30TH ANNIVERSARY EDITION

ADVANCES IN ECONOMETRICS

Series Editors: Thomas B. Fomby, R. Carter Hill, Ivan Jeliazkov, Juan Carlos Escanciano and Eric Hillebrand

Recent Volumes:

Volume 20B:	Econometric Analysis of Financial and Economic Time Series – Edited by Dek Terrell and Thomas B. Fomby
Volume 21:	Modelling and Evaluating Treatment Effects in Econometrics – Edited by Daniel L. Millimet, Jeffrey A. Smith and Edward Vytlacil
Volume 22:	Econometrics and Risk Management – Edited by Jean-Pierre Fouque, Thomas B. Fomby and Knut Solna
Volume 23:	Bayesian Econometrics – Edited by Siddhartha Chib, Gary Koop, Bill Griffiths and Dek Terrell
Volume 24:	Measurement Error: Consequences, Applications and Solutions – Edited by Jane Binner, David Edgerton and Thomas Elger
Volume 25:	Nonparametric Econometric Methods – Edited by Qi Li and Jeffrey S. Racine
Volume 26:	Maximum Simulated Likelihood Methods and Applications – Edited by R. Carter Hill and William Greene
Volume 27A:	Missing Data Methods: Cross-Sectional Methods and Applications – Edited by David M. Drukker
Volume 27B:	Missing Data Methods: Time-Series Methods and Applications – Edited by David M. Drukker
Volume 28:	DSGE Models in Macroeconomics: Estimation, Evaluation and New Developments – Edited by Nathan Balke, Fabio Canova, Fabio Milani and Mark Wynne
Volume 29:	Essays in Honor of Jerry Hausman – Edited by Badi H. Baltagi, Whitney Newey, Hal White and Carter Hill

ADVANCES IN ECONOMETRICS VOLUME 30

30TH ANNIVERSARY EDITION

EDITED BY

DEK TERRELL
Louisiana State University, Baton Rouge, LA

DANIEL MILLIMET
Southern Methodist University, Dallas, Texas

Emerald

United Kingdom – North America – Japan
India – Malaysia – China

Emerald Group Publishing Limited
Howard House, Wagon Lane, Bingley BD16 1WA, UK

First edition 2012

British Library Cataloguing in Publication Data
A catalogue record for this book is available from the British Library

ISBN: 978-1-78190-309-4
ISSN: 0731-9053 (Series)

ISOQAR REGISTERED
UKAS MANAGEMENT SYSTEMS
0026
Certificate Number 1985
ISO 9001
ISO 14001

ISOQAR certified Management Systems, awarded to Emerald for adherence to Quality and Environmental standards ISO 9001:2008 and 14001:2004, respectively

INVESTOR IN PEOPLE

CONTENTS

PART V: SIMULATION

LIST OF CONTRIBUTORS

Lee C. Adkins	Department of Economics, Oklahoma State University, Stillwater, OK, USA
Nathan S. Balke	Department of Economics, Southern Methodist University, Umphrey Lee Center, Dallas, TX, USA
Badi H. Baltagi	Department of Economics and Center for Policy Research, Syracuse University, Syracuse, NY, USA
Randall C. Campbell	Department of Finance and Economics, Mississippi State University, Mississippi State, MS, USA
Jiaqi Chen	Federal Reserve Bank of Dallas, Dallas, TX, USA
Juan Carlos Escanciano	Department of Economics, Indiana University, Bloomington, IN, USA
Thomas B. Fomby	Department of Economics, Southern Methodist University, Dallas, TX, USA
Mary N. Gade	Department of Economics, Oklahoma State University, Stillwater, OK, USA
William Griffiths	Economics Department, University of Melbourne, Victoria, Australia
Jeffery W. Gunther	Federal Reserve Bank of Dallas, Dallas, TX, USA
Joseph H. Haslag	Economics Department, University of Missouri, Columbia, MO, USA
R. Carter Hill	Department of Economics, Louisiana State University, Baton Rouge, LA, USA

Eric Hillebrand	Center for Research in Econometric Analysis of Time Series (CREATES), Department of Economics and Business, Aarhus University, Aarhus C, Denmark
Joseph G. Hirschberg	Department of Economics, University of Melbourne, Victoria, Australia
Yu-Chin Hsu	Department of Economics, University of Missouri, Columbia, MO, USA
Ivan Jeliazkov	Department of Economics, University of California, Irvine, CA, USA
George G. Judge	The Graduate School and Giannini Foundation, University of California, Berkeley, CA, USA
Chihwa Kao	Center for Political Research, Syracuse University, Syracuse NY, USA
Tae-Hwy Lee	Department of Economics, University of California Riverside, Riverside, CA, USA
James P. LeSage	Department of Finance and Economics, Texas State University, San Marcos, TX, USA
Long Liu	Department of Economics, University of Texas at San Antonio, San Antonio, TX, USA
Jenny N. Lye	Department of Economics, University of Melbourne, Victoria, Australia
Michael W. McCracken	Federal Reserve Bank of St. Louis, St. Louis, MO, USA
Daniel Millimet	Department of Economics, Southern Methodist University, Dallas, TX, USA
Ron C. Mittelhammer	School of Economic Sciences, Washington State University, Pullman, WA, USA

Asli Ogunc	Department of Economics and Finance, Texas A&M University – Commerce, TX, USA
R. Kelley Pace	Department of Finance, Louisiana State University, Baton Rouge, LA, USA
Dek Terrell	Department of Economics, Louisiana State University, Baton Rouge, LA, USA
Timothy J. Vogelsang	Department of Economics, Michigan State University, East Lansing, MI, USA
Charley Xia	Department of Mathematics and Statistics, University of Melbourne, Parkville, Victoria, Australia
Jingjing Yang	Department of Economics, College of Business Administration, University of Nevada, Reno, NV, USA
Weiping Yang	Capital One Financial Research, Richmond, VA, USA
Shuang Zhu	Department of Finance, Kansas State University, Manhattan, KS, USA

DEDICATION

This 30th volume of *Advances in Econometrics* is dedicated to Thomas B. Fomby and R. Carter Hill. The first chapter of this volume chronicles the 30-year history of the series and its achievements under the leadership of Thomas Fomby and Carter Hill. Thomas Fomby has edited the series since 1987; Carter Hill joined as co-editor in 1996. The success of this research series over the last 30 years is to a large extent attributable to these two men. Their work with the *Advances* volume, their own research, and their co-authoring of textbooks translated into several languages clearly attest to the deserving dedication of this volume to these two men.

Anyone who has authored a chapter for *Advances in Econometrics*, attended the associated conference, or been associated with the series in any way knows first-hand the time and energy these two individuals have devoted to the series. Each year at the conference dinner, Thomas Fomby discusses the mission of the series, which includes the publication of original articles containing sufficient detail to familiarize non-experts on the topic along with the sharing of computer programs and data used in chapters. While mission statements sound like a rather dry topic, Thomas Fomby's passion to this mission makes his dinner presentation a highlight for everyone in attendance. This passion is apparent in his 25 plus years of editorship as well.

Carter Hill's passion for econometrics and the *Advances in Econometrics* series are apparent in his work as well. In addition to his careful and painstaking work editing volumes, Carter promoted the initial idea of the conference with great vigor. He has also been an ardent advocate of increasing the accessibility of computer code and data to readers. His energy has been crucial in making the conference and the research series a great success.

Like this volume, the conference for the 30th volume of *Advances in Econometrics* was held in honor of Thomas B. Fomby and R. Carter Hill. Friends, family, former students, and both current and past colleagues gathered for a surprise dinner and conference in honor of Thomas and Carter. In addition to a tribute from Emerald Press, the event included a number of entertaining stories beginning with Stan Johnson, the dissertation advisor of both Fomby and Hill, shedding light on their days in graduate school.

One could go on far too long discussing the accomplishments of Thomas Fomby and Carter Hill, both with regard to *Advances in Econometrics* and other aspects of their careers. However, the greatest insight into both men comes from those that have interacted with them most. Stories from their many students, other colleagues such as the guest editors of this volume, authors publishing in the *Advances in Econometrics* series, and countless others using their research or textbooks provide the greatest insight into their accomplishments.

In conclusion, we happily dedicate this 30th volume of *Advances in Econometrics* to its long-standing co-editors, Thomas B. Fomby and R. Carter Hill.

PREFACE

On March 24, 2012, we had one of the greatest surprises of our lives. We walked into the Lod Cook Conference Center on the LSU campus, to find scores of people, including our friend, mentor and major Professor Stanley R. Johnson, and many old colleagues and students. This event, in honor of the 30th volume of *Advances in Econometrics*, was engineered by our friend Dek Terrell, with the aide of his spouse, Dannielle Lewis, our spouses Nancy Fomby and Melissa Waters, Julianna Richard of LSU, Daniel Millimet of SMU, and our former Emerald Press editor Emma Whitfield. True to our nature we were oblivious to the nine months of planning and work that had gone into preparing the event. These cunning and diabolical people, obviously skilled in hiding the truth and keeping poker faces through it all, had organized a surprise conference! Not a birthday party, not a dinner, but a weekend conference with participants from all over the world! We are amazed, and honored.

Econometrics is fun! We have enjoyed a professional and personal friendship for 40 years. We like to talk about econometrics, think about it, do it, internalize it, and write about it. Our passion is to learn, and then facilitate the dissemination of knowledge (the catchy phrase "knowledge transfer" applies). Our friendship began with our graduate studies, under the inspiring guidance of Stan Johnson at the University of Missouri – Columbia. We are a good fit, because we do not mind asking each other "What the heck does that mean?" That curiosity has lasted. We hope that *Advances in Econometrics* will not only continue but improve, keeping the long citation property of the chapters. We are recruiting smart young editors to carry on the tradition.

To watch *Advances* grow and mature through the years has been a wonderful and rewarding experience. The experience has been full of friends, new and old, and full of exciting discoveries in our vast econometrics field. The discipline is now fully engaged in the big data era and there seems to be no end to the potential that exists in econometric theory and application. To us *Advances* has been a "village" and what a wonderful village it is. We look forward to seeing the advances that *Advances* engenders in econometric research and application in the future. We think it has a wonderful future ahead of it.

Once again, we thank Dek Terrell, his co-conspirators, and especially the contributors to this volume for the lasting smiles that memories of this event will bring.

In terms of dedication, Thomas would like to dedicate his 25 plus years of editorship of *Advances* to the memory of his son, Robert Blake Fomby, who passed away shortly after attending the 30th volume of *Advances* conference. He was a fine young man who left us much too early. He is sorely missed.

Thomas B. Fomby
R. Carter Hill
Editors

INTRODUCTION

The collection of chapters in this 30th volume of *Advances in Econometrics* provides a well-deserved tribute to Thomas B. Fomby and R. Carter Hill, who have served as editors of the *Advances in Econometrics* series for 25 and 21 years, respectively. Volume 30 contains a more varied collection of chapters than previous volumes, in essence mirroring the wide variety of econometric topics covered by the series over 30 years. Volume 30 starts with a chapter discussing the history of this series over the last 30 years. The next five chapters can be broadly categorized as focusing on model specification and testing. Following this section are three contributions that examine instrumental variables models in quite different settings. The next four chapters focus on applied macroeconomics topics. The final chapter offers a practical guide to conducting Monte Carlo simulations.

Randy C. Campbell and Asli Ogunc's "A History of the Advances in Econometrics Series" provides an excellent first chapter for this 30th volume of *Advances in Econometrics* in recognition of its longtime editors Thomas B. Fomby and R. Carter Hill. The chapter briefly documents the history of this series, which has included articles by 493 authors. The list of contributors includes four Nobel Prize winners and numerous outstanding econometricians. The chapter summarizes topics, key contributions and highlights beginning with the founding of this research annual in 1982, with Robert L. Basmann and George Rhodes serving as the initial co-editors. The chapter moves quickly through the period where Thomas B. Fomby first entered as co-editor in 1987 to R. Carter Hill's entry in 1996 and then to more recent volumes, which also typically include presentation of the papers at a conference.

In their chapter "Bayesian Unit Root Testing: The Effect of Choice of Prior on Test Outcomes," Charley Xia and William Griffiths focus on the sensitivity of unit root tests to the choice of prior. In particular, they consider using two testing procedures, a credible interval test and a Bayes factor test, using priors attributable to Jeffreys, Lubrano, and Berger and Yang to test for a unit root. These basic testing procedures are extended to allow for Bayesian model averaging and a test with a hierarchical prior for a hyperparameter. Monte Carlo simulations are used to evaluate the testing procedures. The results indicate that conclusions can vary significantly across testing procedure and choice of prior. In terms of both test size and power, the prior recommended by Lubrano performs best in simulations. The chapter concludes with an application to Australian time series.

The chapter by Jenny N. Lye and Joseph G. Hirschberg, "Inverse Test Confidence Intervals for Turning-Points: A Demonstration with Higher Order Polynomials," extends work on a long-standing, but important econometric issue arising in nonlinear regression specifications. Specifically, Lye and Hirschberg are interested in obtaining confidence intervals for the estimated turning points of the outcome with respect to a covariate entering the model nonlinearly. As the turning points are nonlinear functions of the estimated parameters, proper inference, especially in finite samples, is not trivial. However, given the popularity of nonlinear models in many economic contexts, the question is salient. Here, the authors build on prior work by showing how one may derive confidence intervals for turning points in high order polynomial and fractional polynomial regression models by inverting a two-sided test for the location of the value of the regressor where the first derivative function is equal to zero. The model is also extended to perform tests using the second derivative function as well. Lye and Hirschberg illustrate the methods by examining time series data on per capita carbon dioxide emissions and per capita income, testing for the so-called environmental Kuznets curve.

The chapter by Jingjing Yang and Timothy J. Vogelsang, "Serial Correlation Robust LM Type Tests for a Shift in Trend," adds to the literature on testing for structural breaks with an unknown break date. Specifically, Yang and Vogelsang are interested in testing for a break in the trend of a univariate time series. Extending earlier results for models with non-trending data, the authors focus on the class of LM statistics based on nonparametric kernel estimators of the long-run variance and develop a fixed-b asymptotic theory for the nonparametric estimator of the long-run variance and the LM test statistics. The fixed-b theory suggests that, for a given statistic, kernel, and significance level, there usually exists a bandwidth such that the fixed-b asymptotic critical value is the same for both $I(0)$ and $I(1)$ errors. Yang and Vogelsang calculate these robust bandwidths using simulation methods for several well-known kernels. The authors demonstrate that when the robust bandwidth is used, the supremum statistic configured with either the Bartlett or Daniell kernel provides LM tests with good power. Finally, when testing for a slope change, Yang and Vogelsang obtain the surprising finding that less trimming of potential dates for the trend shift leads to higher power, contrasting with the usual relationship between trimming and power.

Michael W. McCracken's contribution, "Consistent Testing for Structural Change at the Ends of the Sample," focuses on testing for structural breaks near the beginning or end of the sample using Chow or Predictive tests for structural change. The asymptotic results in this chapter rely on an innovative parameterization of the location of the break point. Intuitively, this parameterization models breakpoints as local to the beginning or end of the sample. By carefully controlling the rate of convergence of the breakpoint to the endpoint, the number of observations available for the hypothesis tests diverges. The theoretical contribution of the chapter consists of providing asymptotic distributions for the Chow test, a max-Chow test, and variants of the predictive test for structural breaks. Monte Carlo results are used to evaluate the performance of both tests. Results are generally consistent with the theory and suggest that the max-Chow test in particular has some power to detect structural breaks near the end of the sample.

The chapter by Eric Hillebrand and Tae-Hwy Lee, "Stein-Rule Estimation and Generalized Shrinkage Methods for Forecasting Using Many Predictors," provides a very appropriate contribution to the volume, focusing on Stein-rule shrinkage estimates. Their work builds on a number of contributions by Carter Hill, including joint work with Thomas Fomby. The focus of the chapter is to evaluate Stein-rule estimators that shrink an ordinary least squares estimator to a principal component estimator. Monte Carlo simulations evaluate the performance of the three estimators under varying conditions. Results suggest that the Stein-rule estimator can dominate both ordinary least squares and principal components under a relatively large set of simulated conditions.

The contribution by Badi H. Baltagi, Chihwa Kao, and Long Liu, "On the Estimation and Testing of Fixed Effects Panel Data Models with Weak Instruments," considers the asymptotic properties of weak instruments in a within k-class estimator for panel data. In particular, the coefficients of the weak instruments are modeled as shrinking to zero at a rate proportional to $\sqrt{n}T^{\delta}$. The primary results of the chapter show that as $(n, T)\rightarrow\infty$, the within group k-class estimator is consistent if $0<\delta<1/2$ and inconsistent if $1/2<\delta<\infty$.

In the contribution by George G. Judge and Ron C. Mittelhammer, "A Risk Superior Semiparametric Estimator for Overidentified Linear Models," the authors discuss the merits of using a weighted average of ordinary and two-stage least squares (OLS and TSLS), instead of either alone, when one is concerned about the potentially endogeneity of at least some covariates in a linear regression model. Estimation of the optimal weight, in terms of minimized mean squared error, is discussed. The authors present striking Monte Carlo evidence supporting the reduction in MSE from their combined estimator, relative to either TSLS or the usage of OLS or TSLS depending on the results of a Hausman test, in many cases. Finally, Judge and Mittelhammer illustrate the combined estimator in an application to the estimation of the returns to schooling.

The chapter by R. Kelley Pace, James P. LeSage, and Shuang Zhu, "Spatial Dependence in Regressors and its Effect on Performance of Likelihood-Based and Instrumental Variable Estimators," investigates estimators used in applications of the spatial autoregressive (SAR) model and spatial Durbin model (SDM) where there is spatial dependence in the regressors. The chapter contains analysis of the partial R^2 to highlight the fact that spatial dependence of the regressors can lead to weak instruments. It then compares maximum likelihood and restricted maximum likelihood to instrumental variables in both common data sets and simulations. Results of the simulations indicate that instrumental variables techniques may be unreliable in many settings where there is spatial dependence in the regressors, particularly for the SDM model.

The chapter by Nathan S. Balke, "Sectoral Effects of Aggregate Shocks," explores the identification of aggregate monetary and productivity (or technology) factors. The question is an important one as understanding the role of monetary shocks is vital for evaluating the effects of monetary policy, while the ability of productivity shocks to explain business cycle fluctuations remains an open issue. Using sectoral rather than aggregate data, along with a combination of long-run and sign restrictions to identify such factors in a Bayesian common factor model, Balke finds that the monetary factor is responsible for long swings in nominal variables, but has little effect on fluctuations in output, real wage, or labor input growth. In contrast, the productivity factor increases output and real wage growth in the short and long run, as well as increases labor input and decreases in prices. That said, the quantitative effect of the productivity factor on labor input is relatively small. Balke's results prove to be robust to the number of factors included in the model, to alternative priors about the short-run effects of the monetary factor, and to the inclusion of oil prices. Interestingly, oil prices appear to be largely driven by the other aggregate factors in the model.

The chapter by Joseph H. Haslag and Yu-Chin Hsu, "Cyclical Co-movement between Output, the Price Level, and the Inflation Rate," examines two related sets of facts concerning business cycles. First, movements in the price level are countercyclical. Second, movements in the inflation rate are procyclical. One explanation for these stylized facts is that price and output are characterized by two sine waves with the same periodicity. In this case, there exists a phase shift that can generate the observed correlations. Haslag and Hsu examine the validity of this explanation. The authors find that a positive phase shift in the price level can generate these correlations, as well as yield a testable prediction concerning the price level having a causal effect (in the Granger sense) on output. This prediction also holds in the data. Finally, Haslag and Hsu find some changes in the relationships between the price level, inflation, and output over time. Specifically, the authors find that the countercyclical relationship between the price level and output has strengthened over time, while the procyclical relationship between the inflation rate and output has weakened. Further examination suggests that this may be attributable to a diminishing positive phase shift in the price level.

The contribution by Tae-Hwy Lee and Weiping Yang, "Money-Income Granger-Causality in Quantiles," revisits the question of whether money is useful for forecasting future output. While the prior literature has focused on Granger causality tests for money in models for the conditional mean of output, here the authors examine Granger causality in models for conditional quantiles. Using U.S. time series data on real personal income or industrial production, as well as either M1 or M2 to measure money, the analysis indicates that quantile forecasts of output growth are significantly improved by accounting for past money growth, particularly in the tails, whereas conditional mean forecasts are not generally improved by accounting for money. This discovery is intriguing, suggesting that the effectiveness of monetary policy has been masked by the existing focus on only the conditional mean.

The chapter by Jiaqi Chen and Jeffery W. Gunther, "Copula-GARCH Time-Varying Tail Dependence," builds on a growing literature assessing tail dependence in the returns to risky assets. Proper measurement of the joint susceptibility of different assets to the same downside risk is crucial to practitioners and policy-makers, otherwise portfolios that seem sufficiently diversified may, in fact, not be. Here, Chen and Gunther examine the tail dependence between bank returns and insurance. More importantly, the authors use this application to illustrate the value of time-varying copulas in an econometric setting. Specifically, Chen and Gunther apply a parametric copula – GARCH model to estimate time – varying tail dependence between bank and insurance daily stock returns over the previous two decades.

Incorporating data from the recent financial crisis allows the authors to assess whether tail dependence became magnified during the downturn. The authors find that tail dependence did indeed rise during the recent crisis. Moreover, they demonstrate the need to pay close attention to the specification of the time variation in dependence.

The final chapter in the volume, "Monte Carlo Experiments Using Stata: A Primer with Examples," is authored by Lee C. Adkins and Mary N. Gade. Under Carter and Tom's stewardship, a fundamental goal of the *Advances in Econometrics* series has been to bring advanced econometric techniques to students and practitioners. A powerful tool to do this is through Monte Carlo experiments and the sharing of computer code. Adkins and Gade continue this tradition by providing an introduction to the use of Monte Carlo experiments in the software package, Stata. Moreover, the authors generously provide the computer code to facilitate learning, both inside and outside the classroom.

Dek Terrell
Daniel Millimet

PART I
HISTORY

A HISTORY OF THE ADVANCES IN ECONOMETRICS SERIES

Randall C. Campbell and Asli Ogunc

ABSTRACT

Advances in Econometrics *is a series of research annuals first published in 1982 by JAI Press. In this paper, we present a brief history of the series over its first 30 years. We describe key events in the history of the volume, and give information about the key contributors: editors, editorial board members,* Advances in Econometrics *Fellows, and authors who have contributed to the great success of the series.*

Keywords: Advances in Econometrics; history; AIE conference; key contributors

INTRODUCTION

Advances in Econometrics is a series of research annuals first published in 1982 by JAI Press. The *Advances in Econometrics* website (http://faculty.smu.edu/tfomby/aie.htm) lists the mission of the publication, which is, "To annually publish original econometrics papers on designated topics with the intention of expanding the use of newly vintaged econometric techniques throughout the empirical economic literature and beyond."

Throughout its 30 years of existence, this mission has certainly been achieved with outstanding econometricians contributing papers on an

30th Anniversary Edition
Advances in Econometrics, Volume 30, 3–24

ISSN: 0731-9053/doi:10.1108/S0731-9053(2012)0000030006

impressive variety of topics. Through the first 27 volumes, 493 different authors have contributed at least one paper to *Advances in Econometrics*. Among the outstanding econometricians are four Nobel Prize winners and 31 of 100 members of the "Econometricians' Hall of Fame" (see Baltagi, 1998, Table 5). During the period from 1982 to 2002, JAI Press published the first 16 volumes. Elsevier published Volumes 17–21 from 2003 to 2008, while Emerald Publishing has published Volumes 22 to the present. These volumes featured 330 papers with over 7,400 citations.[1] In addition, Volumes 28–30 will soon be published and a call for papers has been issued for Volume 31. Table 1 lists the titles and editors for the first 29 volumes.

The rest of this history is organized as follows. The middle sections summarize the history of the volume and are organized according to important events. The final section summarizes recent developments, lists key contributors, and concludes the paper.

BEGINNINGS OF THE VOLUME: 1982–1986

The founding co-editors of *Advances in Econometrics* (AIE) were Robert L. Basmann (Fig. 1) of Texas A&M University and George F. Rhodes, Jr. (Fig. 2) of Colorado St. University. Professor Basmann received his Ph.D. from Iowa State University. He is well known for his pioneering work on simultaneous equations and two-stage estimators and was named a Fellow of the Econometric Society in 1966. Professor Rhodes received his Ph.D. from Ohio State University. His early work also involved simultaneous equations. In addition, his empirical work in the field of Labor Economics is well known.

The first volume of AIE was published in 1982. Basmann and Rhodes declared the criteria for papers published in AIE as[2]

"Papers shall make original contributions in economics and econometrics that lay major new foundations for continued study," and
"Papers shall be sufficiently long and writing style sufficiently mature and complete that they are self-contained. They should be readable and should not require more than occasional reference to previous works."

A few summary statistics for the AIE volumes and authors are given in Tables 2–5. Tables 2 and 3, respectively, give the top 10 AIE volumes by number of citations and number of papers. Table 4 lists the top 10 most cited papers in the history of AIE. Table 5 lists all contributors who have authored or co-authored at least three papers published in AIE.

Table 1. AIE Titles and Editors: Volumes 1–27.

Volume	Year	Title	Editors
1	1982	Studies of Consumer and Worker Behavior	R.L. Basmann, George F. Rhodes, Jr.
2	1983	Exact Distribution Analysis in Linear Simultaneous Equation Models	R.L. Basmann, George F. Rhodes, Jr.
3	1984	Economic Inequality: Measurement and Policy	R.L. Basmann, George F. Rhodes, Jr.
4	1985	Economic Inequality: Survey Methods and Measurement	R.L. Basmann, George F. Rhodes, Jr.
5	1986	Innovations in Quantitative Economics: Essays in Honor of R.L. Basmann	Daniel J. Slottje
6	1987	Computation and Simulation	Thomas B. Fomby, G. F. Rhodes, Jr.
7	1988	Nonparametric and Robust Inference	Thomas B. Fomby, G. F. Rhodes, Jr.
8	1990	Co-Integration, Spurious Regressions, and Unit Roots	Thomas B. Fomby, G. F. Rhodes, Jr.
9	1991	Econometric Methods and Models for Industrial Organizations	Thomas B. Fomby, G. F. Rhodes, Jr.
10	1994	Simulating and Analyzing Industrial Structure	Thomas B. Fomby, G. F. Rhodes, Jr.
11A	1996	Bayesian Computational Methods and Applications	Thomas B. Fomby, R. Carter Hill
11B	1996	Bayesian Methods Applied to Time Series Data	Thomas B. Fomby, R. Carter Hill
12	1997	Applying Maximum Entropy to Econometric Problems	Thomas B. Fomby, R. Carter Hill
13	1998	Messy Data – Missing Observations, Outliers, and Mixed-Frequency Data	Thomas B. Fomby, R. Carter Hill
14	2000	Applying Kernel and Nonparametric Estimation to Economic Topics	Thomas B. Fomby, R. Carter Hill
15	2000	Nonstationary Panels, Panel Cointegration, and Dynamic Panels	B. Baltagi, T. Fomby, R. Carter Hill
16	2002	Econometric Models in Marketing	P. H. Franses, A. L. Montgomery
17	2003	Maximum Likelihood Estimation of Misspecified Models: Twenty Years Later	Thomas B. Fomby, R. Carter Hill
18	2004	Spatial and Spatiotemporal Econometrics	James P. Lesage, R. Kelley Pace
19	2004	Applications of Artificial Intelligence in Finance and Economics	J. Binner, G. Kendall, S. Chen
20A	2006	Econometric Analysis of Financial and Economic Time Series	Thomas B. Fomby, Dek Terrell
20B	2006	Econometric Analysis of Financial and Economic Time Series	Thomas B. Fomby, Dek Terrell

Table 1. *(Continued)*

Volume	Year	Title	Editors
21	2008	Modeling and Evaluating Treatment Effects in Econometrics	Thomas B. Fomby, R. C. Hill, D. Millimet, J. Smith, E. Vytlacil
22	2008	Econometrics and Risk Management	J-P. Fouque, Thomas B. Fomby, K. Solna
23	2008	Bayesian Econometrics	S. Chib, W. Griffiths, G. Koop, Terrell
24	2009	Measurement Error: Consequences, Applications and Solutions	J. Binner, D. Edgerton, T. Elger
25	2009	Nonparametric Econometric Methods	Qi Li, Jeffrey S. Racine
26	2010	Maximum Simulated Likelihood Methods and Applications	William Greene, R. Carter Hill
27A	2011	Missing Data Methods: Cross-Sectional Methods and Applications	David M. Drukker
27B	2011	Missing Data Methods: Time-Series and Applications	David M. Drukker
28	2012	DSGE Models in Macroeconomics-Estimation, Evaluation, and New Develop.	Balke, Canova, Milani, Wynne
29	2012	Essays in Honor of Jerry Hausman	B. Baltagi and W. Newey

Fig. 1. Dr. Robert L. Basmann.

Fig. 2. Dr. George F. Rhodes, Jr. and his wife, Joan.

Table 2. Top Ten AIE Volumes by Number of Citations.

Volume	Title	Citations
15	Nonstationary Panels, Panel Cointegration, and Dynamic Panels	3,076
8	Co-Integration, Spurious Regressions, and Unit Roots	621
3	Economic Inequality: Measurement and Policy	610
1	Studies of Consumer and Worker Behavior	350
21	Modeling and Evaluating Treatment Effects in Econometrics	232
18	Spatial and Spatiotemporal Econometrics	230
7	Nonparametric and Robust Inference	211
16	Econometric Models in Marketing	209
20B	Econometric Analysis of Financial and Economic Time Series	194
13	Messy Data – Missing Observations, Outliers, and Mixed-Frequency Data	136

The first volume includes papers on consumer and worker behavior in addition to papers on general econometric theory. The total number of citations for Volume 1 is 350, which is the 4th highest number of total citations among AIE volumes, as shown in Table 2. Volume 1 includes the Jorgensen, Lau, and Stoker article entitled "The Transcendental Logarithmic Model of Aggregate Consumer Behavior." This article has 267 total

Table 3. Top Ten AIE Volumes by Number of Papers.

Volume	Title	Number of Papers
23	Bayesian Econometrics	21
5	Innovations in Quantitative Economics: Essays in Honor of R.L. Basmann	16
25	Nonparametric Econometric Methods	15
12	Applying Maximum Entropy to Econometric Problems	14
14	Applying Kernel and Nonparametric Estimation to Economic Topics	14
20B	Econometric Analysis of Financial and Economic Time Series	14
8	Co-Integration, Spurious Regressions, and Unit Roots	13
20A	Econometric Analysis of Financial and Economic Time Series	13
21	Modeling and Evaluating Treatment Effects in Econometrics	13

Table 4. Top Ten AIE Papers by Number of Citations.

Author(s)	Paper	Volume	Citations
Peter Pedroni	Fully Modified OLS for Heterogeneous Cointegrated Panels	15	729
Jorg Breitung	The Local Power of Some Unit Root Tests for Panel Data	15	693
Chihwa Kao and Min-Hsien Chiang	On the Estimation and Inference of a Cointegrated Regression in Panel Data	15	691
Badi H. Baltagi and Chihwa Kao	Nonstationary Panels, Cointegration in Panels, and Dynamic Panels: A Survey	15	434
Richard Blundell, Stephen Bond, and Frank Windmeijer	Estimation in Dynamic Panel Data Models: Improving the Performance of the Standard GMM Estimator	15	434
James E. Foster	On Economic Poverty: A Survey of Aggregate Measures	3	314
Dale W. Jorgensen, Lawrence J. Lau, and Thomas M. Stoker	The Transcendental Logarithmic Model of Aggregate Consumer Behavior	1	267
Joon Y. Park	Testing for Unit Roots and Cointegration by Variable Addition	8	234
Francis X. Diebold and Marc Nerlove	Unit Roots in Economics Time Series: A Selected Survey	8	166
Nanak Kakwani	On Measurement of Taxes, Progressivity and Redistributive Effects of Taxes with Applications to Horizontal and Vertical Equity	3	112

Table 5. Top Authors by Number of Papers Published in AIE.

Author	Number of Papers	Volumes	Total Citations
George F. Rhodes, Jr.	7	4 (Volumes 2, 5, 6, 10)	4
R. Carter Hill	5	5 (Volumes 5, 6, 7, 17, 26)	35
Qi Li	5	5 (Volumes 13, 14, 15, 25, 27)	29
Daniel J. Slottje	5	4 (Volumes 3, 5, 6, 12)	41
Aman Ullah	5	5 (Volumes 6, 7, 14, 15, 25)	60
Badi H. Baltagi	4	4 (Volumes 14, 15, 17, 18)	445
Jane M. Binner	4	2 (Volumes 19, 24)	25
Siddhartha Chib	4	3 (Volumes 11, 16, 23)	82
Thomas B. Fomby	4	4 (Volumes 5, 6, 7, 17)	7
Robert L. Basmann	3	3 (Volumes 2, 3, 4)	52
Zongwu Cai	3	2 (Volumes 20, 25)	30
Shu-Heng Chen	3	2 (Volumes 19, 24)	12
James J. Heckman	3	3 (Volumes 1, 2, 5)	83
Kim P. Huynh	3	2 (Volumes 25, 27)	2
David T. Jacho-Chavez	3	2 (Volumes 25, 27)	2
Dale W. Jorgensen	3	3 (Volumes 1, 3, 5)	271
Nanak Kakwani	3	1 (Volume 3)	220
Gary Koop	3	1 (Volume 23)	19
Douglas Miller	3	2 (Volumes 12,17)	14
Christopher F. Parmeter	3	2 (Volumes 21, 25)	17
Peter C.B. Phillips	3	3 (Volumes 2, 5, 8)	106
Jeffrey S. Racine	3	3 (Volumes 14, 16, 25)	4
Arnold Zellner	3	3 (Volumes 5, 12, 23)	109

citations, which makes it the 7th most cited paper in the history of AIE. The first volume includes papers by two of the authors who have three or more AIE publications, Dale W. Jorgensen and James J. Heckman. The paper "Models of Analysis of Labor Force Dynamics," by Christopher J. Flinn and James J. Heckman (Fig. 3), appeared in the first volume. With this paper, James J. Heckman (Nobel Prize in Economics, 2000) became the first of four future Nobel Prize winners to publish a paper in AIE. Professor Heckman also published papers in Volumes 2 and 5 of AIE.

The second volume, entitled "Exact Distribution Analysis in Linear Simultaneous Equation Models," was edited by Robert L. Basmann, who also contributed one of his sole authored articles. George Rhodes, Jr., who has the highest number of total AIE publications, contributed two articles to the volume.

Volume 3, "Economics Inequality: Measurement and Policy," again edited by Basmann and Rhodes, was published in 1984. In his introduction,

Fig. 3. Dr. James J. Heckman.

Professor Rhodes explains the role of the disciplines and fields that are represented by the papers in this volume in the evolution of welfare economics. The 610 citations of this volume, which is the 3rd highest total in the series, are evidence of the accomplishment of the stated purpose. In addition, this volume contains two of the top 10 most cited papers in AIE: "On Economic Poverty: A Survey of Aggregate Measures," by James E. Foster, which ranks 6th with 314 citations and "On Measurement of Taxes, Progressivity and Redistributive Effect of Taxes with Applications to Horizontal and Vertical Equity," by Nanak Kakwani, which ranks 10th with 112 citations.

Volume 4, edited by Basmann and Rhodes in 1985, continued to develop welfare economics. In the introduction to Volume 4, which is titled "Economic Inequality: Survey Methods and Measurements," Rhodes talks more about the policy implications than the evolution of the field. Volume 4 is also noteworthy in that 7 of the 12 authors have 3 or more AIE publications, as shown in Table 5.

Daniel J. Slottje edited the fifth volume, titled "Innovations in Quantitative Economics: Essays in Honor of Robert L. Basmann." The fifth edition is noteworthy for several reasons. First, this is the first volume that honors a great econometrician, Robert L. Basmann, who was co-founder of AIE. More importantly, this was the volume in which R. Carter Hill and Thomas B. Fomby made their first mark on AIE with their article, "Improved Confidence Sets in a Non-Utopian Setting." Finally, as shown in Table 3, this volume ranks 2nd in total number of papers with 16.

THOMAS B. FOMBY BECOMES CO-EDITOR: 1987–1994

A significant change occurred in 1987 when Robert L. Basmann decided to step down as co-editor of AIE. Daniel Slottje, a former student of Professor Basmann's, was teaching at Southern Methodist University with Thomas B. Fomby. He asked Professor Fomby, an excellent econometrician who had published a paper in Volume 5 of AIE, if he would be interested in working with George Rhodes as co-editor of AIE. Professor Fomby agreed to replace Professor Basmann and remains active as senior co-editor today, 25 years later. When asked what motivated him to become co-editor, Professor Fomby said that he saw the volume as something that could continue well into the future since there was not a research annual of this type in the field of econometrics. In addition, he said that working on a research volume for practitioners was one of the factors that motivated him to become an editor of the volume and that being an editor would provide him the means and motivation to "keep up with all of the new stuff in econometrics." He described AIE as a research volume where new methodology could be quickly disseminated to practitioners, and saw it becoming more methods oriented. This became evident in future volumes in diverse areas such as Bayesian methods, time series, marketing, financial econometrics, and industrial organization to name a few.

The topic for Volume 6 is "Computation and Simulation." In response to rapid advances in high-speed computers during the 1980s, this volume focused on computer intensive techniques in a variety of econometric areas. This was the first of five volumes edited by Thomas B. Fomby and George F. Rhodes, Jr.

Volume 7 is titled "Nonparametric and Robust Inference." It is divided into three sections: Robust Estimation, Robust Inference, and Consumer Demand Studies. The purpose is to connect microeconomic theories to testing and implementation. There are several well-cited papers throughout the volume, with a total of 211 citations to date, 7th highest among AIE volumes.

Volume 8 is titled "Co-Integration, Spurious Regressions, and Unit Roots." This volume is divided into four sections: Survey and the New Concept of Multi-Co-Integration, Developments in Testing, Developments in Theory, and Applications. The volume has 621 total citations, indicating the quality of the papers and high level of interest in this topic and ranking 2nd among AIE volumes. Volume 8 includes two of the top 10 most cited papers in AIE. The volume begins with a survey paper by Francis Diebold and Marc Nerlove, which has generated 166 citations (9th highest in AIE). The most frequently cited paper in the volume is "Testing for Unit Roots and Cointegration by Variable Addition," by Joon Park, which has been cited 234 times (8th highest in AIE). Other frequently cited papers in the volume include Perron (28 citations), Schmidt (54 citations), and Hansen and Phillips (72 citations). Finally, Volume 8 included the paper "Multi-cointegration," by Clive W. J. Granger (Fig. 4) and Tae-Hwy Lee. Clive W.J. Granger (Nobel Prize in Economics, 2003) became the second future Nobel Prize winner to publish in AIE.

Volume 9 is titled "Econometric Methods and Models for Industrial Organization." This is the first of two volumes focusing on econometrics of industrial organization. Papers are divided between the econometric analysis of structure and incentives, regulation, and advertising. This topic was

Fig. 4. Dr. Clive W.J. Granger.

chosen in response to changes in the regulatory environment and other areas during the 1980s and early 1990s.

Volume 10 is the follow-up volume on econometric analysis of industrial organization, and is titled "Simulating and Analyzing Industrial Structure." Volume 10 is somewhat shorter than previous volumes with only six contributed papers. Volume 10 was the final volume edited by Thomas B. Fomby and George F. Rhodes, Jr. as it marked the last volume for which founding co-editor George F. Rhodes, Jr. served as co-editor.

R. CARTER HILL BECOMES CO-EDITOR: 1996–2001

After George F. Rhodes, Jr. stepped down as co-editor of years, having met when AIE in 1991, Thomas B. Fomby asked R. Carter Hill, who had published articles in Volumes 6 and 7, to join him as co-editor for Volume 11. Professors Fomby and Hill had already known each other for 20 years when they were both students of Stan Johnson (Fig. 5) at the University of Missouri-Columbia. Starting with the publication of Volume 11 of AIE in 1996, they have worked together as co-editors of AIE for nearly 20 years now. Fomby noted that since Hill came aboard as co-editor, they have

Fig. 5. Drs. Carter Hill, Stan Johnson, and Thomas B. Fomby.

focused on papers that include more case examples and which have computer code that is fully accessible to readers upon request, allowing for quicker adoption of new methodologies.

Volume 11, which was split into two parts, focuses on Bayesian methods. The popularity of the Bayesian approach to econometrics has grown rapidly due to new computational techniques and fast computers. Markov Chain Monte Carlo (MCMC) techniques have revolutionized the Bayesian methods that were previously considered impractical. The Gibbs sampling and Metropolis–Hastings algorithms made Bayesian methods very operational. Those applications are covered in Volume 11, Part A, entitled "Bayesian Computational Methods and Applications." Part B, "Bayesian Methods Applied to Time Series Data," focuses on the time series applications of Bayesian methods.

Following the volume on Bayesian methods, Fomby and Hill edited Volume 12, entitled "Applying Maximum Entropy to Econometric Problems." In his work "The Bayesian Method of Moments: Theory and Applications," Arnold Zellner explained entropy as a method that can achieve the goals of the Bayesian method with fewer assumptions.

The objective for Volume 13, entitled "Messy Data," was to address the very frequent issue of data problems. The 10 articles in this volume tackled different types and levels of "messy ness" and ways to diagnose and/or deal with these data problems for time series, cross section, and panel platforms.

In 2001, Fomby and Hill edited Volume 14 "Applying Kernel and Nonparametric Estimation to Economic Topics." The volume had two parts, Methodology and Applications. Volume 14 ranks 4th among AIE volumes with 14 contributed papers. Finally, Badi H. Baltagi, published his first AIE paper in Volume 14. He continued to contribute greatly to AIE over the coming years and eventually became one of two inaugural AIE Fellows.

Professor Baltagi joined Fomby and Hill as editor for Volume 15. This volume, entitled "Nonstationary Panels, Panel Cointegration, and Dynamic Panels," includes 11 refereed papers on nonstationary and dynamic panel data models written by 20 authors. Within this context, Blundell et al. discussed the estimation of dynamic panel models using generalized method of moments (GMM) in the dynamic error components framework and reports efficiency gains resulting from the system GMM estimator. This paper has received 434 citations (4th highest in series). Pedroni, in the most frequently cited AIE paper (729 citations), developed methods for estimating and testing hypotheses for cointegrating vectors in dynamic panels. Breitung, in the 2nd most frequently cited AIE paper (693 citations) studied the local power of panel unit root test statistics against a sequence of local power alternatives. Kao and Chiang, in the 3rd most frequently cited paper (691 citations) studied the limiting distributions of ordinary least squares (OLS), fully modified OLS (FMOLS), and dynamic OLS (DOLS) estimators in a panel cointegrated regression model, and showed that the OLS, FMOLS, and DOLS estimators are all asymptotically normally distributed. Finally, co-editor and frequent contributor to AIE, Badi Baltagi contributed a joint paper with Chihwa Kao that has been cited 434 times (tied for 4th highest). Overall, Volume 15 contains the top five most cited papers in AIE history. Due to the large number of heavily cited papers, Volume 15 has a total of 3,076 citations, easily the most of any volume in the series with nearly five times the number of citations as the next most cited volume.

Philip Hans Franses and Alan L. Montgomery served as guest editors for Volume 16, which was titled "Econometric Models in Marketing." This volume focused on the application of econometric methods on marketing theory. The most frequently cited paper in Volume 16 is "Discrete choice models incorporating revealed preferences and psychometric data," by Morikawa, Ben-Akiva, and McFadden. This is also noteworthy in that Daniel McFadden (Fig. 6; Nobel Prize in Economics, 2000) became the third future Nobel Prize winner to publish in AIE.

Fig. 6. Dr. Daniel McFadden.

ADVANCES IN ECONOMETRICS CONFERENCES INAUGURATED: 2002–PRESENT

Another significant event in the history of AIE occurred in November 2002. Volume 17, "Maximum Likelihood Estimation of Misspecified Models: Twenty Years Later," edited by Fomby and Hill, was in recognition of Hal White's seminal work in robust estimation. For this volume, authors were able to present their work in a unique workshop-type setting on the campus of Louisiana State University. R. Carter Hill had for years wanted to have a small conference where contributors to the volume could present their work and receive strong feedback from other contributors. The 1st Annual AIE Conference was held in the Cook Conference Center on the campus of Louisiana State University in November 2002. Hal White attended the conference and contributed a paper to the volume, as did many other outstanding econometricians. The AIE conference is still going strong with the 12th and most recent conference held in March 2012.

The 18th volume was edited by James P. LeSage and R. Kelley Pace and is titled "Spatial and Spatiotemporal Econometrics." LeSage and Pace contributed a lengthy introduction to the volume, which describes spatial regression models and estimation. The contributed papers from this volume generated 230 total citations, making it the 6th most cited volume in AIE history. Authors presented their research at the 2nd Annual AIE Conference, which was held at Louisiana State University in November 2003. Fomby and Hill continued to serve as series editors, however as seen in Table 1, there have been frequent guest editors beginning with Volume 18.

The 19th volume was titled "Applications of Artificial Intelligence in Finance and Economics." Papers for this volume were presented at the 2003 International Conference on Artificial Intelligence. The volume was edited by Jane M. Binner, Graham Kendall, and Shu-Heng Chen.

Volume 20 of *Advances in Econometrics* honored the econometric contributions of 2003 Nobel Prize winners Robert Engle (Fig. 7) and Sir Clive Granger. The volume is titled "Econometric Analysis of Financial and Economic Time Series." Dek Terrell joined Thomas B. Fomby as editor for this volume. Volume 20 was split into two parts. To date, Part 1 has 162 citations and Part 2 has 194 citations. Contributed papers to Volume 20 were presented at the 3rd Annual AIE Conference held at Louisiana State University in November 2004. Professor Granger was not able to attend the conference as he was in New Zealand at the time. However, he did send a "reflections" piece on his career that was included in the volume. Robert F. Engle III (Nobel Prize in Economics, 2003) did attend the conference and gave comments on his career (see Fig. 7). This piece was also included in the volume, marking the fourth Nobel Prize winner to publish in AIE.

Dek Terrell's involvement in Volume 20 is noteworthy for two reasons. First, Professor Terrell would later be recognized as one of two inaugural AIE Fellows. In addition, Professor Terrell holds the Freeport-McMoRan Endowed Chair of Economics at Louisiana State University and is Director of the LSU Division of Economic Development. Generous funding provided through the Freeport-McMoRan Chair has been instrumental in allowing the *Advances in Econometrics* Conference to be held as well as attracting excellent contributors to the volume and conference participants.

To examine the impact of the AIE Conference (and hence the Freeport-McMoRan funding) on the volume, we regress the number of papers per volume on a constant and an indicator variable equal to one for volumes with a conference and zero for volumes without a conference. We obtain the following estimated regression:

$$\underset{(\text{s.e.})}{\widehat{\text{Papers}}} = \underset{(1.25)}{10.778} + \underset{(2.17)}{4.333}\,\text{Conference}$$

Fig. 7. Dr. Robert Engle speaking at Conference dinner.

At the 10% level of significance, the impact of the AIE Conference (and its funding) has been to increase the number of contributed papers from 10.8 to 15.1, an increase of 40%.[3] The funding allows LSU and SMU to bring contributors to the conference. In addition, the ability to receive feedback in a workshop setting is attractive to authors.

Volume 21, "Modeling and Evaluating Treatment Effects in Econometrics," was edited by Thomas B. Fomby, R. Carter Hill, Daniel Millimet, Jeffrey A. Smith, and Edward J. Vytlacil. This volume addressed the estimation of the effects of treatments, which has engaged a lot of discussion during the early 2000s. Thirteen contributed papers and 23 contributing writers had over 232 citations (5th highest among AIE volumes) for the volume that covered most key components of the current evolution in this literature. The papers were presented in the 4th Annual AIE Conference in October 2005. This was the first AIE conference to be held at Southern Methodist University.

The subject of Volume 22, entitled "Econometrics and Risk Management," was credit risk and credit derivatives. This volume, which was edited by Jean-Pierre Fouque, Thomas B. Fomby, and Knut Solna, became especially relevant following the subprime crisis in August 2008. The 5th Annual AIE Conference took place in November 2006 at Cook Conference Center at Louisiana State University.

Following the success of the two parts of Volume 11, Siddhartha Chib, William Griffiths, Gary Koop, and Dek Terrell edited another volume of *Advances in Econometrics* on "Bayesian Econometrics." Volume 23 focuses on the scope and diversity of Bayesian applications as well as the Bayesian inference and computations, both on cross section and time series data. There are 21 contributed papers in this volume, ranking it as the volume with the highest number of papers (see Table 3). The papers were presented at the 6th Annual AIE Conference at Louisiana State University in November 2007.

Volume 24, "Measurement Error: Consequences, Applications and Solutions," was edited by Jane M. Binner, David L. Edgerton, and Thomas Elger. This volume focused on the measurement error in macroeconomic data and their impact on empirical work which impacts macroeconomic policy at the highest levels.

Volume 25 was edited by Qi Li and Jeffrey Racine. This volume focused on nonparametric estimation methods. Authors in the 25th volume presented their research at the 7th Annual AIE Conference at Louisiana State University in November 2008.

Volume 26 is titled "Maximum Simulated Likelihood Methods and Applications." This volume was edited by William Greene and R. Carter

Hill. Papers appearing in the 26th volume were presented at the 8th Annual AIE Conference at Louisiana State University in November 2009.

Volume 27 was edited by David M. Drukker, Director of Econometrics at Stata. The volume is titled "Missing Data Methods: Cross-Sectional Methods and Applications." Due to its length, Volume 27 is split into two parts. Papers appearing in the 27th volume were presented at the 9th Annual AIE Conference at Southern Methodist University in October 2010.

The first of the two forthcoming volumes is edited by Nathan Balke, Fabio Canova, Fabio Milani, and Mark Wynne. Volume 28 is titled "DSGE Models in Macroeconomics – Estimation, Evaluation, and New Developments." Papers for Volume 28 were presented at the 10th Annual AIE Conference at Southern Methodist University in November 2011 (Fig. 8).

Finally, Volume 29 consists of "Essays in Honor of Jerry Hausman." Co-editors for Volume 29 were Badi Baltagi and Whitney Newey. Papers for Volume 29 were recently presented at the 11th Annual AIE Conference at Louisiana State University in February 2012. The conference concluded with a discussion of the papers and final remarks by Jerry Hausman himself.

Fig. 8. 10th AIE Conference at Southern Methodist University.

RECENT DEVELOPMENTS

During the past 30 years, *Advances in Econometrics* has certainly succeeded in "expanding the use of newly vintaged econometric techniques throughout the empirical economic literature and beyond." Numerous outstanding econometricians, including four Nobel Prize winners, have published in the volume. Combined, the papers in AIE have been cited over 7,400 times. Professor Fomby stated that AIE has been able to establish a reputation for quality as a result of the outstanding papers submitted by its authors and their editorial comments and suggestions for improving the publication. Among these many contributors over the years, we acknowledge a few who deserve special mention.

First, the series editors, George F. Rhodes, Jr. and Robert Basmann, who started the series, and Thomas B. Fomby and R. Carter Hill who have been Senior Co-Editors for 26 and 17 years, respectively. In 2009, Ivan Jeliazkov (University of California at Irvine) joined Professors Fomby and Hill as Senior Co-Editor. Professor Jeliazkov earned his Ph.D. at Washington University in St. Louis. He has published extensively, particularly in the area of Bayesian econometrics and MCMC estimation. Professor Jeliazkov had previously published papers in Volumes 23 and 26 of AIE. Dr. Jeliazkov is an international scholar who worked in academic and research institutions around the world including China, Korea, and Australia. He is also the Associate Editor of the *International Journal of Mathematical Modelling and Numerical Optimisation.*

In 2010, Juan Carlos Escanciano (Indiana University) also became a Senior Co-Editor for AIE. Professor Escanciano earned his Ph.D. at Universidad Carlos III de Madrid, Spain. He has published numerous articles on specification tests in semiparametric and nonparametric models and many other areas of econometric theory. Dr. Escanciano also served as the Associate Editor of *Economic Bulletin* and is affiliated with numerous international econometric, statistical, and mathematical associations.

In 2012, Eric T. Hillebrand (Center for Research in Econometric Analysis of Time Series, Aarhus University) became the most recent Senior Co-Editor for AIE. Professor Hillebrand earned his Ph.D. from University of Bremen. Dr. Hillebrand has published extensively in the time series literature, particularly with regard to volatility models and financial econometrics.

In addition to the new Senior Co-Editors, AIE established an Advisory Board in 2011. The three inaugural member of the AIE Advisory Board were Eric Hillebrand (Aarhus University; now a Senior co-Editor),

Doug Miller (University of Missouri-Columbia), and Daniel Millimet (Southern Methodist University). The Advisory Board members are all respected econometricians who have previously contributed to AIE. Professor Hillebrand contributed a paper in Volume 20. Professor Miller contributed one paper in Volume 12 and two papers in Volume 17. Professor Millimet contributed papers in Volumes 21 and 27 and served as guest editor for Volume 21.

Finally, in February 2012 the election of *Advances in Econometrics* Fellows was introduced, and the two inaugural fellows are Badi Baltagi and Dek Terrell. Contributors are eligible to be elected an AIE Fellow provided they have:

(i) Served as editor or co-editor for three or more volumes of *Advances in Econometrics*, OR
(ii) Contributed four papers to the series as author or co-author.

As shown in Table 5, Professor Baltagi of Syracuse University (Fig. 9) ranks 3rd in total articles contributed to the series with 5 articles, which have appeared in Volumes 14, 15, 17, and 18. In addition, he served as editor for Volume 15, which has the highest number of total citations in the series (see Table 2). His contributed paper in Volume 15 is tied for the 4th most total citations (see Table 4). Professor Baltagi is also serving as co-editor for Volume 29 (forthcoming).

Fig. 9. Dr. Badi Baltagi.

Fig. 10. Dr. Dek Terrell.

Professor Terrell (Fig. 10) contributed a paper to Volume 23 and served as editor for Volumes 20A, 20B, 23, and 30 (forthcoming). Volume 20B ranks 9th in total citations (see Table 2). In addition, Table 3 shows that Volume 23 ranks 1st in total number of papers while both parts of Volume 20 rank in the top 10 (the two parts combined would rank 1st with 27 total papers). In addition, Professor Terrell's generous contributions via the Freeport-McMoRan Endowed Chair of Economics have provided necessary funding for the AIE Conference.

The annual AIE conference has further advanced the field of econometrics, allowing a unique forum for authors to share and receive feedback on their research. Professor Fomby expects the volume to become more international over the coming years and would like to see the conference held outside the United States sometime in the coming years.

We feel fortunate to have been students of R. Carter Hill and to have gotten to know Thomas B. Fomby and many of the contributors to the volume through the conferences. Dek Terrell recently organized the 12th Annual AIE Conference in honor of Professors Fomby and Hill (Figs. 11 and 12). Many colleagues showed up to honor these outstanding

Fig. 11. Dr. Thomas B. Fomby, Ms. Emma Whitfield, and Dr. R. Carter Hill.

Fig. 12. 12th AIE Conference at Louisiana State University.

econometricians whose hard work and dedication have been instrumental to the tremendous success of the AIE volumes. AIE celebrated its 30th year in 2012, and given the continuing contributions of many including the newest Senior Co-Editors and Advisory Board members, will be "expanding the use of newly vintage econometric techniques" for many years to come.

NOTES

1. Citation counts were obtained from Google Scholar (www.scholar.google.com) as of May 3, 2012.
2. The AIE website now lists a third criterion, "Papers should, where possible, contain empirical economic examples that illustrate the usefulness of described econometric techniques."
3. We ran a similar regression with citations as the dependent variable and found the coefficient on *conference* to be insignificant. This is not surprising since the conference is relatively recent and older papers have had more time to be cited. In addition, as shown in Table 2, a few volumes, especially Volume 15, tend to dominate in terms of citations.

ACKNOWLEDGMENTS

The authors would like to thank Thomas B. Fomby for taking time to talk with us about AIE and for his comments on the paper. We also thank Carter Hill and Dek Terrell for their comments, and Stephen Barnes for his valuable assistance in collecting data. Any remaining errors are our own.

REFERENCE

Baltagi, B. H. (1998). Worldwide institutional rankings in econometrics: 1989–1995. *Econometric Theory*, *14*, 1–43.

PART II
MODEL SPECIFICATION AND TESTING

BAYESIAN UNIT ROOT TESTING: THE EFFECT OF CHOICE OF PRIOR ON TEST OUTCOMES

Charley Xia and William Griffiths

ABSTRACT

A Monte Carlo experiment is used to examine the size and power properties of alternative Bayesian tests for unit roots. Four different prior distributions for the root that is potentially unity – a uniform prior and priors attributable to Jeffreys, Lubrano, and Berger and Yang – are used in conjunction with two testing procedures: a credible interval test and a Bayes factor test. Two extensions are also considered: a test based on model averaging with different priors and a test with a hierarchical prior for a hyperparameter. The tests are applied to both trending and non-trending series. Our results favor the use of a prior suggested by Lubrano. Outcomes from applying the tests to some Australian macroeconomic time series are presented.

Keywords: Test size and power; Monte Carlo simulation; model averaging; hierarchical prior

30th Anniversary Edition
Advances in Econometrics, Volume 30, 27–57

ISSN: 0731-9053/doi:10.1108/S0731-9053(2012)0000030007

INTRODUCTION

Whether or not macroeconomic time series contain unit roots continues to play an important role in econometrics and macroeconomic theory. The statistical properties of a unit root process differ substantially from those of its stationary counterpart, and the existence of unit roots in macroeconomic variables has implications for both empirical studies and theoretical modeling. From an econometrics perspective, the order of integration of time series variables has implications for modeling the data in levels or first differences. It is also the platform upon which we consider co-integrating or long-run equilibrium relationships between nonstationary time series. Business cycle theory, which describes fluctuations in output as transitory deviations from a long-run deterministic trend, loses its empirical support if macroeconomic variables have unit roots. The simple permanent income model predicts consumption is a random walk and co-integrated with total income. Models of persistent unemployment have formed around whether there is a natural rate of unemployment or there has been a unit root hysteresis effect (Blanchard & Summers, 1987). Whether stock prices are mean reverting has implications for predicting future prices based on historical data. The emphasis on unit roots in these and other examples has acted as catalyst for developing sophisticated testing procedures.

In a seminal paper detailing the importance of the unit root distinction in macroeconomic data, Nelson and Plosser (1982) used an augmented Dickey–Fuller test to find overwhelming support for the unit root hypothesis in 13 out of 14 long-run annual time series from the United States. However, beginning from Sims (1988), Bayesian authors have highlighted inadequacies in classical methods. Their major point was that the asymptotic distribution of classical test statistics changes discontinuously between the stationary and the unit root cases, leading to disconnected confidence intervals. The Bayesian framework was argued to be a more logically sound and consistent basis for testing the unit root hypothesis. The choice between classical or Bayesian methods was further fuelled when Dejong and Whiteman (1991), using a Bayesian approach, found significantly less evidence for unit root processes in the Nelson and Plosser dataset.

These results were questioned by Phillips (1991), who illustrated that Bayesian test conclusions were not robust to prior specifications. He argued that a uniform prior for time series coefficients was inappropriate as a non-informative prior, and that, with a Jeffrey's prior, Bayesian and classical procedures yield similar results. Lubrano (1995) extended the refinement of Bayesian testing procedures by showing that the specification of the test model and an adequate treatment of the initial observation are essential to making sensible inferences.

In this paper we investigate further the properties of Bayesian tests for unit roots. We examine the size and power properties of tests derived from alternative prior densities for the (potential) unit root, and we apply the tests to some Australian macroeconomic time series. The testing model is developed in the second section. In the third section we specify the marginalized likelihood for the hypothesized unit root and discuss the merits of alternative assumptions about initial observations. In the fourth section we describe two Bayesian testing procedures: a credible interval test and a Bayes factor test. The alternative priors that we consider are outlined in the fifth section; these include four basic priors, a test based on model averaging with different priors, and a hierarchical prior for a hyperparameter. Monte Carlo results on the size and power of the test procedures are presented in the sixth section for both trending and non-trending data. The results from applying the tests to macroeconomic data are given in the seventh section.

FORMULATION OF THE TESTING MODEL

When setting up a model to test for unit roots, it is convenient to consider two cases. The first is where the series is trending and we wish to use a unit root test to distinguish between a trend stationary process with a deterministic trend, and a random walk with drift. The second is where the series has no definite trend and we wish to distinguish between a stationary autoregressive (AR) process and a random walk. A simple model that illustrates both these cases is

$$y_t = \mu + \delta t + u_t \qquad u_t = \rho u_{t-1} + \varepsilon_t \tag{1}$$

where we assume the ε_t are independent $N(0, \sigma^2)$. This parameterization is often labeled a components model since it makes clear distinctions between the deterministic and stochastic components. Combining the two equations in Eq. (1) gives

$$y_t = (1-\rho)\mu + \rho\delta + (1-\rho)\delta t + \rho y_{t-1} + \varepsilon_t \tag{2}$$

The trending and non-trending cases correspond to $\delta \neq 0$ and $\delta = 0$, respectively. The unit root hypothesis is $H_0 : \rho = 1$. Under this null hypothesis, and the trending case, the intercept μ and time trend t disappear and δ becomes the drift parameter in the random walk with drift model $y_t = \delta + y_{t-1} + \varepsilon_t$. The alternative hypothesis $H_1 : |\rho| < 1$ yields a trend stationary model, characterizing a process that is stable and mean reverting around a deterministic trend. If y_t is logarithmic, then δ has a direct interpretation as the rate of growth. In the non-trending case the model under the null is the random walk $y_t = y_{t-1} + \varepsilon_t$; under the alternative the process is stable and mean reverting around the mean μ. When $|\rho| > 1$, we have an explosive process. While asymptotic theory excludes the possibility of an explosive infinite series, $|\rho| > 1$ can be modeled in the Bayesian framework under the assumption of a finite time frame. The above components model has been adopted by Zivot (1994) and Lubrano (1995) for unit root testing, and promoted by Schotman and van Dijk (1991a, 1991b) and DeJong, Nankervis, Savin, and Whiteman (1992).

An alternative parameterization that has been used in unit root testing is the so-called reduced form of the components model

$$y_t = \beta_0 + \beta_1 t + \rho y_{t-1} + \varepsilon_t \tag{3}$$

where the parameters in Eqs. (1) and (3) are related by

$$\beta_0 = (1 - \rho)\mu + \rho\delta \qquad \beta_1 = (1 - \rho)\delta \tag{4}$$

In this formulation, the convenience of a linear model in Eq. (3) is complicated by the nonlinear dependence of β_0 and β_1 upon ρ in Eq. (4). To establish the same dynamics as exist for the unit root hypothesis test using the components model, the constraints $\beta_0 = \delta$ and $\beta_1 = 0$ are required under the null. If we consider β_0 and β_1 as independent of ρ, then y_t will not be of the same order of magnitude under the null and alternative hypotheses. In particular, if the time trend coefficient $\beta_1 \neq 0$, then y_t will be $O(t)$ under the alternative hypothesis and $O(t^2)$ under the null (see Bauwens, Lubrano, & Richard, 1999, pp. 164–168). This behavior, where a time trend is always present, will be undesirably captured by a likelihood function that is based on the reduced form. Consequently, the null and alternative hypotheses will no longer discriminate between a random walk with drift model and a trend stationary model. In the components model the time trend disappears in the presence of a unit root and whether $\delta = 0$ then governs the order of magnitude of y_t. There is no discrepancy in the behavior of the data under the null and alternative hypotheses. For this reason the components model is considered the better parameterization for a Bayesian framework; it is the one that we consider in this paper.

Generalizing to an AR(p) Process

Eqs. (1) and (2) are relevant for an AR(1) process. To incorporate a more sophisticated serial correlation structure, we rewrite the error process in Eq. (1) as $A(L)u_t = \varepsilon_t$ where $A(L)$ is a polynomial of order p in the lag operator; that is, $A(L) = 1 - a_1L - a_2L^2 - \cdots - a_pL^p$. A stationary property of y_t is equivalent to the roots of the polynomial function $A(L)$ lying outside the unit circle. The possibility of a unit root is isolated by the convenient representation (see Hamilton, 1994, pp. 516–518)

$$A(L) = (1 - \rho L) + (1 - L)A * (L) \tag{5}$$

where $\rho = 1 - A(1)$, $A * (L) = a_1^*L + \cdots + a_{p-1}^*L^{p-1}$, and $a_j^* = \sum_{i=j}^{p-1} a_{i+1}$, $j = 1, 2, \ldots, p - 1$. A data process has a unit root if $\rho = 1$. Using Eq. (5) and substituting $A(L)u_t = \varepsilon_t$ into $y_t = \mu + \delta t + u_t$ yields

$$y_t = (1 - \rho)\mu + \rho\delta + (1 - \rho)\delta t + \rho y_{t-1} + A * (L)(\Delta y_t - \delta) + \varepsilon_t \tag{6}$$

Here one can see that the parameter ρ is concerned with the long-run dynamics of the series, while the a_j^* parameters are concerned with the short term. Relative to the AR(1) model, p–1 lags of Δy_t have been added.

The lags of Δy_t include an extra nonlinearity in δ. Lubrano (1995) proposed replacing $(\Delta y_t - \delta)$ by $(\Delta y_t - \hat{\delta})$, where $\hat{\delta}$ is an estimator of δ from the empirical mean of Δy_t. This simplification can be theoretically justified as linearizing $A * (1)\delta$ around $\widehat{A*(1)\delta} = 0$ and $\hat{\delta}$. Its advantage is that it leads to simpler calculations and a marginal likelihood function of ρ that retains the same structure irrespective of the autoregressive order. Rewriting Eq. (6) with this simplification yields

$$y_t = (1 - \rho)\mu + \rho\delta + (1 - \rho)\delta t + \rho y_{t-1} + A * (L)\Delta y_t^* + \varepsilon_t \tag{7}$$

where $\Delta y_t^* = \Delta y_t - \hat{\delta}$.

Conditioning on the parameter of interest ρ, the components model in Eq. (7) is linear in the nuisance parameters $\beta = (\mu, \delta, a_1^*, \ldots, a_{p-1}^*)'$, which allows us to write it in a regression form given by

$$y_t(\rho) = x_t'(\rho)\beta + \varepsilon_t \qquad \varepsilon_t \sim N(0, \sigma^2) \tag{8}$$

To obtain Eq. (8), we define $x_t' = (1, t)$, $y_t(\rho) = (1 - \rho L)y_t$, and $x_t'(\rho) = (x_t' - \rho x_{t-1}', \Delta y_{t-1}^*, \ldots, \Delta y_{t-p+1}^*)$. Since an AR($p$) model reserves the first p observations, we let the sample size be $T+p$. Then, stacking the last T observations, we obtain the matrix regression format (conditional on ρ)

$$y(\rho) = X(\rho)\beta + \varepsilon \tag{9}$$

where $y(\rho)$ is a $(T \times 1)$ vector function of ρ, $X(\rho)$ is a $(T \times (p+1))$ matrix function of ρ, and $\varepsilon \sim N(0, \sigma^2 I_T)$. For an AR(1) process, we have $x_t'(\rho) = (x_t' - \rho x_{t-1}')$ and $\beta = (\mu, \delta)'$. This notation can be found in Zellner and Tiao (1964), Lubrano (1995), and Bauwens et al. (1999).

MARGINALIZED LIKELIHOOD FOR ρ

To test for a unit root under the Bayesian framework, the marginal posterior density $f(\rho|y)$ is required. From Bayes' theorem, it is given by

$$f(\rho|y) \propto L(\rho; y)\pi(\rho)$$

where $\pi(\rho)$ is a prior density for ρ, and $L(\rho;\, y)$ is the marginalized likelihood function for ρ given by

$$L(\rho; y) = \int_{\phi(\rho)} L(\rho, \beta, \sigma^2; y)\pi(\beta, \sigma^2|\rho)d\beta d\sigma^2 \tag{10}$$

In Eq. (10), $L(\rho, \beta, \sigma^2; y)$ is the joint likelihood function, and $\pi(\beta, \sigma^2|\rho)$ is a (conditional) prior density with support $\varphi(\rho)$.

The nature of both the joint likelihood and the marginalized likelihood depends on the treatment of the initial observations $y_0, y_1, y_2, \ldots, y_{p-1}$. If these observations are treated as fixed, then the joint likelihood function is an approximate one, given by

$$L_A(\beta, \sigma^2, \rho; y) \propto (\sigma^2)^{-T/2} \exp\left\{-\frac{1}{2\sigma^2}[y(\rho) - X(\rho)\beta]'[y(\rho) - X(\rho)\beta]\right\} \tag{11}$$

The corresponding marginalized likelihood function, obtained by expressing ignorance using the diffuse prior $\pi(\beta, \sigma^2|\rho) \propto \sigma^{-2}$, and then integrating out $(\beta,\, \sigma^2)$ as depicted in Eq. (10), is

$$L_A(\rho; y) \propto |\tilde{M}(\rho)|^{-1/2}\tilde{S}(\rho)^{-[T-(p+1)]/2} \tag{12}$$

where $\tilde{M}(\rho) = X'(\rho)X(\rho)$ and $\tilde{S}(\rho) = y'(\rho)[I_T - X(\rho)\tilde{M}^{-1}(\rho)X'(\rho)]y(\rho)$. It turns out that $L_A(\rho;\, y)$ is not suitable for posterior inference about ρ because it is unbounded at $\rho = 1$ when the constant term μ is present (Lubrano, 1995). To conduct meaningful posterior analysis, we require the marginalized likelihood function to be finite and positive at $\rho = 1$ and to be integrable over the parameter space of ρ.

This problem can be overcome by making a distributional assumption about the initial observation y_0. It is often reasonable to assume that the observed sample is a segment of a data generating process, one which began long ago or in the infinite past. The initial observation is the closest reference to this unobservable past, and can bring in valuable information. As noted in Uhlig (1994), one can compare the distance of the initial observation from the time trend to form an impression of the size of the root. To see this, consider an AR(1) process and the components model in Eq. (1), and assume some starting date at $t = -s$ such that $u_{-s} = \phi$ is fixed. The initial observation y_0 can be shown to have the distribution

$$y_0 | (\mu, \sigma^2, \phi, s) \sim N(\mu + \rho^s \phi, \sigma^2 q(\rho, s))$$

with

$$q(\rho, s) = \sum_{i=0}^{s} \rho^{2i} = \frac{1 - \rho^{2(s+1)}}{1 - \rho^2}$$

Since $q(\rho, s)$ is a geometric series of nonnegative ρ^2, it is well defined for all $\rho \in \mathbb{R}$, and there is no implicit truncation of the parameter space. Under the stationarity assumption, and an infinite time horizon, $y_0 \sim N(\mu, \sigma^2/(1 - \rho^2))$. For small ρ the variance of y_0 is small and we should observe the initial observation close to the time trend.

To incorporate the density of the initial observation we simplify calculations by assuming $\phi = 0$; then,

$$y_0 | (\mu, \sigma^2, s) \sim N(\mu, \sigma^2 q(\rho, s)) \tag{13}$$

The exact likelihood function of an AR(p) process considers the first p observations as random. However, this joint density is complex (see Hamilton, 1994, pp. 123–125). Following Schotman and van Dijk (1991a, 1991b), and Lubrano (1995), we consider only the distribution of y_0 and let $(y_1, y_2, \ldots y_{p-1})$ remain fixed.

The exact likelihood function for (β, σ^2, ρ) is then calculated as the product of the conditional data density in Eq. (11) and the marginal distribution of y_0 in Eq. (13)

$$L_E(\beta, \sigma^2, \rho; y) \propto q(\rho, s)^{-1/2} (\sigma^2)^{-T/2} \times \exp\left\{ -\frac{1}{2\sigma^2} \left([y(\rho) - X(\rho)\beta]'[y(\rho) - X(\rho)\beta] + \frac{1}{q(\rho, s)} (y_0 - \mu)^2 \right) \right\} \tag{14}$$

It is convenient to rewrite the term in the exponent as

$$-\frac{1}{2\sigma^2}([y(\rho) - X(\rho)\beta]'[y(\rho) - X(\rho)\beta] + (y_0^* - x_0'\beta)'(y_0^* - x_0'\beta))$$

where $y_0^* = y_0\, q(\rho, s)^{-1/2}$ and $x_0' = (q(\rho, s)^{-1/2}, 0)$. Then, combining Eq. (14) with the diffuse prior $\pi(\beta, \sigma^2|\rho) \propto \sigma^{-2}$, and integrating out (β, σ^2), yields the marginalized exact likelihood function

$$L_E(\rho; y, s) \propto q(\rho, s)^{-1/2}\left|M(\rho)\right|^{-1/2} S(\rho)^{-(T-p)/2} \tag{15}$$

where

$$M(\rho) = X'(\rho)X(\rho) + \begin{pmatrix} q(\rho, s)^{-1} & 0 \\ 0 & 0 \end{pmatrix}$$

$$\tilde{\beta}(\rho) = M(\rho)^{-1}\left[X'(\rho)y(\rho) + \begin{pmatrix} y_0 q(\rho, s)^{-1} \\ 0 \end{pmatrix}\right]$$

$$S(\rho) = y'(\rho)y(\rho) + y_0^2 q(\rho, s)^{-1} - \tilde{\beta}'(\rho)M(\rho)\tilde{\beta}(\rho)$$

A function $q(\rho, v)$ which behaves similarly to $q(\rho, s)$ for reference values of s and v, but is easier to use is (Bauwens et al., 1999, pp. 182–183)

$$q(\rho, v) = \frac{1 + v}{1 + v - \rho^2} \tag{16}$$

To ensure the variance of y_0 in Eq. (11) is nonnegative, $q(\rho, v)$ implicitly truncates the domain of ρ to $\rho \in (-\sqrt{1 + v}, \sqrt{1 + v})$. Both s and v are subjective and affect the likelihood function, but the impact of choosing v which controls the range of values of ρ is more conspicuous than the impact on ρ of choosing a starting date s. Also, calculating the Jeffreys prior (considered below) is easier for $q(\rho, v)$. In what follows our calculations are based on replacing $q(\rho, s)$ with $q(\rho, v)$. For testing, Bauwens et al. recommended a value of $v = 1/3$ which was sufficiently large in the sense that the truncation point $\sqrt{1 + v} = 1.155$ never actively affected the likelihood function.

TESTING THE UNIT ROOT HYPOTHESIS

The ingredients for testing the unit root hypothesis are the marginalized likelihood function $L_E(\rho; y, v)$ as defined in Eq. (15), but with $q(\rho, s)$ replaced by $q(\rho, v)$, and a prior density $\pi(\rho)$. Before considering a number of possible choices for $\pi(\rho)$, we consider two Bayesian testing procedures, one based on a Bayes factor, and one based on a credible interval.

For the Bayes factor approach, we consider the posterior probability for the point null hypotheses $H_0 : \rho = 1$ against the alternative $H_1 : \rho \neq 1$. The starting point is the assigning of a probability mass ω to $H_0 : \rho = 1$, and defining the prior as

$$\pi(\rho) = \begin{cases} \omega & \text{for } \rho = 1 \\ (1 - \omega)\pi^*(\rho) & \text{for } \rho \neq 1 \end{cases}$$

where $\omega \in [0, 1]$ and $\pi^*(\rho)$ is required to be a proper density integrating to one. The posterior probability of H_0 is then given by

$$P(H_0|y, v) = \left[1 + \frac{1 - \omega}{\omega B}\right]^{-1}$$

where B is the Bayes factor

$$B = \frac{L_E(\rho = 1; y, v)}{\int_{\rho \neq 1} L_E(\rho; y, v)\pi^*(\rho)d\rho}$$

If we set $\omega = 0.5$, implying equal prior weights to the null and alternative hypotheses, then the posterior probability $P(H_0|y, v)$ depends only on the Bayes factor. The decision rule: reject the unit root hypothesis if $P(H_0|y, v) < 0.5$ is equivalent to rejecting H_0 when $B < 1$. Posterior odds ratios for point null hypotheses have been considered by Sims (1988), Schotman and van Dijk (1991a, 1991b), and Koop (1991).

For the credible interval approach for testing $H_0 : \rho = 1$, we reject H_0 if $\rho = 1$ lies outside the 95% credible interval $(-\sqrt{1 + v}, \rho_{\text{sup}})$, where ρ_{sup} is defined by

$$\int_{-\sqrt{1+v}}^{\rho_{sup}} f(\rho|y)d\rho = 0.95$$

This approach is equivalent to calculating a nonstationary probability

$$P(\rho \geq 1|y, v) = \int_{1}^{\sqrt{1+v}} f(\rho|y)d\rho$$

and rejecting the unit root hypothesis if $P(\rho \geq 1|y, v) < 0.05$. The above testing procedure has been adopted by Phillips (1991) and Lubrano (1995).

For any continuous proper density $\kappa(\rho)$ defined over the domain of ρ, it is possible to calculate a Bayes factor and a credible interval. Berger and Delampady (1987) recommend reporting both for hypothesis testing problems with a point null hypothesis. The Bayes factor provides the data evidence against the point null, while the credible interval reports the size of the discrepancy. As conclusions reached under the two testing procedures may not always agree for the same set of observations, it is of interest to study the consequences of using both. Our Monte Carlo simulations will attempt to address this problem along with the choice of prior for ρ.

ALTERNATIVE PRIORS FOR ρ

The prior in Bayesian analysis serves as an elicitation of the information available to the decision maker before observing the data. When there is insufficient information or in situations which desire objective inference, the search has been for a non-informative prior to convey some notion of "ignorance" or "emphasizing" data evidence. One central focus of the Bayesian literature on unit root testing has been the appropriate non-informative prior for ρ. Since the early work of Bayes (1764), the uniform distribution has been seen as non-informative for location parameters. A uniform prior for ρ was used by Sims (1988) to illustrate how Bayesian procedures can reject the unit root hypothesis systematically more than classical procedures. However, this foundation has been deeply questioned by Phillips (1991) who raised concerns that ρ is not akin to a location parameter and its impact upon sample moments (e.g., mean, variance, autocorrelations) changes drastically for different values. As an objective framework, Phillips proposed a Jeffreys prior based on the principles of Jeffreys (1946), and the arguments of Perks (1947). The Jeffrey's prior is proportional to the square root of the determinant of the information matrix. It leads to posterior inferences that are invariant under one-to-one reparameterizations, and, since it favors parameter values where the information is large, it has been interpreted as emphasizing data evidence. For ρ the Jeffreys prior turns out to be steeply increasing as ρ approaches unity and beyond. This finding is then interpreted to mean that a process generated in these intervals is itself more "informative" in the sense of high Fisher's information. Within such views the flat uniform prior is then subjective, since it implicitly down-weighs the sample information for large ρ values by imposing a bias toward the stationary region. However, there have been criticisms of Jeffreys rule, and, given that the asymptotic distribution for ρ hits a discontinuity at unity, with explosive values inconsiderable, Perks' justifications can only be confined to the stationary region of the parameter space.

Following the non-informative prior debate for unit root testing, Berger and Yang (1994) and Lubrano (1995) proposed alternative solutions. In the remainder of this section, we describe the priors used in our Monte Carlo experiment: the uniform, Jeffreys, Berger–Yang and Lubrano priors, a model averaging solution to the choice of prior, and a prior for v that can be used as an alternative to fixing v.

Jeffreys Prior

The Jeffreys prior based on the exact likelihood function of an AR(1) process with no time trend and the structural form in Eq. (2) is (see Bauwens et al., 1999, pp. 195–196)

$$\pi_J(\rho|v,T) \propto \begin{cases} \sqrt{\frac{1+v+\rho^2}{1+v-\rho^2}\left((1+v)\frac{1-\rho^{2T}}{1-\rho^2}-1\right)+\frac{1}{1-\rho^2}\left(T-\frac{1-\rho^{2T}}{1-\rho^2}\right)} & \rho \neq 1 \\ \sqrt{\frac{2+v}{v}((1+v)T-1)+\frac{T(T-1)}{2}} & \rho = 1 \end{cases} \tag{17}$$

This prior has an integrable infinite singularity at $\rho = \pm\sqrt{1+v}$ and is therefore proper by truncating $\rho \in (-\sqrt{1+v}, \sqrt{1+v})$; T is the sample size.

Lubrano Prior

This prior is an arcsine distribution proposed by Lubrano (1995) after observing the similarity of this distribution with the Jeffreys prior in Eq. (17), and removing the unintuitive property of having a prior dependent on the sample size T.

$$\pi_L(\rho|v) = \frac{1}{\pi\sqrt{1+v-\rho^2}} \tag{18}$$

Berger–Yang Reference Prior

Reference priors originated with Bernardo (1979) under the principle of maximizing the missing information in an experiment, where distance between densities is measured by the Kullback–Leibler divergence. The Berger and Yang (1994) reference prior for the AR(1) process with no time trend and truncation $\rho \in (-\sqrt{1+v}, \sqrt{1+v})$ is

$$\pi_{BY}(\rho|v) = \begin{cases} \frac{1}{2\pi\sqrt{1-\rho^2}} & \rho<1 \\ \frac{\text{arcsec}^{-1}(\sqrt{1+v})}{4|\rho|\sqrt{\rho^2-1}} & 1<|\rho|<\sqrt{1+v} \end{cases} \tag{19}$$

This density has an integrable infinite singularity at $\rho = 1$.

Uniform Prior

A uniform or diffuse prior for $\rho \in (-\sqrt{1+v}, \sqrt{1+v})$ is given by

$$\pi_U(\rho|v) = \frac{1}{2\sqrt{1+v}} \tag{20}$$

Plots of the above four functions are given in Figs. 1(a) and 1(b) for $v = 1/3$ and $T = 25$. Two different graphs are needed to accommodate a change

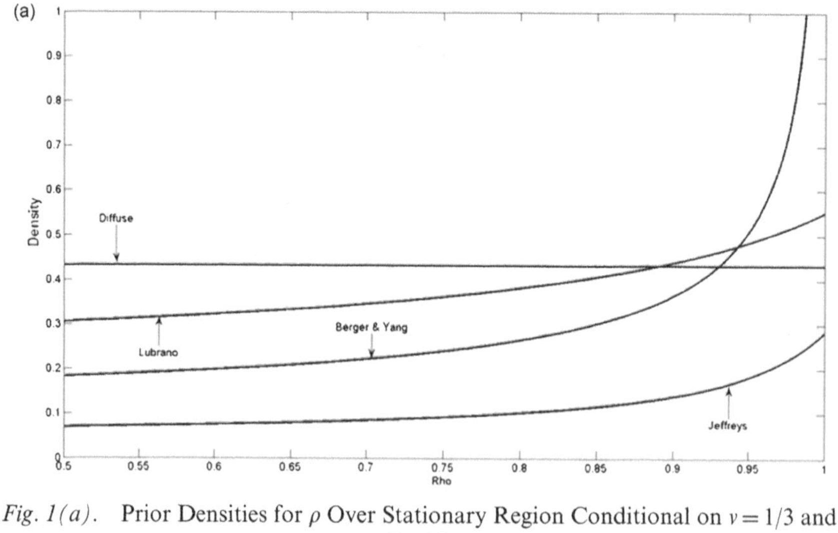

Fig. 1(a). Prior Densities for ρ Over Stationary Region Conditional on $v = 1/3$ and $T = 25$.

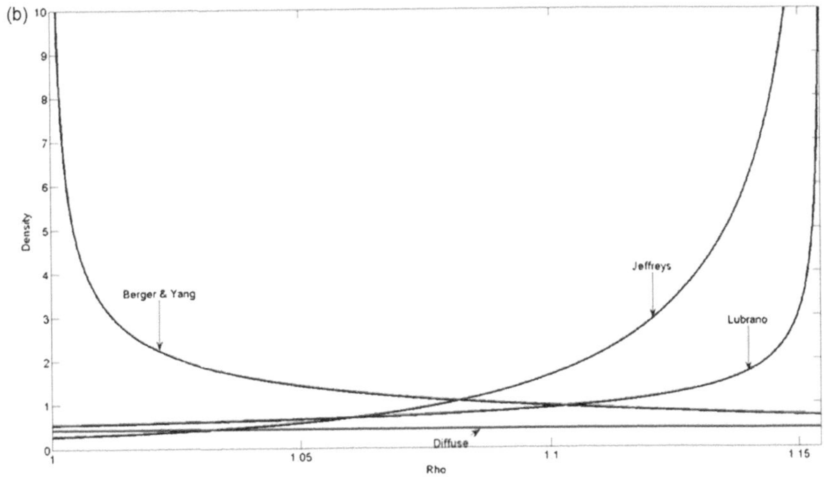

Fig. 1(b). Prior Densities for ρ Over Nonstationary Region Conditional on $v = 1/3$ and $T = 25$.

of scale for $\rho > 1$. All priors are quite diffuse over the stationary region up to approximately 0.95, but their treatment over the nonstationary region differs substantially. The Jeffreys prior explodes exponentially over the nonstationary region at a rate governed by the sample size T. The explosion is more controlled for the Lubrano prior, while the Berger–Yang prior is decreasing over the nonstationary region.

Model Averaging with Different Priors

If a decision maker recognizes uncertainty in the selection process for a prior, then a natural extension is to accommodate this uncertainty through Bayesian model averaging. Specifically, we may recognize four different Bayesian models that are described by the same marginalized exact likelihood function for ρ, the same belief for the domain $\rho \in (-\sqrt{1+v}, \sqrt{1+v})$, but four different priors given by Eqs. (17)–(20). If $M_i = \{L_E(\rho; y, v), \pi_i(\rho|v)\}, i = J, L, BY, U$, then the posterior probability for model i is

$$P(M_i|y, v) = \frac{P(y|M_i, v)P(M_i)}{P(y|v)}$$

where

$$P(y|M_i, v) = \int_{-\sqrt{1+v}}^{\sqrt{1+v}} L_E(\rho; y, v)\, \pi_i(\rho|v) d\rho$$

$$P(y|v) = \sum_{i=1}^{4} P(y|M_i, v) P(M_i)$$

The model-averaged posterior density for ρ is

$$\begin{aligned} f(\rho|y, v) &= \sum_{i=1}^{4} f(\rho|y, v, M_i) P(M_i|y, v) \\ &\propto L_E(\rho; y, v) \sum_{i=1}^{4} \pi_i(\rho|v) P(M_i|y, v) \\ &\propto L_E(\rho; y, v) \sum_{i=1}^{4} \pi_i(\rho|v) P(y|M_i, v) \quad \text{if } P(M_i) = 1/4 \end{aligned} \tag{21}$$

Assuming the four priors are given equal prior probabilities, from Eq. (21) we can see that, a posteriori, the prior that carries the most weight is the one with largest marginal likelihood $P(y|M_i, v)$. Using this framework, the Bayes factor is given by $B = L_E(\rho = 1; y, v)/P(y|v)$, and the posterior probability of nonstationarity is

$$P(\rho \geq 1|y, v) = \sum_{i=1}^{4} P(\rho \geq 1|y, v, M_i) P(M_i|y, v)$$

A Prior for v

In the application of unit root testing it is often sensible to assume that ρ values implying explosive series are unlikely. The Bayesian procedures described above can accommodate this subjective belief through two channels: the first is to specify a prior distribution for ρ such that its density is decreasing for $\rho \geqslant 1$. However, if one is to accept the arguments of Phillips (1991), and use the Jeffreys and Lubrano priors that explode over the nonstationary interval, reflecting increasing sample information, an alternative must be found. The alternative that we have suggested for the specifications considered so far is to restrict the parameter space to $\rho \in (-\sqrt{1+v}, \sqrt{1+v})$, and to follow the recommendation of $v = 1/3$ as a possible choice (Bauwens et al., 1999). One way to generalize this approach, and make it more flexible, is to assign a prior distribution to v. We chose to assign an exponential distribution with mean $E(v)=\lambda^{-1}$, where λ is chosen such that $P(v>1/3)=0.05$. This constitutes a hierarchical approach where the choice of a domain for ρ is made separately from the choice of prior density in this uncertain domain. The exponential distribution can be justified as that which maximizes entropy subject to the constraint that the mean is equal to λ^{-1}, where λ is a constant ensuring that $P(v>1/3)=0.05$ ex post.

Under this set of assumptions, the joint posterior density for (ρ, v) is

$$f(\rho, v|y) \propto L_E(\rho, v; y)\pi(\rho|v)\pi(v) \tag{22}$$

where $L_E(\rho, v; y)$ is given by Eqs. (15) and (16), $\pi(\rho|v)$ is one of the priors in Eqs. (17)–(20), and $\pi(v)$ is the exponential prior density. To perform the credible interval test using this set up, we need to evaluate

$$P(\rho \geq 1|y) = \int_0^\infty \int_1^{\sqrt{1+v}} f(\rho, v|y)d\rho dv$$

For the Bayes factor test, the likelihood function at $\rho=1$ can be derived as

$$L_E(\rho = 1; y) = \int_0^\infty L_E(\rho = 1, v; y)\pi(v)dv$$

The marginal likelihood $P(y)$ needed for the Bayes factor (and Bayesian averaging if it used) is

$$P(y) = \int_0^\infty \int_{-\sqrt{1+v}}^{\sqrt{1+v}} L_E(\rho, v; y)\pi(\rho|v)\pi(v)d\rho dv$$

MONTE CARLO SIMULATIONS

A Monte Carlo simulation can be used to study the repeated-sample performance of the Bayesian testing procedures, the choice of priors, and of Bayesian methods in general. A desirable test should minimize test size and maximize test power, although the Bayesian test procedures we have described are not such that they would necessarily conform to the sampling theory convention of a 0.05 size. Moreover, a "pure" Bayesian, who makes a decision conditional on one observed sample, may not be interested in sampling theory criteria for test evaluation. Nevertheless, there are enough sampling-theorists out there to warrant a test evaluation within their framework.

Two AR(1) processes were used in the simulation: one with a constant and no time trend, and one with a constant and a time trend. These two models capture the dynamic properties of a random walk and a random walk with drift, respectively. The true parameters used were

$$\mu = 1 \qquad \sigma^2 = 1 \qquad \delta = \{0, 1\} \quad \rho = [0.75 : 0.01 : 1.00]$$
$$y_0|v \sim N(1, q(\rho, v)) \qquad\qquad v = \left\{\tfrac{1}{6}, \tfrac{1}{3}\right\}$$

The sample size was $T = 50$, and $N = 10{,}000$ replications were used for each ρ value. Three rejection criteria for the unit root hypothesis were considered: (i) $B < 1$, (ii) $P(\rho \geq 1|y, v) < 0.05$, and (iii) both $B < 1$ and $P(\rho \geq 1|y, v) < 0.05$ hold. When an exponential prior was assigned to v, (ii) was replaced by $P(\rho \geq 1|y) < 0.05$. For the exponential prior we set $\lambda = 8.985$, implying $P(v > 1/3) = 0.05$. The Bayesian tests were further benchmarked against the relevant Dickey–Fuller test for each model.

Numerical integration was handled using the package Matlab and the in-built adaptive Gauss–Kronrod quadrature function (quadgk). This algorithm allows for integration of integrable singularities and pretesting was done to ensure the calculated integrating constants multiplied by the kernel densities integrated to one. Double integrals required for the case where a prior is assigned to v used Matlab's quad2d function.

(a)

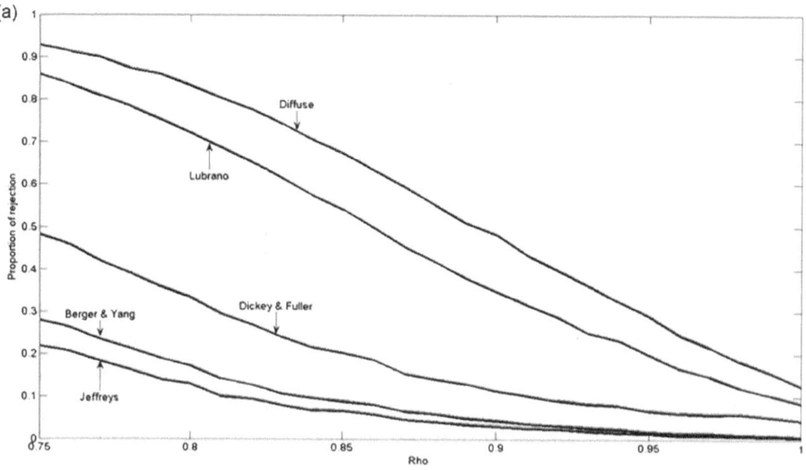

Fig. 2(a). Power Functions for Credible Interval Test on Non-Trending Data with $v = 1/3$.

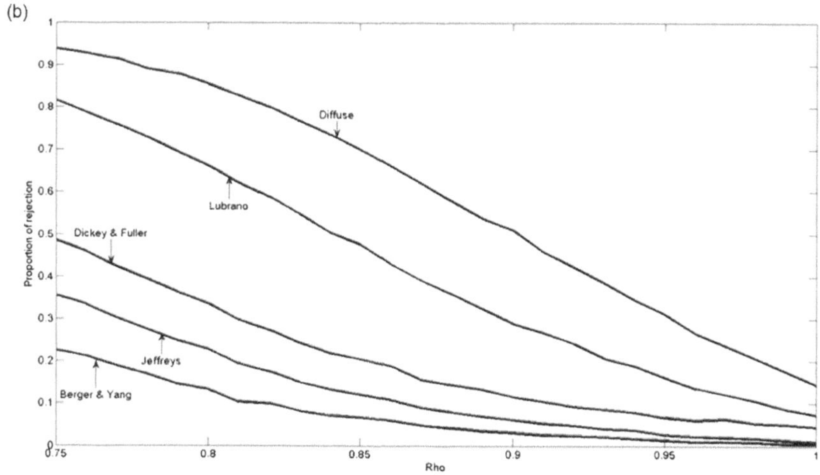

Fig. 2(b). Power Functions for Credible Interval Test on Non-Trending Data with $v = 1/6$.

Results for Non-Trending Case

The simulated power functions for the credible interval tests applied to the non-trending AR(1) case are presented in Figs. 2(a) and 2(b) for $v = 1/3$ and $v = 1/6$, respectively. Perhaps the most striking point is the relatively large test size for the uniform (diffuse) prior case. It is 13.4% for $v = 1/3$ and 15.2% for $v = 1/6$. This result supports the case argued in Phillips (1991), that the uniform prior is biased toward the stationary region. The Lubrano prior also results in a high test size relative to the Dickey–Fuller 5% test, although at 8.7% for $v = 1/3$ and 6.1% for $v = 1/6$, it is not as marked as that for the uniform prior. The controlled explosion of Lubrano's prior over the nonstationary region only occurs close to the truncation point $\sqrt{1+v}$. A large v means that this explosion has less chance to affect the likelihood function than the Jeffreys prior which explodes exponentially. Its power at $\rho = 0.95$ is 16.8% for $v = 1/6$ and 19.2% for $v = 1/3$. The test powers for the uniform and Lubrano cases are also much larger than those from the other priors, perhaps reflecting the greater test size, although the difference is dramatic. The Lubrano prior seems to be well suited for a credible interval test, with the Monte Carlo simulation suggesting that its size and power can be uniformly controlled by the choice of v, and that the size and power function from a uniform prior can be uniformly brought down by an appropriate choice of v.

The sizes of the tests from the Jeffreys and Berger–Yang priors are approximately equal and very low, and these tests also have relatively low power. Their relative power performance is sensitive to v: the Jeffreys prior leads to a more powerful test for small v, and the Berger–Yang prior does better for larger v. The Berger–Yang prior assigns prior probabilities of 0.25 to each of the intervals $\rho \in [0, 1)$ and $\rho \in (1, \sqrt{1+v})$. A small v implies more of the density is concentrated around explosive values closer to unity. Conditional on a smaller v having no systematic effect on the likelihood function, as v gets smaller this prior will increasingly amplify the sample probabilities for the nonstationary region. The observed low power for the credible interval test suggests the likelihood function tends to have significant tail probabilities extending to $\rho \geqslant 1$. The lower power for $v = 1/6$ compared to $v = 1/3$ occurs because the prior's amplification effect in the nonstationary region is larger for smaller v. The opposite effect occurs with the Jeffreys prior which explodes exponentially over the nonstationary region. For $\rho = 0.95$, it leads to a power of 3.9% when $v = 1/6$ and 1.6% when $v = 1/3$.

The different sizes of the tests make it difficult it compare the Bayesian tests with the Dickey–Fuller test, but, from the steepness of the power functions, it would appear that size-corrected versions of the tests from the uniform and Lubrano priors would be more powerful than the Dickey–Fuller test, but those from the Jeffreys and Berger–Yang priors would not.

We now consider the Bayes factor tests whose power functions are graphed in Figs. 3(a) and 3(b). The following observations can be made:

1. All Bayesian test sizes are approximately the same (slightly less than the 5% of the Dickey–Fuller test), with the exception of that from the Jeffreys prior. Particularly noteworthy is the 3.9% test size from the uniform prior. In contrast to the credible interval test, the Bayes factor test with a uniform prior does not overly favor the stationary region. This finding is consistent with a point made by Berger and Delampady (1987) who caution against using credible regions corresponding to diffuse priors as a test for a point null hypothesis. They note that the likelihood of a special point ρ_0 that lies outside a credible interval is often not too much smaller than the average likelihood of ρ within the interval. They recommended the credible interval be taken as a discrepancy measure of the true parameter ρ from a point null hypothesis value ρ_0.
2. The tests from the uniform and Lubrano priors maintain their relative positions as the most powerful tests.

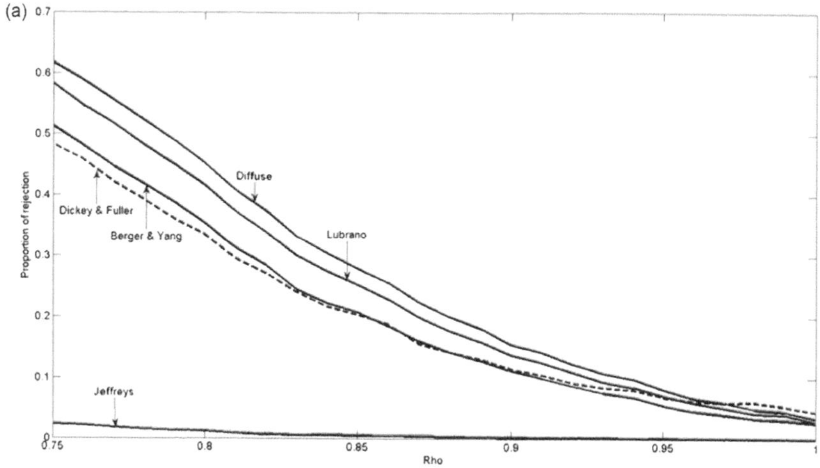

Fig. 3(a). Power Functions for Bayes Factor Test on Non-Trending Data with $\nu = 1/3$.

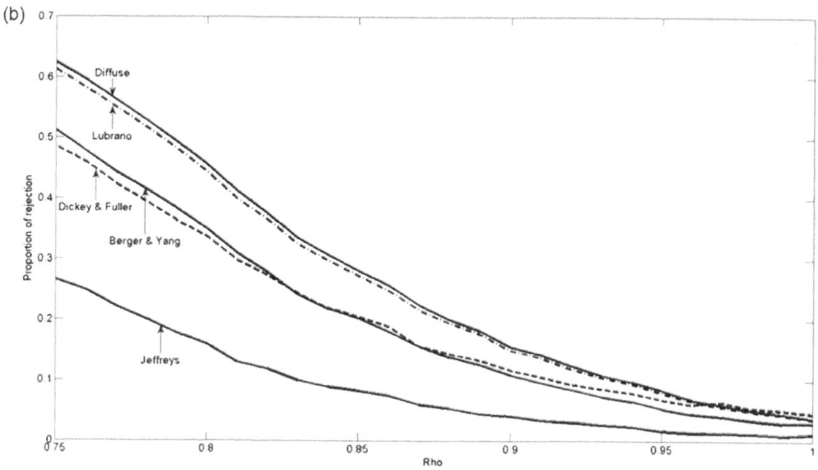

Fig. 3(b). Power Functions for Bayes Factor Test on Non-Trending Data with $\nu = 1/6$.

3. The performance of the test from the Berger–Yang prior improved dramatically to make it comparable to that of the Dicky–Fuller test.
4. The test from the Jeffreys prior has very low power and size, particularly for $\nu = 1/3$.

5. With the exceptions of the Dickey–Fuller test (which stayed the same) and the Berger–Yang based test (which improved), the power functions and sizes of the Bayes factor tests were lower than their corresponding credible interval tests.

The size and power of each of the dual criterion tests that reject when both $B<1$ and $P(\rho \geq 1|y, v)<0.05$ will, by construction, be no greater than the smaller size and power of the two corresponding separate tests. Thus, the power functions for the dual criterion tests all turn out to be very similar to those for the Bayes factor tests, with the exception of the Berger–Yang based tests whose power functions tend to mimic those for the corresponding credible interval tests.

Results for Trending Case

The power functions for the credible interval and Bayes factor tests for the trending case are presented in Figs. 4(a) and 4(b) for $v=1/3$. Those for $v=1/6$ are omitted since the effects of changing v are similar to those for the non-trending case. It is worth noting that both the Jeffreys prior in Eq. (17) and Berger and Yang's reference prior in Eq. (19) are tailored for an AR(1) process with no time trend and may not be justified as priors in this case. The power functions for these two priors for both the credible interval test

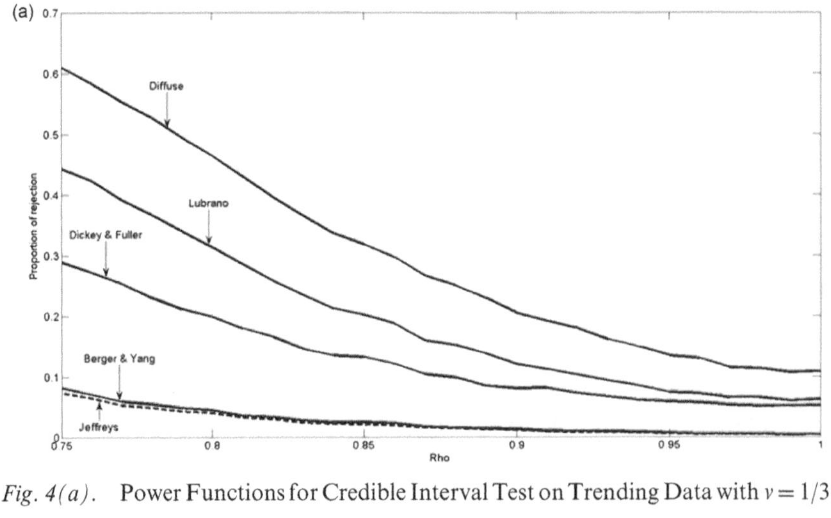

Fig. 4(a). Power Functions for Credible Interval Test on Trending Data with $v=1/3$.

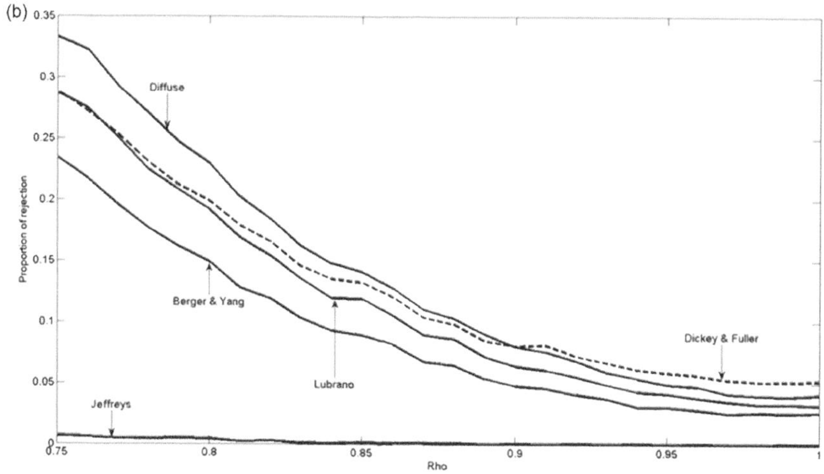

Fig. 4(b). Power Functions for Bayes Factor Test on Trending Data with $\nu = 1/3$.

and the Bayes factor test are dominated by the Dickey–Fuller test in the stationary region. A Bayes factor test with the Berger–Yang prior is viable with a size at 2.6%, but both tests based on the Jeffreys prior are not, with very low power in both cases.

The test from the uniform prior exhibits the same large size for the credible interval test at 11.5%, which may again be indicative of the unsatisfactory nature of credible interval tests based on diffuse priors. On the other hand, the credible interval test from the Lubrano prior has size only slightly above 0.05, and is much more powerful than the Dickey–Fuller, Jeffreys and Berger–Yang alternatives. For the Bayes factor tests relative performance is less clear cut. The test from the Lubrano prior no longer dominates the Dickey–Fuller test. It's size is 3.2%, compared with 5% for the Dickey–Fuller test, and its power only matches that of the Dickey–Fuller test for $\rho < 0.8$. However, changing ν to size correct the Lubrano-based test is likely to make it comparable. The test based on the uniform prior performs well. It has a size of 4.3%, less than that of the Dickey–Fuller test, but still dominates the Dickey–Fuller test in terms of power for $\rho < 0.9$.

Results from Bayesian Model Averaging

Figs. 5(a)–5(d) exhibit the results from the same Monte Carlo simulation but also including the results from Bayesian model averaging; the Dickey–Fuller

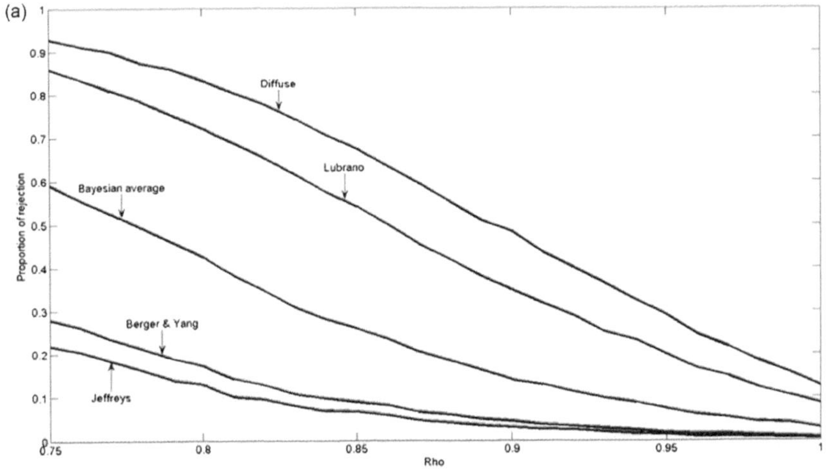

Fig. 5(a). Bayesian Averaging: Power Functions for Credible Interval Test on Non-Trending Data with $\nu = 1/3$.

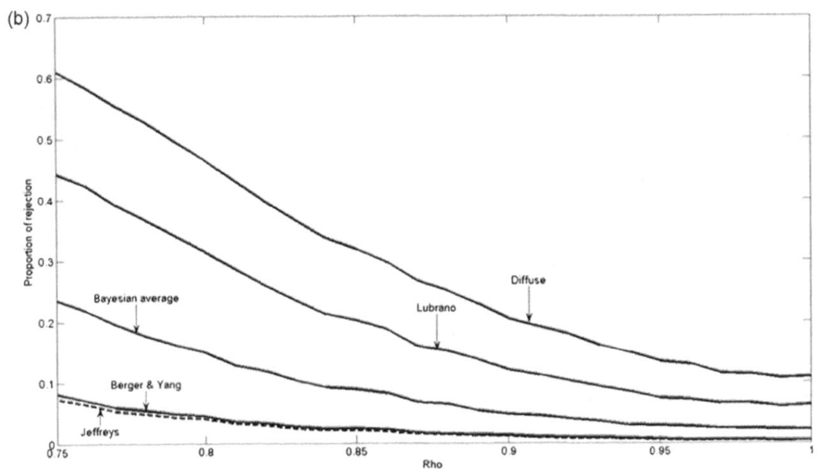

Fig. 5(b). Bayesian Averaging: Power Functions for Credible Interval Test on Trending Data with $\nu = 1/3$.

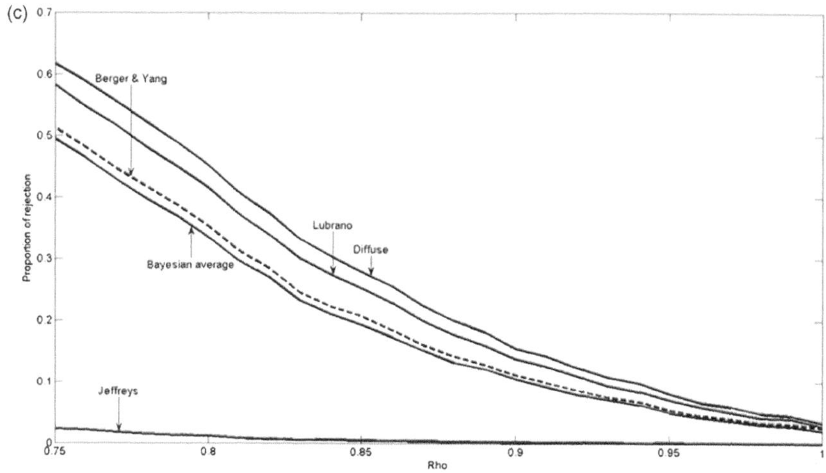

Fig. 5(c). Bayesian Averaging: Power Functions for Bayes Factor Test on Non-Trending Data with $v = 1/3$.

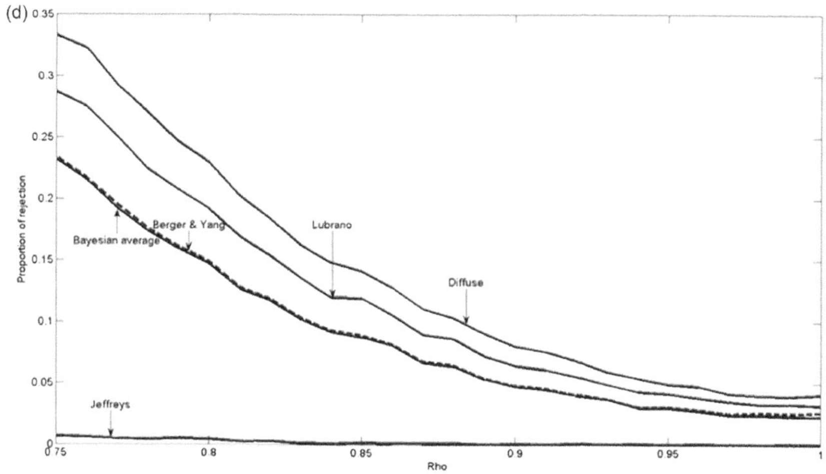

Fig. 5(d). Bayesian Averaging: Power Functions for Bayes Factor Test on Trending Data with $v = 1/3$.

test results are removed for better comparison. In all cases, we see that Bayesian averaging with equal $P(M_i)$ does quite well to reflect the relative average positions of the results from the four priors. For the credible interval tests one could argue that it is a good compromise between power and size. However, it is hard to recommend it over the Lubrano prior which has consistently good power and whose size can be corrected by changing v.

Results from Including a Prior for v

The power functions for the Lubrano prior tests with $v = 1/3$ and $v = 1/6$ are compared with those from a Lubrano prior, coupled with an exponential prior for v, in Figs. 6(a)–6(d). The power function for the latter is labeled "Lubrano hierarchy." We set $\lambda = 8.985$ to give an a priori probability of $P(v > 1/3) = 0.05$. For both the trending and non-trending data the power function for the credible interval test version of the Lubrano hierarchy is located between those for $v = 1/3$ and $v = 1/6$. The Bayes factor test version dominates those for $v = 1/3$ and $v = 1/6$. These results suggest that assigning a prior to v is a good strategy. In the credible interval case, it avoids having to specify a value for v and it provides a test that is no worse than $v = 1/6$. In

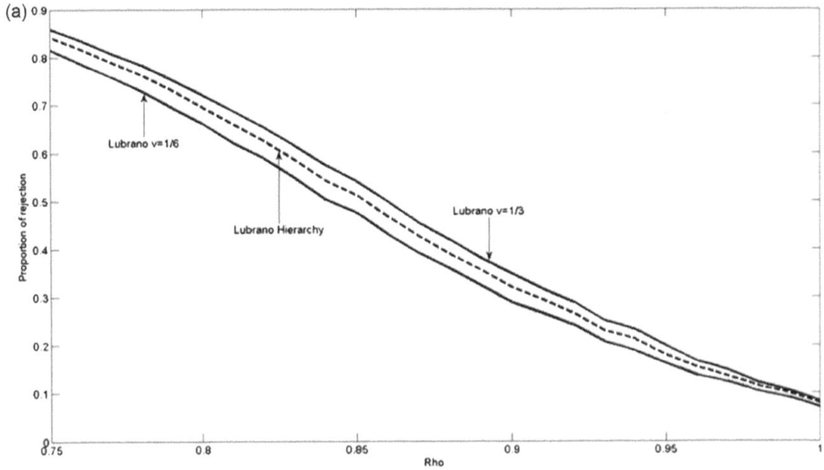

Fig. 6(a). Hierarchical Prior: Power Functions for Credible Interval Test on Non-Trending Data Using Lubrano Prior with $v = 1/3$, $v = 1/6$, and $\lambda = 8.985$.

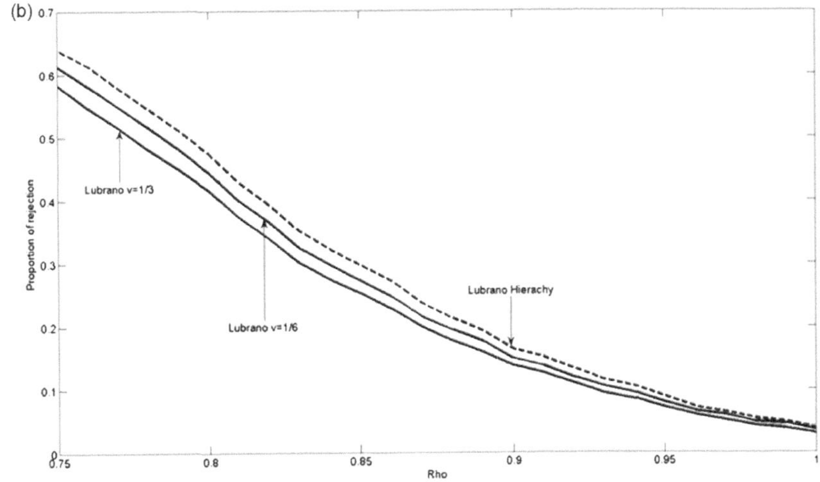

Fig. 6(b). Hierarchical Prior: Power Functions for Bayes Factor Test on Non-Trending Data Using Lubrano Prior with $\nu = 1/3$, $\nu = 1/6$, and $\lambda = 8.985$.

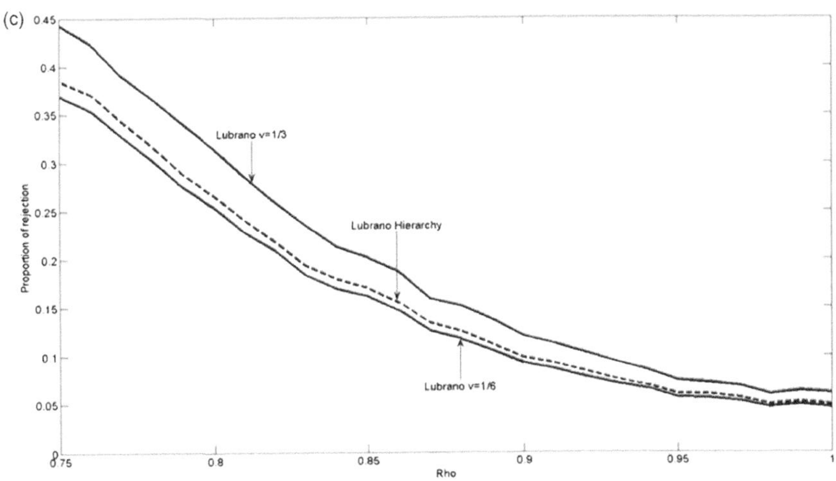

Fig. 6(c). Hierarchical Prior: Power Functions for Credible Interval Test on Trending Data Using Lubrano Prior with $\nu = 1/3$, $\nu = 1/6$, and $\lambda = 8.985$.

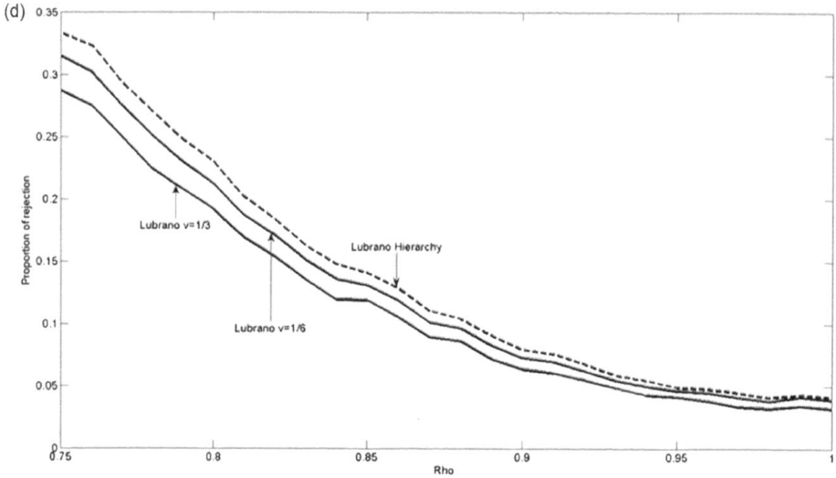

Fig. 6(d). Hierarchical Prior: Power Functions for Bayes Factor Test on Trending Data Using Lubrano Prior with $v = 1/3$, $v = 1/6$, and $\lambda = 8.985$.

the Bayes factor case it not only avoids the need to specify v, it also has better performance than the conventional settings of $v = 1/3$ and $v = 1/6$.

UNIT ROOTS IN AUSTRALIAN MACROECONOMIC TIME SERIES

Few papers have adopted a Bayesian framework to examine the unit root hypothesis in Australian macroeconomic time series. Mayadunne, Evans, and Inder (1995) studied 11 Australian macroeconomic variables using the ADF, Phillips–Perron, and KPSS tests, rejecting a unit root only for the unemployment rate. Narayan and Smyth (2005) looked at 16 Australian macroeconomic variables and found the unit root hypothesis was less likely once structural breaks were accounted for. We chose to apply Bayesian procedures to test for a unit root in 12 key Australian macroeconomic time series. The variables chosen and the results are given in Tables 1 and 2. The data were obtained from Australian Bureau of Statistics. All data were converted to a quarterly time span and natural logs were taken of all variables except the unemployment rate, the CPI, and the 10-year Treasury

Table 1. Unit Root Tests of Australian Macroeconomic Variables.

Series	Start	p	Trend	ADF	Lubrano & $v \sim \exp(8.985)$		BA & $v \sim \exp(8.985)$	
					$p(\rho \geq 1/y)$	Bayes factor	$p(\rho \geq 1/y)$	Bayes factor
Nominal GDP	1959:03	1	Yes	−2.29	0.43	17.40	0.74	10.27
Real GDP	1959:03	1	Yes	−2.24	0.33	24.26	0.66	14.97
GDP per capita	1973:03	1	Yes	−2.27	0.23	12.40	0.53	10.00
Household consumption	1959:03	1	Yes	−2.19	0.47	28.10	0.75	15.09
Investment	1959:03	3	Yes	−3.70[a]	0.009[a]	0.74[a]	0.04[a]	1.06
Employment	1978:02	2	Yes	−2.76	0.038[a]	3.56	0.16	4.33
Unemployment Rate	1978:02	2	No	−2.02	0.037[a]	5.38	0.18	6.11
M1	1975:02	2	Yes	−0.99	0.41	17.35	0.69	11.33
Money base	1975:02	1	Yes	−3.10	0.27	11.17	0.55	9.10
CPI	1970:03	5	No	−1.70	0.09	8.43	0.32	8.55
Treasury bond yield	1970:01	2	No	−1.13	0.19	20.84	0.52	15.17
ASX index	1875:01	2	Yes	−2.99	0.14	28.31	0.51	21.17

[a]Denotes unit root hypothesis rejected by their respective tests.

Table 2. Unit Root Tests of First Difference of Australian Macroeconomic Variables.

Series	Start	p	Trend	ADF	Lubrano & $\nu \sim \exp(8.985)$		BA & $\nu \sim \exp(8.985)$	
					$p(\rho \geq 1/y)$	Bayes factor	$p(\rho \geq 1/y)$	Bayes factor
Nominal GDP	1959:04	1	No	-15.21^a	0^a	0^a	0^a	0^a
Real GDP	1959:04	1	No	-13.08^a	0^a	0^a	0^a	0^a
GDP per capita	1973:04	1	No	-12.17^a	0^a	0^a	0^a	0^a
Household consumption	1959:04	1	No	-13.15^a	0^a	0^a	0^a	0^a
Investment	1959:04	1	No	-11.99^a	0^a	0^a	0^a	0^a
Employment	1978:03	1	No	-5.65^a	$(1.3 \times 10^{-6})^a$	$(9.2 \times 10^{-5})^a$	$(5.1 \times 10^{-6})^a$	$(1.1 \times 10^{-4})^a$
Unemployment Rate	1978:03	1	No	-5.31^a	$(2.6 \times 10^{-6})^a$	$(1.8 \times 10^{-4})^a$	$(9.9 \times 10^{-6})^a$	$(2.3 \times 10^{-4})^a$
M1	1975:03	1	No	-9.48^a	0^a	0^a	0^a	0^a
Money base	1975:03	2	No	-11.45^a	0^a	0^a	0^a	0^a
CPI	1970:04	4	No	-7.76^a	0^a	0^a	0^a	0^a
Treasury bond yield	1970:02	1	No	-10.59^a	0^a	0^a	0^a	0^a
ASX index	1875:02	1	No	-17.57^a	0^a	0^a	0^a	0^a

[a]Denotes unit root hypothesis rejected by their respective tests.

bond yield. Based on favorable results in the simulation study, two procedures were adopted:

- Credible interval and Bayes factor tests using a Lubrano prior for ρ and an exponential prior for ν.
- Credible interval and Bayes factor tests using a Bayesian averaging framework and an exponential prior for ν.

A trend was included whenever a graph of the data suggested one was present, and the lag length was chosen using a Schwarz criterion. In theory there is no reason to pretest for lag length using a Bayesian averaging framework, since the likelihood functions for different lag lengths can be treated as possible alternative models. However, doing so is computationally expensive. From Table 1, we note that classical and Bayesian results are in fair agreement at a 5% significance level. A unit root for investment is rejected by the Dickey–Fuller test and by all the Bayesian tests except for the Bayes factor test used with model averaging. The credible interval test used with the Lubrano prior rejects unit roots for employment and unemployment, but this decision is not supported by the Dickey–Fuller tests or the other Bayesian tests. This outcome is consistent with the results of the simulation experiment where the credible interval Lubrano-based test was more powerful than all other tests with the exception of that from a uniform prior.

In Table 2 we report the results from applying the same tests to first differences of the variables. The presence of a unit root is rejected by all tests for all variables.

CONCLUSION

In this paper we adopt a Bayesian framework to examine the unit root hypothesis in time series analysis. Monte Carlo simulations were used to study the power and size of Bayesian procedures in small samples using different objective priors. The results indicate that both testing procedures (posterior odds or posterior credible intervals) as well as the choice of prior have a significant impact on performances. In light of these observations, a Bayesian average over priors was introduced as a compromise to allow all priors a role in rejecting the unit root hypothesis. While this approach does not in any way address the issue of a "robust" and non-informative prior for time series coefficients, it is a natural basis for averaging across results from different models. We further looked at ways to implement subjective beliefs over explosive values by considering a maximum entropy density on the truncation variable v. Both Bayesian averaging and introducing a hyperparameter as if it was a random sample from some underlying distribution are hierarchical model approaches. This methodology is most natural to the Bayesian setup, since frequentist methods do not treat models as random variables and therefore cannot give a rigorous justification for model averaging.

The prior which led to the best performance in terms of test size and power was the Lubrano prior. Introducing a hierarchical prior for v improved performance for the Bayes factor tests and proved to be a good risk averse strategy for credible interval tests. Finally, we implemented Bayesian procedures to test the unit root hypothesis in Australian macroeconomic variables. While the test results were similar in most cases, examining the magnitudes of the test values suggests that discrepancies can arise.

REFERENCES

Bauwens, L., Lubrano, M., & Richard, J. F. (1999). *Bayesian inference in dynamic econometric models*. Oxford: Oxford University Press.

Bayes, T. (1764). An essay toward solving a problem in the doctrine of chances. Reprinted in Pearson, E. S., & Kendall, M. G. (Eds.). (1970). *Studies in the history of statistics and probability*. London: Charles Griffin.

Berger, J. O., & Delampady, M. (1987). Testing precise hypotheses. *Statistical Science, 2*, 317–352.

Berger, J. O., & Yang, R. Y. (1994). Non informative priors and Bayesian testing for the AR(1) model. *Econometric Theory, 10*, 461–482.

Bernardo, J. M. (1979). Reference prior distributions for Bayesian inference. *Journal of the Royal Statistical Society B, 41*, 113–147.

Blanchard, O. J., & Summers, L. H. (1987). Hysteresis in unemployment. *European Economic Review, 31*, 288–295.

DeJong, D. N., Nankervis, J. C., Savin, N. E., & Whiteman, C. H. (1992). Integration versus trend stationarity in time series. *Econometrica, 60*, 423–433.

Dejong, D. N., & Whiteman, C. H. (1991). Reconsidering trends and random walks in macroeconomic time series. *Journal of Monetary Economics, 28*, 221–254.

Hamilton, J. (1994). *Time series analysis*. Princeton, NJ: Princeton University Press.

Jeffreys, H. (1946). An invariant form for the prior probability in estimation problems. *Proceedings of the Royal Society of London, Series A, 186*, 453–461.

Koop, G. (1991). Intertemporal properties of real output: A Bayesian analysis. *Journal of Business and Economic Statistics, 9*, 253–266.

Lubrano, M. (1995). Testing for unit roots in a Bayesian framework. *Journal of Econometrics, 69*, 81–109.

Mayadunne, G., Evans, M., & Inder, B. (1995). An empirical investigation of shock persistence in economic time series. *Economic Record, 71*, 145–156.

Narayan, P., & Smyth, R. (2005). Structural breaks and unit roots in Australian macroeconomic time series. *Pacific Economic Review, 10*, 421–437.

Nelson, C. R., & Plosser, C. I. (1982). Trends and random walks in macroeconomic time series: Some evidence and implications. *Journal of Monetary Economics, 10*, 139–162.

Perks, F. J. A. (1947). Some observations on inverse probability including a new indifference rule. *Journal of the Institute of Actuaries, 73*, 285–334.

Phillips, P. C. B. (1991). To criticize the critics: An objective Bayesian analysis of stochastic trends. *Journal of Applied Econometrics, 6*, 333–364.

Schotman, P. C., & van Dijk, H. K. (1991a). A Bayesian analysis of the unit root in real exchange rates. *Journal of Econometrics, 49*, 195–238.

Schotman, P. C., & van Dijk, H. K. (1991b). On Bayesian routes to unit roots. *Journal of Applied Econometrics, 49*, 387–401.

Sims, C. A. (1988). Bayesian skepticism on unit root econometrics. *Journal of Applied Econometrics, 6*, 423–434.

Uhlig, H. (1994). On Jeffreys prior when using the exact likelihood function. *Econometric Theory, 10*, 633–644.

Zellner, A., & Tiao, G. C. (1964). Bayesian analysis of the regression model with autocorrelated errors. *Journal of the American Statistical Association, 59*, 763–778.

Zivot, E. (1994). A Bayesian analysis of the unit root hypothesis within an unobserved components model. *Econometric Theory, 10*, 552–578.

INVERSE TEST CONFIDENCE INTERVALS FOR TURNING-POINTS: A DEMONSTRATION WITH HIGHER ORDER POLYNOMIALS

Jenny N. Lye and Joseph G. Hirschberg

ABSTRACT

In this chapter we demonstrate the construction of inverse test confidence intervals for the turning-points in estimated nonlinear relationships by the use of the marginal or first derivative function. First, we outline the inverse test confidence interval approach. Then we examine the relationship between the traditional confidence intervals based on the Wald test for the turning-points for a cubic, a quartic, and fractional polynomials estimated via regression analysis and the inverse test intervals. We show that the confidence interval plots of the marginal function can be used to estimate confidence intervals for the turning-points that are equivalent to the inverse test. We also provide a method for the interpretation of the confidence intervals for the second derivative function to draw inferences for the characteristics of the turning-point.

This method is applied to the examination of the turning-points found when estimating a quartic and a fractional polynomial from data used for the estimation of an Environmental Kuznets Curve. The Stata *do files*

30th Anniversary Edition
Advances in Econometrics, Volume 30, 59–95

ISSN: 0731-9053/doi:10.1108/S0731-9053(2012)0000030008

used to generate these examples are listed in Appendix A along with the data.

Keywords: Inverse test confidence intervals; likelihood profile; quartic; fractional polynomials; *Stata*; marginal effect function; Fieller method

INTRODUCTION

Regression models in many applications employ a functional specification which allows for a nonlinear relationship between a regressor and the dependent variable. The most common of these is the quadratic specification with a wide application of more complex higher order polynomials and semi-parametric methods where part of the specification is linear and part is allowed to be flexible functions of x. Typically these models are of the form

$$y_t = \beta(x_t) + \left(\alpha_0 + \sum_{j=1}^{M} \alpha_j z_{jt}\right) + \varepsilon_t$$

In this specification we assume that the nonlinear specification is in the variable x while the other regressors (z) are assumed to have a linear influence on the dependent variable.[1]

An important characteristic of these models is that they possess the property of a nonconstant marginal effect of the x variable over the values of x. In this chapter we demonstrate that the confidence intervals of the marginal effect function with respect to x can be used for inferences concerning the turning-points (sometimes referred to as the stationary points). We also show that these intervals are equivalent to a class of previously defined inferential methods. In addition we demonstrate how the marginal of the marginal effects function, or more directly the second order function can be used for inferences in high order models. In this discussion we follow the implications of the consideration of the derivative functions implied by these nonlinear specifications.

Marginal effects or derivative functions of nonlinear regression specifications have a significant history in the interpretation of regression results. The interpretation of the marginal effects is central to many econometric applications and is extensively covered in texts such as Wooldridge (2010, Chap. 2). Recently, the statistics package *Stata* has introduced a series of post-estimation procedures to compute marginal effects and functions based on the parameters estimated from a wide variety of models. The original version of the currently available *margins* routine was a user written routine as described in Bartus (2005).

An early exposition of the marginal effects for polynomial regressions can be found in Williams (1959, Chap. 6), however his treatment has been rarely referenced and it is only recently that the estimation of high order polynomials has become widespread. In the case of the quadratic specification the location of the maximum or minimum is defined as $-1/2$ times the ratio of the parameter on the linear term divided by the parameter on the squared term. Thus to draw inferences concerning the location of the turning-point requires the construction of confidence intervals for the ratio of regression parameters. However, Dufour (1997) in his survey of "Impossibility Theorems in Econometrics" observes that the direct application of the Wald test to the ratio of parameters "deviates arbitrarily from any approximating distribution" because it is a function of the data to which it is applied. He then proposes that a valid interval can be found using the Fieller (1944) approach as the inversion of the appropriate likelihood ratio test and demonstrates that the interval is defined as the roots to a quadratic expression. In defining the general form of Fieller's method for the construction of confidence intervals for ratios of linear functions of regression parameters Rao (1973) and Zerbe (1978) obtain equivalent expressions (Eqs. 4b.2.16 and 2.5–2.7 respectively). Applications of the Fieller technique for the inferences for varying income elasticities from Engle curve estimation can be found in Hirschberg, Lye, and Slottje (2008).

The inversion of a two sided test for a linear combination of the regression parameters is not only limited to the consideration of ratios of parameter values. In the examination of the existence of turning-points in nonlinear functions estimated by polynomial regressions, Williams (1959, Chap. 6) and Hellbronner (1979) demonstrated that the confidence interval of the location of the turning-points for a cubic regression could be constructed by inverting the test for the location of the value of the regressor where the first derivative function was equal to zero. Because the first derivative function of the cubic is a quadratic that implies two turning-points, the inversion test used to construct their confidence intervals, involves the solution of the four roots for a quartic to define the intervals of interest.[2] In the particular application considered by Hellbronner, he limited his interest to only the interval that lay in the range of the x variable. Williams also employs the second derivative to locate the point of maximum rate of change and although he alludes to the application of inverse tests, he does not discuss the details of its implementation.

In this chapter we demonstrate the uses for the inversion of the test for a single linear combination of parameters to draw inferences for a set of related nonlinear specifications. Vandaele (1981) has shown that these inverse t-tests are equivalent to the profile likelihood tests for these cases since these tests are equivalent to the corresponding likelihood ratio tests.

First we will outline the concept behind the inversion test. Then we discuss the application of the Fieller/likelihood profile test that involves the inversion of the t-test for the location of the turning-points (in this case an extremum) defined by the estimated quadratic regression. We then introduce the cubic and demonstrate that inferences concerning the location of turning-points can be found via the inversion tests. We then extend the cubic to consider the quartic specification that has recently been found in more applications in economics. This discussion is followed by the consideration of an example of the use of a quartic. The last section introduces a more general specification referred to as a fractional polynomial. This specification differs from a traditional polynomial in that all the higher order derivatives are not zero by assumption. We then demonstrate how this method can be applied to the same data examined using the quartic. The *Stata* routines for these analyses are given in the appendices along with a listing of the data used in the applications.

THE INVERSE TEST CONFIDENCE INTERVAL

The inverse test confidence interval is found by first defining a relationship for a critical value of a test statistic for a variable of interest and then solving for the values of the variable that define the limits of the interval. For example, a test of a linear combination of the mean vector of normally distributed random k dimensional vector $\mathbf{c}_{k\times 1} \sim N[\boldsymbol{\mu}_{k\times 1}, \boldsymbol{\Omega}_{k\times k}]$ defined as $\mathbf{R}_{1\times k}\boldsymbol{\mu} = r$ we can propose the following 2-sided hypothesis test

$$H_0 : \mathbf{R}_{1\times k}\boldsymbol{\mu} = r$$
$$H_1 : \mathbf{R}_{1\times k}\boldsymbol{\mu} \neq r$$

And the usual confidence interval for the scalar r is given by

$$CI = \mathbf{Rc} \pm z_{\alpha/2}\sqrt{\mathbf{R\Omega R}'} \tag{1}$$

where $z_{\alpha/2}$ is the critical standard normal value for a $(1-\alpha)$ confidence interval. Under H_0 we can construct the test statistic defined by

$$T = \frac{\mathbf{Rc} - r}{\sqrt{\mathbf{R\Omega R'}}} \tag{2}$$

To establish if the H_0 can be rejected or not we need to determine if $T \geq |z_{\alpha/2}|$.

Alternatively, one could "invert the problem" by determining the critical values for r referred to as $\tilde{r}$, where the null hypothesis would be rejected at the $(1-\alpha)$ level of significance as defined by $z_{\alpha/2}$. Thus the implied value of the test statistic ($\tilde{T}^2$) is given by

$$\tilde{T}^2 = \frac{(\mathbf{Rc} - \tilde{r})^2}{\mathbf{R\Omega R'}} = z^2_{\alpha/2} \tag{3}$$

The inversion involves the solution to a quadratic in $\tilde{r}$ defined as

$$(\mathbf{Rc} - \tilde{r})^2 - z^2_{\alpha/2}\mathbf{R\Omega R'} = 0 \tag{4}$$

The two solutions for $\tilde{r}$ define the interval.[3]

$$\{\tilde{r}_U, \tilde{r}_L\} = \hat{r} \pm z_{\alpha/2}\sqrt{\mathbf{R\Omega R'}}, \quad \text{where } \hat{r} = \mathbf{Rc} \tag{5}$$

In this case the solution is trivial. To paraphrase Molière's *Bourgeois Gentilhomme* who discovers that he is already speaking prose, we can now observe that we have been using inverse confidence intervals all along. However, in a number of cases the solution is not equivalent to the traditional Wald test approach where we add and subtract a multiple of the standard error of the parameter estimate to the estimate.

Consider a related case where the multiplicative matrix $\mathbf{R}$ is a function of an unknown parameter θ and r is given. In this case we wish to test the value(s) of θ for which $\mathbf{R}(\theta)\boldsymbol{\mu} = r$ using the two-sided test defined by

$$H_0 : \mathbf{R}(\theta)\boldsymbol{\mu} = r$$
$$H_1 : \mathbf{R}(\theta)\boldsymbol{\mu} \neq r$$

One way to proceed is to use estimates of μ defined by $\mathbf{c}$ and to solve $\mathbf{R}(\hat{\theta})\mathbf{c} = r$ for estimate(s) of θ referred to as $\hat{\theta}$ when it is possible to solve $\mathbf{R}^{-1}(r, \mathbf{c}) = \hat{\theta}$. We could then construct a Wald test with an estimate of the variance of the $\hat{\theta}$ derived from the application of the delta method to find a linear approximation for $\mathbf{R}^{-1}(r, \mathbf{c})$.[4]

In this case the inverse test provides an alternative solution. Again specify the test statistic under the null (note that we expect up to twice the number of values for the value of θ to define the interval(s) for all the solutions for $\mathbf{R}^{-1}(r, \mathbf{c})$).

$$\tilde{T}^2 = \frac{(\mathbf{R}(\tilde{\theta})\mathbf{c} - r)^2}{\mathbf{R}(\tilde{\theta})\mathbf{\Omega}\mathbf{R}(\tilde{\theta})'} = z^2_{\alpha/2} \tag{6}$$

Thus we can write this expression as a quadratic in $\mathbf{R}(\tilde{\theta})$ given by

$$\mathbf{R}(\tilde{\theta})(\mathbf{cc}' - z^2_{\alpha/2}\mathbf{\Omega})\mathbf{R}(\tilde{\theta})' - 2r\mathbf{c}'\mathbf{R}(\tilde{\theta})' + r^2 = 0 \tag{7}$$

which can be solved for the confidence interval(s). Since this is a quadratic form there are now potentially two solutions for every $\hat{\theta}$. Note that for any value of α, there is no guarantee that they will be real valued.

In this chapter we investigate a particular set of cases where the null hypothesis of interest can be defined as $\mathbf{R}(\theta)\mathbf{c} = 0$ and where the restriction vector can be written as a polynomial function in θ and $\mathbf{c}$ is the vector of estimated regression parameters. Typically we have the linear function vector defined as

$$\mathbf{R}(\theta) = (\, R_0 \quad R_1\theta \quad R_2\theta^2 \quad \cdots \quad R_j\theta^j \quad \cdots \quad R_k\theta^k \,) \tag{8}$$

Thus in the cases dealt with here there are $2k$ potential confidence bounds defined by the roots to the polynomial defined by the solution to

$$\mathbf{R}(\tilde{\theta})(\mathbf{cc}' - z^2_{\alpha/2}\mathbf{\Omega})\mathbf{R}(\tilde{\theta})' = 0 \tag{9}$$

The graphical equivalent that we demonstrate is found from the plot of the confidence interval defined by

$$\mathrm{CI}(f(\theta)) = \mathbf{R}(\theta)\mathbf{c} \pm z_{\alpha/2}\sqrt{\mathbf{R}(\theta)\mathbf{\Omega}\mathbf{R}(\theta)'} \tag{10}$$

From which we can locate the values of θ where the upper and lower bounds cross the zero reference line to define the $\tilde{\theta}$.[5]

THE QUADRATIC SPECIFICATION

The quadratic functional form is widely used in regression to identify a U or inverted U-shape relationship between a regressor and the dependent variable. Many authors claim to have found such a relationship by focussing on the significance and signs of the estimated coefficients corresponding to the quadratic (Clark, Oswald, & Warr, 1996) and possibly also examining whether the position of the extremum falls within the range of the data (see Richmond & Kaufmann, 2006). Dufour's (1997) admonition notwithstanding, a number of authors continue to use the delta method to calculate a standard error for the extremum and then construct symmetric confidence intervals (e.g., Kearsley & Riddel, 2010).[6] Although there has been a considerable number of Monte Carlo studies that record the similarity of the confidence intervals obtained using the Fieller and the delta it has been shown that the coverage of the delta confidence interval is related to the characteristics of the particular sample used in estimation (Hirschberg & Lye, 2010a). More recently, authors have used confidence intervals for the extremum based on the Fieller method as well as the Sasabuchi (1980) intersection-union test (see Huang, Lin, & Yeh, 2012; Schnabel & Wagner, 2008).[7]

The quadratic specification is typically written as

$$y_t = \beta_1 x_t + \beta_2 x_t^2 + \left(\alpha_0 + \sum_{j=1}^{M} \alpha_j z_{jt} \right) + \varepsilon_t \tag{11}$$

Again we assume that there is a single regressor x entered as a quadratic function, M linearly related regressors denoted by z_j, and N observations. By setting the marginal effect or the first derivative of $E(y|x,z)$ with respect to x, $((\partial E(y|x,z))/\partial x) = \beta_1 + 2\beta_2 x_t$, equal to zero, we can derive the value of x that defines the extremum of the relationship between y and x given values for the other regressors, as $\theta = (-\beta_1/2\beta_2)$. θ is either the maximum (when $\beta_2 < 0$) or the minimum (when $\beta_2 > 0$). These conditions are established via the signs of the second order derivative of the function where $x = \theta$. In this case $(\partial^2 E(y|\theta, z)/\partial x^2) = 2\beta_2$, thus the sign of β_2 will establish the nature of the extremum.[8] The usual estimate of the extremum is given by $\hat{\theta} = (-\hat{\beta}_1/2\hat{\beta}_2)$, where$\hat{\beta}_1$ and $\hat{\beta}_2$ are the OLS estimates of β_1 and β_2 in Eq. (11) respectively. The point defined by $\hat{\theta}$ also defines the value of x where the marginal effect of x on y changes sign.

The Delta Method Applied to the Quadratic

The delta method for drawing inferences about a nonlinear function such as θ, is to estimate its variance by use of a first order Taylor series expansion to linearize the nonlinear function. In this case the estimated variance of $\hat{\theta}$ is given by (see, e.g., Rao, 1973, pp. 385–389)

$$\frac{1}{4}\left[\begin{pmatrix}\frac{-1}{\hat{\beta}_2}\end{pmatrix} \quad \begin{pmatrix}\frac{\hat{\beta}_1}{\hat{\beta}_2^2}\end{pmatrix}\right]\begin{bmatrix}\hat{\sigma}_1^2 & \hat{\sigma}_{12}\\ \hat{\sigma}_{12} & \hat{\sigma}_2^2\end{bmatrix}\begin{bmatrix}\begin{pmatrix}\frac{-1}{\hat{\beta}_2}\end{pmatrix}\\ \begin{pmatrix}\frac{\hat{\beta}_1}{\hat{\beta}_2^2}\end{pmatrix}\end{bmatrix} = \frac{\hat{\sigma}_1^2\hat{\beta}_2^2 - 2\hat{\beta}_1\hat{\beta}_2\hat{\sigma}_{12} + \hat{\beta}_1^2\hat{\sigma}_2^2}{4\hat{\beta}_2^4} \tag{12}$$

where $\hat{\sigma}_i^2$ is the estimated variance of $\hat{\beta}_i$ and $\hat{\sigma}_{ij}$ is the estimated covariance between $\hat{\beta}_i$ and $\hat{\beta}_j$. A $100(1-\alpha)\%$ confidence interval for θ is defined by

$$\{\tilde{\theta}_U, \tilde{\theta}_L\}_{\text{delta}} = \hat{\theta} \pm t_{\alpha/2}\sqrt{\frac{\hat{\sigma}_1^2\hat{\beta}_2^2 - 2\hat{\beta}_1\hat{\beta}_2\hat{\sigma}_{12} + \hat{\beta}_1^2\hat{\sigma}_2^2}{4\hat{\beta}_2^4}} \tag{13}$$

where $t_{\alpha/2}$ is the value from the t distribution with an $(\alpha/2)\%$ level of significance and $N-k$ degrees of freedom. This is commonly referred to as a delta method confidence interval and can readily be computed using post-regression software routines available in most statistics packages.

The Inversion Method Applied to the Quadratic

An alternative approach to construct a confidence interval for θ is found by inverting the t-test associated with the null hypothesis

$$\mathrm{H}_0 : \beta_1 + 2\beta_2\theta = 0$$
$$\mathrm{H}_1 : \beta_1 + 2\beta_2\theta \neq 0$$

A $100(1-\alpha)\%$ confidence inter val for θ is obtained by solving[9]

$$\tilde{T}^2 = \frac{(\hat{\beta}_1 + 2\hat{\beta}_2\tilde{\theta})^2}{(\sigma_1^2 + 4\sigma_{12}\tilde{\theta} + 4\sigma_2^2\tilde{\theta}^2)} = t_{\alpha/2}^2 \tag{14}$$

The solution is found by solving $a\tilde{\theta}^2 + b\tilde{\theta} + c = 0$ where $a = 4\hat{\sigma}_2^2(\hat{t}_2^2 - t_{\alpha/2}^2)$, $b = 4(\hat{\beta}_1\hat{\beta}_2 - t_{\alpha/2}^2\hat{\sigma}_{12})$, and $c = \hat{\sigma}_1^2(\hat{t}_1^2 - t_{\alpha/2}^2)$, where we define the t-ratios for the parameter estimates as: $\hat{t}_1 = \hat{\beta}_1/\hat{\sigma}_1$and $\hat{t}_2 = \hat{\beta}_2/\hat{\sigma}_2$. Then, the roots can be defined as

$$\{\tilde{\theta}_U, \tilde{\theta}_L\}_{\text{Fieller}} = \frac{t_{\alpha/2}^2\hat{\sigma}_{12} - \hat{\beta}_1\hat{\beta}_2 \pm \sqrt{(\hat{\beta}_1\hat{\beta}_2 - t_{\alpha/2}^2\hat{\sigma}_{12})^2 - \hat{\sigma}_2^2\hat{\sigma}_1^2(\hat{t}_2^2 - t_{\alpha/2}^2)(\hat{t}_1^2 - t_{\alpha/2}^2)}}{2\hat{\sigma}_2^2(\hat{t}_2^2 - t_{\alpha/2}^2)} \tag{15}$$

If $\hat{t}_2^2 > t_{\alpha/2}^2$, then Eq. (15) has two real roots and a finite confidence interval is constructed. This condition corresponds to being able to reject $H_0 : \beta_2 \neq 0$ when α is the level of significance (Buonaccorsi, 1979). In addition to the finite interval case, the resulting confidence interval may be the complement of a finite interval when (b^2–$4ac > 0$, $a < 0$) or of the whole real line when b^2–$4ac < 0$, $a < 0$. These conditions are discussed in Scheffé (1970) and Zerbe, Laska, Meisner, and Kushner (1982). This approach results in a confidence interval that is equivalent to the Fieller method (Fieller, 1944) which provides a general procedure for constructing confidence limits for the ratio of two parameters, estimated by normally distributed random variables.

Graphically (as shown in Hirschberg & Lye, 2010b) the 100(1−α)% confidence interval for θ following the inversion method, is obtained by plotting an estimate of the first derivative as a function of x ($\hat{\beta}_1 + 2\hat{\beta}_2 x$) with its corresponding 100(1−α)% confidence interval defined as

$$\mathrm{CI}\left(\frac{\partial E(y|x,z)}{\partial x}\right) = (\hat{\beta}_1 + 2\hat{\beta}_2 x) \pm t_{\alpha/2}\sqrt{(\hat{\sigma}_1^2 + 4x\hat{\sigma}_{12} + 4x^2\hat{\sigma}_2^2)} \quad (16)$$

An estimate of the extremum value,$x = \hat{\theta}$, is found by solving the equation $\hat{\beta}_1 + 2\hat{\beta}_2\hat{\theta} = 0$. Similarly, the bounds that define a 100(1−α)% confidence interval on θ are found by solving for x in the relationship

$$(\hat{\beta}_1 + 2\hat{\beta}_2 x) \pm t_{\alpha/2}\sqrt{(\hat{\sigma}_1^2 + 4x\hat{\sigma}_{12} + 4x^2\hat{\sigma}_2^2)} = 0 \quad (17)$$

which is equivalent to solving the roots of the equation

$$(\hat{\beta}_1 + 2\hat{\beta}_2 x)^2 - t_{\alpha/2}^2(\hat{\sigma}_1^2 + 4x\hat{\sigma}_{12} + 4x^2\hat{\sigma}_2^2) = 0 \quad (18)$$

By rearranging the terms in Eq. (18), this can be written as the quadratic equation, $ax^2 + bx + c = 0$, where a, b, and c are the terms defined above.

THE CUBIC SPECIFICATION

Aside from the quadratic the next most commonly employed polynomial specification is the cubic. Many applications that use the cubic are accompanied by a plot of the estimated function along with descriptive comments on its shape (e.g., Bagliani, Bravo, & Dalmazzone, 2008). In a limited number of cases, the definition of the turning-points implied by the roots of the first derivative function and the determination of the signs of the second order derivative are used to determine the nature of the turning-points (Hellbronner, 1979; Naghshpour, 2005; Vitaliano & Held, 1991). In some cases Wald tests are employed to find standard errors for the turning-points as in French and Zarkin (1995). Alternative techniques for the construction of confidence intervals have been applied in the application of the cubic; Plassmann and Khanna (2007) use a Bayesian approach for the establishment of turning-points.

The cubic polynomial specification is defined as

$$y_t = \beta_1 x_t + \beta_2 x_t^2 + \beta_3 x_t^3 + \left(\alpha_0 + \sum_{j=1}^{M} \alpha_j z_{jt} \right) + \varepsilon_t \tag{19}$$

In this case the first derivative function is given by a quadratic equation defined as

$$\frac{\partial E(y/x, z)}{\partial x} = \beta_1 + 2\beta_2 x + 3\beta_3 x^2 \tag{20}$$

The solution of this equation when set to zero is given by the solution to a quadratic. Thus there are two turning-points from the two roots (given $\beta_3 \neq 0$) defined as

$$\theta_i = -\frac{1}{3\beta_3}(\beta_2 + \phi_i \gamma), \quad \text{where } \gamma = \sqrt{\beta_2^2 - 3\beta_1\beta_3}, \ \phi_1 = 1, \quad \text{and} \quad \phi_2 = -1 \tag{21}$$

The second derivative of the cubic evaluated at the ith root of the quadratic in Eq. (21) defined by θ_i is given as $(\partial^2 y/\partial x^2) = 2\beta_2 + 6\beta_3\theta_i$. Both roots of Eq. (21) are real if $\gamma > 0$. If this is the case the second derivative test indicates that θ_i is a relative (or local) minimum if $(2\beta_2 + 6\beta_3\theta_i > 0)$ and θ_i is a relative maximum when $(2\beta_2 + 6\beta_3\theta_i < 0)$.

The Delta Method Applied to the Cubic

The approximate variance of $\hat{\theta}_i$ based on the delta method is

$$V(\hat{\theta}_i) = d_1\hat{\sigma}_1^2 + 2d_1d_2\hat{\sigma}_{12} + 2d_1d_3\hat{\sigma}_{13} + d_2\hat{\sigma}_2^2 + 2d_2d_3\hat{\sigma}_{23} + d_3\hat{\sigma}_3^2 \tag{22}$$

where $[d_{i1} \ \ d_{i2} \ \ d_{i3}] = [\frac{\partial\theta_i}{\partial\beta_1} \ \frac{\partial\theta_i}{\partial\beta_2} \ \frac{\partial\theta_i}{\partial\beta_3}]$ evaluated at the estimates, thus $d_{i1} = (\phi_i/2\hat{\gamma})$, $d_{i2} = (-(\phi_i\hat{\beta}_2 + \hat{\gamma})/3\hat{\beta}_3\hat{\gamma})$, and $d_{i3} = ((2\hat{\beta}_2\hat{\gamma} - \phi_i(3\hat{\beta}_1\hat{\beta}_3 - 2\hat{\beta}_2^2))/$

$6\hat{\beta}_3^2\hat{\gamma}$) for $i = 1, 2$. A 100(1−α)% delta method based confidence interval for θ_i is given by: $\{\tilde{\theta}_{iU}, \tilde{\theta}_{iL}\}_{\text{delta}} = \hat{\theta}_i \pm t_{\alpha/2}\sqrt{V(\hat{\theta}_i)}$. A number of programs in computer packages can be used to calculate the delta variance estimates such as *nlcom* in *Stata*.

The Inversion Method Applied to the Cubic

A confidence interval for θ_i can be constructed by inverting the t-test associated with the null hypothesis

$$\begin{aligned} H_0 &: \beta_1 + 2\beta_2\theta + 3\beta_3\theta^2 = 0 \\ H_1 &: \beta_1 + 2\beta_2\theta + 3\beta_3\theta^2 \neq 0 \end{aligned}$$

A 100(1−α)% confidence interval for θ_i is obtained by solving for the values of θ at which the following expression holds.

$$T^2 = \frac{(\hat{\beta}_1 + 2\hat{\beta}_2\tilde{\theta} + 3\beta_3\tilde{\theta}^2)^2}{(\sigma_1^2 + 4\tilde{\theta}\sigma_{12} + 4\tilde{\theta}^2\sigma_2^2 + 6\tilde{\theta}^2\sigma_{13} + 12\tilde{\theta}^3\sigma_{23} + 9\tilde{\theta}^4\sigma_3^2)} = t_{\alpha/2}^2 \tag{23}$$

By rearrangement we define a quartic equation whose roots are the limits of the implied confidence intervals for the turning-points of the original cubic function. This quartic expression that is equivalent to Eq. (23) is given below.

$$\begin{aligned} &9(\hat{\beta}_3^2 - t_{\alpha/2}^2{}^2\hat{\sigma}_3^2)\tilde{\theta}^4 + 12(\hat{\beta}_2\hat{\beta}_3 - t_{\alpha/2}^2\hat{\sigma}_{23})\tilde{\theta}^3 + 2((3\hat{\beta}_1\hat{\beta}_3 + 2\hat{\beta}_2^2) - t_{\alpha/2}^2(3\hat{\sigma}_{13} + 2\hat{\sigma}_2^2))\tilde{\theta}^2 \\ &\qquad + 4(\hat{\beta}_1\hat{\beta}_2 - t_{\alpha/2}^2\hat{\sigma}_{12})\tilde{\theta} + (\hat{\beta}_1^2 - t_{\alpha/2}^2\hat{\sigma}_1^2) = 0 \end{aligned} \tag{24}$$

The corresponding four roots to this quartic define the confidence intervals for the turning-points and can be found numerically using widely available software such as the *polyroots* function in *Stata* and other widely used software that implement the algorithm proposed by Jenkins and Traub (1970) for the numerical derivation of the roots for high order polynomials. The roots can then be associated with the corresponding confidence bounds for θ_i.

These can also be obtained graphically by plotting an estimate of the first derivative as a function of x $(\hat{\beta}_1 + 2\hat{\beta}_2 x_t + 3\hat{\beta}_3 x_t^2)$ and a 100(1−α) confidence interval.

$$\mathrm{CI}\left(\frac{\partial E(y/x,z)}{\partial x_t}\right) = (\hat{\beta}_1 + 2\hat{\beta}_2 x_t + 3\hat{\beta}_3 x_t^2) \pm t_{\alpha/2}\sqrt{\begin{pmatrix} \hat{\sigma}_1^2 + 4x_t\hat{\sigma}_{12} + 4x_t^2\hat{\sigma}_2^2 \\ +6x_t^2\hat{\sigma}_{13} + 12x_t^3\hat{\sigma}_{23} + 9x_t^4\hat{\sigma}_3^2 \end{pmatrix}} \tag{25}$$

The estimated values of θ_i correspond to the values at which $\hat{\beta}_1 + 2\hat{\beta}_2 x_t + 3\hat{\beta}_3 x_t^2 = 0$ and the corresponding bounds are given by the values where the confidence interval intersects the zero reference line. Note that the CI is defined for all possible values of x_t however if the interval never cuts the zero reference line we can conclude that for this value of α that the interval is infinite. The possibility of an infinite bound is one of the characteristics of the inversion test and corresponds to the case where we obtain imaginary roots for Eq. (24).

In order to clearly identify the nature of the root as either a maximum or a minimum we invert the test that the second derivative is equal to zero as we have done for the confidence intervals of the turning-points. The function of the parameters defined by the second derivative function is given as $((\partial^2 y/\partial x^2) = 2\beta_2 + 6\beta_3\theta_i)$ thus we can write the test as[10]

$$T^2 = \frac{(2\hat{\beta}_2 + 6\hat{\beta}_3\breve{\theta})^2}{(36\breve{\theta}^2\hat{\sigma}_3^2 + 24\hat{\sigma}_{23}\breve{\theta} + 4\hat{\sigma}_2^2)} = t_{\alpha/2}^2 \tag{26}$$

We can solve for the roots of the quadratic defined below in Eq. (27) to establish the interval over which the turning-points can be identified as a maximum or a minimum.

$$36(\hat{\beta}_3^2 - t_{\alpha/2}^2\hat{\sigma}_3^2)\breve{\theta}^2 + 24(\hat{\beta}_2\hat{\beta}_3 - t_{\alpha/2}^2\hat{\sigma}_{23})\breve{\theta} + 4(\hat{\beta}_2^2 - t_{\alpha/2}^2\hat{\sigma}_2^2) = 0 \tag{27}$$

In analogous fashion to the intervals found for the first derivative function, we also find that this interval is equivalent to plotting the second derivative function with the corresponding 100(1−α) confidence interval evaluated at different values of x is given by

$$\mathrm{CI}\left(\frac{\partial^2(y/x,z)}{\partial x_t^2}\right) = (2\hat{\beta}_2 + 6\hat{\beta}_3 x_t) \pm t_{\alpha/2}\sqrt{36x_t^2\hat{\sigma}_3^2 + 24\hat{\sigma}_{23}x_t, + 4\hat{\sigma}_2^2} \tag{28}$$

Thus, the plot of the confidence interval for the second derivative function can be used to define the characteristics of the roots of the first derivative function.

THE QUARTIC SPECIFICATION

Following the trend in the applied economics literature to estimate higher order polynomials a number of studies report the estimation of a quartic or a fourth-order polynomial (Coate & Fry, 2012; Kumar & Viswanathan, 2007; Wong, 2011). As in the application of lower order polynomials, the most common approach to describing the estimated function is to plot the estimated functional form. In this section we demonstrate that the inversion test confidence interval for the turning-points can be found from the confidence interval of the first derivative function. We follow the approach outlined by Williams (1959, Section 6.14).For the quartic or fourth order polynomial the regression is specified as

$$y_t = \beta_1 x_t + \beta_2 x_t^2 + \beta_3 x_t^3 + \beta_4 x_t^4 + \left(\alpha_0 + \sum_{j=1}^{M} \alpha_j z_{jt}\right) + \varepsilon_t \tag{29}$$

where the first derivative function is given by a cubic equation defined as

$$\frac{\partial y}{\partial x} = \beta_1 + 2\beta_2 x + 3\beta_3 x^2 + 4\beta_4 x^3 \tag{30}$$

The turning-points can be found from the solution of this equation when set to zero

$$\theta^3 + a_1\theta^2 + a_2\theta + a_3 = 0 \tag{31}$$

where $a_1 = (3\beta_3/4\beta_4)$, $a_2 = (\beta_2/2\beta_4)$, and $a_3 = (\beta_1/4\beta_4)$ and define the intermediate terms: $Q = (a_1^2 - 3a_2/9)$ and $R = (2a_1^3 - 9a_1a_2 + 27a_3/54)$. If $Q^3 - R^2 \geq 0$ then there are 3 real roots. Defining $\lambda = \arccos(R/\sqrt{Q^3})$ then the roots are given by the equations

$$\theta_1 = -2\sqrt{Q}\cos\left(\frac{\lambda}{3}\right) - \frac{a_1}{3} \tag{32}$$

$$\theta_2 = -2\sqrt{Q}\cos\left(\frac{\lambda + 2\pi}{3}\right) - \frac{a_1}{3} \tag{33}$$

$$\theta_3 = -2\sqrt{Q}\cos\left(\frac{\lambda + 4\pi}{3}\right) - \frac{a_1}{3} \tag{34}$$

If $R^2 - Q^3 > 0$ then there is only 1 real root given by the equation

$$\theta_1 = -\text{sgn}(R)[(\sqrt{R^2 - Q^3} + |R|)^{1/3} + Q(\sqrt{R^2 - Q^3} + |R|)^{-1/3}] - \frac{a_1}{3} \tag{35}$$

For the quartic the second order derivative given the value of the roots as defined by the quadratic equation

$$\left[\frac{\partial^2 y}{\partial x^2}\bigg|\theta_i\right] = 2\beta_2 + 6\beta_3\theta_i + 12\beta_4\theta_i^2 \tag{36}$$

where θ_i represents the root being examined as estimated by $\hat{\theta}_i$ based on the estimates of the parameters of the model.

To determine whether a particular root is a relative maximum or minimum point we would test if the value is <0, or >0 respectively. Note that although the sign of the second derivative function is sufficient to determine if a turning-point is a relative maximum or minimum, the second derivative being equal to zero is only a necessary condition for the location of an inflection point.

The Delta Method Applied to the Quartic

The expressions defined by Eqs. (32)–(35) for the roots of the quartic though they are complex can be used to derive delta standard errors when a program is available to obtain the approximate standard errors for nonlinear functions of regression parameters. An example is shown in the *Stata* routine provided in section A.1 for the application in section "Example of a Quartic." Note that due to the complexity of these expressions it may be necessary to derive the estimated variance covariance matrix for the intermediate expression such as the a_i defined above.

The delta confidence interval of the second derivative function can also be determined by estimating the variance of the roots to Eq. (36) evaluated at the turning-points defined by $\hat{\theta}_i$ derived in Eqs. (32)–(35) above. An example of this is described in "Example of a Quartic."

The Inversion Method Applied to the Quartic

Alternatively we can construct the intervals from the inversion of the test using the tests for the first derivatives and the second derivatives. Again the first derivative function for the quartic is defined as

$$\frac{dy_t}{dx_t} = \beta_1 + 2\beta_2 x_t + 3\beta_3 x_t^2 + 4\beta_4 x_t^3 \tag{37}$$

A confidence interval for the intervals θ^i can be constructed by inverting the *F*-test associated with the null hypothesis that the first derivative is equal to zero versus the two-sided alternative

$$\begin{aligned} \mathrm{H}_0 &: \beta_1 + 2\beta_2\theta + 3\beta_3\theta^2 + 4\beta_4\theta^3 = 0 \\ \mathrm{H}_1 &: \beta_1 + 2\beta_2\theta + 3\beta_3\theta^2 + 4\beta_4\theta^3 \neq 0 \end{aligned}$$

A 100(1−α)% confidence interval for θ_i is obtained by solving for the solutions to

$$T^2 = \frac{(\hat{\beta}_1 + 2\hat{\beta}_2\tilde{\theta} + 3\hat{\beta}_3\tilde{\theta}^2 + 4\hat{\beta}_4\tilde{\theta}^3)^2}{(\hat{\sigma}_1^2 + 4\tilde{\theta}\hat{\sigma}_{12} + 2(2\hat{\sigma}_2^2 + 3\hat{\sigma}_{13})\tilde{\theta}^2 + 4(3\hat{\sigma}_{23} + 2\hat{\sigma}_{14})\tilde{\theta}^3 + (9\hat{\sigma}_3^2 + 16\hat{\sigma}_{24})\tilde{\theta}^4 + 24\hat{\sigma}_{34}\tilde{\theta}^5 + 16\hat{\sigma}_4^2\tilde{\theta}^6)} = t_{\alpha/2}^2 \tag{38}$$

The solution can be found by finding the roots of the 6th order polynomial defined by

$$\begin{aligned} &16(\hat{\beta}_4^2 - t^2\hat{\sigma}_4^2)\tilde{\theta}^6 + 24(\hat{\beta}_3\hat{\beta}_4 - t^2\hat{\sigma}_{34})\tilde{\theta}^5(+9\hat{\beta}_3^2 + 16\hat{\beta}_2\hat{\beta}_4 - t^2(9\hat{\sigma}_3^2 + 16\hat{\sigma}_{24}))\tilde{\theta}^4 \\ &+ (12\hat{\beta}_2\hat{\beta}_3 + 8\hat{\beta}_1\hat{\beta}_4 - t^2(12\hat{\sigma}_{23} + 8\hat{\sigma}_{14}))\tilde{\theta}^3 + 2((3\hat{\beta}_1\hat{\beta}_3 + 2\hat{\beta}_2^2) - t^2(3\hat{\sigma}_{13} + 2\hat{\sigma}_2^2))\tilde{\theta}^2 \\ &+ 4(\hat{\beta}_1\hat{\beta}_2 - t^2\hat{\sigma}_{12})\tilde{\theta} + (\hat{\beta}_1^2 - t^2\hat{\sigma}_1^2) = 0 \end{aligned} \tag{39}$$

Note that although a general analytic solution for these roots is not available we can obtain numerical solutions as discussed above.

An estimate of the first derivative as a function of x can be plotted with a $1-\alpha$ confidence interval as

$$\mathrm{CI}\left(\frac{\partial Y_t}{\partial x_t}\right) = (\hat{\beta}_1 + 2\hat{\beta}_2 x_t + 3\hat{\beta}_3 x_t^2 + 4\hat{\beta}_4 x_t^3) \pm t_{\alpha/2}^2 \sqrt{\begin{pmatrix} \hat{\sigma}_1^2 + 4x_t\hat{\sigma}_{12} + 2(2\hat{\sigma}_2^2 + 3\hat{\sigma}_{13})x_t^2 \\ +4(3\hat{\sigma}_{23} + 2\hat{\sigma}_{14})x_t^3 + (9\hat{\sigma}_3^2 + 16\hat{\sigma}_{24})x_t^4 \\ +24\hat{\sigma}_{34}x_t^5 + 16\hat{\sigma}_4^2 x_t^6 \end{pmatrix}} \tag{40}$$

and the equivalent bounds for the first derivative function can be located where the confidence interval includes zero.

In this case the second order derivative function can also be used as a test of which solutions are relative extremum points. Thus we are interested in the region in which the null hypothesis can be rejected.

$$\mathrm{H}_0 : 2\beta_2 + 6\beta_3\theta + 12\beta_4\theta^2 = 0$$
$$\mathrm{H}_1 : 2\beta_2 + 6\beta_3\theta + 12\beta_4\theta^2 \neq 0$$

Using the test for the second order derivative function at the estimated root we find that the solution becomes the roots of a quartic.

$$T^2 = \frac{(2\hat{\beta}_2 + 6\hat{\beta}_3\breve{\theta} + 12\hat{\beta}_4\breve{\theta}^2)^2}{(4\hat{\sigma}_{22} + 24\hat{\sigma}_{23}\breve{\theta} + (48\hat{\sigma}_{24} + 36\hat{\sigma}_{33})\breve{\theta}^2 + 144\hat{\sigma}_{34}\breve{\theta}^3 + 144\hat{\sigma}_{44}\breve{\theta}^4)} = t_{\alpha/2}^2 \tag{41}$$

For the limits of the second derivative function where we are unable to reject the null hypothesis we can solve for the solutions defined by $\breve{\theta}$ of the quartic defined in Eq. (42)

$$\begin{aligned} &144[\hat{\beta}_4^2 - t_{\alpha/2}^2\hat{\sigma}_{44}]\breve{\theta}^4 + 144[\hat{\beta}_3\hat{\beta}_4 - t_{\alpha/2}^2\hat{\sigma}_{34}]\breve{\theta}^3 \\ &+ [36(\hat{\beta}_3^2 - t_{\alpha/2}^2\hat{\sigma}_{33}) + 48(\hat{\beta}_2\hat{\beta}_4 - t_{\alpha/2}^2\hat{\sigma}_{24})]\breve{\theta}^2 \\ &+ 24[\hat{\beta}_2\hat{\beta}_3 - t_{\alpha/2}^2\hat{\sigma}_{23}]\breve{\theta} + 4[\hat{\beta}_2^2 - t_{\alpha/2}^2\hat{\sigma}_{22}] = 0 \end{aligned} \tag{42}$$

The four roots of Eq. (42) define the limits of the regions over which we cannot reject the hypothesis that the sign of the second derivative is neither positive nor negative.

Again the alternative to the solution to this polynomial is to inspect the plot of the (100−α)% confidence interval of second derivative function defined by

$$\mathrm{CI}\left(\frac{\partial^2 Y_t}{\partial x_t^2}\right) = 2\hat{\beta}_2 + 6\hat{\beta}_3 x_t + 12\hat{\beta}_4 x_t^2 \pm t_{\alpha/2}^2 \sqrt{\begin{pmatrix} 4\hat{\sigma}_{22} + 24\hat{\sigma}_{23} x_t \\ +(48\hat{\sigma}_{24} + 36\hat{\sigma}_{33}) x_t^2 \\ +144(\hat{\sigma}_{34} x_t^3 + \hat{\sigma}_{44} x_t^4) \end{pmatrix}} \tag{43}$$

By establishing if the CI for the second derivative function at the location of estimated roots of the first derivative function (37) defined by the values of $\hat{\theta}_i$ is greater or less than zero we can infer the sign of the second order derivative at the turning-point.

Example of a Quartic[11]

Utt, Hunter, and McCormick (2001) estimate an Environmental Kuznets Curve (EKC) for the United States using annual observations on CO_2 per capita emissions from 1929 to 1996. Using a quartic polynomial they estimate an EKC with two peaks at about US$8,000 and US$28,000 and a trough at about US$20,000 (as measured in constant 1990 dollars).[12] In this case the regression is specified with only one independent variable which is included as a quartic of the form

$$y_t = \beta_0 + \beta_1 x_t + \beta_2 x_t^2 + \beta_3 x_t^3 + \beta_4 x_t^4 + \varepsilon_t \tag{44}$$

where y is the CO_2 emissions per capita as reported in Utt et al. (2001) and x is the real GDP per capita (in constant US$2000) from Winer, Tofias, Grofman, and Aldrich (2008). Although these data are time series in nature and they exhibit the characteristics of potential unit-roots we appeal to the analysis by Liu, Skjerpen, and Telle (2009) in which they demonstrate that the properties of the asymptotic inferential methods work well when such a high order polynomial is estimated in the levels. Consequently, for this example we will apply OLS to this sample and employ the asymptotic standard errors.

Table 1 presents the resulting parameter estimates and their standard errors. In Fig. 1 we have plotted the scatter plot of the observations along with the predicted function. Note that we have included additional observations with x values from $30,000 to $35,000 that are outside the range of the observations.[13]

Table 1. The Parameter Estimates for the Quartic Function Applied to the Utt et al. Data Series as Defined in Appendix B.

Coefficient	Estimate	SE	*t*-Stat.
$\hat{\beta}_0$	2.98	0.10	28.42
$\hat{\beta}_1$	8.49	0.69	12.34
$\hat{\beta}_2$	−9.12	1.12	−8.16
$\hat{\beta}_3$	3.76	0.63	5.95
$\hat{\beta}_4$	−0.53	0.11	−4.62
Adjusted R^2	0.86	N	68

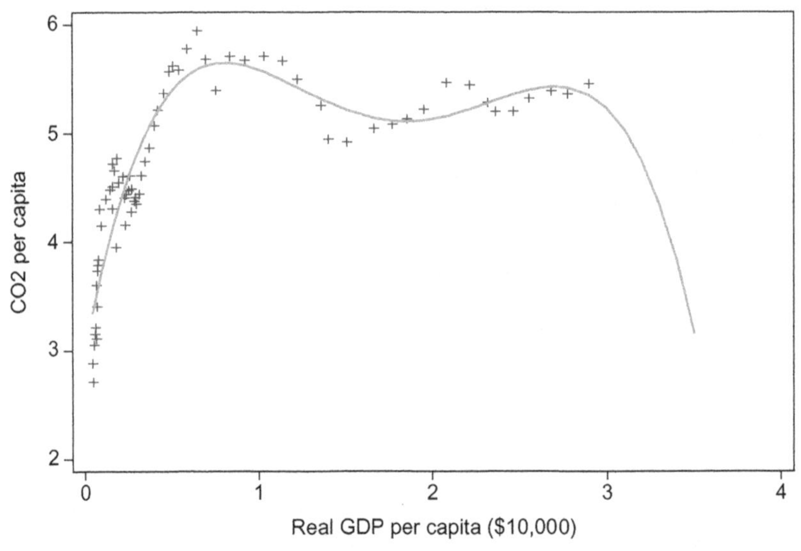

Fig. 1. The Scatter Plot of the Observations and the Quartic Function Fit by the Regression.

Using the first derivative approach we estimate three turning-points. Two maxima at US$8,076 and US$27,045 and a minimum at $US18,476. Listed in Table 2 are the estimated standard errors for these nonlinear functions of the data as estimated by the use of the delta method applied to the nonlinear equations defined in Eqs. (32)–(34). Also listed in Table 2 are the confidence intervals obtained using both the standard errors estimated via the delta

Table 2. The Estimated Turning-Points and the Standard Error Estimated via the Application of the Delta Method with the Confidence Intervals Based on the Two Methods.

Turning-Point	Estimate	SE	95% Confidence Interval			
			Delta		Inverse test	
$\hat{\theta}_1$	0.81	0.04	0.72	0.89	0.73	0.91
$\hat{\theta}_2$	1.85	0.11	1.62	2.07	1.66	2.10
$\hat{\theta}_3$	2.70	0.18	2.35	3.06	2.46	3.41

All values are in $10,000.

method and using the inversion test approach based on the solution to the sixth order polynomial as defined in Eq. (39). The latter estimates were obtained numerically through the application of the *Stata* routine *polyroots*.

From Table 2 we note that the most of the bounds are similar except for the upper bound on the last turning-point that appears to be far out of the sample range for the x variable. The maximum value for real GDP per capita in the sample is $29,000 which is well within the interval over which we are unable to reject the null hypothesis that the first derivative function is zero thus implying a flat relationship up to $30,600 implied by the delta CI or $34,100 found via the inversion test.

Alternatively, the inverse test confidence interval from the CI of the first derivative function defined by Eq. (40) can be found from Fig. 2.[14] We have included the location of the roots obtained from the numeric solution to demonstrate that they coincide with the values found by locating the points where the 95% CI of the 1st derivative function cut the zero reference line.

In the case that the polynomial defined by Eq. (39) does not have 6 real roots we would find that the equivalent figure to Fig. 2 would have CIs that would not cross the zero line for these imaginary roots. Thus it may be possible for the equivalent plot to display open ended intervals for some roots in which case the upper or lower limit may be ∞ or $-\infty$ respectively.

In order to determine the nature of these turning-points we now can investigate the 2nd derivative function of these estimated values. The second order derivative function for the quartic is given by Eq. (36) and we can now evaluate this function at the turning-points defined by the roots of the cubic implied by the first derivative function defined in Eq. (31). The delta method would involve the substitution of the estimates for the roots ($\hat{\theta}_i$) into the second derivative function to determine the sign. The *Stata* routine in

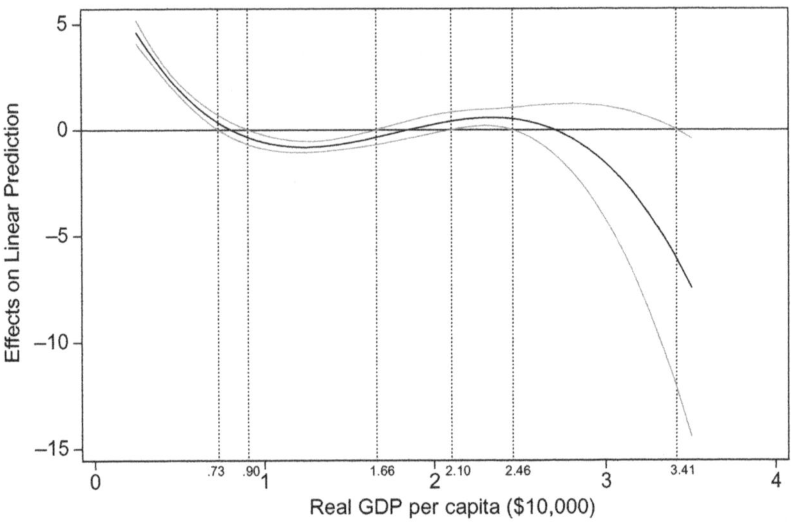

Fig. 2. The First Derivative Function with a 95% Confidence Interval and the Confidence Intervals Found Using the Inverse Test.

Table 3. The 95% CI for the Second Order Derivative Evaluated at Each of the Estimated Turning-Points with the Corresponding Delta and Inverse Confidence Intervals.

Turning-Point	$(\partial^2 y/\hat{\partial} x^2)\|\hat{\theta}_i$	SE	95% Confidence Interval ($10,000)			
			Delta		Inverse	
$\hat{\theta}_1 = 0.81$	−4.15	0.31	−4.76	−3.54	−4.75	−3.55
$\hat{\theta}_2 = 1.85$	1.87	0.33	1.22	2.53	1.24	2.51
$\hat{\theta}_3 = 2.70$	−3.42	2.10	−7.61	0.77	−7.53	0.69

Appendix A has been set to evaluate the second derivatives at these points. The resulting intervals for each turning-point are given in Table 3. From this table we can infer that the first turning-point has a negative second derivative indicating a local maximum, the second turning-point has a positive second derivative indicating a local minimum, and the third has a positive second order derivative which implies another local maximum. In addition, we also find that the estimated standard errors for the first two turning-points indicate that each of these values of $(\partial^2 y/\partial x^2)|\hat{\theta}_i$ are significantly different from zero. However for the third turning-point we are unable to reject the null hypothesis that the second derivative is equal to zero for this point which is necessary but not sufficient to indicate that the last turning-point (located at \$27,045) is not a relative extrema.

Alternately, we can use the inverse test to compute the confidence interval for the implied turning-points by computing the CI for the second derivative function evaluated at each turning-point. The plot of the 95% confidence interval for the second derivative function can be used to obtain equivalent intervals for Table 3. Fig. 3 plots the first derivative function as the solid line and the 95% CI for the second derivative function as defined in Eq. (36) as the dotted lines. Note that the turning-points have been located by the vertical lines. The intervals for the second derivative as shown in Fig. 3, can be found by locating the values along the reference lines defined at the turning-points of the 1st derivative function, that are cut by the CI of the 2nd derivative functions. These intervals are listed in Table 3. Fig. 3 provides an indication of the regions for which the sign of the second derivative is positive, negative, and indeterminate. Thus one can locate the CI for the 2nd derivative at those values of the regressor that coincide with the roots of the quartic defined in Eq. (42).

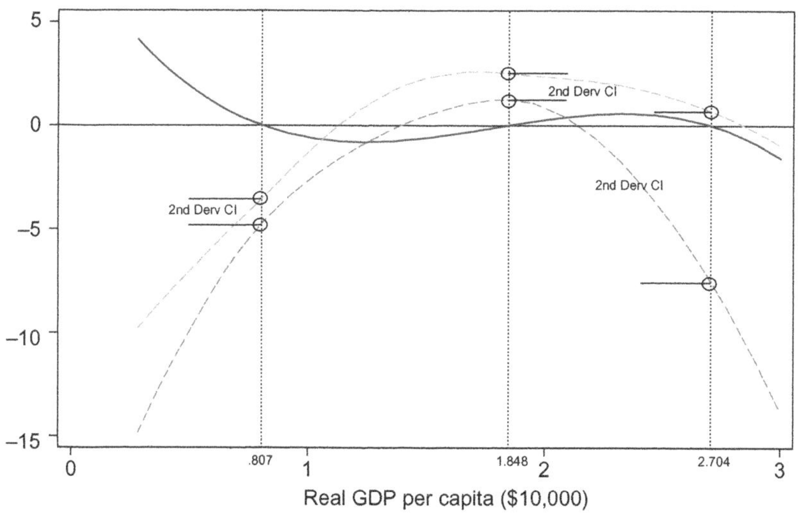

Fig. 3. 1st Derivative Function with 2nd Derivative 95% CI. The CI for the 2nd Derivative Function Can be Read at Each Turning-Point.

In sum the location and classification of the turning-points can be established by use of the appropriate plot of the first and second derivatives of the estimated function along with the corresponding confidence intervals. Thus it is possible to implement the inversion test by the use of the appropriate plots of the 1st and 2nd derivatives and their associated CIs.

FRACTIONAL POLYNOMIALS

In order to demonstrate the general nature of the inverse test intervals we consider an extension of standard polynomials referred to as "fractional polynomials" proposed by Royston and Altman (1994). In the economics literature these specifications are similar to the application of the minflex Laurent expansion employed by Barnett (1983) to generalize the use of a second order Taylor series for demand system estimation. Fractional polynomials involve the use of the positive and negative integer powers as well as the log, square-root, and inverse of the square root of the variable of interest. A primary application of this approach has been where regression models cannot be fit using the usual least squares approach (e.g., limited dependent variables and other maximum likelihood models). Due to the more complex nature of the estimation needed for such methods as smoothing splines (as discussed in Ruppert, Wand, & Carroll, 2003) and other "true" nonparametric techniques, fractional polynomials have been used where linear specifications can be accommodated by existing software such as in the application of the Cox proportional hazards model (see Kristiansen, Grønbæk, Becker, & Tolstrup, 2008).

A typical specification for a 3rd order fractional polynomial as identified by the powers $\{-3, -2, -1, -1/2, 0, 1/2, 1, 2, 3\}$ would be defined as

$$y_t = \beta_0 + x_t^{-3}\beta_1 + x_t^{-2}\beta_2 + x_t^{-1}\beta_3 + x_t^{-1/2}\beta_4 + \beta_5 \ln x_t + x_t^{1/2}\beta_6 + x_t\beta_7 + x_t^2\beta_8 + x_t^3\beta_9 + \varepsilon_t \tag{45}$$

As in the cases for the regular polynomials we can define a vector of regressors that are functions of the covariate of interest (x).[15]

$$y_t = X_t \mathrm{B} + \varepsilon_t$$

where

$$X_t = (1 \quad x_t^{-3} \quad x_t^{-2} \quad x_t^{-1} \quad x_t^{-1/2} \quad \ln(x_t) \quad x_t^{1/2} \quad x_t \quad x_t^2 \quad x_t^3) \tag{46}$$

and $\mathrm{B} = (\beta_0 \;\; \beta_1 \;\; \cdots \;\; \beta_9)'$. We can define $w_t^2 = x_t$ so that the vector of regressors can be written in terms of integer powers

$$X_t = (1 \quad w_t^{-6} \quad w_t^{-4} \quad w_t^{-2} \quad w_t^{-1} \quad 2\ln(w) \quad w_t \quad w_t^2 \quad w_t^4 \quad w_t^6) \tag{47}$$

To determine the locations of the extrema for this function we can evaluate the first order and second order derivatives using the vectors for the first and second derivatives defined as

$$\frac{\partial X_t}{\partial w_t} = [0 \quad -6w_t^{-7} \quad -4w_t^{-5} \quad -2w_t^{-3} \quad -w_t^{-2} \quad 2w_t^{-1} \quad 1 \quad 2w_t \quad 4w_t^3 \quad 6w_t^5] \tag{48}$$

$$\frac{\partial^2 X_t}{\partial w_t^2} = [0 \quad 42w_t^{-8} \quad 20w_t^{-6} \quad 6w_t^{-4} \quad 2w_t^{-3} \quad -2w_t^{-2} \quad 0 \quad 2 \quad 12w_t^2 \quad 30w_t^4] \tag{49}$$

Thus the estimated first derivative function is defined as $[\partial X_t/\partial w_t]\hat{\mathrm{B}}$ can be written as a 12th order polynomial defined as

$$\left[\frac{\partial X_t}{\partial w_t}\right]\hat{\mathrm{B}} = w_t^{-7}\begin{pmatrix} 6\hat{\beta}_9 w_t^{12} + 4\hat{\beta}_8 w_t^{10} + 2\hat{\beta}_7 w_t^8 + \hat{\beta}_6 w_t^7 + 2\hat{\beta}_5 w_t^6 \\ -\hat{\beta}_4 w_t^5 - 2\hat{\beta}_3 w_t^4 - 4\hat{\beta}_2 w_t^2 - 6\hat{\beta}_1 \end{pmatrix} \tag{50}$$

Thus, Eq. (50) can be used to solve for the locations of the extrema in terms of the estimated parameters $\hat{\beta}_i$ as $\hat{\theta}$ when $[\partial X_t/\partial w_t]\hat{\mathrm{B}} = 0$ thus we have

$$(6\hat{\beta}_9\hat{\theta}^{12} + 4\hat{\beta}_8\hat{\theta}^{10} + 2\hat{\beta}_7\hat{\theta}^8 + \hat{\beta}_6\hat{\theta}^7 + 2\hat{\beta}_5\hat{\theta}^6 - \hat{\beta}_4\hat{\theta}^5 - 2\hat{\beta}_3\hat{\theta}^4 - 4\hat{\beta}_2\hat{\theta}^2 - 6\hat{\beta}_1) = 0 \tag{51}$$

However, the roots of a 12th order polynomial have no analytic solution, thus the application of the traditional delta method would require the use of numeric derivatives for the approximation of the variance of the real valued roots. Or alternatively, one could derive confidence intervals for the roots via the use of a parametric bootstrap where one draws values of the coefficients from a multivariate normal pseudo-random number generator defined by using the estimated parameters and covariance.[16]

The test that we invert to locate the confidence intervals for the extrema is the test that the value of the first derivative to be equal to zero is defined as

$$\frac{[\partial X_t/\partial w_t]\hat{\mathrm{B}}}{\sqrt{[\partial X_t/\partial w_t]\hat{\mathrm{cov}}(\hat{\mathrm{B}})[\partial X_t/\partial w_t]'}} = t_{\alpha/2} \tag{52}$$

The value of the covariate (w) that satisfies this expression can be found by the solution of the polynomial in w defined by

$$\left[\frac{\partial X_t}{\partial w_t}\right](\hat{\mathrm{B}}\hat{\mathrm{B}}' - t_{\alpha/2}^2 \hat{\mathrm{cov}}(\hat{\mathrm{B}}))\left[\frac{\partial X_t}{\partial w_t}\right]' = 0 \tag{53}$$

In this case it would be necessary to solve the 24th order polynomial. We list part of this 24-order polynomial in Eq. (54) to demonstrate that such applications can lead to highly complex expressions that would rarely be used for computation.[17]

$$\begin{aligned}
&(36\hat{\beta}_9^2 - 36t_{\alpha/2}^2\hat{\sigma}_{99})\tilde{\theta}^{24} \\
&\quad + (48\hat{\beta}_8\hat{\beta}_9 - 48t_{\alpha/2}^2\hat{\sigma}_{89})\tilde{\theta}^{22} \\
&\quad + (16\hat{\beta}_8^2 + 24\hat{\beta}_7\hat{\beta}_9 - 24t_{\alpha/2}^2\hat{\sigma}_{79} - 16t_{\alpha/2}^2\hat{\sigma}_{88})\tilde{\theta}^{20} \\
&\quad + (12\hat{\beta}_6\hat{\beta}_9 - 12t_{\alpha/2}^2\hat{\sigma}_{69})\tilde{\theta}^{19} \\
&\vdots \\
&\vdots \\
&\quad + (16\hat{\beta}_2^2 + 24\hat{\beta}_1\hat{\beta}_3 - 24t_{\alpha/2}^2\hat{\sigma}_{13} - 16t_{\alpha/2}^2\hat{\sigma}_{22})\tilde{\theta}^{4} \\
&\quad + (48\hat{\beta}_1\hat{\beta}_2 - 48t_{\alpha/2}^2\hat{\sigma}_{12})\tilde{\theta}^{2} \\
&\quad + (36\hat{\beta}_1^2 - 36t_{\alpha/2}^2\hat{\sigma}_{11})
\end{aligned} \tag{54}$$

Again the roots of this polynomial could be determined by using a numeric solution as was done for the case of the quartic. Furthermore, it may be difficult to interpret the implications for imaginary roots.

The alternative graphic solution can be found by defining the confidence intervals from the plots of the estimated 1st derivative function for any nonlinear specification. We only require the estimation of a linear combination of the estimated parameters to define the first derivative function as $[\partial X_t/\partial w_t]\hat{\mathrm{B}}$ and the confidence interval is given by:

$$\mathrm{CI}\left(\frac{\partial Y_t}{\partial x_t}\right) = \left[\frac{\partial X_t}{\partial w_t}\right]\hat{\mathrm{B}} \pm t_{\alpha/2}\sqrt{\left[\frac{\partial X_t}{\partial w_t}\right]\mathrm{c\hat{o}v}(\hat{\mathrm{B}})\left[\frac{\partial X_t}{\partial w_t}\right]'} \tag{55}$$

This interval can be used to locate the confidence bounds for the turning-points of the polynomial by locating the points where the CI of the first derivative cut the zero reference line.

The second order derivative function can also be plotted in order to establish the intervals for the classification of the turning-points. In this case we would use the confidence intervals for the second derivative model defined by

$$\mathrm{CI}\left(\frac{\partial^2 Y_t}{\partial x_t^2}\right) = \left[\frac{\partial^2 X_t}{\partial w_t^2}\right]\hat{\mathrm{B}} \pm t_{\alpha/2}\sqrt{\left[\frac{\partial^2 X_t}{\partial w_t^2}\right]\mathrm{c\hat{o}v}(\hat{\mathrm{B}})\left[\frac{\partial^2 X_t}{\partial w_t^2}\right]'} \tag{56}$$

In the same manner as we have discussed in the case of the application to the quartic.

An Application of the Fractional Polynomial[18]

Following in analogous fashion to the application for the quartic we reanalyze the Utt et al. (2001) data using a 3rd order fractional polynomial as defined in Eq. (45). The resulting estimated function with the data for estimation is plotted in Fig. 4.

By comparing Figs. 1 and 4 one can note the two primary differences in the estimated functions. First, the fractional polynomial estimates a local maximum and a minimum between .2 and .4 that was missed by the quartic. Second, the fractional estimate does not exhibit the pronounced fall in the extrapolated values past the maximum value of the data that was present in the estimated quartic function. The estimated parameters for the 3rd order fractional polynomial are listed in Table 4. Note that not all are significantly different from zero but the fit is better than the quartic.

Due to the complications in the estimation of the turning-points and the confidence intervals based on the delta method and the inversion test, we will only use plots of the first and second derivatives and their corresponding confidence intervals to establish the nature of the turning-points in this case.

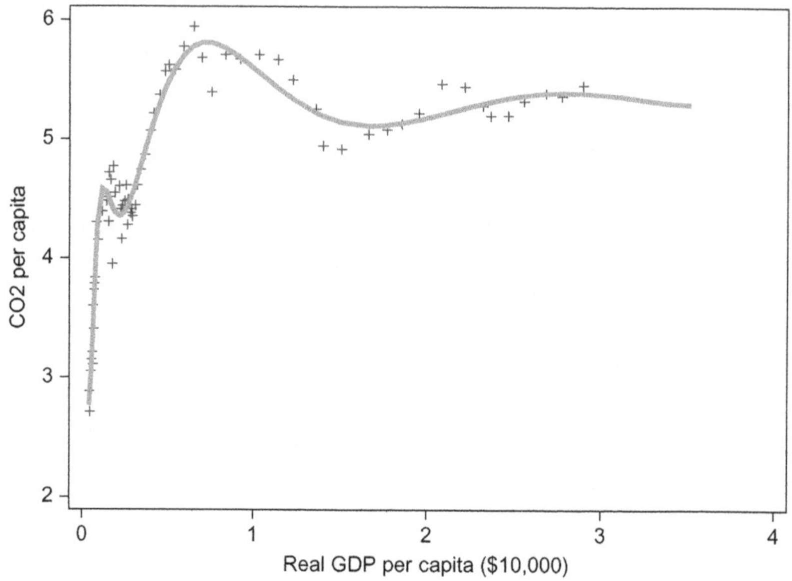

Fig. 4. The Scatter Plot and Fitted Function Based on 3rd Order Fractional Polynomial.

Table 4. Parameter Estimates for the 3rd Order Fractional Polynomial Regression.

Coefficient	Estimate	SE	t-Stat.
$\hat{\beta}_0$	1721.771	651.417	2.64
$\hat{\beta}_1$	−0.014	0.005	−2.87
$\hat{\beta}_2$	1.598	0.503	3.18
$\hat{\beta}_3$	−166.668	50.490	−3.30
$\hat{\beta}_4$	1750.692	541.094	3.24
$\hat{\beta}_5$	1952.256	635.487	3.07
$\hat{\beta}_6$	−4217.859	1496.489	−2.82
$\hat{\beta}_7$	967.629	385.969	2.51
$\hat{\beta}_8$	−54.342	29.278	−1.86
$\hat{\beta}_9$	2.785	2.129	1.31
Adjusted R^2	0.95	N	68

Fig. 5 is a plot of the first derivative function with the corresponding 95% confidence interval.[19] From this plot we can locate the 5 points where the first derivative function is equal to zero which correspond to the first maximum at \$1,200 then a minimum at approximately \$2,200 followed by a maximum at \$7,300, a minimum at \$16,700 and a maximum at \$28,000. From this plot we note that the confidence intervals for the first derivative indicate that once the level of GDP per capita exceeds \$22,200 the first derivative is never significantly different from zero, thus we can conclude that for values of GDP per capita over \$22,200 there is no further variation in CO_2 with respect to GDP per capita. The second order function can be used to establish the nature of the turning-points identified by the first derivative.

In order to more accurately determine the location of the turning-points and the confidence intervals we have plotted the first derivative function and the confidence interval for the second derivative function in two figures defined for two different ranges of the regressor. Fig. 6 shows the second derivative confidence interval and the first derivative function for values of the per capita income under \$7,000. Fig. 7 has the same plot from \$7,000 to \$30,000. From Fig. 6 we find that the second derivative test indicates that the first turning-point at \$1,200 is a relative maximum, the second turning-point at \$2,200 is a relative minimum.

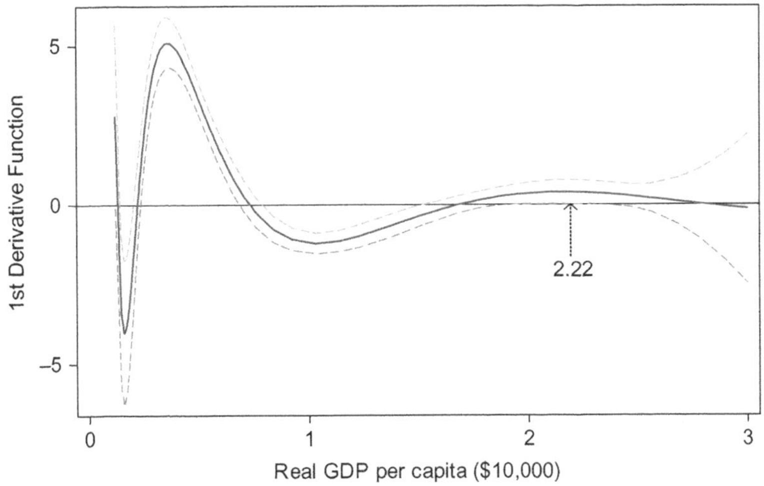

Fig. 5. The First Derivative Function of the 3rd Order Fractional Polynomial Function with the Corresponding 95% CI.

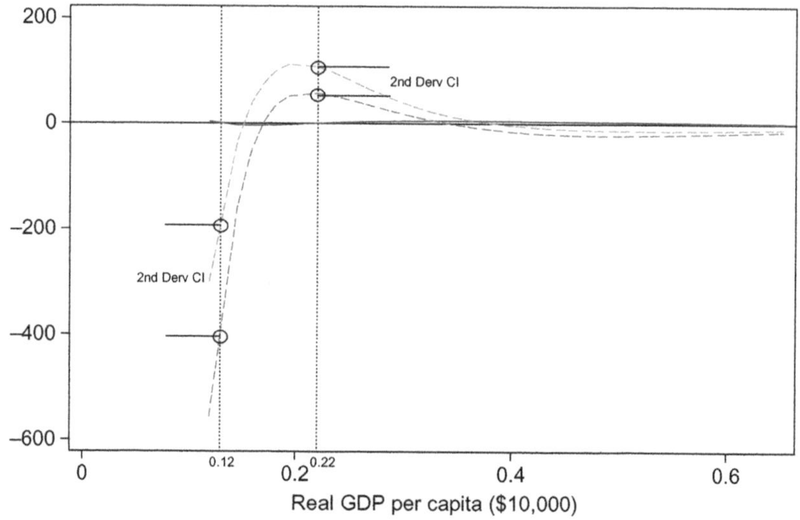

Fig. 6. The First Derivative Function and the Second Derivative Confidence Interval for GDP Per Capita from 0 to $7,000.

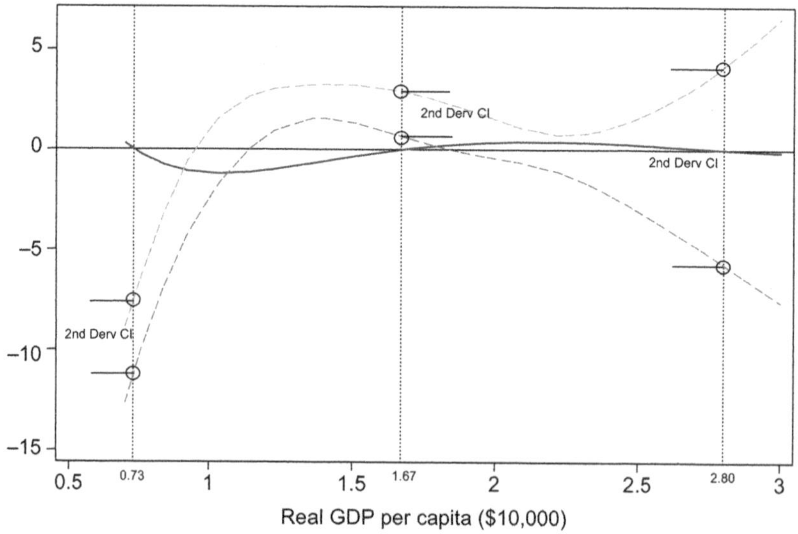

Fig. 7. The First Derivative Function and the Second Derivative Confidence Interval for GDP Per Capita from $7,000 to $30,000.

From Fig. 7 we note that the third turning-point at \$7,000 is a maximum and the fourth turning-point at \$16,700 is a minimum. However, we can also read from Fig. 7 that the confidence interval for the second derivative function evaluated at the position of the last turning-point at \$28,000 includes zero thus we cannot reject the null hypothesis that the turning-point is neither a maximum nor a minimum.

CONCLUSIONS

In this chapter we have presented a method for the determination of the location and properties of turning-points estimated for nonlinear models specified in regressions. By the construction of the inverse test we can locate the limits for confidence intervals of high order polynomials as well as for fractional polynomials as the roots to high order polynomials. We then demonstrate that the results for the first derivative function can be extended to the second derivative function which can provide tests for the determination of the nature of the turning-points identified by the first derivative. Alternatively, we have shown that graphic methods can be used to provide equivalent results and that they can be much more easily employed especially for models that may be specified that do not have an analytical solution found from higher order polynomials.

Future research in this area could be directed to the consideration of the more complex results obtained when multiple regressors are allowed to interact in an extension of the general response surface model. In addition to the fractional polynomials it would also be possible to use the methods proposed here to evaluate the turning-points for other polynomial related specifications such as the "Fourier Flexible Form" specification as proposed by Gallant (1981) and the class of Chebyshev polynomials (see Smyth, 2005 for an example). These specifications are similar to the fractional polynomial in that they also allow for the definition of higher order derivative functions that are linear in the regression parameters thus they can be plotted in a similar manner to the fractional polynomial examined here.

NOTES

1. In this discussion we assume that there is only one variable to be entered in a nonlinear form. In most cases the extension to more than one can be accommodated in a similar manner to the techniques we discuss here.

2. Williams refers to the inverse test confidence intervals as fiducial limits.

3. We will follow the convention where we use the circumflex (ˆ) to denote the estimated location and the tilda (∼) to denote the estimated interval limits.

4. Note that the analytic solution may not be available as we find in our discussion below. As an alternative to the delta method one could use a simulation or empirical bootstrap constructed by draws from an estimate of $N[\boldsymbol{\mu}, \Omega]$.

5. The case where they do not cross the zero reference line will indicate the situation where no real root for the equivalent polynomial defined by Eq. (9).

6. We refer to the delta as the Wald test confidence interval derived by combination of the delta approximation for the variance of the ratio with the definition of the confidence interval by adding and subtracting a multiple of the estimated standard error.

7. Lind and Mehlum (2010) observe that the Sasabuchi (1980) test at the α level of significance is equivalent to determining if the $(1-2\alpha)$ confidence interval for the extremum is within the range of the x variable.

8. This also justifies the use of the test of the null hypothesis that $\beta_2 = 0$ to establish whether an inflection point is present or not and to justify the consideration of this functional form.

9. We will substitute t^2 for the F with one degree of freedom in the numerator.

10. Here we use $\breve{\theta}$ to denote confidence bounds for the 2nd derivative function.

11. The *Stata* routine to generate the results in this section is given in section A.1. Note that the figures have been edited to add comments.

12. The data used in this analysis are listed in Appendix B.

13. Here we exploit a common feature in many regression software packages to predict out-of-sample when the dependent variable is missing but the regressors are available.

14. The first derivative plot can be provided by the *marginplot* routine in *Stata* as shown in the program provided in Appendix A. Alternatively one can use the *nlcom* routine with a little more programming this method is also shown in the *Stata* routine in the appendix.

15. Here we drop all other covariates to simplify our expressions. They would not influence the remainder of this section as long as they were not interacted with the x variable.

16. This technique may prove problematic when roots are only derived as real for a subset of simulations.

17. This expression was obtained using the *MuMath* computer algebra routine.

18. The *Stata* routine to generate the results in this section is given in section A.2.

19. The *Stata* program used to create these results is listed in section A.2.

ACKNOWLEDGMENTS

We would like to thank the anonymous referee for a number of comments that proved very helpful in the preparation of this version of the chapter. We also wish to thank participants at the Econometric Society Australasian Meeting held in July 2012. This research is supported by research funds supplied by the Faculty of Business and Economics, University of Melbourne. The responsibility for any errors is entirely ours.

REFERENCES

Bagliani, M., Bravo, G., & Dalmazzone, S. (2008). A consumption-based approach to environmental Kuznets curves using the ecological footprint indicator. *Ecological Economics*, *65*, 650–661.

Barnett, W. A. (1983). New indices of money supply and the flexible Laurent demand system. *Journal of Business and Economic Statistics*, *1*, 7–23.

Bartus, T. (2005). Estimation of marginal effects using Margeff. *The Stata Journal*, *5*, 309–329.

Buonaccorsi, J. P. (1979). On Fieller's theorem and the general linear model. *The American Statistician*, *33*, 162.

Clark, A., Oswald, A., & Warr, P. (1996). Is job satisfaction U-shaped in age? *Journal of Occupation and Organizational Psychology*, *69*, 57–81.

Coate, B., & Fry, T. (2012). *Better off dead? Prices realised for Australian paintings sold at auction*. ACEI Working Paper, AWP-02-2012. Association for Cultural Economics, http://www.culturaleconomics.org

Dufour, J. (1997). Some impossibility theorems in econometrics with applications to structural and dynamic models. *Econometrica*, *65*, 1365–1387.

Fieller, E. C. (1944). A fundamental formula in the statistics of biological assay and some applications. *Quarterly Journal of Pharmacy and Pharmacology*, *17*, 117–123.

French, M., & Zarkin, G. (1995). Is moderate alcohol use related to wages? Evidence from four worksites. *Journal of Health Economics*, *14*, 319–344.

Gallant, A. R. (1981). On the bias in flexible functional forms and an essentially unbiased form: The Fourier flexible form. *Journal of Econometrics*, *15*, 211–245.

Hellbronner, E. (1979). Position and confidence limits of an extremum. *Journal of Chemical Education*, *56*, 240–243.

Hirschberg, J. G., & Lye, J. N. (2010a). A geometric comparison of the delta and Fieller confidence intervals. *The American Statistician*, *64*, 234–241.

Hirschberg, J. G., & Lye, J. N. (2010b). Two geometric representations of confidence intervals for ratios of linear combinations of regression parameters: An application to the NAIRU. *Economics Letters*, *108*, 73–76.

Hirschberg, J. G., Lye, J. N., & Slottje, D. J. (2008). Inferential methods for elasticity estimates. *Journal of Econometrics*, *147*, 299–315.

Huang, H., Lin, Y., & Yeh, C. (2012). An appropriate test of the Kuznets hypothesis. *Applied Economics Letters*, *19*, 47–51.

Jenkins, M. A., & Traub, J. F. (1970). A three-stage algorithm for real polynomials using quadratic iteration. *SIAM Journal on Numerical Analysis*, *7*, 545–566.

Kearsley, A., & Riddel, M. (2010). A further inquiry into the pollution haven hypothesis and the environmental Kuznets curve. *Ecological Economics*, *69*, 905–919.

Kristiansen, L., Grønbæk, M., Becker, U., & Tolstrup, J. S. (2008). Risk of pancreatitis according to drinking habits: A population-based cohort study. *American Journal of Epidemiology*, *168*, 932–937.

Kumar, K., & Viswanathan, B. (2007). Changing structure of income indoor air pollution relationships in India. *Energy Policy, 35*, 5496–5504.

Lind, J., & Mehlum, H. (2010). With or without U? The appropriate test for a U-shaped relationship. *Oxford Bulletin of Economics and Statistics, 72*, 109–118.

Liu, G., Skjerpen, T., & Telle, K. (2009). Unit roots, polynomial transformations and the environmental Kuznets curve. *Applied Economics Letters, 16*, 285–288.

Naghshpour, S. (2005). The cyclical nature of family income distribution in the United States: An empirical note. *Journal of Economics and Finance, 29*, 138–143.

Plassmann, F., & Khanna, N. (2007). Assessing the precision of turning-point estimates in polynomial regression functions. *Econometric Reviews, 26*, 503–528.

Rao, C. R. (1973). *Linear statistical inference and its applications* (2nd ed.). New York, NY: Wiley.

Richmond, A., & Kaufmann, R. (2006). Is there a turning-point in the relationship between income and energy use and/or carbon emissions?. *Ecological Economics, 56*, 176–189.

Royston, P., & Altman, D. G. (1994). Regression using fractional polynomials of continuous covariates: Parsimonious parametric modelling. *Journal of the Royal Statistical Society. Series C (Applied Statistics), 43*, 429–467.

Ruppert, D., Wand, M. P., & Carroll, R. J. (2003). *Semiparametric regression.* Cambridge, UK: Cambridge University Press.

Sasabuchi, S. (1980). A test of a multivariate normal mean with composite hypotheses determined by linear inequalities. *Biometrika, 67*, 429–439.

Scheffé, H. (1970). Multiple testing versus multiple estimation. Improper confidence sets. Estimation of directions and ratios. *The Annals of Mathematical Statistics, 41*, 1–29.

Schnabel, C., & Wagner, J. (2008). *Union membership and age: The inverted U-shape hypothesis under test.* Institute for the Study of Labor (IZA) Discussion Paper No. 3842. University of Bonn, Bonn, Germany.

Smyth, G. K. (2005). Polynomial approximation. *Encyclopaedia of biostatistics* (2nd ed.). Wiley. On line edition. http://onlinelibrary.wiley.com/book/10.1002/0470011815

Utt, J. A., Hunter, W. W., & McCormick, R. E. (2001). On the relationship between net carbon emissions and income. Mimeo, Washington State University.

Vandaele, W. (1981). Wald, likelihood ratio and Lagrange multiplier tests as an *F* test. *Economics Letters, 8*, 361–365.

Vitaliano, D., & Held, J. (1991). Road accident external effects: An empirical assessment. *Applied Economics, 23*, 373–378.

Williams, E. J. (1959). *Regression analysis.* New York, NY: Wiley.

Winer, S. L., Tofias, M. W., Grofman, B., & Aldrich, J. H. (2008). Trending economic factors and the structure of Congress in the growth of government, 1930–2002. *Public Choice, 135*, 415–448.

Wong, M. (2011). Estimating the impact of the ethnic housing quotas in Singapore. Retrieved from http://maisy.wharton.upenn.edu/research. Accessed on March 8, 2012.

Wooldridge, J. M. (2010). *Econometric analysis of cross section and panel data.* Cambridge, MA: MIT Press.

Zerbe, G. O. (1978). On Fieller's theorem and the general linear model. *The American Statistician, 32*, 103–105.

Zerbe, G. O., Laska, E., Meisner, M., & Kushner, A. B. (1982). On multivariate confidence regions and simultaneous confidence limits for ratios. *Communications in Statistics, Theory and Methods, 11*, 2401–2425.

APPENDIX A

A.1 : Stata *Program to Generate the Results in Section "Example of a Quartic"*

```
/*
        Read the Carbon data from Utt etal and run the quartic regression
        Save the estimated parameters, the predicted values and plot the function and the data.
        Use the mono colour scheme for all graphs.
*/
        set scheme s1mono
        use "C:\data\old_v\cubic\carbon_stock\utt.dta" , clear

        label variable co2 "CO2 per capita"
        label variable g "Real GDP per capita ($10,000)"

        regress co2 g c.g#c.g c.g#c.g#c.g c.g#c.g#c.g#c.g
        estimates store eq_1
        predict p_co2
        label variable p_co2 "Predicted CO2"

        twoway (scatter co2 g, msymbol(+)) (line p_co2 g, sort lwidth(1)), ylabel(,nogrid) ///
        ytitle("CO2 per capita")   legend(off) name(Functionfrac, replace)

/*
        Define the roots of the quartic and apply the delta method to estimate confidence intervals
        Using nlcom to compute the non-linear function of the parameters to find the roots of the quartic
        and the delta standard errors with the confidence interval.  By using sequential calls to nlcom
        with the post option we can build up the complex functions for these roots.
*/

        quietly nlcom       (a1: (3*_b[c.g#c.g#c.g])/(4*_b[c.g#c.g#c.g#c.g])) ///
                                      (a2:(_b[c.g#c.g])/(2*_b[c.g#c.g#c.g#c.g]) ) ///
                                      (a3:(_b[g])/(4*_b[c.g#c.g#c.g#c.g]) ) ///
                                      (b2:_b[c.g#c.g] ) (b3:_b[c.g#c.g#c.g]) (b4:_b[c.g#c.g#c.g#c.g]), post

        quietly nlcom       (a1: (_b[a1]) )                                          ///
                                      (q1: ( (_b[a1]^2-3*_b[a2])/9) ) ///
                                      (q2: (2*_b[a1]^3 - 9*_b[a1]*_b[a2] + 27*_b[a3])/54 ) ///
                                      (b2:_b[b2] ) (b3:_b[b3]) (b4:_b[b4]),  post

        nlcom     (x1: -2*sqrt(_b[q1])*cos( acos((_b[q2])/sqrt((_b[q1]^3)))   /3) - (_b[a1]/3)) ///
                  (x2: -2*sqrt(_b[q1])*cos((acos((_b[q2])/sqrt((_b[q1])^3))+ 2*_pi)/3) - (_b[a1]/3)) ///
                  (x3: -2*sqrt(_b[q1])*cos((acos((_b[q2])/sqrt((_b[q1])^3))+ 4*_pi)/3) - (_b[a1]/3)) ///
                  (t1a: 2*_b[b2] + 6*_b[b3]*.8076 + 12*_b[b4]*.8076^2 )  ///
                  (t2b: 2*_b[b2] + 6*_b[b3]*2.7045 + 12*_b[b4]*2.7045^2 )  ///
                  (t3c: 2*_b[b2] + 6*_b[b3]*1.8476 + 12*_b[b4]*1.8476^2 )

/*
        Retrieve the original quartic regression parameter and covariance estimates
*/
        estimates restore eq_1
/*
        Store the parameters from the quartic as b1-b4, and the elements of the covariance matrix as Vij
*/

        scalar b1=_b[g]
        scalar b2=_b[c.g#c.g]
        scalar b3=_b[c.g#c.g#c.g]
        scalar b4=_b[c.g#c.g#c.g#c.g]

        matrix V = get(VCE)
        scalar v34 = V[3,4]
        scalar v24 = V[2,4]
        scalar v14 = V[1,4]
        scalar v23 = V[2,3]
        scalar v13 = V[1,3]
```

```
        scalar v12 = V[1,2]
        scalar v11 = V[1,1]
        scalar v22 = V[2,2]
        scalar v33 = V[3,3]
        scalar v44 = V[4,4]
/*
        Define the 6th order polynomial with roots that are the equivalent 95% inverse CIs for
        each inflexion point.  Set t_s to the square of the critical value for two tailed t-test.
*/
        scalar t_ = 1.96
        scalar t_s = t_^2

        scalar c0 = b1^2 - t_s*v11
        scalar c1 = 4*b1*b2 - 4*t_s*v12
        scalar c2 = 6*b1*b3 + 4*b2^2 - 6*t_s*v13- 4*t_s*v22
        scalar c3 = 12*b2*b3 + 8*b1*b4 - 12*t_s*v23 - 8*t_s*v14
        scalar c4 = 9*b3^2 + 16*b2*b4 - 9*t_s*v33 - 16*t_s*v24
        scalar c5 = 24*b3*b4 - 24*t_s*v34
        scalar c6 = 16*b4^2 - 16*t_s*v44

/*
        Use the mata program polyroots to compute the roots of this polynomial.
        st_numscalar inserts the scalar values to the polyroot program
*/
        mata polyroots((st_numscalar(("c0")),st_numscalar(("c1")),st_numscalar(("c2")),  ///
        st_numscalar(("c3")), st_numscalar(("c4")),st_numscalar(("c5")),st_numscalar(("c6"))))

/*
        Use the margins program to estimate the first derivative function and the estimated 95%CI then
        plot with marginsplot along with the indictors for the solution found from the polyroots command
        to show they coincide with the Fieller type bounds.
*/
        quiet margins, dydx(g) at(g=(0.25(0.05)3.5))

        marginsplot, recast(line) recastci(rline) ciopts(color(*.7)) yline(0, lstyle(foreground)) ///
        name(FirstDerv1, replace) plotregion(fcolor(gs16)) ylabel(,nogrid) ///
        xline(3.41144302 2.45864884 2.09915614 1.65867625 .904947489 .732782011 , lpattern(".") lwidth( .2 ))

/*
        Plot out the 1st and 2nd derivative functions to find the regions for inflexion points as opposed to
        maxima and minima using predictnl
*/
        predictnl d1 = _b[c.g] + 2 *_b[c.g#c.g] * g + 3 *_b[c.g#c.g#c.g] * g^2  ///
        + 4 *_b[c.g#c.g#c.g#c.g] * g^3 , se(d1_se)
        generate up1 = d1 + t_ * d1_se
        generate lw1 = d1 - t_ * d1_se

        line  d1 up1 lw1 g, lwidth( .4 .3 .3) lpattern( "1" "-#-" "-#-") sort  ///
        xline(0.8076194 2.704491 1.847637, lpattern(".") lwidth( .2 ) ) ///
        name(d1, replace)yline(0, lstyle(foreground)) ///
        ylabel(,nogrid) ytitle("1st Derivative Function") legend(off)

        predictnl d2 = 2 *_b[c.g#c.g] + 6 *_b[c.g#c.g#c.g] * g + 12 *_b[c.g#c.g#c.g#c.g] * g^2 , se(d2_se)
        generate up2 = d2 + t_ * d2_se
        generate lw2 = d2 - t_ * d2_se

        line  d1 up2 lw2 g if lw2 > -15 , lwidth( .4 .3 .3) lpattern( "1" "-#-" "-#-") sort  ///
        xline(0.8076194 2.704491 1.847637, lpattern(".") lwidth( .2 ) ) ///
        name(d1_d2, replace)yline(0, lstyle(foreground))       ///
        ylabel(,nogrid) title("1st Derivative Function with 2nd Derivative 95% CI") legend(off)

        twoway (scatter co2 g) (line p_co2 g, sort), name(Function2, replace) ///
        xline(3.41144302 2.45864884 2.09915614 1.65867625 .904947489 .732782011, lpattern(".") lwidth( .2 )) ///
        ylabel(,nogrid) title("Function with turning-point intervals") legend(off)

        line p_co2 d1 d2 up2 lw2 g,        ///
        name(all, replace) lstyle(p4 p1 p2 p2) yline(0,lstyle(foreground))          ///
        xline(3.41144302 2.45864884 2.09915614 1.65867625 .904947489 .732782011, lpattern(".") lwidth( .2 )) ///
        ylabel(,nogrid) title("Function with turning-point intervals") legend(off)
```

A.2 : Stata *Program to Generate the Results in the Example Application of the Fractional Polynomial*

```
/*
        Read the Carbon data from Utt etal and run a fractional polynomial
        with powers -3 -2 -1 -1/2 0 1/2 1 2 3
        Save the estimated parameters, the predicted values and plot the function and the data
        Use the mono-colour scheme for all graphs.
*/
        use "C:\data\old_v\cubic\carbon_stock\utt.dta" , clear
        label variable co2 "CO2 per capita"
        label variable g "Real GDP per capita ($10,000)"
        set scheme s1mono

        generate ig3 = 1/g^3
        generate ig2 = 1/g^2
        generate ig1 = 1/g
        generate igs = 1/sqrt(g)
        generate g0 = log(g)
        generate gs = sqrt(g)
        generate g1 = g
        generate g2 = g^2
        generate g3 = g^3

        regress co2 ig3 ig2 ig1 igs g0 gs g1 g2 g3
        estimates store eq_1
        predict p_co2
        label variable p_co2 "Predicted CO2"
        twoway (scatter co2 g, msymbol(+)) (line p_co2 g, sort lwidth(1)), ylabel(,nogrid) ///
        ytitle("CO2 per capita")   legend(off) name(Functionfrac, replace)
/*
        Plot out the 1st and 2nd derivative functions to find the regions for inflexion points as opposed to
        maxima and minima
*/
        generate t_ = 1.96
        predictnl d1 =  -2*_b[ig2] / g^3 - _b[ig1] / g^2 + _b[g0] / g + _b[g1] + 2* _b[g2] * g ///
                                         - .5 * _b[igs]*ig1*igs + .5 * _b[gs]* igs       ///
                                         - 3 * _b[ig3] / g^4 + 3* _b[g3] * g^2 , se(d1_se)
        generate up1 = d1 + t_ * d1_se
        generate lw1 = d1 - t_ * d1_se

        line  d1 up1 lw1 g if ((g > .095 )* (g < 3.1)), lwidth( .4 .3 .3) lpattern( "1" "-#-" "-#-") sort , ///
        ylabel(,nogrid) ytitle("1st Derivative Function") legend(off)  ///
        name(d1frac, replace)yline(0, lstyle(foreground))

        predictnl d2 =  6*_b[ig2] / g^4 + 2* _b[ig1] / g^3 - _b[g0] / g^2 + 2* _b[g2]  ///
                                         + .75 * _b[igs]*ig2*igs - .25 * _b[gs]*ig1* igs  ///
                                         + 12 * _b[ig3] / g^5 + 6 * _b[g3] * g  , se(d2_se)
        generate up2 = d2 + t_ * d2_se
        generate lw2 = d2 - t_ * d2_se

        line  d1 up2 lw2 g if ((g > .7 )* (g < 3.1)) , lwidth( .4 .3 .3) lpattern( "1" "-#-" "-#-") sort  ///
        xline(.73 1.67 2.8 , lpattern(".") lwidth( .3 ) ) ///
        name(d1_d2frac1, replace)yline(0, lstyle(foreground)) ///
        ylabel(,nogrid) title("1st Derivative Function with 2nd Derivative 95% CI") legend(off)

        line  d1 up2 lw2 g if ((g < .7 )* (g > .12)) , lwidth( .4 .3 .3) lpattern( "1" "-#-" "-#-") sort  ///
        xline(.13 .22 , lpattern(".") lwidth( .3 ) ) ///
        name(d1_d2frac2, replace)yline(0, lstyle(foreground)) ///
        ylabel(,nogrid) title("1st Derivative Function with 2nd Derivative 95% CI") legend(off)
```

APPENDIX B: DATA USED FOR EXAMPLE FROM UTT ET AL. (2001)

CO_2	g	CO_2	g
4.3	0.085003	5.068	0.400714
3.786	0.074033	5.211	0.418914
3.212	0.06162	5.368	0.453311
2.709	0.046979	5.567	0.485656
2.882	0.044872	5.619	0.506366
3.05	0.05218	5.582	0.542679
3.151	0.057553	5.774	0.58988
3.601	0.065376	5.943	0.652405
3.732	0.071262	5.681	0.701269
3.109	0.066247	5.392	0.758539
3.404	0.070367	5.707	0.836963
3.835	0.076747	5.669	0.921925
4.149	0.094976	5.707	1.030728
4.393	0.12005	5.664	1.138708
4.479	0.14524	5.496	1.224937
4.716	0.158819	5.25	1.360127
4.51	0.159439	4.943	1.4017
4.305	0.157226	4.915	1.509263
4.656	0.169435	5.041	1.663832
4.771	0.18359	5.078	1.769473
3.949	0.17917	5.127	1.854223
4.545	0.193692	5.216	1.951673
4.603	0.219915	5.464	2.082665
4.408	0.228283	5.442	2.216931
4.439	0.237771	5.278	2.319561
4.156	0.234249	5.195	2.364967
4.476	0.250976	5.197	2.46678
4.609	0.260075	5.318	2.557764
4.488	0.269218	5.386	2.684405
4.277	0.268288	5.361	2.774956
4.378	0.286005	5.455	2.898218
4.409	0.291215		2.9
4.35	0.296448		3

Appendix B. (*Continued*)

CO_2	g	CO_2	g
4.442	0.313843		3.1
4.611	0.326307		3.2
4.74	0.345757		3.3
4.863	0.370008		3.4
			3.5

SERIAL CORRELATION ROBUST *LM* TYPE TESTS FOR A SHIFT IN TREND

Jingjing Yang and Timothy J. Vogelsang

ABSTRACT

We analyze Lagrange Multiplier (LM) *tests for a shift in trend of a univariate time series at an unknown date. We focus on the class of* LM *statistics based on nonparametric kernel estimates of the long run variance. Extending earlier work for models with nontrending data, we develop a fixed-*b *asymptotic theory for the statistics. The fixed-*b *theory suggests that, for a given statistic, kernel, and significance level, there usually exists a bandwidth such that the fixed-*b *asymptotic critical value is the same for both* I(0) *and* I(1) *errors. These "robust" bandwidths are calculated using simulation methods for a selection of well-known kernels. We find when the robust bandwidth is used, the supremum statistic configured with either the Bartlett or Daniell kernel gives* LM *tests with good power. When testing for a slope change, we obtain the surprising finding that less trimming of potential shift dates leads to higher power, which contrasts the usual relationship between trimming and power. Finite sample simulations indicate that the robust* LM *statistics have stable size with good power.*

Keywords: Fixed-*b* asymptotics; structural change; long run variance; nonmonotonic power; kernel; robust bandwidth

30th Anniversary Edition
Advances in Econometrics, Volume 30, 97–131

ISSN: 0731-9053/doi:10.1108/S0731-9053(2012)0000030009

INTRODUCTION

A recent paper by Yang and Vogelsang (2011) carried out an analysis of *LM* tests for a shift in the mean of a nontrending time series. We extend the approach of Yang and Vogelsang (2011) to tests for a shift in the trend of a univariate time series. The robust tests for a shift in trend developed in this chapter add to the recent literature on tests for a shift in trend that are robust to the strength of serial correlation and that focus on Wald-type tests; see Harvey, Leyboune, and Taylor (2009), Perron and Yabu (2009), Vogelsang (1997), and Sayginsoy and Vogelsang (2011). Here we focus on the class of *LM* statistics based on nonparametric kernel estimators of the long run variance. We treat the shift date as unknown. The main theoretical contribution of the chapter is to develop a fixed-*b* asymptotic theory for the long run variance estimator and the *LM* test statistics. Completely analogous to the findings in Yang and Vogelsang (2011) we show that the fixed-*b* limit of the *LM* statistics depends on the kernel and bandwidth needed to implement the long run variance estimator. The fixed-*b* limit also depends on the magnitude of the trend shift under the alternative. The fixed-*b* limiting random variables allow us to theoretically capture the impact of the choice of bandwidth on both the size and the power of the tests.

We derive results for both the case of stationary $I(0)$ errors and integrated/nearly integrated $I(1)$ errors. We focus on five well-known kernels: Bartlett, Parzen, Bohman, Daniell, and quadratic spectral (QS). As in Yang and Vogelsang (2011) we find that, with very few exceptions, there exist bandwidth choices for the long run variance estimator that result in critical values being the same in both the $I(0)$ and $I(1)$ error cases. These serial correlation "robust" bandwidths depend on the kernel, significance level, and amount of trimming applied to the grid of potential shift dates. Throughout the chapter, an *LM* test implemented with its $I(0)/I(1)$ robust bandwidth and corresponding fixed-*b* critical value is labeled "robust *LM*" test. The robust *LM* tests are desirable in practice because they provide a simple solution to the bandwidth choice problem while simultaneously giving robustness to the strength of serial correlation in the errors.

Using an asymptotic power analysis we explore the impact of choice of kernel on the robust *LM* statistics. We find that when testing for an intercept shift, the Bartlett kernel usually gives the most power, whereas when testing for a slope change, the Daniell kernel has the best power. We treat the date of the shift as unknown and we apply the well-known mean, exponential mean, and supremum statistics. These statistics require that the set of possible shift dates be constrained away from the endpoints of the sample. This constraint on possible shift dates is called trimming. Our power analysis suggests that the supremum statistic delivers tests with the highest power. In standard settings, more trimming tends to increase power because it restricts the range of possible shift dates. When testing for an intercept shift, we find that increasing trimming does increase power although only slightly. When testing for a slope change, we find that the opposite is true; more trimming reduces power, often substantially so. In practice we recommend that the Bartlett supremum robust *LM* statistic be used to test for a shift in intercept and the Daniell supremum robust *LM* statistic be used to test for a slope change. In the latter case we recommend that as little trimming be used as possible because this enhances power and extends the range of dates over which potential shifts can be detected.

While we believe this is the first study in the literature that focuses on *LM* tests for a shift in trend, there is a recent literature on *LM* tests for a shift in mean. Vogelsang (1999) examined several *LM*-type tests for a shift in mean and documented the possibility and sources of nonmonotonic power. Nonmonotonic power occurs when a power curve initially rises as the alternative moves away from the null but power eventually falls as the alternative moves farther away from the null. In same cases power can fall to zero for extremely large deviations away from the null. The reason that power can be nonmonotonic when using *LM* statistics is intuitive to explain. *LM* statistics use long run variance estimators based on residuals from the model estimated under the null hypothesis of no shift. Therefore, when there is a shift, the long run variance estimator is not invariant to the magnitude of the shift. A large shift can cause the denominator of an *LM* statistic to be large and this can cripple power. Yang and Vogelsang (2011) showed that the possibility of nonmonotonic power in *LM* tests for a shift in mean is directly linked to the bandwidth used for the long run variance estimator. Small bandwidths tend to give *LM* tests with monotonic power, whereas larger bandwidths tend to give *LM* tests with nonmonotonic power. This relationship between the bandwidth and nonmonotonic power can explain the monotonic power of the hybrid *LM* tests for a shift in mean recently proposed by Kejriwal (2009). Not surprisingly, we find that the relationship between the bandwidth and the possibility of nonmonotonic power carries over to tests for a shift in trend.

The remainder of the chapter is organized as follows. In the next section we describe the model and assumptions and we develop fixed-*b* asymptotic theory for the *LM* tests. We use the fixed-*b* theory to show that, for a given percentage point, there exist bandwidths such that $I(0)$ and $I(1)$ critical values are the same. We use the fixed-*b* theory to explore the power properties of the robust

LM statistics. In the Section "Finite sample performance and comparisons" we provide results from simulations that compare the finite sample properties of the robust *LM* tests. Null rejection probabilities are close to the nominal level. Power of the robust *LM* tests relative to existing Wald-type tests depends on the strength of the serial correlation. When serial correlation is weak, the robust *LM* tests tend to have lower power, whereas with strong serial correlation, power of the robust *LM* tests tends to be higher although in some cases the power curves cross and neither *LM* nor *Wald* tests dominate. Section "Conclusions" concludes and the proof of the main theoretical result of the chapter is given in the appendix.

TESTING FOR A SHIFT IN TREND

The Models and Assumptions

We assume that one of the following three models generates a time series process y_t:

$$y_t = \mu + \beta t + \delta_1 DU_t^0 + u_t \tag{1}$$

$$y_t = \mu + \beta t + \delta_2 DT_t^0 + u_t \tag{2}$$

$$y_t = \mu + \beta t + \delta_1 DU_t^0 + \delta_2 DT_t^0 + u_t \tag{3}$$

where

$$DU_t^0 = \mathbf{1}(t > T_b^0), \quad DT_t^0 = \mathbf{1}(t > T_b^0)(t - T_b^0),$$

$\mathbf{1}(\cdot)$ is the indicator function and T_b^0 denotes the date of the shift in trend. Data generating process (DGP) (1) allows for a shift in the intercept of a trending time series; (2) allows only for a shift in the slope of a trending time series and requires that segments of the trend before and after the shift date be joined, and finally; (3) allows for a shift in the intercept as well as the slope of a trending time series. Throughout the chapter we assume that T_b^0 is unknown. Following Yang and Vogelsang (2011) we assume the errors are of the form

$$u_t = \rho u_{t-1} + \varepsilon_t, \quad t = 1, \cdots, T \tag{4}$$

$$\varepsilon_t = d(L)e_t, \;\; d(L) = \sum_{i=0}^{\infty} d_i L^i, \quad \sum_{i=0}^{\infty} i|d_i| < \infty, \quad d(1)^2 > 0 \tag{5}$$

where L is the lag operator, $\{e_t\}$ is a martingale difference sequence with $\sup_t E(e_t^4) < \infty$, $E(e_t|e_{t-1}, e_{t-2}, \ldots) = 0$ and $E(e_t^2|e_{t-1}, e_{t-2}, \ldots) = 1$. The errors are $I(0)$ when $|\rho| < 1$, and the errors are nearly $I(1)$ when $\rho = 1 - c/T$, where c is a constant. The pure unit root error case corresponds to $c = 0$. When (4) and (5) hold, some well-known results follow[1]:

$$T^{-1/2} \sum_{t=1}^{[rT]} u_t \Rightarrow \sigma W(r) = \sigma \int_0^r dW(s) \quad \text{if } |\rho| < 1$$

$$T^{-1/2} u_{[rT]} \Rightarrow d(1) V_c(r) \quad \text{if } \rho = 1 - \frac{c}{T}$$

where $\sigma^2 = d(1)^2/(1-\rho)^2$, $W(r)$ is a standard Wiener process, $V_c(r) = \int_0^r \exp\{-c(r-s)\} dW(s)$, $[rT]$ is the integer part of rT where $r \in [0, 1]$, and $\Rightarrow$ denotes weak convergence.

Tests for a shift in trend are constructed using the following three regression models:

$$y_t = \mu + \beta t + \delta_1 DU_t + \xi_t \tag{6}$$

$$y_t = \mu + \beta t + \delta_2 DT_t + \xi_t \tag{7}$$

$$y_t = \mu + \beta t + \delta_1 DU_t + \delta_2 DT_t + \xi_t \tag{8}$$

where

$$DU_t = \mathbf{1}(t > T_b), \quad DT_t = \mathbf{1}(t > T_b)(t - T_b)$$

and T_b is the shift date used in the estimated model. When $T_b = T_b^0$, the estimated models are correctly specified and $\xi_t = u_t$, but when $T_b \neq T_b^0$, the estimated models are misspecified and ξ_t is a function of u_t and both T_b and T_b^0. Following the structural change literature, we make the standard assumption that the shift points, defined as $\lambda_0 = T_b^0/T$ and $\lambda = T_b/T$, remain fixed as the sample size increases. In our asymptotic theory the deterministic trend regressors become scaled and mapped into functions defined on the unit interval [0, 1] as follows. Rewrite the three regressions (6)–(8) in generic notation as

$$y_t = f_t' \theta + g_t(T_b)' \delta + \xi_t \tag{9}$$

where $f_t' = [1, t]'$, $\theta = [\mu, \beta]'$, and

$$g_t(T_b) = \begin{cases} [DU_t] & \text{for} \quad (6) \\ [DT_t] & \text{for} \quad (7) \\ \begin{bmatrix} DU_t \\ DT_t \end{bmatrix} & \text{for} \quad (8), \end{cases} \qquad \delta = \begin{cases} [\delta_1] & \text{for} \quad (6) \\ [\delta_2] & \text{for} \quad (7) \\ \begin{bmatrix} \delta_1 \\ \delta_2 \end{bmatrix} & \text{for} \quad (8). \end{cases}$$

We can write the DGP models in generic notation as

$$y_t = f_t'\theta + g_t(T_b^0)'\delta + u_t. \tag{10}$$

Define the scaling matrices $\tau_f = \begin{bmatrix} 1 & 0 \\ 0 & T^{-1} \end{bmatrix}$ and

$$\tau_g = \begin{cases} 1 & \text{for} \quad (6) \\ T^{-1} & \text{for} \quad (7) \\ \tau_f & \text{for} \quad (8). \end{cases}$$

Then it follows directly that

$$\tau_f f_{[rT]} \to F(r), \quad \tau_g g_{[rT]}(T_b) \to G(r, \lambda)$$

where $F(r)' = [1, r]'$ and

$$G(r, \lambda) = \begin{cases} [\mathbf{1}(r > \lambda)] & \text{for} \quad (6) \\ [\mathbf{1}(r > \lambda)(r - \lambda)] & \text{for} \quad (7) \\ \begin{bmatrix} \mathbf{1}(r > \lambda) \\ \mathbf{1}(r > \lambda)(r - \lambda) \end{bmatrix} & \text{for} \quad (8). \end{cases}$$

Define the "detrended" versions of $dW(r)$, $V_c(r)$, and $G(r, \lambda)$ as

$$\widetilde{dW}(r) = dW(r) - F(r)' \left(\int_0^1 F(s)F(s)' ds \right)^{-1} \int_0^1 F(s)\, dW(s)$$

$$\tilde{V}_c(r) = V_c(r) - F(r)' \left(\int_0^1 F(s)F(s)' ds \right)^{-1} \int_0^1 F(s) V_c(s)\, ds$$

$$\tilde{G}(r, \lambda) = G(r, \lambda) - F(r)' \left(\int_0^1 F(s)F(s)' ds \right)^{-1} \int_0^1 F(s) G(s, \lambda)\, ds.$$

The LM statistics studied here use an estimator of σ^2 from the class of nonparametric spectral density estimators given by

$$\tilde{\sigma}^2(m) = \tilde{\gamma}_0 + 2\sum_{j=1}^{T-1} k(j/m)\tilde{\gamma}_j, \qquad \tilde{\gamma}_j = T^{-1}\sum_{t=j+1}^{T} \tilde{\xi}_t\tilde{\xi}_{t-j}$$

where $\tilde{\xi}_t$ are the OLS residuals from the regression

$$y_t = f_t'\theta + \xi_t \tag{11}$$

which is the regression model obtained under the null hypothesis of no shift in the trend function. The function $k(x)$ is the kernel function and m is the bandwidth (or truncation lag for kernels that truncate). A kernel is labeled type 1 if $k(x)$ is twice continuously differentiable everywhere, and type 2 if $k(x)$ is continuous, twice continuously differentiable everywhere except at $|x| = 1$ and $k(x) = 0$ for $|x| \geq 1$. For type 2 kernels define the derivative from the left at $x = 1$ as $k'_-(1) = \lim_{h\to 0}[(k(1) - k(1-h))/h]$.

LM Tests for a Shift in Trend

In the respective models we focus on testing the null hypothesis that there is no shift in trend:

$$H_0 : \delta = 0$$

against the alternative

$$H_1 : \delta \neq 0$$

For a given shift date, T_b, used in regression (9), define the *LM* statistic as

$$LM(T_b) = \frac{SSR_0 - SSR(T_b)}{\tilde{\sigma}^2(m)}$$

where $SSR_0 = \sum_{t=1}^{T}\tilde{\xi}_t^2$ is the sum of squared residuals under the null hypothesis and $SSR(T_b)$ is the sum of squared residuals from regression (9). Because we treat the shift date as unknown, we follow Andrews (1993) and Andrews and Ploberger (1994) and consider mean, exponential mean, and supremum statistics of the form

$$MeanLM = T^{-1} \sum_{T_b \in \Lambda^*} LM(T_b)$$

$$ExpLM = \log\left(T^{-1} \sum_{T_b \in \Lambda^*} \exp\left(\tfrac{1}{2} LM(T_b)\right)\right)$$

$$SupLM = \sup_{T_b \in \Lambda^*} LM(T_b)$$

where $\Lambda^* = \{T_b^*, T_b^* + 1, \ldots, T - T_b^*\}$ is the set of possible shift dates. The parameter $\lambda^* = T_b^*/T$ is held fixed as T increases and λ^* determines the amount of "trimming" used in computing the statistics.

Fixed-b Asymptotic Theory

In this subsection we describe fixed-b asymptotic results for the LM tests. These results complement the fixed-b results derived by Sayginsoy and Vogelsang (2011) for the case of nonparametric long run variance *Wald* statistics for testing for a shift in trend. We provide results under sequences of alternatives that depend on whether the error, u_t, is $I(0)$ or $I(1)$. The alternative is given by

$$H_A : \delta = \tau_i \bar{\delta}$$

where $\bar{\delta} = \bar{\delta}_1$ for (1), $\bar{\delta} = \bar{\delta}_2$ for (2), $\bar{\delta}' = [\bar{\delta}_1, \bar{\delta}_2]'$ for (3), $\tau_0 = T^{-1/2}\tau_g$, and $\tau_1 = T^{1/2}\tau_g$. Note that τ_0 gives the appropriate local alternative rate(s) when u_t is $I(0)$, whereas τ_1 gives the appropriate local alternative rate(s) when u_t is $I(1)$. Because the numerator of $LM(T_b)$, $SSR_0 - SSR(T_b)$, is identical to the *Wald* statistic, its limit follows directly from the results of Sayginsoy and Vogelsang (2011) (Theorems 1 and 2). Our theoretical contribution is obtaining the fixed-b limit of $\tilde{\sigma}^2(m)$ under H_A. Results for the null distribution of the LM statistics follow directly by setting $\bar{\delta} = 0$. The following theorem gives the limiting distributions of the LM statistics under H_A.

Theorem 1. Suppose the true model is given by (10) with shift date $T_b^0 = \lambda_0 T$. Suppose the LM statistic is computed using the regression (9) with shift date $T_b = \lambda T$. Let $m = bT$, where $b \in (0, 1]$ is fixed as T increases. Under H_A, as $T \to \infty$,

$$LM(T_b) \Rightarrow \frac{P_i(\lambda, \lambda_0)' \left(\int_0^1 \tilde{G}(s, \lambda)\tilde{G}(s, \lambda)' ds \right)^{-1} P_i(\lambda, \lambda_0)}{\Phi_i(b)} \equiv LM_i^\infty(\lambda, \lambda_0, b, \delta^*)$$

$$MeanLM_m \Rightarrow \int_{\lambda^*}^{1-\lambda^*} LM_i^{\infty}(\lambda, \lambda_0, b, \delta^*)\, d\lambda \equiv MeanLM_b^{\infty}$$

$$ExpLM_m \Rightarrow \log\left(\int_{\lambda^*}^{1-\lambda^*} \exp\left(\frac{1}{2} LM_i^{\infty}(\lambda, \lambda_0, b, \delta^*)\right) d\lambda\right) \equiv ExpLM_b^{\infty}$$

$$SupLM_m \Rightarrow \sup_{\lambda \in [\lambda^*, 1-\lambda^*]} LM_i^{\infty}(\lambda, \lambda_0, b, \delta^*) \equiv SupLM_b^{\infty}$$

where $i = 0$ if $\{u_t\}$ is $I(0)$ and $i = 1$ if $\{u_t\}$ is $I(1)$ and

$$\delta^* = \begin{cases} \bar{\delta}\sigma^{-1}, & \text{if } \{u_t\} \text{ is } I(0), \\ \bar{\delta}d(1)^{-1}, & \text{if } \{u_t\} \text{ is } I(1). \end{cases}$$

$$P_0(\lambda, \lambda_0) = \left(\int_0^1 \widetilde{G}(s, \lambda)\widetilde{G}(s, \lambda)' ds\right)^{-1} \times \left[\int_0^1 \widetilde{G}(s, \lambda) dW(s) + \bar{\delta}\sigma^{-1} \int_0^1 \widetilde{G}(s, \lambda)\widetilde{G}(s, \lambda_0)' ds\right]$$

$$P_1(\lambda, \lambda_0) = \left(\int_0^1 \widetilde{G}(s, \lambda)\widetilde{G}(s, \lambda)' ds\right)^{-1} \times \left[\int_0^1 \widetilde{G}(s, \lambda) V_c(s) + \bar{\delta}d(1)^{-1} \int_0^1 \widetilde{G}(s, \lambda)\widetilde{G}(s, \lambda_0)' ds\right]$$

$$\Phi_i(b) = \begin{cases} \int_0^1 \int_0^1 -\frac{1}{b^2} k''\left(\frac{r-s}{b}\right) Q_i(r) Q_i(s)\, dr\, ds, & \text{if } k(\cdot) \text{ is of type 1} \\ \int\int_{|r-s|<b} -\frac{1}{b^2} k''\left(\frac{r-s}{b}\right) Q_i(r) Q_i(s)\, dr\, ds & \\ \quad + \frac{2}{b} k'_{-}(1) \int_0^{1-b} Q_i(r+b) Q_i(r)\, dr, & \text{if } k(\cdot) \text{ is of type 2} \\ \frac{2}{b} \int_0^1 Q_i(r)^2 dr - \frac{2}{b} \int_0^{1-b} Q_i(r+b) Q_i(r)\, dr, & \text{if } k(\cdot) \text{ is Bartlett} \end{cases}$$

$$Q_0(r) = \int_0^r \widetilde{dW}(s) + \int_0^r \widetilde{G}(s, \lambda_0)' ds \bar{\delta}\sigma^{-1},$$

$$Q_1(r) = \int_0^r \widetilde{V}_c(s) ds + \int_0^r \widetilde{G}(s, \lambda_0)' ds \bar{\delta} d(1)^{-1}.$$

We provide a proof of Theorem 1 in the appendix. Notice that the asymptotic random variables given by Theorem 1 depend on the persistence of the errors, the kernel, the bandwidth and, under the alternative, the magnitude of the intercept and/or slope change. We can explore the impact of the choice of kernel and bandwidth on the null distributions and power of the *LM* tests by simulating from these asymptotic distributions.

Robust Bandwidths

By setting $\bar{\delta} = 0$ Theorem 1 gives fixed-b asymptotic limits of the *LM* statistics under the null hypothesis of no shift in trend. Notice that the limiting distributions given by Theorem 1 are different for $I(0)$ and $I(1)$ errors. Therefore, the *LM* statistics are, in general, not robust to the strength of the serial correlation in u_t. Theoretically exploring the relationship between critical values and the strength of serial correlation is difficult given the nonstandard nature of the limits in Theorem 1. Alternatively we use numerical methods to explore the relationship between critical values and the strength of serial correlation. We simulated the fixed-b limits given by Theorem 1 using 10,000 replications with the Wiener processes in the limits approximated by scaled partial sums of 1,000 i.i.d. $N(0, 1)$ random deviates. We simulated critical values for both the $I(0)$ and the $I(1)$ limits and for the $I(1)$ case we simulated the limits for a range of values for $c \geq 0$. Simulations were carried out for the Bartlett, Parzen, Bohman, Daniell, and QS kernel over the relatively fine bandwidth grid $b = 0.001, 0.002, 0.0003, \ldots, 0.998, 0.999, 1.0$. We found that for all three models, each type of *LM* statistic, and most kernels there exists, for a given percentage point, a value of b such that the $I(0)$ and the $I(1)(c = 0)$ critical values are the same. We use b^* to denote these robust bandwidths, and we denote the corresponding fixed-b critical value by $cv(b^*)$. We found that b^* exists for significance levels ranging from 1% to 20% and across a wide range of trimming values. In addition we also found that for $c > 0$, critical values corresponding to b^* tend to be below or very close to $cv(b^*)$ making $cv(b^*)$ robust to $I(1)$ errors more generally. This suggests that in practice use of b^* and $cv(b^*)$ will give rejection probabilities close to the nominal level for weak, strong, and unit root serial correlation in the errors. The existence of these robust bandwidths is, on the one hand, remarkable given that such robust bandwidths do not exist for Wald type trend function tests. On the other hand, the existence of these robust bandwidths is not surprising given that Yang and Vogelsang (2011) found the existence of robust bandwidths when using LM statistics to test for a shift in mean of a nontrending series.

While it would be impractical to plot critical value curves for all models, statistics, kernels, significance levels, and trimming values, it is useful to depict a few cases to show the existence of the robust bandwidths. Here we focus on Model 2. We report results for the Daniell kernel (our recommended kernel based on power considerations discussed in the next section) and the Bartlett kernel. We focus on the 5% significance level and 10% trimming ($\lambda^* = 0.1$). Figs. 1–3 plot asymptotic critical values for the $MeanLM_b^\infty$, $ExpLM_b^\infty$, and $SupLM_b^\infty$ statistic for the Daniell kernel. The x-axis depicts the value of b. We plot critical values for the $I(0)$ error case and for the $I(1)$ error case for $c = 0, 12, 36, 60$. When b is small, the $I(1)$ critical values are much larger than the $I(0)$ critical value. This is a typical finding for models with deterministic regressors. As b increases, the critical values in the $I(1)$ cases rapidly decrease, hit a minimum, and then slowly rise. In the $I(0)$ case the critical values start much lower when b is small,

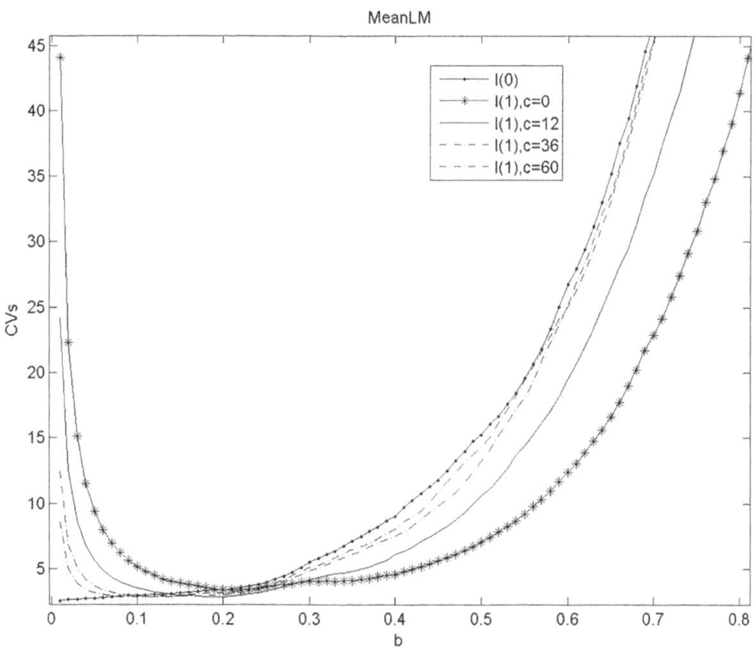

Fig. 1. 5% Fixed-b Asymptotic Critical Values of $MeanLM_b^\infty$. Daniell Kernel, 10% Trimming, Model 2.

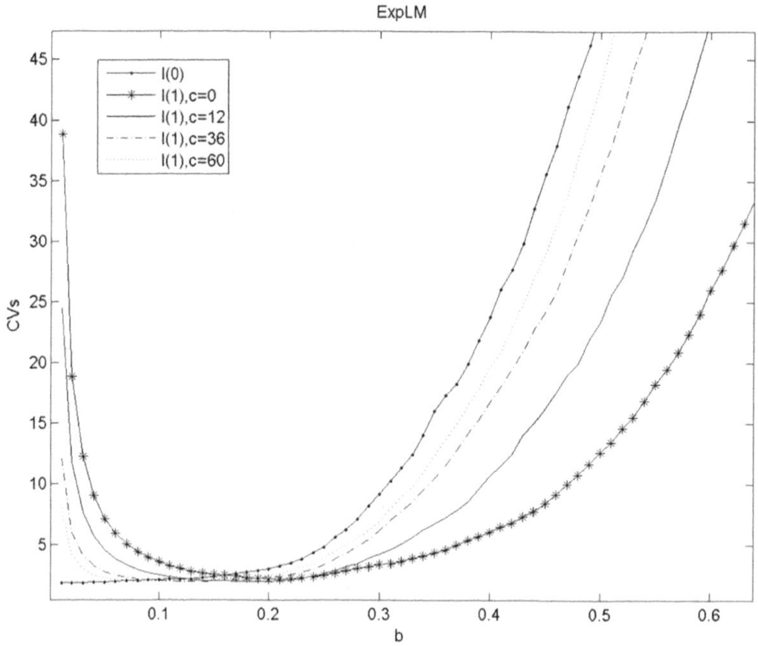

Fig. 2. 5% Fixed-b Asymptotic Critical Values of $ExpLM_b^{\infty}$. Daniell Kernel, 10% Trimming, Model 2.

decrease slowly as b increases, hit a minimum, and then increase relatively quickly as b increases further. Notice that there is an intersection point where the $I(0)$ and $I(1)(c = 0)$ critical values are the same and the other $I(1)$ critical values are smaller. The bandwidth where the intersection occurs is the robust bandwidth, b^*. Fig. 4 plots critical values for the $MeanLM_b^{\infty}$ statistic using the Bartlett kernel. Here we see that the $I(0)$ and $I(1)(c = 0)$ critical value curves do not intersect. Therefore, b^* does not exist in this case. We found this exception to be rare and in most cases the robust bandwidths exist.

That the robust bandwidths exist is useful in practice because the robust bandwidths allow configurations of the *LM* statistics that are robust to the strength of serial correlation while simultaneously solving the bandwidth choice problem. As the power analysis in the next section shows, the *SupLM* statistic configured with the Bartlett kernel tends to deliver robust tests with the highest power in Models 1 and 3 (intercept shift) regardless of

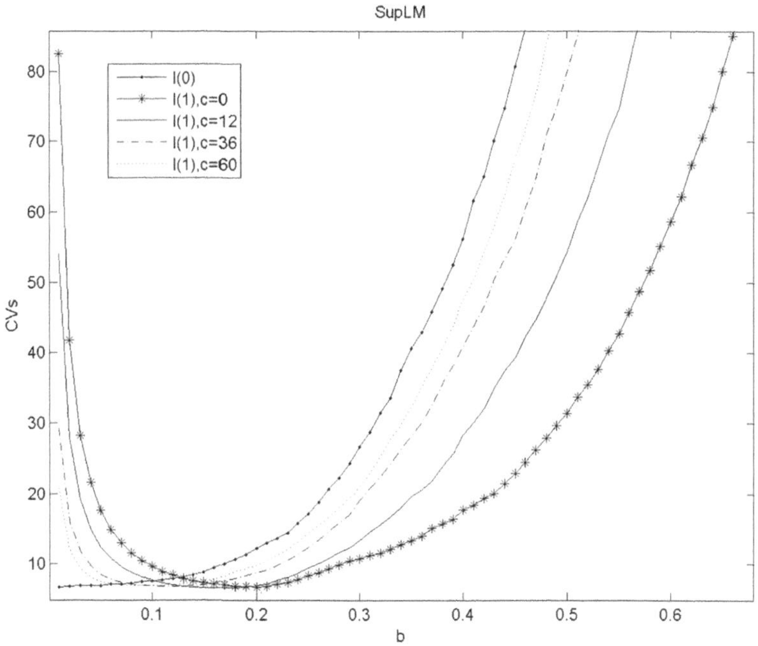

Fig. 3. 5% Fixed-*b* Asymptotic Critical Values of $SupLM_b^\infty$. Daniell Kernel, 10% Trimming, Model 2.

significance level or trimming. The *SupLM* statistic configured with the Daniell kernel tends to deliver robust tests with the highest power in Models 2 and 3 (slope change). We computed the Bartlett and Daniell kernel robust bandwidths for *SupLM* for significance levels 1%, 2%,..., 20% and trimming values 1%, 5%, 10%, 15%, and 20%. In Tables 1 and 3 we tabulate b^* and $cv(b^*)$ for the Bartlett $SupLM_{b^*}^\infty$ statistic for Models 1 and 3, whereas in Tables 2 and 4 we tabulate b^* and $cv(b^*)$ for the Daniell $SupLM_{b^*}^\infty$ statistic for Models 2 and 3. Asymptotic critical values for other kernels and for the $MeanLM_{b^*}^\infty$ and $ExpLM_{b^*}^\infty$ statistics are available upon request.

Asymptotic Power

Because Theorem 1 provides an approximation to the sampling behavior of the *LM* statistics under both the null of no shift trend and the alternative of

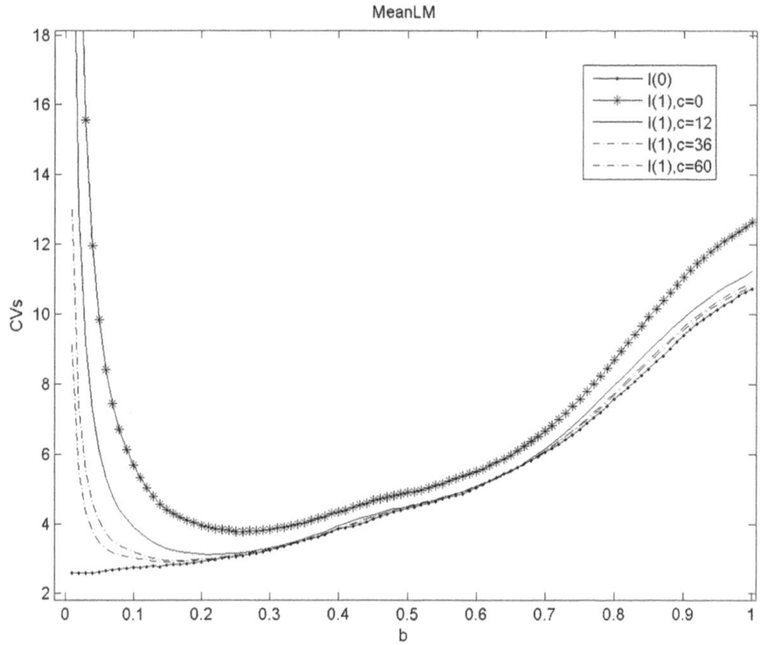

Fig. 4. 5% Fixed-*b* Asymptotic Critical Values of $MeanLM_b^\infty$. Bartlett Kernel, 10% Trimming, Model 2.

a shift in trend, we can use the asymptotic random variables from the theorem to compute local asymptotic power, which provides predictions about finite sample power of the tests. For the three models we simulated asymptotic power of the robust *LM* statistics for the set of five kernels, significance levels 1%, 2%,..., 20% and trimming values 1%, 5%, 10%, 15%, and 20% using the same simulation methods as used for the critical values. The trend shift was located in the middle of the sample, that is, $\lambda_0 = 0.5$. Similar results were obtained for $\lambda_0 = 0.25, 0.75$. Results are only reported for the Bartlett and Daniell kernels as those are the two kernels that deliver highest power depending on the model. Results for other kernels are available upon request. Results are only reported for Models 1 and 2. Results for Model 3 in the case of an intercept shift are similar to Model 1 and results for Model 3 in the case of a slope change are similar to Model 2.

Figs. 5–8 plot asymptotic power for detecting a shift in intercept using the Bartlett and Daniell kernel *LM* statistics at significance level 5% with

Table 1. $I(0)/I(1)$ Robust Bandwidths and Asymptotic Critical Values, $SupLM_{b^*}^{\infty}$. Bartlett Kernel, Model 1 (Regressors $[1, t, DU_t]$).

Level (%)	1% Trimming		5% Trimming		10% Trimming		15% Trimming		20% Trimming	
	b^*	$cv(b^*)$	b^*	$cv(b^*)$	b^*	$cv(b^*)$	b^*	$cv(b^*)$	b^*	$cv(b^*)$
80	0.0840	9.2645	0.0920	8.7001	0.0960	8.3900	0.0980	8.1728	0.0980	7.9787
81	0.0830	9.3745	0.0910	8.7845	0.0960	8.4438	0.0980	8.2296	0.0980	8.0409
82	0.0830	9.4517	0.0910	8.8491	0.0950	8.5447	0.0970	8.3161	0.0970	8.1294
83	0.0820	9.5557	0.0900	8.9565	0.0950	8.6038	0.0960	8.4208	0.0970	8.2079
84	0.0820	9.6460	0.0900	9.0135	0.0940	8.7162	0.0960	8.4863	0.0960	8.3003
85	0.0810	9.7539	0.0890	9.1390	0.0940	8.7840	0.0950	8.5892	0.0950	8.4088
86	0.0800	9.9146	0.0880	9.2659	0.0930	8.8947	0.0950	8.6734	0.0950	8.4900
87	0.0800	9.9988	0.0880	9.3597	0.0930	8.9705	0.0940	8.7806	0.0940	8.5887
88	0.0790	10.1480	0.0870	9.4675	0.0920	9.0697	0.0930	8.8980	0.0940	8.6922
89	0.0780	10.3220	0.0870	9.5651	0.0910	9.1941	0.0930	8.9825	0.0930	8.8140
90	0.0770	10.4930	0.0860	9.6988	0.0900	9.3468	0.0920	9.0974	0.0920	8.9270
91	0.0760	10.6780	0.0850	9.8434	0.0890	9.4885	0.0910	9.2420	0.0910	9.0626
92	0.0760	10.8320	0.0840	10.0230	0.0890	9.6152	0.0900	9.3902	0.0900	9.2112
93	0.0740	11.1240	0.0840	10.1580	0.0880	9.7887	0.0890	9.5467	0.0890	9.3807
94	0.0730	11.3310	0.0820	10.3990	0.0870	9.9711	0.0880	9.7381	0.0880	9.5471
95	0.0720	11.5990	0.0810	10.6030	0.0860	10.1440	0.0860	9.9933	0.0870	9.7736
96	0.0700	11.9780	0.0790	10.9310	0.0840	10.4300	0.0850	10.2190	0.0850	10.0450
97	0.0690	12.3620	0.0780	11.2630	0.0820	10.7820	0.0830	10.5720	0.0830	10.3790
98	0.0660	12.9780	0.0760	11.6610	0.0790	11.2760	0.0800	11.0810	0.0800	10.8930
99	0.0620	14.0720	0.0720	12.4390	0.0760	11.9590	0.0760	11.7720	0.0770	11.5880

Table 2. $I(0)/I(1)$ Robust Bandwidths and Asymptotic Critical Values, $SupLM_{b^*}^{\infty}$. Daniell Kernel, Model 2 (Regressors $[1, t, DT_t]$).

Level (%)	1% Trimming		5% Trimming		10% Trimming		15% Trimming		20% Trimming	
	b^*	$cv(b^*)$	b^*	$cv(b^*)$	b^*	$cv(b^*)$	b^*	$cv(b^*)$	b^*	$cv(b^*)$
80	0.1220	7.2539	0.1480	6.4718	0.1650	6.0510	0.1870	5.6209	0.2080	5.4290
81	0.1210	7.3922	0.1460	6.5662	0.1640	6.1435	0.1840	5.7018	0.2060	5.4773
82	0.1190	7.5163	0.1440	6.6548	0.1620	6.2352	0.1820	5.7773	0.2020	5.5205
83	0.1170	7.6796	0.1420	6.7544	0.1610	6.3217	0.1810	5.8516	0.2000	5.5707
84	0.1160	7.8422	0.1410	6.8384	0.1590	6.4137	0.1780	5.9496	0.1970	5.6319
85	0.1140	7.9629	0.1380	6.9726	0.1580	6.5151	0.1760	6.0387	0.1940	5.6991
86	0.1120	8.1375	0.1350	7.1215	0.1560	6.6094	0.1740	6.1590	0.1920	5.7772
87	0.1100	8.2983	0.1330	7.2739	0.1540	6.7188	0.1720	6.2795	0.1890	5.8696
88	0.1080	8.5285	0.1300	7.4304	0.1530	6.8136	0.1700	6.3831	0.1870	5.9681
89	0.1050	8.7361	0.1280	7.5812	0.1500	6.9176	0.1670	6.5025	0.1840	6.0722
90	0.1030	8.9560	0.1260	7.7643	0.1470	7.0390	0.1650	6.6270	0.1810	6.1957
91	0.1010	9.2455	0.1240	7.9489	0.1450	7.1475	0.1630	6.7631	0.1780	6.3433
92	0.0980	9.5267	0.1210	8.1962	0.1410	7.3166	0.1610	6.9011	0.1750	6.4650
93	0.0960	9.8249	0.1170	8.4354	0.1370	7.5117	0.1580	7.0257	0.1730	6.6144
94	0.0930	10.1400	0.1140	8.6846	0.1330	7.7410	0.1560	7.1489	0.1700	6.7910
95	0.0900	10.5320	0.1100	9.0108	0.1300	7.9796	0.1520	7.3137	0.1670	6.9442
96	0.0860	11.0740	0.1050	9.4789	0.1260	8.3209	0.1460	7.5247	0.1630	7.1386
97	0.0820	11.7940	0.1000	9.9709	0.1200	8.7277	0.1400	7.8183	0.1580	7.3678
98	0.0770	12.5720	0.0950	10.5590	0.1130	9.2207	0.1320	8.3071	0.1500	7.6430
99	0.0680	14.3020	0.0870	11.5230	0.1010	10.2990	0.1230	8.9591	0.1360	8.2330

Table 3. $I(0)/I(1)$ Robust Bandwidths and Asymptotic Critical Values, $SupLM_{b^*}^{\infty}$. Bartlett Kernel, Model 3 (Regressors $[1, t, DU_t, DT_t]$).

Level (%)	1% Trimming		5% Trimming		10% Trimming		15% Trimming		20% Trimming	
	b^*	$cv(b^*)$	b^*	$cv(b^*)$	b^*	$cv(b^*)$	b^*	$cv(b^*)$	b^*	$cv(b^*)$
80	0.0710	12.6680	0.0800	11.6420	0.0870	10.8950	0.0940	10.3400	0.1000	9.8923
81	0.0710	12.7690	0.0790	11.7450	0.0870	10.9620	0.0930	10.4440	0.0990	9.9929
82	0.0700	12.9540	0.0790	11.8490	0.0860	11.0940	0.0930	10.5020	0.0990	10.0500
83	0.0700	13.0330	0.0780	11.9700	0.0860	11.1700	0.0920	10.6270	0.0990	10.1390
84	0.0690	13.2340	0.0780	12.0510	0.0850	11.3080	0.0920	10.6940	0.0980	10.2200
85	0.0690	13.3430	0.0780	12.1920	0.0850	11.3770	0.0920	10.7890	0.0970	10.3330
86	0.0680	13.5590	0.0770	12.3240	0.0850	11.4950	0.0910	10.8880	0.0970	10.4350
87	0.0680	13.7070	0.0770	12.4720	0.0840	11.6160	0.0900	11.0350	0.0960	10.5350
88	0.0670	13.9020	0.0760	12.6140	0.0840	11.7270	0.0900	11.1290	0.0950	10.6540
89	0.0660	14.1750	0.0750	12.8140	0.0830	11.8520	0.0890	11.2670	0.0950	10.7580
90	0.0660	14.3210	0.0750	12.9930	0.0820	12.0440	0.0880	11.4190	0.0940	10.8970
91	0.0650	14.5790	0.0740	13.1730	0.0820	12.1670	0.0870	11.5920	0.0930	11.0410
92	0.0640	14.8450	0.0730	13.4080	0.0810	12.3060	0.0860	11.7660	0.0920	11.2000
93	0.0630	15.1990	0.0720	13.6810	0.0800	12.5170	0.0850	11.9470	0.0910	11.3970
94	0.0620	15.6080	0.0700	14.0120	0.0790	12.7330	0.0850	12.1140	0.0890	11.6210
95	0.0600	16.0840	0.0700	14.2350	0.0780	13.0000	0.0840	12.3410	0.0880	11.8320
96	0.0590	16.5040	0.0680	14.6400	0.0770	13.2930	0.0820	12.6320	0.0870	12.1050
97	0.0570	17.1800	0.0660	15.1720	0.0740	13.8420	0.0810	12.9120	0.0850	12.4190
98	0.0540	18.2180	0.0630	15.9890	0.0720	14.3270	0.0780	13.5010	0.0830	12.8160
99	0.0500	19.8360	0.0600	17.0280	0.0690	15.1250	0.0740	14.2370	0.0780	13.6690

Table 4. $I(0)/I(1)$ Robust Bandwidths and Asymptotic Critical Values, $SupLM_{b^*}^{\infty}$. Daniell Kernel, Model 3 (Regressors $[1, t, DU_t, DT_t]$).

Level (%)	1% Trimming		5% Trimming		10% Trimming		15% Trimming		20% Trimming	
	b^*	$cv(b^*)$	b^*	$cv(b^*)$	b^*	$cv(b^*)$	b^*	$cv(b^*)$	b^*	$cv(b^*)$
80	0.0620	13.2490	0.0680	12.1870	0.0740	11.4160	0.0780	10.8730	0.0820	10.4790
81	0.0610	13.4570	0.0680	12.3370	0.0730	11.5410	0.0780	10.9990	0.0810	10.5860
82	0.0610	13.5710	0.0670	12.4880	0.0730	11.6430	0.0770	11.1060	0.0810	10.7110
83	0.0600	13.8240	0.0670	12.5800	0.0720	11.8190	0.0770	11.2330	0.0800	10.8280
84	0.0600	13.9180	0.0670	12.7310	0.0720	11.8910	0.0760	11.3730	0.0800	10.9360
85	0.0600	14.1240	0.0660	12.8920	0.0720	12.0360	0.0760	11.4850	0.0790	11.0780
86	0.0590	14.3190	0.0660	13.0790	0.0710	12.2110	0.0750	11.6400	0.0790	11.1970
87	0.0590	14.5660	0.0650	13.2740	0.0710	12.3130	0.0750	11.7390	0.0780	11.3500
88	0.0580	14.7740	0.0650	13.4350	0.0700	12.5450	0.0740	11.9490	0.0770	11.5470
89	0.0580	14.9710	0.0640	13.6480	0.0700	12.6600	0.0740	12.0680	0.0770	11.6680
90	0.0570	15.2580	0.0640	13.9070	0.0690	12.8830	0.0730	12.2680	0.0760	11.8600
91	0.0560	15.6060	0.0630	14.0830	0.0690	13.0280	0.0720	12.5190	0.0750	12.0700
92	0.0560	15.9310	0.0620	14.3810	0.0680	13.2710	0.0720	12.6580	0.0750	12.2310
93	0.0550	16.2640	0.0610	14.7290	0.0670	13.5520	0.0710	12.9010	0.0740	12.4620
94	0.0540	16.6800	0.0600	15.0990	0.0660	13.8560	0.0700	13.1740	0.0730	12.7390
95	0.0530	17.1670	0.0600	15.4220	0.0650	14.2020	0.0690	13.5290	0.0720	13.0540
96	0.0510	17.8510	0.0580	15.8730	0.0640	14.5400	0.0670	13.9710	0.0700	13.4560
97	0.0500	18.5370	0.0570	16.5510	0.0630	14.9800	0.0660	14.4280	0.0680	13.9780
98	0.0480	19.6860	0.0550	17.2250	0.0610	15.7300	0.0640	14.9700	0.0660	14.6350
99	0.0440	21.7370	0.0520	18.4100	0.0580	16.7770	0.0600	16.1440	0.0620	15.6700

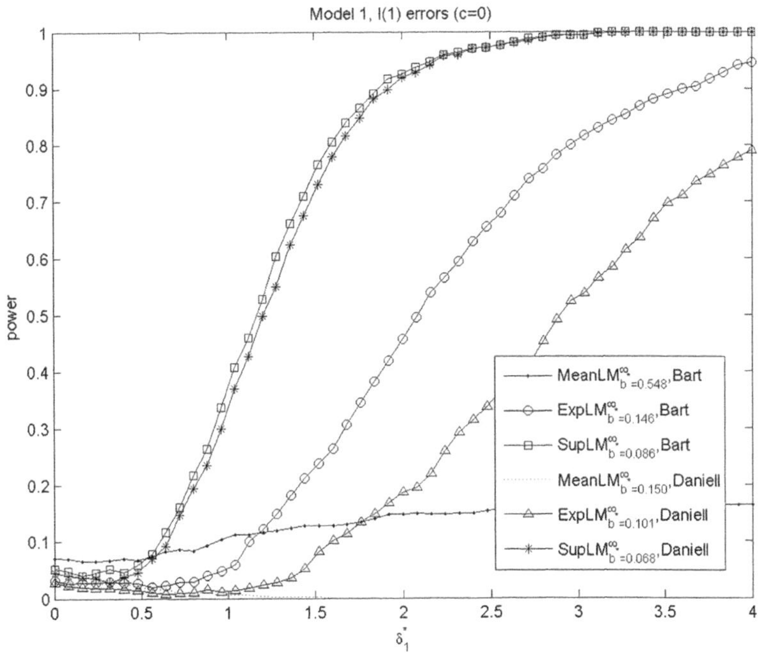

Fig. 5. Local Asymptotic Power Using b^* and $cv(b^*)$. Model 1, $I(1)$ Errors with $c = 0$.

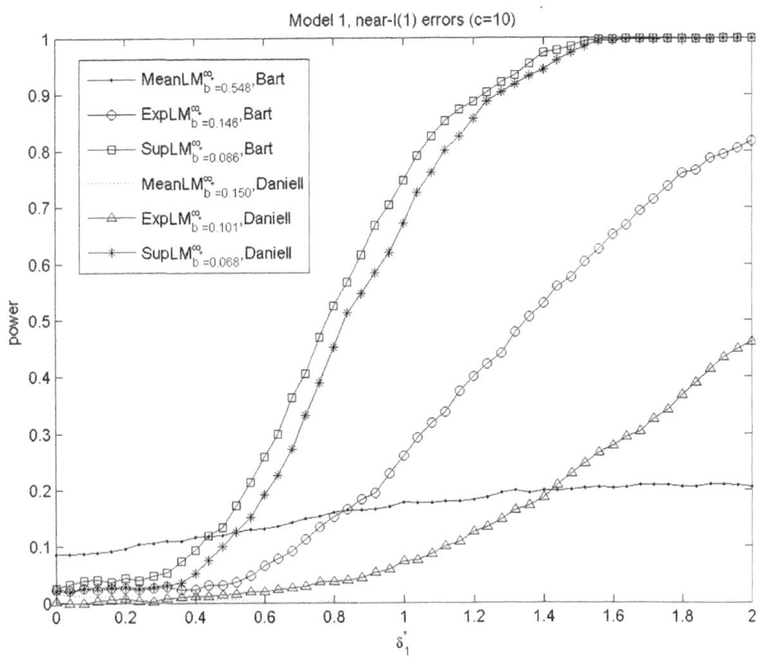

Fig. 6. Local Asymptotic Power Using b^* and $cv(b^*)$. Model 1, $I(1)$ Errors with $c = 10$.

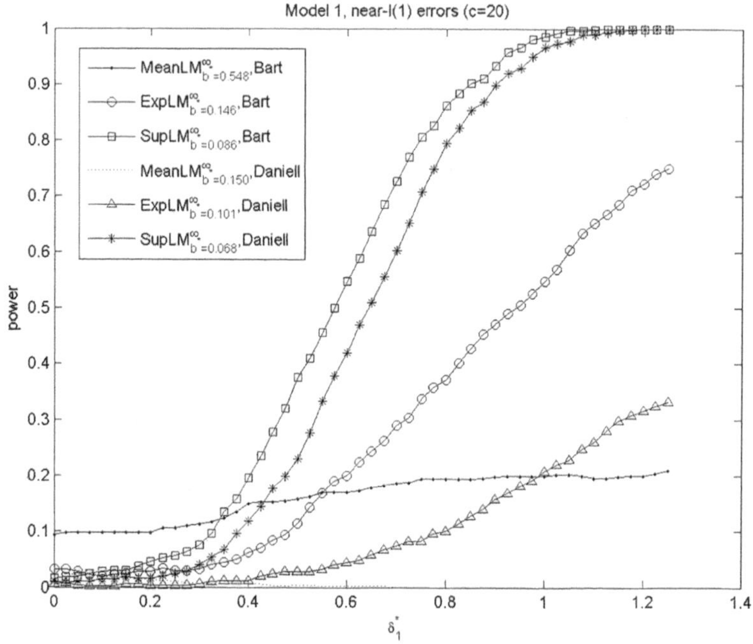

Fig. 7. Local Asymptotic Power Using b^* and $cv(b^*)$. Model 1, $I(1)$ Errors with $c = 20$.

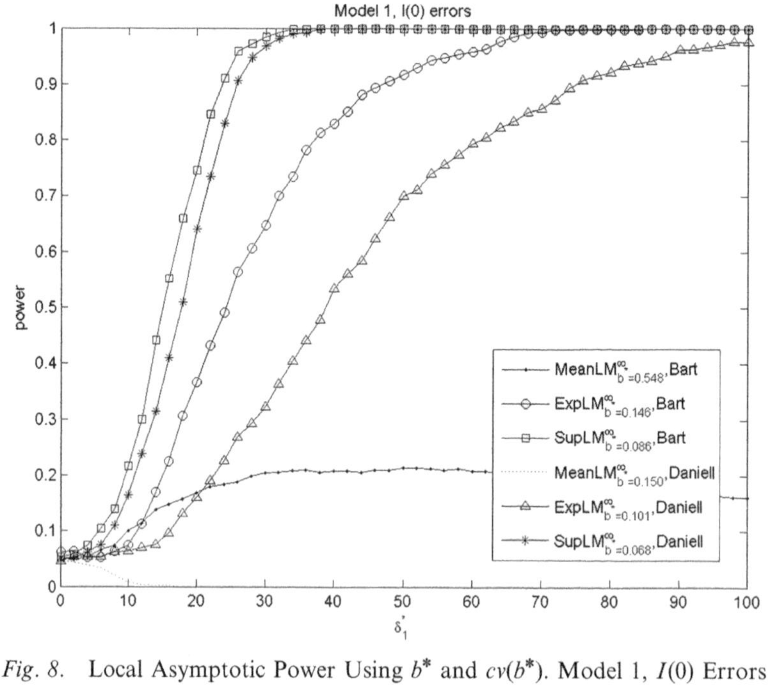

Fig. 8. Local Asymptotic Power Using b^* and $cv(b^*)$. Model 1, $I(0)$ Errors.

trimming of 10%. Power patterns are similar for other significance levels and trimming. Figs. 5–7 give results for $I(1)$ errors with $c = 0, 10, 20$, whereas Fig. 8 gives results for $I(0)$ errors. Figs. 9–12 have the same configurations as Figs. 5–8 and plot asymptotic power for detecting a shift in slope in Model 2.

Focusing on Figs. 5–8 it is apparent that the Bartlett $SupLM_{b^*}$ statistic has the highest power for all four serial correlation configurations. Power of the Daniell $SupLM_{b^*}$ statistic is similar but slightly lower. Power of the $ExpLM_{b^*}$ statistics is substantially lower, whereas power of the $MeanLM_{b^*}$ statistics is either zero (Daniell) or flat (Bartlett). The Bartlett $MeanLM_{b^*}$ statistic has rejections above 5% when $\delta_1^* = 0$ and the errors are $I(1)$ because there is no robust bandwidth for this statistic.

Figs. 9–12 tell a different story for Model 2. Regardless of the error configuration, the Daniell $SupLM_{b^*}$ statistic has, by far, the highest power. Many of the statistics depict nonmonotonic power, that is, power that

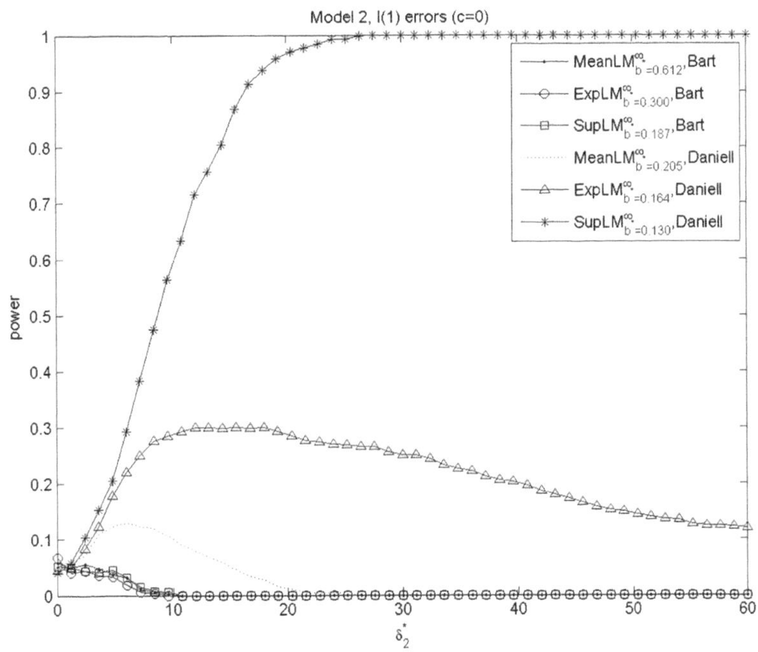

Fig. 9. Local Asymptotic Power Using b^* and $cv(b^*)$. Model 2, $I(1)$ Errors with $c = 0$.

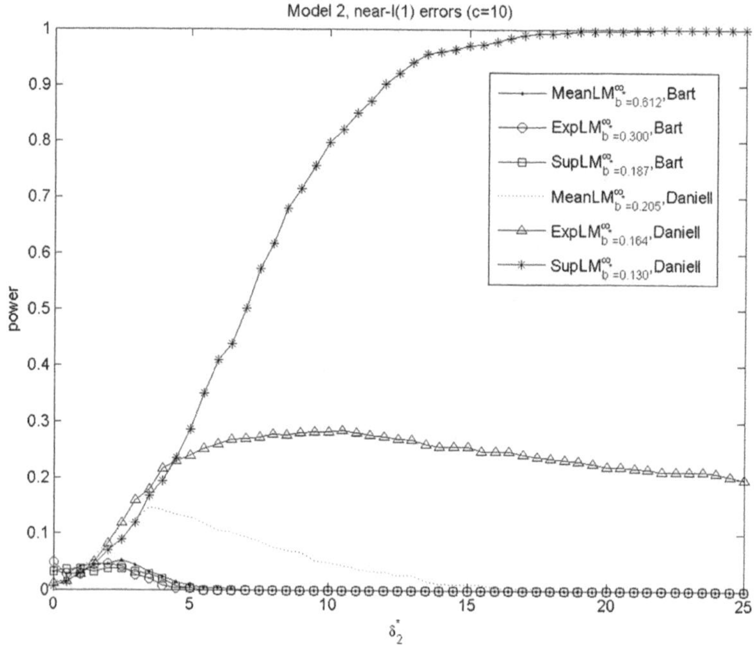

Fig. 10. Local Asymptotic Power Using b^* and $cv(b^*)$. Model 2, $I(1)$ Errors with $c = 10$.

initially increases as the level shift grows but eventually starts to fall as the level shift becomes large. Yang and Vogelsang (2011) documented the relationship between the bandwidth and nonmonotonic power of LM tests for a shift in mean in a nontrending series. They found that smaller bandwidths tend to give tests with monotonic power, whereas larger bandwidths tend to give tests with nonmonotonic power. In unreported simulations, we found similar patterns in the relationship between power and the bandwidth in Models 1–3. Notice in Fig. 12 that the Daniell $SupLM_{b^*}$ statistic has the smallest value of b^* and it is the only statistic that has monotonic power. In Model 1 power of the robust statistics tends to be monotonic because the robust bandwidths tend to be smaller in Model 1 compared to Model 2.

We now focus on the impact of trimming on power. In standard settings power tends to fall as trimming decreases because less trimming increases the possible range of shift dates. In Models 1 and 3 (intercept shift), power

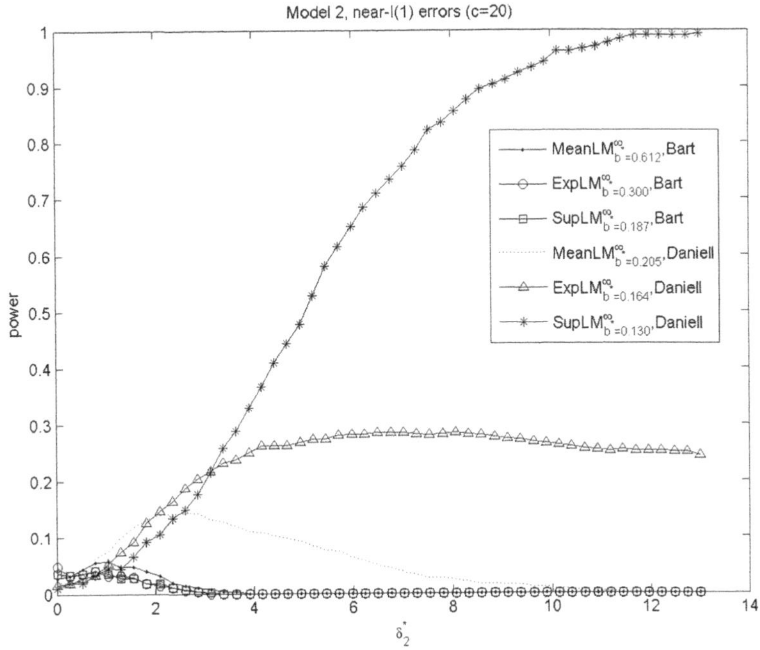

Fig. 11. Local Asymptotic Power Using b^* and $cv(b^*)$. Model 2, $I(1)$ Errors with $c = 20$.

of the Bartlett $SupLM_{b^*}$ statistic is not that sensitive to the amount of trimming but we did find that reducing trimming tends to reduce power slightly when the errors are $I(0)$. See Figs. 13 and 14, which plot power for the Bartlett $SupLM_{b^*}$ statistic in Model 1 for the five trimming values of 1%, 5%, 10%, 15%, and 20%. Similar results hold in Model 3 (slope change) for the Daniell $SupLM_{b^*}$ statistic. The relationship between trimming and power is much more interesting in Model 2. Figs. 15 and 16 plot power of the Daniell $SupLM_{b^*}$ across trimming values. Fig. 15 gives results for $I(1)$ errors with $c = 0$, and Fig. 16 gives results for $I(0)$ errors. Results for $I(1)$ errors with $c = 10, 20$ are similar and are omitted. The results are striking. Power is highest when 1% trimming is used. As trimming increases, power steadily decreases and eventually becomes nonmonotonic for trimming of 20%. These patterns are the opposite of what one would expect and our findings clearly point to a recommendation that 1% trimming be used in Model 2.

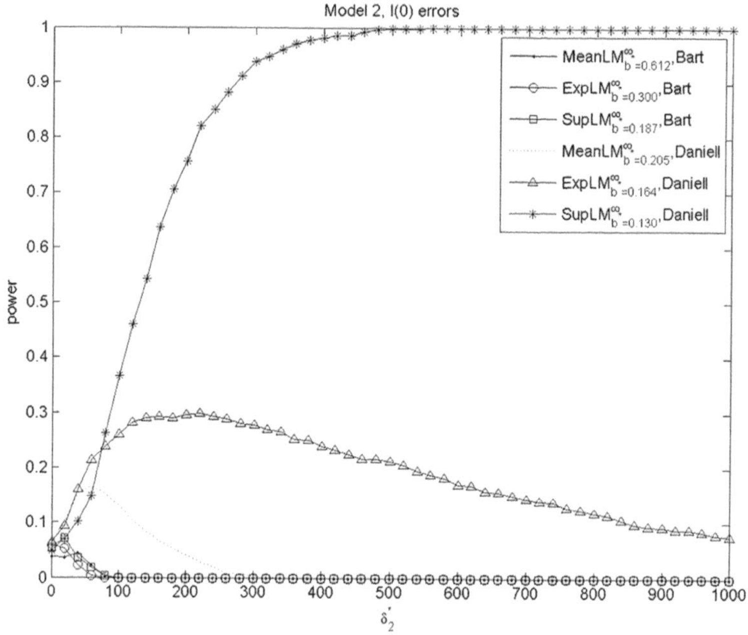

Fig. 12. Local Asymptotic Power Using b^* and $cv(b^*)$. Model 2, $I(0)$ Errors.

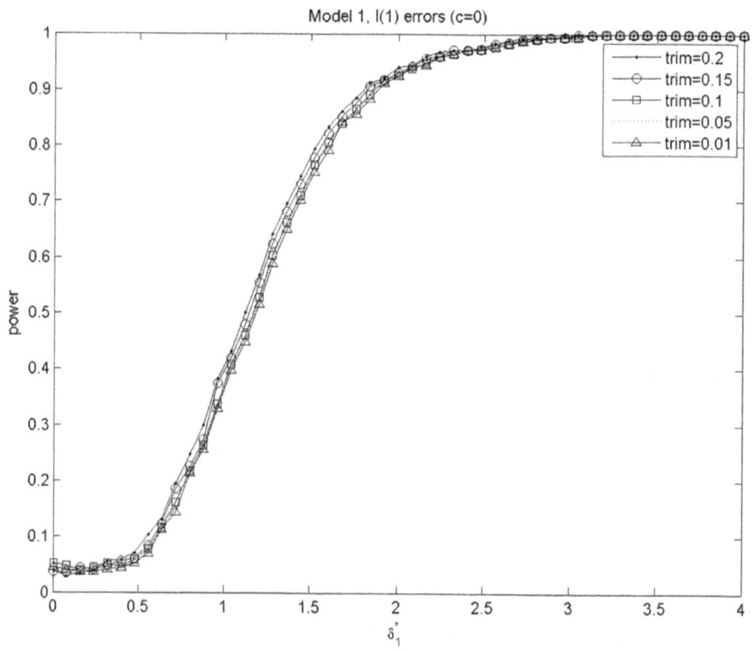

Fig. 13. Local Asymptotic Power Using b^* and $cv(b^*)$. Bartlett $SupLM_{b^*}$ Statistic. Model 1, $I(1)$ Errors with $c = 0$.

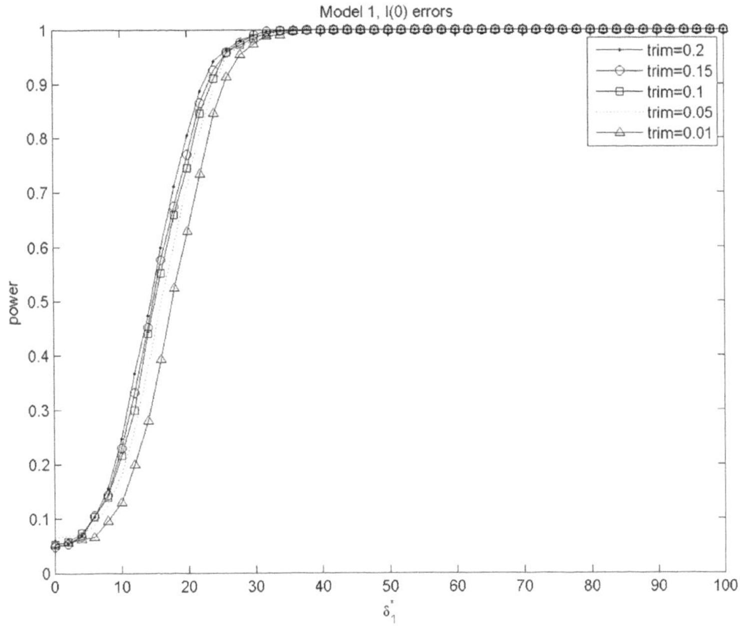

Fig. 14. Local Asymptotic Power Using b^* and $cv(b^*)$. Bartlett $SupLM_{b^*}$ Statistic. Model 1, $I(0)$ Errors.

The relationships between trimming and power can be explained intuitively by close examination of the critical values Tables 1–4. For Models 1 and 3, as trimming increases, the robust bandwidths do not change much and so we would expect power to increase as trimming increases. In contrast, we see for Model 2 that as trimming increases the robust bandwidths increase substantially and it is the increase in the robust bandwidths that leads to significantly lower power. Why the robust bandwidths change so much in Model 2 as trimming increases and change so little in Models 1 and 3 remains an interesting question.

The practical implications of our power analysis are clear. In Model 1 the Bartlett $SupLM_{b^*}$ statistic gives the test with the highest power. Trimming does not matter much except when errors are $I(0)$ in which case power decreases as trimming decreases. In Model 2 the Daniell $SupLM_{b^*}$ statistic gives the highest power. Contrary to standard intuition, we find that less trimming leads to higher power and trimming of 1% is recommended. In Model 3, the recommended statistic depends on whether one is looking to

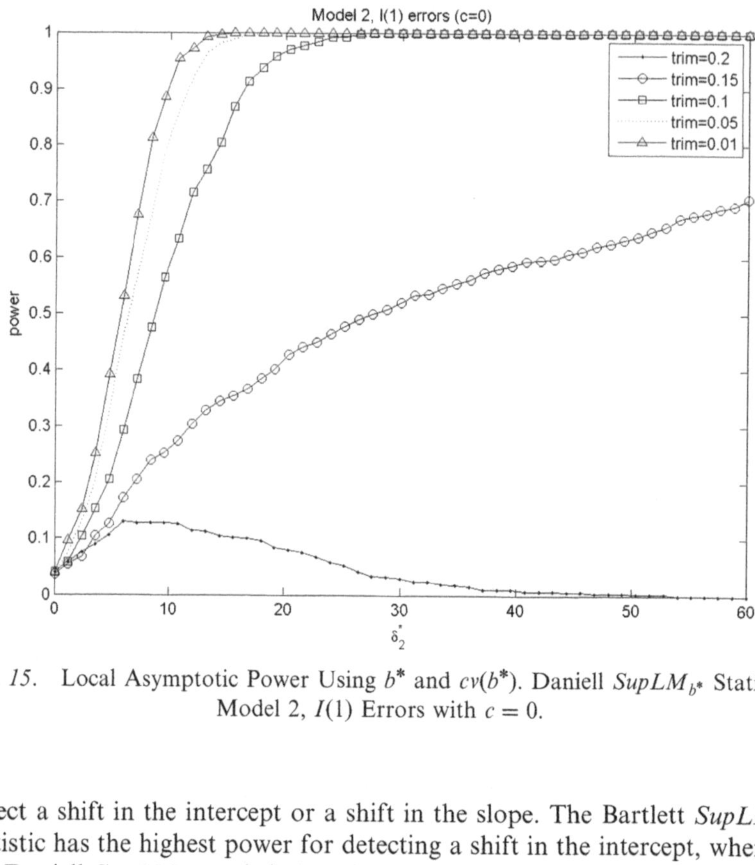

Fig. 15. Local Asymptotic Power Using b^* and $cv(b^*)$. Daniell $SupLM_{b^*}$ Statistic. Model 2, $I(1)$ Errors with $c = 0$.

detect a shift in the intercept or a shift in the slope. The Bartlett $SupLM_{b^*}$ statistic has the highest power for detecting a shift in the intercept, whereas the Daniell $SupLM_{b^*}$ statistic has the highest power for detecting a shift in slope. As in Model 1, power is not that sensitive to trimming, although less trimming tends to lower power when the errors are $I(0)$.

FINITE SAMPLE PERFORMANCE AND COMPARISONS

We use a small simulation study to investigate the finite sample performance of the robust *LM* statistics. For brevity, we only report results for Model 2. We use the same simulation design as Sayginsoy and Vogelsang (2011) to allow direct comparisons to results in their Tables 1 and 2. We focus on the Daniell and Bartlett $SupLM_{b^*}$ statistics given their robustness and desirable

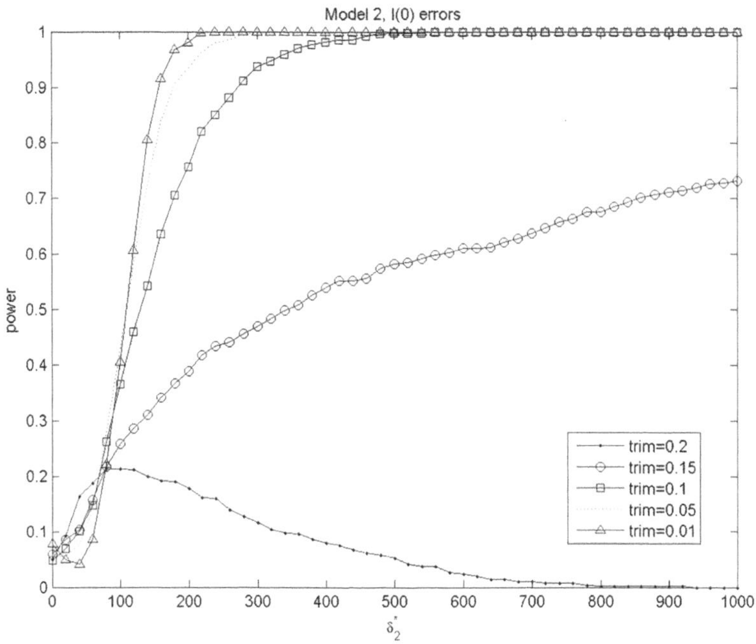

Fig. 16. Local Asymptotic Power Using b^* and $cv(b^*)$. Daniell $SupLM_{b^*}$ Statistic. Model 2, $I(0)$ Errors.

power properties. Results for the $MeanLM_{b^*}$ and $ExpLM_{b^*}$ statistics and other kernels are available upon request. For comparison purposes, we include results for *SupWald* statistic proposed by Sayginsoy and Vogelsang (2011) implemented using the J-statistic scaling factor, the Daniell kernel, the feasible integrated power optimal bandwidth, fixed-b critical values and 10% trimming (see Sayginsoy and Vogelsang (2011) for details). We denote this statistic $SupW_J$. Sayginsoy and Vogelsang (2011) report results for the tests of Harvey et al. (2009) and Perron and Yabu (2009), which are directly comparable to the results we report here.

Following Sayginsoy and Vogelsang (2011) we generated data for the regression error using the ARMA(1,1) model

$$u_t = \rho u_{t-1} + e_t + \theta e_{t-1},$$
$$u_0 = e_0 = 0,$$

Table 5. Finite Sample Null Rejection Probabilities, 5% Nominal Level, $\lambda_0 = 0.5$. Model 2 (Regressors $[1, t, DT_t]$).

		$SupLM_{b^*}$ Bartlett			$SupLM_{b^*}$ Daniell			
		Trimming			Trimming			
θ	ρ	0.05	0.1	0.2	0.05	0.1	0.2	$SupW_J$
−0.4	0	0.062	0.051	0.018	0.062	0.098	0.076	0.009
	0.3	0.048	0.054	0.032	0.048	0.060	0.059	0.016
	0.5	0.031	0.048	0.040	0.031	0.039	0.049	0.020
	0.7	0.018	0.038	0.045	0.018	0.025	0.037	0.032
	0.9	0.023	0.035	0.044	0.023	0.026	0.034	0.061
	0.95	0.034	0.042	0.048	0.034	0.040	0.043	0.081
	1	0.046	0.051	0.052	0.046	0.054	0.057	0.108
0	0	0.040	0.053	0.036	0.040	0.050	0.056	0.018
	0.3	0.027	0.048	0.043	0.027	0.034	0.045	0.021
	0.5	0.017	0.040	0.048	0.017	0.023	0.037	0.024
	0.7	0.015	0.031	0.046	0.015	0.016	0.029	0.024
	0.9	0.023	0.034	0.044	0.023	0.021	0.029	0.032
	0.95	0.034	0.040	0.048	0.034	0.035	0.041	0.038
	1	0.047	0.051	0.053	0.047	0.055	0.053	0.048
0.4	0	0.030	0.050	0.042	0.030	0.037	0.047	0.024
	0.3	0.020	0.044	0.045	0.020	0.028	0.039	0.025
	0.5	0.016	0.036	0.046	0.016	0.018	0.034	0.019
	0.7	0.014	0.028	0.046	0.014	0.014	0.026	0.015
	0.9	0.022	0.030	0.044	0.022	0.021	0.028	0.022
	0.95	0.034	0.040	0.048	0.034	0.035	0.039	0.026
	1	0.049	0.050	0.053	0.049	0.055	0.054	0.032
0.8	0	0.024	0.047	0.041	0.024	0.034	0.045	0.029
	0.3	0.018	0.039	0.045	0.018	0.024	0.039	0.022
	0.5	0.015	0.033	0.045	0.015	0.018	0.032	0.017
	0.7	0.014	0.028	0.046	0.014	0.013	0.025	0.013
	0.9	0.022	0.031	0.043	0.022	0.021	0.028	0.019
	0.95	0.034	0.041	0.048	0.034	0.035	0.039	0.023
	1	0.048	0.050	0.053	0.048	0.054	0.053	0.026
−0.4	0	0.067	0.045	0.021	0.067	0.080	0.068	0.003
	0.3	0.061	0.052	0.030	0.061	0.060	0.057	0.007
	0.5	0.044	0.051	0.037	0.044	0.044	0.048	0.017
	0.7	0.026	0.044	0.044	0.026	0.028	0.037	0.036
	0.9	0.016	0.029	0.045	0.016	0.016	0.026	0.049
	0.95	0.024	0.035	0.047	0.024	0.023	0.030	0.066
	1	0.049	0.047	0.049	0.049	0.052	0.053	0.115
0	0	0.055	0.052	0.035	0.055	0.051	0.053	0.012
	0.3	0.040	0.052	0.038	0.040	0.042	0.046	0.014
	0.5	0.032	0.048	0.041	0.032	0.032	0.040	0.018
	0.7	0.019	0.038	0.044	0.019	0.022	0.030	0.022

Table 5. *(Continued)*

		$SupLM_{b^*}$ Bartlett			$SupLM_{b^*}$ Daniell			
		Trimming			Trimming			
θ	ρ	0.05	0.1	0.2	0.05	0.1	0.2	$SupW_J$
	0.9	0.016	0.027	0.045	0.016	0.013	0.023	0.027
	0.95	0.022	0.034	0.047	0.022	0.021	0.030	0.038
	1	0.050	0.047	0.049	0.050	0.051	0.054	0.060
0.4	0	0.045	0.053	0.037	0.045	0.045	0.048	0.010
	0.3	0.035	0.050	0.039	0.035	0.036	0.044	0.018
	0.5	0.027	0.044	0.041	0.027	0.030	0.038	0.027
	0.7	0.018	0.036	0.043	0.018	0.018	0.028	0.029
	0.9	0.015	0.026	0.044	0.015	0.013	0.022	0.023
	0.95	0.022	0.034	0.048	0.022	0.021	0.029	0.027
	1	0.048	0.047	0.050	0.048	0.051	0.054	0.047
0.8	0	0.041	0.051	0.037	0.041	0.041	0.047	0.016
	0.3	0.032	0.048	0.039	0.032	0.035	0.042	0.024
	0.5	0.025	0.043	0.041	0.025	0.028	0.037	0.033
	0.7	0.016	0.035	0.043	0.016	0.018	0.027	0.027
	0.9	0.015	0.026	0.044	0.015	0.012	0.022	0.020
	0.95	0.022	0.034	0.048	0.022	0.021	0.029	0.025
	1	0.049	0.047	0.050	0.049	0.051	0.055	0.043

where $\{e_t\}$ is an i.i.d. $N(0, 1)$ time series. We generated data for $\{y_t\}$ using Model 2 where we set $\mu = 0$ and $\beta = 0$ without loss of generality. We give results for the sample sizes $T = 50, 100$ for null rejection probabilities and $T = 100$ for power. Ten thousand replications were used in all cases and the nominal level is 5%.

Empirical null ($\delta_2 = 0$) rejection probabilities are given in Table 5 for $\rho = 0, 0.5, 0.7, 0.9, 1.0$, and $\theta = -0.4, 0, 0.4, 0.8$. For both sample sizes it is clear from the table that rejection probabilities are close to 5%. The robust bandwidth does deliver tests that are robust to both weak and very strong serial correlation in the data. Table 6 reports finite sample power. Power is not size-adjusted and is a reflection of actual power that would be obtained in practice when using asymptotic critical values to determine rejections. We report results for three levels of trimming: 0.05, 0.1, and 0.2. We do not report results for trimming of 0.01 because this level of trimming is not feasible with $T = 100$.

First compare power between the two kernels for the $SupLM_{b^*}$ statistic. Regardless of trimming the Daniell kernel has much higher power and this is

Table 6. Finite Sample Power, 5% Nominal Level, $\lambda_0 = 0.5$. Model 2 (Regressors $[1, t, DT_t]$), $T = 100$.

		$T = 50$						
		$SupLM_{b^*}$ Bartlett			$SupLM_{b^*}$ Daniell			
		Trimming			Trimming			
ρ	δ	0.05	0.1	0.2	0.05	0.1	0.2	$SupW_J$
0	0	0.055	0.052	0.035	0.051	0.051	0.053	0.012
	0.012	0.049	0.056	0.036	0.052	0.058	0.070	0.031
	0.024	0.046	0.055	0.036	0.053	0.079	0.113	0.120
	0.036	0.038	0.047	0.023	0.065	0.105	0.145	0.334
	0.048	0.034	0.030	0.010	0.091	0.135	0.167	0.603
	0.06	0.036	0.016	0.003	0.143	0.175	0.185	0.829
	0.072	0.044	0.009	0.001	0.213	0.224	0.194	0.946
	0.084	0.058	0.004	0.000	0.304	0.280	0.203	0.984
	0.096	0.074	0.002	0.000	0.398	0.337	0.207	0.995
0.5	0	0.032	0.048	0.041	0.021	0.032	0.040	0.018
	0.03	0.034	0.052	0.048	0.025	0.042	0.067	0.056
	0.06	0.037	0.052	0.034	0.038	0.070	0.116	0.210
	0.09	0.036	0.032	0.014	0.072	0.110	0.146	0.429
	0.12	0.043	0.016	0.003	0.140	0.162	0.166	0.642
	0.15	0.060	0.006	0.000	0.245	0.231	0.180	0.797
	0.18	0.079	0.002	0.000	0.375	0.309	0.189	0.882
	0.21	0.099	0.001	0.000	0.507	0.387	0.189	0.929
	0.24	0.117	0.000	0.000	0.635	0.468	0.188	0.955
0.8	0	0.015	0.032	0.046	0.005	0.016	0.025	0.025
	0.2	0.043	0.030	0.017	0.086	0.099	0.123	0.281
	0.4	0.093	0.001	0.000	0.492	0.336	0.153	0.539
	0.6	0.141	0.000	0.000	0.874	0.624	0.154	0.691
	0.8	0.166	0.000	0.000	0.988	0.831	0.134	0.783
	1	0.174	0.000	0.000	1.000	0.943	0.111	0.842
	1.2	0.173	0.000	0.000	1.000	0.985	0.086	0.868
	1.4	0.167	0.000	0.000	1.000	0.997	0.062	0.908
	1.6	0.157	0.000	0.000	1.000	1.000	0.043	0.928
1	0	0.050	0.047	0.049	0.051	0.051	0.054	0.060
	0.2	0.066	0.052	0.047	0.103	0.086	0.068	0.087
	0.4	0.091	0.040	0.025	0.252	0.180	0.103	0.151
	0.6	0.115	0.019	0.007	0.465	0.307	0.127	0.224
	0.8	0.139	0.006	0.001	0.688	0.462	0.134	0.292
	1	0.163	0.002	0.000	0.846	0.616	0.133	0.350
	1.2	0.174	0.001	0.000	0.936	0.741	0.125	0.403
	1.4	0.184	0.000	0.000	0.977	0.841	0.115	0.452
	1.6	0.189	0.000	0.000	0.994	0.906	0.104	0.493

consistent with the predictions of the asymptotic analysis for Model 2. Second, compare power across the level of trimming. We see that, opposite to standard settings, increasing the amount of trimming leads to lower power exactly as predicted by the asymptotic power analysis. Less trimming, rather than more, should be used in Model 2. Third, compare power of the 5% trimming Daniell $SupLM_{b^*}$ statistic with the $SupW_J$ statistic. When serial correlation is weak ($\rho = 0, 0.5$), $SupW_J$ dominates in terms of power, whereas when serial correlation is very strong ($\rho = 1.0$), $SupLM_{b^*}$ has higher power. When serial correlation is moderately strong ($\rho = 0.8$), neither statistic dominates in terms of power. The power curves cross for a value of δ between 0.4 and 0.6. $SupW_J$ has higher power for detecting small slope changes, whereas $SupLM_{b^*}$ has higher power for detecting medium-to-large slope changes.

CONCLUSIONS

In this chapter we analyze *LM* tests for a shift in the trend function of a univariate time series where we treat the date of the shift as unknown. We focus on *LM* statistics constructed using nonparametric kernel estimators of the long run variance. We approximate the sampling null distributions of the *LM* tests using fixed-*b* asymptotic theory, which allows us to capture the impact of the kernel and bandwidth choice of the long run variance estimator on the *LM* statistics. As Yang and Vogelsang (2011) found for *LM* tests for a shift in the mean of a nontrending series, we find that, given a significance level, there exist bandwidths such that the null critical values are the same for both stationary and unit root errors in the data. Use of these "robust" bandwidths solves the bandwidth choice problem while simultaneously making the tests robust to the strength of the serial correlation in the data.

Using a local asymptotic power approximation we investigate the impact on power of the choice of kernel (Bartlett, Parzen, Bohman, Daniell, QS) and the method of handling the unknown shift date (*MeanLM*, *ExpLM*, *SupLM*). When testing for an intercept shift, the Bartlett kernel tends to deliver the highest power, whereas when testing for a slope change, the Daniell kernel tends to deliver the highest power. We find that the *SupLM* statistic generally does best in terms of power. The *MeanLM*, *ExpLM*, and *SupLM* statistics require "trimming" of the possible range of shift dates. In standard settings, using more trimming, that is, restricting the range of possible shift dates, tends to increase power in detecting a trend shift assuming the unknown shift date is in the range of possible shift dates. In a model with only an intercept shift dummy or a model with both an intercept shift dummy and a slope change dummy, we find that more trimming does lead to higher power although the sensitivity in power is small. In an *a priori* unexpected finding, we find that in a model with only a slope change dummy, less trimming leads to tests with much higher power than more trimming. This counter-intuitive finding is not surprising *ex post* when one examines the values of the robust bandwidths, which depend on the level of trimming. In most cases, the values of the robust bandwidths change but only slightly as the trimming varies. But, in the model with only a slope change dummy, the robust bandwidths increase substantially as trimming increases. Larger bandwidths tend to reduce power of the *LM* statistics and can even induce nonmonotonic power. In a model with only a slope change dummy, it is best to use very little trimming.

We report finite sample simulations that show that the robust *LM* statistics have empirical rejections probabilities close to the nominal level. The finite sample power properties of the tests closely follow the predictions of the asymptotic analysis. Power rankings of the robust *LM* tests relative to existing *Wald* tests depend on the strength of the serial correlation. When serial correlation is weak, the robust *LM* tests are dominated by the *Wald* tests. When serial correlation is very strong, the opposite is true and the robust *LM* tests dominate. When serial correlation is moderately strong, power curves cross and neither the *LM* nor the *Wald* tests dominate the other in terms of power.

In practice we can recommend the Bartlett $SupLM_{b^*}$ statistic when testing for an intercept shift. We recommend the Daniell $SupLM_{b^*}$ statistic when testing for a slope change and that minimal trimming be used in this case.

ACKNOWLEDGMENT

We thank a referee and participants at the 30th Anniversary Advances in Econometrics Conference for helpful comments and suggestions.

NOTE

1. See Phillips (1987).

REFERENCES

Andrews, D. W. K. (1993). Tests for parameter instability and structural change with unknown change point. *Econometrica, 61*, 821–856.

Andrews, D. W. K., & Ploberger, W. (1994). Optimal tests when a nuisance parameter is present only under the alternative. *Econometrica, 62*, 1383–1414.

Harvey, D., Leybourne, S., & Taylor, A. (2009). Simple, robust, and powerful tests of the breaking trend hypothesis. *Econometric Theory, 25*(04), 995–1029.

Hashimzade, N., & Vogelsang, T. J. (2008). Fixed-*b* asymptotic approximation of the sampling behavior of nonparametric spectral density estimators. *Journal of Time Series Analysis, 29*, 142–162.

Kejriwal, M. (2009). Tests for a mean shift with good size and monotonic power. *Economics Letters, 102*, 78–82.

Kiefer, N. M., & Vogelsang, T. J. (2005). A new asymptotic theory for heteroskedasticity-autocorrelation robust tests. *Econometric Theory, 21*, 1130–1164.

Perron, P., & Yabu, T. (2009). Testing for shifts in trend with an integrated or stationary noise component. *Journal of Business and Economic Statistics, 27*(3), 369–396.

Phillips, P. C. B. (1987). Time series regression with unit roots. *Econometrica, 55*, 277–302.

Sayginsoy, O., & Vogelsang, T. J. (2011). Powerful tests of structural change that are robust to strong serial correlation. *Econometric Theory, 27*, 992–1025.

Vogelsang, T. J. (1997). Wald-type tests for detecting shifts in the trend function of a dynamic time series. *Econometric Theory, 13*, 818–849.

Vogelsang, T. J. (1999). Sources of nonmonotonic power when testing for a shift in mean of a dynamic time series. *Journal of Econometrics, 88*, 283–300.

Yang, J., & Vogelsang, T. (2011). Fixed-*b* analysis of LM-type tests for a shift in mean. *The Econometrics Journal, 14*(3), 438–456.

APPENDIX

Proof of Theorem 1. The result for the numerator of $LM(T_b)$ follows directly from Sayginsoy and Vogelsang (2011) although we use different notation here. To complete the proof we need to derive the fixed-b limit of $\widetilde{\sigma}^2(m)$ in the case of $I(0)$ errors and the fixed-b limit of $T^{-2}\widetilde{\sigma}^2(m)$ in the case of $I(1)$ errors. Because the fixed-b algebra for $\widetilde{\sigma}^2(m)$ is identical to the algebra used by Hashimzade and Vogelsang (2008), we can express $\widetilde{\sigma}^2(m)$ as a function of the partial sums of $\widetilde{\xi}_t$ (the OLS residuals from regression (11)) and appeal to the continuous mapping theorem to obtain the fixed-b limits using arguments in Kiefer and Vogelsang (2005), Hashimzade and Vogelsang (2008), and Sayginsoy and Vogelsang (2011). Therefore, it is sufficient to derive the limits of the scaled partial sums of $\widetilde{\xi}_t$. Define

$$\widetilde{S}_{[rT]} = \sum_{t=1}^{[rT]} \widetilde{\xi}_t$$

where

$$\begin{aligned}
\widetilde{\xi}_t &= y_t - f_t'\left(\sum_{s=1}^{T} f_s f_s'\right)^{-1} \sum_{s=1}^{T} f_s y_s \\
&= u_t - f_t'\left(\sum_{s=1}^{T} f_s f_s'\right)^{-1} \sum_{s=1}^{T} f_s u_s + g_t(T_b^0)'\delta - f_t'\left(\sum_{s=1}^{T} f_s f_s'\right)^{-1} \sum_{s=1}^{T} f_s g_s(T_b^0)'\delta
\end{aligned}$$

giving

$$\begin{aligned}
\widetilde{S}_{[rT]} = &\sum_{t=1}^{[rT]} u_t - \sum_{t=1}^{[rT]} f_t'\left(\sum_{s=1}^{T} f_s f_s'\right)^{-1} \sum_{s=1}^{T} f_s u_s \\
&+ \sum_{t=1}^{[rT]} g_t(T_b^0)'\tau_i\bar{\delta} - \sum_{t=1}^{[rT]} f_t'\left(\sum_{s=1}^{T} f_s f_s'\right)^{-1} \sum_{s=1}^{T} f_s g_s(T_b^0)'\tau_i\bar{\delta}
\end{aligned}$$

The appropriate rate of scaling depends on whether the errors are $I(0)$ or $I(1)$. For the $I(0)$ case we have

$$\begin{aligned}
T^{-1/2}\widetilde{S}_{[rT]} = &T^{-1/2}\sum_{t=1}^{[rT]} u_t - T^{-1}\sum_{t=1}^{[rT]} f_t'\tau_f\left(T^{-1}\sum_{s=1}^{T} \tau_f f_s f_s'\tau_f\right)^{-1} T^{-1/2}\sum_{s=1}^{T} \tau_f f_s u_s \\
&+ T^{-1}\sum_{t=1}^{[rT]} g_t(T_b^0)'\tau_g\bar{\delta} - T^{-1}\sum_{t=1}^{[rT]} f_t'\tau_f\left(T^{-1}\sum_{s=1}^{T} \tau_f f_s f_s'\tau_f\right)^{-1} \\
&\times T^{-1}\sum_{s=1}^{T} \tau_f f_s g_s(T_b^0)'\tau_g\bar{\delta}
\end{aligned}$$

$$\Rightarrow \sigma \int_0^r d\widetilde{W}(s) + \int_0^r \widetilde{G}(s,\lambda_0)' ds \bar{\delta}$$
$$= \sigma\left(\int_0^r d\widetilde{W}(s) + \int_0^r \widetilde{G}(s,\lambda_0)' ds \bar{\delta}\sigma^{-1}\right) \equiv \sigma Q_0(r)$$

For the $I(1)$ case we have

$$\begin{aligned}
T^{-3/2}\widetilde{S}_{[rT]} &= T^{-3/2}\sum_{t=1}^{[rT]} u_t - T^{-1}\sum_{t=1}^{[rT]} f_t'\tau_f \left(T^{-1}\sum_{s=1}^{T} \tau_f f_s f_s' \tau_f\right)^{-1} \\
&\quad \times T^{-3/2}\sum_{s=1}^{T} \tau_f f_s u_s \\
&\quad + T^{-1}\sum_{t=1}^{[rT]} g_t(T_b^0)'\tau_g\bar{\delta} - T^{-1}\sum_{t=1}^{[rT]} f_t'\tau_f \left(T^{-1}\sum_{s=1}^{T} \tau_f f_s f_s' \tau_f\right)^{-1} \\
&\quad \times T^{-1}\sum_{s=1}^{T} \tau_f f_s g_s(T_b^0)'\tau_g\bar{\delta} \\
&\Rightarrow d(1)\int_0^r \widetilde{V}_c(s) ds + \int_0^r \widetilde{G}(s,\lambda_0)' ds\bar{\delta} \\
&= d(1)\left(\int_0^r \widetilde{V}_c(s) ds + \int_0^r \widetilde{G}(s,\lambda_0)' ds\bar{\delta} d(1)^{-1}\right) \equiv d(1)Q_1(r)
\end{aligned}$$

CONSISTENT TESTING FOR STRUCTURAL CHANGE AT THE ENDS OF THE SAMPLE

Michael W. McCracken

ABSTRACT

In this chapter we provide analytical and Monte Carlo evidence that Chow and Predictive tests can be consistent against alternatives that allow structural change to occur at either end of the sample. Attention is restricted to linear regression models that may have a break in the intercept. The results are based on a novel reparameterization of the actual and potential break point locations. Standard methods parameterize both of these locations as fixed fractions of the sample size. We parameterize these locations as more general integer-valued functions. Power at the ends of the sample is evaluated by letting both locations, as a percentage of the sample size, converge to 0 or 1. We find that for a potential break point function, the tests are consistent against alternatives that converge to 0 or 1 at sufficiently slow rates and are inconsistent against alternatives that converge sufficiently quickly. Monte Carlo evidence supports the theory though large samples are sometimes needed for reasonable power.

30th Anniversary Edition
Advances in Econometrics, Volume 30, 133–169

ISSN: 0731-9053/doi:10.1108/S0731-9053(2012)0000030010

Keywords: Structural change; Chow test; Predictive test; intercept correction

JEL Classification: C53; C12; C52

INTRODUCTION

In this chapter we establish the consistency of tests designed to detect structural breaks in the intercept of a linear regression at the beginning and end of a sample of observations. Our results contrast with comments made within the structural break literature wherein it seems to be common knowledge that these tests are inconsistent against alternatives that allow breaks at the ends of the sample (e.g., Dufour, Ghysels, & Hall, 1994). Whether or not structural break tests are consistent is a serious issue since, as documented by Stock and Watson (1996), a large percentage of economic variables exhibit structural breaks across time. Whether or not tests for structural breaks in an intercept are consistent is of particular interest for forecasting agents since as noted by Clements and Hendry (1996), structural breaks in the intercept is one of the most common reasons for real-time predictive failure. It is for this reason they recommend the use of intercept-corrections when constructing forecasts.

In this chapter we apply novel asymptotics to standard Chow (1960) and Predictive (Ghysels & Hall, 1990) tests for structural change in the intercept of linear regression models. To derive the asymptotic behavior of these tests, one has to specify the location of both the actual break and the potential break. The standard approach is to define the location of the actual break as $T_B = [\lambda_B T]$ and the potential break as $R = [\lambda T]$ for fixed fractions $0 < \lambda_B, \lambda < 1$ of the sample size T. Asymptotics are derived by letting T diverge holding these fractions fixed. Power at the ends of the sample is derived by allowing λ_B to approach either 0 or 1.

The approach we take to detecting structural breaks at the ends of the sample is methodologically distinct. We parameterize the actual and potential break points as more general integer-valued functions. As an example, suppose that we parameterize the location of the actual break as $T_B = T - [(1-\lambda)T^b]$, $0 < b < 1$ and $0 < \lambda < 1$. By using this parameterization we are able to refine the notion of the "end" of the sample to the notion of "local to the end" of the sample. Note that using this parameterization, the ratio T_B/T is allowed to converge to 1 in the same fashion as considered in previous work. The difference is that we can control the rates of convergence more delicately by allowing b to vary. Similarly, we can parameterize the location of the potential break as $R = T - [(1-\lambda)T^a]$ $0 < a \leq 1$ and $0 < \lambda < 1$. By taking this approach we also allow the location of the potential break to be "local to the end" of the sample. Note that by letting $a = 1$ we retain the standard method of selecting the potential break point.

Allowing for these more general integer-valued functions we first derive the limiting null distributions of Chow and Predictive tests of structural change. We show that these tests are both asymptotically chi-square. As a corollary we are able to derive the limiting distribution of a max-Chow test designed to detect structural breaks at either the beginning or end of the sample.

We then derive the limiting behavior of these tests under the "local to the end" alternatives discussed above. We are able to show that Chow, max-Chow, and Predictive tests can be consistent against such alternatives if the choice of potential break is chosen appropriately. We obtain the intuitive result that power increases as the distance between the actual and potential break decreases. Our results make clear that whether or not a test for structural change is consistent depends crucially on the particular definition of the "end" of the sample.

We conclude by examining the finite sample size and power of the tests using Monte Carlo experiments. As the theory suggests power increases the closer the potential break point is to the actual break point. In accordance with that result, for a fixed choice of the potential break we see that power of the test decreases as the actual break gets closer to the ends of the sample.

THEORY

In this section we provide analytical results for Chow, max-Chow, and Predictive tests for structural change. Throughout we maintain that there are no breaks on the interior of the sample and hence if a break occurs it does so at a location T_B satisfying $\lim T_B/T \in \{0, 1\}$. In this way we consider our results as complementary to existing results on testing for structural breaks over the interior of the sample by Chow (1960), Kramer, Ploberger, and Alt (1988), Ploberger, Kramer, and Kontrus (1989), Ghysels and Hall (1990), Andrews (1993), Sowell (1996), and Bai (1997).

The following notation will be used. Forecasts and/or predictions of the scalar y_{t+1}, $t = 1, \ldots, T$, are generated using a $(k_1 + 1 = k \times 1)$ vector of covariates $x_{2,t} = (x'_{1,t}, d_{t,R})' = (1, z'_t, d_{t,R})'$. For the potential break location R

the scalar $d_{t,R}$ denotes an indicator function that takes the value 1 if $t>R$ and 0 otherwise. A dummy variable for the actual break location d_{t,T_B} is defined similarly. The $((k_1-1)\times 1)$ vector z_t denotes the subset of predictors (other than a constant) that do not depend upon the sample size T. Since, under the alternative, we treat the actual break location T_B as distinct from the potential break location R, it is useful to define the $(k_1+1=k\times 1)$ vector $x_{3,t}=(x'_{1,t},d_{t,T_B})'=(1,z'_t,d_{t,T_B})'$.

In all results we allow T_B, R, $P\equiv T-R$ and $P_B\equiv T-T_B$ to diverge as the sample size T diverges. By using this asymptotic approximation we distinguish our results from others that treat P and P_B as finite while still allowing R and T_B to diverge. For example, using this approximation Andrews (2003) and Andrews and Kim (2006) establish the asymptotic behavior of several Predictive tests for structural change. Asymptotically valid critical values are constructed using parametric subsampling methods. Since they treat P and P_B as finite their tests are inconsistent. Even so, Monte Carlo experiments show the tests can have reasonable power.

We address the Chow and Predictive tests in separate sub-sections. For each we first show that the test can be asymptotically chi-square under the null hypothesis even when the potential break point location is allowed to be at the end of the sample in the sense that either $\lim R/T=0$ or 1. We then proceed to show that the test can be consistent against alternatives that are at the end of the sample in the sense that $\lim T_B/T=0$ or 1. Before doing so we need first provide a set of assumptions sufficient for the results.

Assumption 1. (a) The DGP satisfies $y_{t+1}=x'_{3,t}\beta^*_3+u_{t+1}$ with $Ex_{3,t}u_{t+1}\equiv Eh_{3,t+1}=0$ for all t with $\beta^*_3=(\beta^{*'}_{3,1},\beta^*_{3,2})'$, (b) under the null hypothesis $\beta^*_{3,1}=\beta^*_1$ and $\beta^*_{3,2}=0$, (c) under the alternative hypothesis $\beta^*_{3,2}\neq 0$, and (d) the parameters are estimated by ordinary least squares (OLS).

Assumption 2. Maintain the null hypothesis. (a) $U_t=[u_{t+1},u_{t+1}z_t,z'_t]'$ is covariance stationary with $Eu^2_{t+1}=\sigma^2$, (b) $E(u_{t+1}|x_{1,t},\ u_{t+1-j}\ \mathrm{j}\geq 1)=0$, (c) for some $r>8$ U_t is uniformly L^r bounded, (d) for some $r>d>2$, U_t is strong mixing with coefficients of size $-rd/(r-d)$, and (e) $\lim_{T\to\infty}T^{-1}E(\sum_{t=1}^T U_t-EU_t)(\sum_{t=1}^T U_t-EU_t)'=V<\infty$ is positive definite.

Assumption 2′. Maintain the alternative hypothesis and let $U_t=[u_{t+1},u_{t+1}z_t,z'_t]'$. (a) $E(u_{t+1}|x_{1,t},\ u_{t+1-j},\ j\geq 1)=0$, (b) for some $r>8$, U_t is uniformly L^r bounded, (c) for some $r>d>2$, U_t is strong mixing with coefficients of size $-rd/(r-d)$, and (d) $\lim_{T\to\infty}T^{-1}\ E(\sum_{t=1}^T U_t-EU_t)(\sum_{t=1}^T U_t-EU_t)'=V<\infty$ is p.d.

Assumption 1 is largely notational but is stated explicitly in order to make the relevant environment clear. We restrict attention to breaks in the intercept of OLS-estimated linear regression models. As such we can map the alternative into testing whether or not the scalar $\beta^*_{3,2}$ takes the value 0. Assumptions 2 and 2′ are more substantive but are standard. The primary difference between these is that under the null we require that the subset of predictor z_t be covariance stationary. Under the alternative we do not. We make this distinction explicit because we want to handle environments where lagged dependent variables are used as predictors and hence, when a break occurs, they will fail to be covariance stationary. The moment and mixing conditions are sufficient for application of weak convergence results in Hansen (1992). The assumption that the population forecast errors are martingale differences insures that the asymptotic null distribution is pivotal but is not needed for consistency of the test. Note that we allow for the forecast errors to be conditionally heteroskedastic.

Chow Test

For the Chow test two linear models, $x'_{i,t}\beta^*_i$, $i = 1, 2, \ldots$, are each estimated using OLS. We denote the residuals associated with models 1 and 2 as $\hat{v}_{1,t+1} = y_{t+1} - x'_{1,t}\hat{\beta}_{1,T}$ and $\hat{v}_{2,t+1} = y_{t+1} - x'_{2,t}\hat{\beta}_{2,T}$, respectively. The actual statistic takes the form

$$W = T \times \frac{(T^{-1}\sum_{s=1}^{T}\hat{v}^2_{1,s+1}) - (T^{-1}\sum_{s=1}^{T}\hat{v}^2_{2,s+1})}{(T^{-1}\sum_{s=1}^{T}\hat{v}^2_{2,s+1})}. \tag{1}$$

The Theorem 1 provides the null limiting distribution of the Chow test allowing the potential break location to satisfy $\lim R/T \in [0, 1]$.

Theorem 1. Maintain Assumptions 1 and 2 and let z_1 and z_2 denote independent standard normal variates. (i) If $\lim R/T = \lambda \in (0,1)$ then $W \rightarrow_d [(1-\lambda)^{1/2} z_1 - \lambda^{1/2} z_2]^2$, (ii) if $\lim R/T = 0$, then $W \rightarrow_d z_1^2$, and (iii) if $\lim R/T = 1$, then $W \rightarrow_d z_2^2$.

Note that in cases (i)–(iii) of Theorem 1, the test statistic is asymptotically chi-square (Eq. (1)). Case (i) is the standard case in which both z_1 and z_2 contribute to the limiting distribution but since the quadratic form is reduced rank we retain only 1 degree of freedom. Cases (ii) and (iii) differ in that only z_1 or z_2 contributes to the limiting distribution. Regardless, asymptotically valid critical values are readily obtained.

Cases (ii) and (iii) lead to another potentially useful result. Note that these two cases imply that two Chow tests, one constructed at the beginning of the sample and a second constructed at the end of the sample, are asymptotically independent of one another. This allows for the simple construction of a max-Chow test for structural breaks at either the beginning or end of the sample. In the following let W_1 and W_2 denote Chow tests as in Eq. (1) with corresponding potential break point location R_1 and R_2, respectively.

Corollary 1. Maintain Assumptions 1 and 2 and let z_1 and z_2 denote independent standard normal variates. If $\lim R_1/T=0$ and $\lim R_2/T=1$, then $W_{\max} = \max[W_1, W_2] \rightarrow_d \max[z_1^2, z_2^2]$.

As we will see in Theorem 2, the ability of these tests to detect alternatives at the ends of the sample depends crucially upon the distance between the potential break point and the actual break point. In the notation of Corollary 1, W_2 has little power to detect alternatives at the beginning of the sample, while W_1 has little power to detect alternatives at the end of the sample. Clearly the max-Chow test overcomes the problem of testing for breaks when one is unwilling to assume that the hypothesized break is known to have occurred at a particular end of the sample.

Choosing asymptotically valid critical values for the max-Chow test is particularly simple using standard methods. Let $F_z(.)$ denote the c.d.f. associated with a chi-square (Eq. (1)) variate and let θ denote the chosen size of the test. Algebra reveals that the $(1-\theta)$-percentile associated with $W_{\max}$ equals $F_z^{-1}((1-\theta)^{1/2})$. These values are easily approximated using chi-square tables for $\theta \in \{0.10, 0.05, 0.01\}$. Constructing asymptotically valid estimates of the p-value associated with the test are also readily constructed using the approximation $1.0-F_z(W_{\max})^2$.

We now turn attention to the power of Chow tests. As previously noted we only consider power against alternatives that are local to the ends of the sample. We derive our results on the power of these tests in two steps. In the first we provide general propositions that show that the relationship between the location of the potential break and the actual break determines whether or not these tests will detect an alternative that is local to the end of the sample. In the second we specialize these results to a particular parameterization of these locations.

Theorem 2. Maintain Assumptions 1 and 2′. (i) If $R \leq T_B$, then $W = O_p(RP_B^2/PT)$, (ii) If $R \geq T_B$, then $W = O_p(PT_B^2/RT)$.

Corollary 2. Maintain Assumptions 1 and 2′ and let $\lim R_1/T=0$ and $\lim R_2/T=1$. (i) If $R_1 \geq T_B$, then $W_{\max} = O_p(P_1 T_B^2/R_1 T)$, (ii) if $R_2 \leq T_B$, then $W_{\max} = O_p(R_2 P_B^2/P_2 T)$, (iii) if $R_2 \geq T_B$ and $R_1 \leq T_B$, then $W_{\max} = O_p(\max[R_1 P_B^2/P_1 T, P_2 T_B^2/R_2 T])$.

Theorem 2 and its corollary show that consistency of the tests depends crucially upon the relationships among R, P, T_B, and P_B. The details of these relationships depend, however, on the parameterization of the location of potential and actual breaks R and T_B. Consider the parameterization mentioned in the Introduction and hence let's focus attention on detecting breaks at the end of the sample. There we considered using $T_B = T - [(1-\lambda)T^b]$ and $R = T - [(1-\lambda)T^a]$ for scalars a and b satisfying $0 < a \leqslant 1$, $0 < b < 1$ and $0 < \lambda < 1$. Note that by choosing this parameterization we are imposing the constraint that $\lim R/T$ is nonzero. We do so since the power associated with such a potential break point location is uniformly lower than that for all other choices of R satisfying $\lim R/T \in (0,1]$ and hence is uninteresting.

Using this parameterization suppose that $R \leqslant T_B$ with a equal to 1. This corresponds to case (i) in Theorem 2 with $\lim \mathrm{R/T} \in (0,1)$. To determine consistency note that

$$\frac{RP_B^2}{PT} = \frac{T - [(1-\lambda)T]}{[(1-\lambda)T]} \frac{[(1-\lambda)T^b]^2}{T} \approx \lambda(1-\lambda)T^{2b-1}$$

and hence the test is consistent against alternatives for which $1/2 < b < 1$ but is inconsistent against alternatives for which $0 < b \leqslant 1/2$. This is a somewhat surprising result since the case where $\lim R/T \in (0,1)$ is the standard one within the literature but within that literature it is "known" (see Dufour et al., 1994 for a discussion) that the test is inconsistent against alternatives at the ends of the sample. Our results show that result depends crucially upon the parameterization of the actual break point.

Now suppose that we again use the same parameterization but with $a \in (0,1)$. Moreover, suppose we use a conservative potential break point location so that $R \leqslant T_B$ and hence $0 < b \leqslant a < 1$. This also corresponds to case (i) in Theorem 2. Similar arguments to that above reveal that $RP_B^2/PT \approx (1-\lambda)T^{2b-a}$ and hence the test is consistent against alternatives for which $a/2 < b \leqslant a$ but is inconsistent against alternatives for which $0 < b \leqslant a/2$. This result makes clear that for any actual break point characterized by b there exists choices of potential break point parameter a such that the test is consistent. Note, however, that we also obtain the result that for any potential break point location parameter a there exist alternatives that the test will not detect. Since we have assumed here that $b \leqslant a$ we obtain the intuitive result that the optimal choice of potential break point location is to equate it to the actual break point location and hence $a = b$.

Suppose again that we are using this parameterization and $a \in (0,1)$. Moreover suppose we use an aggressive potential break point location so that $R \geqslant T_B$ and hence $0 < a \leqslant b < 1$. This corresponds to case (ii) in Theorem 2. Similar arguments to that above reveal that $PT_B^2/RT \approx (1 - \lambda)T^a$ and hence the test is consistent against all alternatives for which $a \leqslant b < 1$. It is important to note here that this result essentially requires knowledge of the location of the break and hence consistency is not so surprising. The downside of this result is that the power of the test is increasing in a. That is, as we choose smaller values of a in order to ensure consistency we are simultaneously lowering the rate at which the test diverges. Since we have assumed here that $a \leqslant b$ we once again obtain the intuitive result that the optimal choice of potential break point location is to equate it to the actual break point location and hence $a = b$.

Predictive Test

Consider the Predictive (Ghysels & Hall, 1990) test for structural change in the intercept of a linear regression. This test is equivalent to a test of zero mean prediction error as discussed in the literature on out-of-sample testing.[1] For that reason we are able to apply many existing results from West (1996) and West and McCracken (1998) when discussing the null asymptotics of these tests. Not all results follow from this previous work, however. In particular, new null asymptotics are derived here for cases in which $\lim R/T = 0$.

For the Predictive tests only model 1 is estimated. We denote the one-step ahead forecast error from this model as $\hat{u}_{1,t+1} = y_{t+1} - x'_{1,t}\hat{\beta}_{1,t}, \quad t = R, \ldots, T$. We consider three distinct means of constructing the forecasts: the recursive, rolling, and fixed schemes. Under the recursive scheme, the regression parameters are reestimated with added data as forecasting moves forward through time and hence for $t = R, \ldots, T$, $\hat{\beta}_{1,t}$ depends upon observations $s = 1, \ldots, t$. Under the rolling scheme, only a fixed window of the past R observations is used and hence for $t = R, \ldots, T$, $\hat{\beta}_{1,t}$ depends upon observations $s = t - R + 1, \ldots, t$. Under the fixed scheme, forecasts are constructed in the same fashion as considered by Ghysels and Hall (1990).

That is, $\hat{\beta}_{1,t}$ is estimated only once using observations $s=1, \ldots, R$ so that $\hat{\beta}_{1,t} = \hat{\beta}_{1,R}$ for $t=R, \ldots, T$.

Below we provide the formulas for the Predictive tests. Following suggestions made in West (1996) and West and McCracken (1998) we consider Predictive tests of the form

$$V_R = \frac{(P^{-1/2}\sum_{t=R}^{T}\hat{u}_{1,t+1})^2}{(P^{-1}\sum_{t=R}^{T}\hat{u}_{1,t+1}^2)} \quad \text{Recursive, } 0 \le \lim R/T \le 1, \tag{2a}$$

$$V_L = \frac{(P^{-1/2}\sum_{t=R}^{T}\hat{u}_{1,t+1})^2}{(2R/3P)(P^{-1}\sum_{t=R}^{T}\hat{u}_{1,t+1}^2)} \quad \text{Rolling with } 0 \le \lim R/T \le 1/2 \tag{2b}$$

$$V_L = \frac{(P^{-1/2}\sum_{t=R}^{T}\hat{u}_{1,t+1})^2}{(1-(P/R)^2/3)(P^{-1}\sum_{t=R}^{T}\hat{u}_{1,t+1}^2)} \quad \text{Rolling with } 1/2 \le \lim R/T \le 1 \tag{2c}$$

$$V_F = \frac{(P^{-1/2}\sum_{t=R}^{T}\hat{u}_{1,t+1})^2}{(1+P/R)(P^{-1}\sum_{t=R}^{T}\hat{u}_{1,t+1}^2)} \quad \text{Fixed (Split Sample)} \tag{2d}$$

West (1996) considers the null asymptotics for the recursive scheme when $0 \leqslant \lim R/T \leqslant 1$, while West and McCracken (1998) consider null asymptotics for the fixed and rolling schemes when $0 < \lim R/T \leqslant 1$. In Theorem 3 we provide new results for the rolling and fixed schemes when $\lim R/T=0$. This is important since the previous results indicate that the power of Chow tests to detect breaks at the beginning of the sample is improved if we consider potential break points at the beginning of the sample. The same may hold for Predictive tests.

Theorem 3. Maintain Assumptions 1 and 2 and let $\lim R/T=0$. When the rolling scheme is used make the additional assumption that $\lim P/R^{3/2}=0$. For the rolling and fixed schemes, $V \to_d$ chi-square (1).

Theorem 3 fills a gap in the literature on Predictive tests by showing that even if we let the potential break point be chosen close to the beginning of the sample we can still obtain an asymptotically chi-square test of the null. Since this is not a surprising result we now turn attention to the power of Predictive tests. As before we only consider power against single-break alternatives that are local to the ends of the sample. We first derive general results on the power of these tests for arbitrary actual break locations T_B

and then we specialize these results to a particular parameterization of these locations.

Theorem 4. Maintain Assumptions 1 and 2′ and consider the recursive scheme. (i) If $R \leqslant T_B$, then $V_R = O_p(T_B^2(\sum_{t=T_B}^{T} t^{-1})^2/P)$, (ii) if $R \geqslant T_B$, then $V_R = O_p(T_B^2(\sum_{t=R}^{T} t^{-1})^2/P)$.

Theorem 5. Maintain Assumptions 1 and 2′ and consider the rolling scheme.

(a) Let $0 \leqslant \lim R/T \leqslant 1/2$: (i) if $T_B \leqslant R$, then $V_L = O_p(\max[P/R, T_B^4/R^3])$, (ii) If $T_B > \max[R, P]$, then $V_L = O_p(\max[P/R, P_B^2(2R - P_B)^2/R^3])$, (iii) If $R < T_B < P$ then $V_L = O_p(\max[P/R, R])$.
(b) Let $1/2 < \lim R/T \leqslant 1$: (i) If $T_B \geqslant R$, then $V_L = O_p(P_B^2(2R - P_B)^2/PR^2)$, (ii) if $T_B < \min[R, P]$, then $V_L = O_p(T_B^4/PR^2)$, and (iii) if $P < T_B < R$, then $V_L = O_p(\mathrm{P}(2T_B - P)^2/R^2)$.

Theorem 6. Maintain Assumptions 1 and 2′ and consider the fixed scheme. (i) If $R \geqslant T_B$, then $V_F = O_p(PT_B^2/RT)$, and (ii) if $R \leqslant T_B$, then $V_F = O_p(RP_B^2/PT)$.

The results for each forecasting scheme again clearly indicate that the relationships among R, P, T_B, and P_B are crucial for determining whether the test will be consistent and at what rate it will diverge if it is. For the fixed scheme the orders of magnitude match those of the Chow test and as such are easily interpretable. Those for the rolling scheme are distinct but straightforward to calculate. This is not the case for the recursive scheme since closed form (asymptotic) approximations for terms like $T_B^2(\sum_{t=T_B}^{T} t^{-1})^2/P$ and $T_B^2(\sum_{t=R}^{T} t^{-1})^2/P$ are not easily obtained. However, one can show that these are at least of the same order as those associated with the fixed schemes. For example, consider case (i) and let $\lim T_B/T = 1$. Algebra reveals that $T_B^2(\sum_{t=T_B}^{T} t^{-1})^2/P \geq T_B^2(\sum_{t=T_B}^{T} T^{-1})^2/P = T_B^2 P_B^2/\ PT^2 \approx (1 - \lambda)T^{2b-a}$ and hence the test is consistent against alternatives for which $a/2 < b \leqslant a$ but inconsistent against those for which $0 < b \leqslant a/2$.

MONTE CARLO EVIDENCE

In this section we provide Monte Carlo evidence on the usefulness of the Chow and Predictive tests when breaks occur at the ends of the sample. For a range of sample sizes and break point locations we evaluate the finite sample size and power of the tests. For brevity we restrict attention to results for the Chow, max-Chow, and recursive Predictive test. Unreported simulations indicate that the fixed Predictive test behaves similar to the Chow test under the null,[2] while its power is usually lower than that using the recursive scheme. Results for the rolling Predictive test are subject to large size distortions (e.g., rejection frequencies near 60% for a nominally 5% test!). That the rolling test for zero mean prediction error is subject to large size distortions is also found in West and McCracken (1998).

Monte Carlo Design

The data-generating process has the representation

$$\begin{pmatrix} y_t \\ x_t \end{pmatrix} = \begin{pmatrix} \beta_{0t}^* \\ 0 \end{pmatrix} + \begin{pmatrix} 0.3 & 0.1 \\ 0 & 0.5 \end{pmatrix} \begin{pmatrix} y_{t-1} \\ x_{t-1} \end{pmatrix} + \begin{pmatrix} u_{y,t} \\ u_{x,t} \end{pmatrix}$$

for i.i.d. mutually independent standard normal disturbance $u_{y,t}$ and $u_{x,t}$. The intercept β_{0t}^* takes the value 0 prior to the time of the break and takes the value 1 afterward – an increase of a little less than one standard deviation. In the size experiments the parameter β_{0t}^* takes the value 0 throughout the entire sample of $T \in \{100;\ 200;\ 400;\ 800;\ 1{,}600;\ 3{,}200\}$ observations. Starting values for y_t and x_t were drawn from the unconditional distribution. The rejection frequency is estimated based upon 5,000 replications.

Throughout the experiments we consider actual and potential break point locations using the parameterization suggested in the text. That is, when a break occurs at the end of the sample, it does so at time $T_B = T - [(1-\lambda)T^b]$ for $b \in \{0.1, 0.2, \ldots, 0.8, 0.9\}$ and $\lambda = 0.85$. The choice of parameter $\lambda = 0.85$ is motivated by suggestions in Andrews (1993), while the choice of parameter b was made simply to span the unit interval. For brevity we only provide Monte Carlo evidence on breaks that occur at the end of the sample.

When the potential break location is at the end of the sample it does so at time $R = T - [(1-\lambda)T^a]$ for $a \in \{0.1, 0.2, \ldots, 0.8, 0.9, 1.0\}$ and $\lambda = 0.85$. Note that although we allow the potential break location to lie on the interior of the sample (i.e., $a = 1.0$), we do not allow the same for the actual break location. When the potential break location occurs at the beginning of the sample it does so at time $R = [(1-\lambda)T^a]$ for similar values of λ and a. When the max-Chow test is constructed we assume that the choice of potential break locations is symmetric in the sense that $R_1 = T - R_2$ and hence the same value of b can be associated with each of the two potential break points. Note that as a practical matter, we model the potential breaks as indicating that a break occurs between time R and $R+1$ and hence any dummy variables used in the construction of the test take the value 0 at time R and take the value 1 at time $R+1$.

Table 1 reports the numerical values for the actual and potential break locations associated with the parameterization discussed above. These locations vary with the sample size T but do so more slowly than they would if we were using the standard parameterization $[\lambda T]$ for the actual and potential break points. Note that for a fixed value of T, there is no variation in the location of the actual and potential break point locations for small values of a and b. This, along with the finite sample sizes, implies that there are instances in which the tests cannot be numerically

Table 1. Locations of Actual and Potential Breaks.

$\lambda=0.85$						
End of sample $T-[(1-\lambda)T^c]$						
$c\backslash T$	100	200	400	800	1,600	3,200
1	85	170	340	680	1,360	2,720
0.9	91	183	367	739	1,486	2,986
0.8	95	190	382	769	1,546	3,105
0.7	97	194	391	784	1,574	3,158
0.6	98	197	395	792	1,588	3,181
0.5	99	198	397	796	1,594	3,192
0.4	100	199	399	798	1,598	3,197
0.3	100	200	400	799	1,599	3,199
0.2	100	200	400	800	1,600	3,200
0.1	100	200	400	800	1,600	3,200
Beginning of sample $[(1-\lambda)T^c]$						
$c\backslash T$	100	200	400	800	1,600	3,200
1	15	30	60	120	240	480
0.9	9	17	33	61	114	214
0.8	5	10	18	31	54	95
0.7	3	6	9	16	26	42
0.6	2	3	5	8	12	19
0.5	1	2	3	4	6	8
0.4	0	1	1	2	2	3
0.3	0	0	0	1	1	1
0.2	0	0	0	0	0	0
0.1	0	0	0	0	0	0

constructed. For instance to construct the recursive Predictive test we require that P is at least equal to 2 and R is at least equal to 1. Similarly, for the Chow test we need both R and P to be at least equal to 1. As we will see in the tables there are times where the tests cannot be constructed and these are denoted "N.A."

When the tests can be constructed we do so using an OLS-estimated linear regression model for the variable y_{t+1}. The linear model always includes an intercept and first lags of both y_{t+1} and x_t as predictors. At times the linear model includes the dummy variable $d_{t,R}$ indicating the location of the potential break as necessary for construction of the relevant test statistic.

Size Results

Table 2 reports the actual size of 5% nominal tests for each of the three test statistics when the potential break location is at the end of the sample. In the first panel we see that for a fixed value of a, the recursive Predictive tests tend to have higher size distortions at the smaller sample sizes but as the sample size increases the size distortions decrease. For example when $a = 0.6$ the size of the test varies across 0.306, 0.187, 0.118, 0.088, 0.074, and 0.065 as the sample size T increases from 100 to 3,200.

In the second and third panels of Table 2 we find that the Chow and max-Chow tests are more reasonably sized than are the recursive Predictive tests. These too are subject to some size distortions but are much less pronounced than those above. For example when $a = 0.9$ the size of the max-Chow test varies across 0.076, 0.060, 0.056, 0.047, 0.050, and 0.055 as the sample size T increases from 100 to 3,200.

It is important to emphasize that the Chow and max-Chow tests tend to be reasonably well sized across all potential break locations for any fixed sample size. This is not the case for the recursive Predictive test. The Predictive test becomes increasingly oversized as the potential break location gets closer to the end of the sample. For example when $T = 1{,}600$ the size of the test varies across 0.050, 0.053, 0.064, 0.065, 0.074, 0.111, and 0.313 as the potential break location parameter a decreases from 1.0 to 0.4. In the same experiment the Chow test remains near 5% regardless of the parameter a.

Table 3 reports the actual size of 5% nominal tests for each of the three test statistics when the potential break location is at the beginning of the sample. In each of the three panels the actual size of the test remains near

Table 2. Size of Tests with Potential Break Location at the End.

Nominal 5% critical values										
Recursive	1.000	0.900	0.800	0.700	0.600	0.500	0.400	0.300	0.200	0.100
$T=100$	0.073	0.095	0.131	0.192	0.306	NA	NA	NA	NA	NA
$T=200$	0.066	0.071	0.092	0.113	0.187	0.304	NA	NA	NA	NA
$T=400$	0.060	0.062	0.075	0.094	0.118	0.188	NA	NA	NA	NA
$T=800$	0.050	0.049	0.056	0.064	0.088	0.146	0.295	NA	NA	NA
$T=1{,}600$	0.050	0.053	0.064	0.065	0.074	0.111	0.313	NA	NA	NA
$T=3{,}200$	0.047	0.056	0.052	0.052	0.065	0.090	0.201	NA	NA	NA
Chow										
$T=100$	0.069	0.076	0.064	0.067	0.058	0.053	NA	NA	NA	NA
$T=200$	0.062	0.060	0.062	0.055	0.056	0.052	0.052	NA	NA	NA
$T=400$	0.058	0.056	0.062	0.059	0.055	0.058	0.049	N.A.	NA	NA
$T=800$	0.049	0.047	0.047	0.047	0.052	0.050	0.050	0.047	NA	NA
$T=1{,}600$	0.050	0.050	0.060	0.055	0.047	0.051	0.053	0.045	NA	NA
$T=3{,}200$	0.048	0.055	0.048	0.048	0.047	0.049	0.056	0.050	NA	NA
max-Chow										
$T=100$	0.074	0.076	0.070	0.066	0.062	NA	NA	NA	NA	NA
$T=200$	0.060	0.061	0.061	0.055	0.054	0.054	NA	NA	NA	NA
$T=400$	0.056	0.055	0.054	0.060	0.053	0.054	NA	NA	NA	NA
$T=800$	0.054	0.049	0.048	0.049	0.056	0.051	0.049	NA	NA	NA
$T=1{,}600$	0.051	0.049	0.055	0.052	0.046	0.049	0.050	NA	NA	NA
$T=3{,}200$	0.045	0.053	0.049	0.051	0.048	0.053	0.053	NA	NA	NA

5% for all sample sizes and potential break locations.[3] This is to be expected for the Chow and max-Chow tests since the test is constructed in a symmetric fashion to those from Table 2. Since the recursive test is constructed differently, and we now have more out-of-sample observations to construct the test (i.e., P is large), it is not surprising that the test is more reasonably sized.

Power Results

Tables 4–12 provide evidence on the power of the three tests when we allow the locations of the potential and actual break points to vary while holding the sample sizes fixed at $T=200$, 800, 3,200. Tables 4, 7, and 10 correspond to the recursive Predictive test, Tables 5, 8, and 11 correspond to the Chow test, and Tables 6, 9, and 12 correspond to the max-Chow test. For each of the Predictive and Chow tests we consider power of the test when the

Table 3. Size of Tests with Potential Break Location at the Beginning.

Nominal 5% critical values										
Recursive	1.000	0.900	0.800	0.700	0.600	0.500	0.400	0.300	0.200	0.100
$T=100$	0.054	0.059	0.056	0.047	0.038	0.032	NA	NA	NA	NA
$T=200$	0.053	0.050	0.054	0.052	0.041	0.036	0.033	NA	NA	NA
$T=400$	0.052	0.051	0.051	0.054	0.055	0.044	0.041	NA	NA	NA
$T=800$	0.052	0.050	0.051	0.050	0.050	0.049	0.045	0.043	NA	NA
$T=1,600$	0.054	0.051	0.053	0.051	0.052	0.053	0.050	0.049	NA	NA
$T=3,200$	0.050	0.052	0.051	0.052	0.052	0.050	0.049	0.049	NA	NA
Chow										
$T=100$	0.065	0.069	0.062	0.061	0.056	NA	NA	NA	NA	NA
$T=200$	0.056	0.058	0.058	0.057	0.053	0.052	NA	NA	NA	NA
$T=400$	0.053	0.047	0.054	0.057	0.049	0.052	NA	NA	NA	NA
$T=800$	0.053	0.050	0.051	0.051	0.049	0.046	0.047	NA	NA	NA
$T=1,600$	0.052	0.055	0.053	0.050	0.049	0.049	0.048	NA	NA	NA
$T=3,200$	0.045	0.047	0.051	0.051	0.047	0.049	0.049	NA	NA	NA
max-Chow										
$T=100$	0.074	0.076	0.070	0.066	0.062	NA	NA	NA	NA	NA
$T=200$	0.060	0.061	0.061	0.055	0.054	0.054	NA	NA	NA	NA
$T=400$	0.056	0.055	0.054	0.060	0.053	0.054	NA	NA	NA	NA
$T=800$	0.054	0.049	0.048	0.049	0.056	0.051	0.049	NA	NA	NA
$T=1,600$	0.051	0.049	0.055	0.052	0.046	0.049	0.050	NA	NA	NA
$T=3,200$	0.045	0.053	0.049	0.051	0.048	0.053	0.053	NA	NA	NA

potential break point location is correctly chosen at the end of the sample and incorrectly chosen at the beginning of the sample. Below we discuss several of the most important observations.

Location of Actual Break Relative to Potential Break

For each test and for each choice of potential and actual break location, power of the test increases with the sample size. When the potential break location is correctly chosen so that $a=b=0.5$ the power of the recursive Predictive test varies across 0.511, 0.557, and 0.800 as T increases from 200 to 3,200. Similarly, the power of the Chow test varies across 0.295, 0.505, and 0.805 as T increases from 200 to 3,200. To conclude that the power of the recursive Predictive test can dominate that of the Chow test is misleading due to the previously discussed size distortions.

Recall that the theory suggests that when the actual and potential break points are the same, the power of the test should increase as the actual break

Table 4. Power of Recursive Test for Break at End.

$T = 200$, Nominal Size = 5%

Potential break correctly specified at end										
Actual break	1.000	0.900	0.800	0.700	0.600	0.500	0.400	0.300	0.200	0.100
0.900	0.797	0.960	0.766	0.548	0.392	0.436	NA	NA	NA	NA
0.800	0.407	0.649	0.835	0.616	0.439	0.469	NA	NA	NA	NA
0.700	0.208	0.324	0.510	0.667	0.475	0.489	NA	NA	NA	NA
0.600	0.108	0.145	0.217	0.358	0.499	0.505	NA	NA	NA	NA
0.500	0.087	0.112	0.156	0.237	0.504	0.511	NA	NA	NA	NA
0.400	0.073	0.085	0.114	0.160	0.286	0.518	NA	NA	NA	NA
0.300	0.066	0.070	0.088	0.116	0.191	0.301	NA	NA	NA	NA
0.200	0.066	0.070	0.088	0.116	0.191	0.301	NA	NA	NA	NA
Potential break incorrectly specified at beginning										
Actual break	1.000	0.900	0.800	0.700	0.600	0.500	0.400	0.300	0.200	0.100
0.900	0.207	0.194	0.191	0.181	0.157	0.150	0.146	NA	NA	NA
0.800	0.112	0.109	0.106	0.105	0.088	0.082	0.077	NA	NA	NA
0.700	0.079	0.082	0.077	0.075	0.061	0.058	0.054	NA	NA	NA
0.600	0.062	0.062	0.064	0.062	0.050	0.045	0.042	NA	NA	NA
0.500	0.059	0.057	0.058	0.058	0.047	0.043	0.040	NA	NA	NA
0.400	0.057	0.053	0.055	0.054	0.046	0.042	0.036	NA	NA	NA
0.300	0.056	0.053	0.053	0.052	0.043	0.039	0.035	NA	NA	NA
0.200	0.056	0.053	0.053	0.052	0.043	0.039	0.035	NA	NA	NA

point approaches the interior of the sample. For example, in the first panel of Table 7 the power of the recursive Predictive test varies across 0.498, 0.557, 0.785, 0.972, 0.999, and 1.000 as the parameter a ($= b$) varies from 0.4 to 1.0. In the first panel of Table 8 the power of the Chow test varies across 0.167, 0.276, 0.505, 0.803, 0.978, 0.999, and 1.000 as the parameter a varies from 0.3 to 1.0.

Similarly, recall that the theory suggests that the power of the test should increase as the potential break point approaches the location of the actual break point. This holds for all sample sizes and all tests. This implies that for any given row we often see power increase and then decrease in a hump-shaped fashion with the hump occurring when $a = b$. For example in the first panel of Table 4 when $b = 0.8$ the power of the recursive Predictive test varies across 0.407, 0.649, 0.835, 0.616, 0.439, and 0.469 as the parameter a varies from 1.0 to 0.5. In the first panel of Table 8 when $b = 0.5$ the power of the Chow test varies across 0.061, 0.091, 0.133, 0.233, 0.405, 0.505, 0.276, and 0.166 as the parameter a varies from 1.0 to 0.3.

Since the actual breaks are at the end of the sample the theory predicts that when the potential break is at the beginning of the sample we should

Table 5. Power of Chow Test for Break at End.

$T = 200$, Nominal Size = 5%

Potential break correctly specified at end										
Actual break	1.000	0.900	0.800	0.700	0.600	0.500	0.400	0.300	0.200	0.100
0.900	0.827	0.973	0.790	0.526	0.265	0.189	0.110	NA	NA	NA
0.800	0.420	0.706	0.857	0.618	0.329	0.231	0.129	NA	NA	NA
0.700	0.209	0.360	0.573	0.676	0.370	0.262	0.141	NA	NA	NA
0.600	0.111	0.159	0.248	0.373	0.404	0.287	0.152	NA	NA	NA
0.500	0.088	0.118	0.167	0.235	0.412	0.295	0.157	NA	NA	NA
0.400	0.072	0.086	0.108	0.136	0.211	0.295	0.163	NA	NA	NA
0.300	0.063	0.066	0.076	0.076	0.095	0.115	0.165	NA	NA	NA
0.200	0.063	0.066	0.076	0.076	0.095	0.115	0.165	NA	NA	NA
Potential break incorrectly specified at beginning										
Actual break	1.000	0.900	0.800	0.700	0.600	0.500	0.400	0.300	0.200	0.100
0.900	0.047	0.040	0.035	0.032	0.036	0.048	NA	NA	NA	NA
0.800	0.046	0.043	0.041	0.040	0.040	0.049	NA	NA	NA	NA
0.700	0.045	0.049	0.045	0.045	0.043	0.049	NA	NA	NA	NA
0.600	0.050	0.051	0.050	0.051	0.048	0.050	NA	NA	NA	NA
0.500	0.051	0.053	0.053	0.053	0.050	0.051	NA	NA	NA	NA
0.400	0.053	0.055	0.055	0.054	0.051	0.051	NA	NA	NA	NA
0.300	0.054	0.058	0.056	0.056	0.053	0.052	NA	NA	NA	NA
0.200	0.054	0.058	0.056	0.056	0.053	0.052	NA	NA	NA	NA

Table 6. Power of max-Chow Test for Break at End.

$T = 200$, Nominal Size = 5%

Potential break correctly specified at end										
Actual break	1.000	0.900	0.800	0.700	0.600	0.500	0.400	0.300	0.200	0.100
0.900	0.742	0.948	0.693	0.420	0.193	0.143	NA	NA	NA	NA
0.800	0.323	0.612	0.798	0.514	0.246	0.175	NA	NA	NA	NA
0.700	0.158	0.284	0.483	0.574	0.283	0.199	NA	NA	NA	NA
0.600	0.091	0.127	0.196	0.290	0.319	0.225	NA	NA	NA	NA
0.500	0.077	0.097	0.133	0.184	0.327	0.233	NA	NA	NA	NA
0.400	0.070	0.077	0.094	0.111	0.166	0.237	NA	NA	NA	NA
0.300	0.063	0.064	0.071	0.068	0.082	0.098	NA	NA	NA	NA
0.200	0.063	0.064	0.071	0.068	0.082	0.098	NA	NA	NA	NA

observe substantially reduced power. In the second panel of Table 10 when $b = 0.8$ the power of the recursive Predictive test varies across 0.418, 0.391, 0.371, 0.365, 0.363, 0.364, 0.352, and 0.352 as the parameter a varies from 1.0 to 0.3. In the second panel of Table 11 when $b = 0.8$ the power of the

Table 7. Power of Recursive Test for Break at End.

$T = 800$, Nominal Size = 5%

Potential break correctly specified at end										
Actual break	1.000	0.900	0.800	0.700	0.600	0.500	0.400	0.300	0.200	0.100
0.900	0.999	1.000	0.997	0.922	0.647	0.436	0.425	NA	NA	NA
0.800	0.753	0.962	0.999	0.960	0.724	0.496	0.460	NA	NA	NA
0.700	0.275	0.488	0.780	0.972	0.766	0.530	0.477	NA	NA	NA
0.600	0.105	0.172	0.288	0.515	0.785	0.548	0.488	NA	NA	NA
0.500	0.060	0.089	0.121	0.198	0.374	0.557	0.496	NA	NA	NA
0.400	0.053	0.062	0.073	0.103	0.177	0.347	0.498	NA	NA	NA
0.300	0.051	0.055	0.060	0.079	0.119	0.203	0.498	NA	NA	NA
0.200	0.049	0.053	0.057	0.067	0.085	0.139	0.285	NA	NA	NA
Potential break incorrectly specified at beginning										
Actual break	1.000	0.900	0.800	0.700	0.600	0.500	0.400	0.300	0.200	0.100
0.900	0.548	0.516	0.497	0.490	0.483	0.456	0.453	0.451	NA	NA
0.800	0.207	0.194	0.188	0.182	0.180	0.171	0.166	0.163	NA	NA
0.700	0.094	0.092	0.089	0.088	0.086	0.083	0.078	0.076	NA	NA
0.600	0.060	0.057	0.059	0.062	0.060	0.056	0.053	0.050	NA	NA
0.500	0.054	0.051	0.055	0.052	0.056	0.052	0.047	0.044	NA	NA
0.400	0.053	0.049	0.052	0.050	0.052	0.050	0.045	0.042	NA	NA
0.300	0.052	0.050	0.050	0.050	0.051	0.050	0.045	0.043	NA	NA
0.200	0.051	0.050	0.051	0.049	0.051	0.049	0.046	0.043	NA	NA

Chow test varies across 0.082, 0.052, 0.046, 0.041, 0.036, 0.040, and 0.043 as the parameter a varies from 1.0 to 0.4. In each of these cases the power of the test is reduced relative to when the potential break was correctly chosen at the end of the sample rather than the beginning. The reduction in power is most severe for the Chow test with power essentially equal to the size of the test. This is of course the exact situation for which the max-Chow was designed. In Table 12 when $b = 0.8$ the power of the max-Chow test varies across 0.955, 1.000, 1.000, 1.000, 0.972, 0.647, and 0.258 as the parameter a varies from 1.0 to 0.4.

Comparisons among the Three Tests

There are some clear relationships among the three test statistics in terms of power. Recall that in each of the power simulations the actual break occurs at the end of the sample. In such an environment we expect that the Chow tests, with potential break points chosen at the end of the sample, will have greater power than the max-Chow tests. Since Chow tests constructed at the beginning of the sample have no power to detect breaks at the end of the

Table 8. Power of Chow Test for Break at End.

$T=800$, Nominal Size = 5%										
Potential break correctly specified at end										
Actual break	1.000	0.900	0.800	0.700	0.600	0.500	0.400	0.300	0.200	0.100
0.900	0.999	1.000	0.999	0.930	0.643	0.338	0.180	0.118	NA	NA
0.800	0.737	0.973	0.999	0.965	0.736	0.419	0.220	0.138	NA	NA
0.700	0.260	0.525	0.838	0.978	0.777	0.466	0.244	0.153	NA	NA
0.600	0.096	0.189	0.338	0.591	0.803	0.491	0.263	0.160	NA	NA
0.500	0.061	0.091	0.133	0.233	0.405	0.505	0.271	0.164	NA	NA
0.400	0.051	0.063	0.076	0.110	0.175	0.315	0.276	0.166	NA	NA
0.300	0.049	0.056	0.059	0.078	0.104	0.165	0.279	0.167	NA	NA
0.200	0.049	0.049	0.049	0.055	0.064	0.075	0.104	0.169	NA	NA
Potential break incorrectly specified at beginning										
Actual break	1.000	0.900	0.800	0.700	0.600	0.500	0.400	0.300	0.200	0.100
0.900	0.089	0.051	0.037	0.034	0.034	0.032	0.043	NA	NA	NA
0.800	0.059	0.043	0.038	0.038	0.041	0.037	0.045	NA	NA	NA
0.700	0.052	0.043	0.043	0.042	0.044	0.043	0.046	NA	NA	NA
0.600	0.052	0.046	0.047	0.047	0.046	0.044	0.047	NA	NA	NA
0.500	0.052	0.048	0.049	0.049	0.048	0.045	0.047	NA	NA	NA
0.400	0.054	0.048	0.050	0.049	0.048	0.045	0.047	NA	NA	NA
0.300	0.054	0.049	0.051	0.049	0.048	0.046	0.047	NA	NA	NA
0.200	0.053	0.050	0.051	0.050	0.049	0.046	0.047	NA	NA	NA

Table 9. Power of max-Chow Test for Break at End.

$T=800$, Nominal Size = 5%										
Potential break correctly specified at end										
Actual break	1.000	0.900	0.800	0.700	0.600	0.500	0.400	0.300	0.200	0.100
0.900	0.997	1.000	0.996	0.878	0.524	0.247	0.129	NA	NA	NA
0.800	0.629	0.947	0.999	0.934	0.638	0.321	0.166	NA	NA	NA
0.700	0.197	0.419	0.762	0.956	0.692	0.367	0.192	NA	NA	NA
0.600	0.081	0.138	0.252	0.496	0.723	0.390	0.202	NA	NA	NA
0.500	0.060	0.073	0.103	0.178	0.325	0.405	0.207	NA	NA	NA
0.400	0.057	0.054	0.067	0.094	0.138	0.239	0.211	NA	NA	NA
0.300	0.056	0.052	0.056	0.068	0.089	0.121	0.210	NA	NA	NA
0.200	0.055	0.050	0.051	0.054	0.064	0.060	0.086	NA	NA	NA

sample it seems likely that the power of max-Chow tests will be reduced regardless of the fact that asymptotically appropriate critical values are being used. This is usually the case when the actual and potential break point parameters are large. For example, in the first panel of Table 5 and

Table 10. Power of Recursive Test for Break at End.

T = 3,200, Nominal Size = 5%										
Potential break correctly specified at end										
Actual break	1.000	0.900	0.800	0.700	0.600	0.500	0.400	0.300	0.200	0.100
0.900	1.000	1.000	1.000	1.000	0.964	0.666	0.417	NA	NA	NA
0.800	0.983	1.000	1.000	1.000	0.983	0.743	0.469	NA	NA	NA
0.700	0.471	0.780	0.984	1.000	0.987	0.777	0.494	NA	NA	NA
0.600	0.134	0.246	0.466	0.801	0.988	0.793	0.508	NA	NA	NA
0.500	0.067	0.088	0.138	0.237	0.445	0.800	0.514	NA	NA	NA
0.400	0.049	0.060	0.064	0.089	0.128	0.250	0.517	NA	NA	NA
0.300	0.048	0.057	0.053	0.057	0.079	0.116	0.284	NA	NA	NA
0.200	0.048	0.056	0.051	0.052	0.064	0.091	0.180	NA	NA	NA
Potential break incorrectly specified at beginning										
Actual break	1.000	0.900	0.800	0.700	0.600	0.500	0.400	0.300	0.200	0.100
0.900	0.958	0.945	0.936	0.933	0.930	0.929	0.915	0.916	NA	NA
0.800	0.418	0.391	0.371	0.365	0.363	0.364	0.352	0.352	NA	NA
0.700	0.119	0.115	0.111	0.109	0.107	0.109	0.107	0.105	NA	NA
0.600	0.060	0.059	0.061	0.059	0.061	0.060	0.058	0.058	NA	NA
0.500	0.048	0.052	0.052	0.051	0.051	0.050	0.049	0.049	NA	NA
0.400	0.047	0.051	0.050	0.050	0.050	0.050	0.049	0.048	NA	NA
0.300	0.049	0.051	0.050	0.051	0.052	0.049	0.047	0.049	NA	NA
0.200	0.050	0.052	0.051	0.052	0.052	0.050	0.049	0.049	NA	NA

when $b = 0.9$ we find that the power of the Chow test varies across 0.827, 0.973, 0.790, 0.526, 0.265, 0.189, and 0.110 as a changes from 1.0 to 0.4 while in a similar experiment the power of the max-Chow test varies across 0.742, 0.948, 0.693, 0.420, 0.193, and 0.143.

Comparing the power of the Chow and the recursive Predictive test is a bit more complicated. For each of the Chow and max-Chow tests the actual size of the test was always close to the nominal size of 5%. That is not the case for the recursive Predictive test. As the potential break point location gets closer to the end of the sample, the size distortions become quite large. In such a situation it is not clear how to evaluate whether one test is more powerful than the other. On the other hand, when the break location is not so close to the end of the sample ($a = 1.0$, 0.9, 0.8) the size of the Predictive tests is reasonable. In these situations the power of the Chow test dominates that of the Predictive test so long as the actual break point is not too near the end of the sample. In Table 4 when $a = 0.9$ the power of the Predictive test varies across 0.797, 0.960, 0.766, 0.548, 0.392, and 0.436 as b varies from 1.0 to 0.5. In Table 5 the Chow test varies across 0.827, 0.973,

Table 11. Power of Chow Test for Break at End.

$T = 3{,}200$, Nominal Size = 5%

Potential break correctly specified at end										
Actual break	1.000	0.900	0.800	0.700	0.600	0.500	0.400	0.300	0.200	0.100
0.900	1.000	1.000	1.000	1.000	0.970	0.653	0.274	0.119	NA	NA
0.800	0.975	1.000	1.000	1.000	0.984	0.745	0.336	0.140	NA	NA
0.700	0.431	0.795	0.991	1.000	0.989	0.782	0.375	0.151	NA	NA
0.600	0.125	0.251	0.514	0.865	0.990	0.797	0.391	0.158	NA	NA
0.500	0.065	0.088	0.151	0.267	0.528	0.805	0.399	0.161	NA	NA
0.400	0.049	0.059	0.068	0.094	0.152	0.281	0.402	0.162	NA	NA
0.300	0.048	0.057	0.052	0.059	0.077	0.102	0.208	0.163	NA	NA
0.200	0.049	0.055	0.050	0.051	0.055	0.060	0.091	0.162	NA	NA
Potential break incorrectly specified at beginning										
Actual break	1.000	0.900	0.800	0.700	0.600	0.500	0.400	0.300	0.200	0.100
0.900	0.251	0.115	0.059	0.040	0.028	0.033	0.037	NA	NA	NA
0.800	0.082	0.052	0.046	0.041	0.036	0.040	0.043	NA	NA	NA
0.700	0.048	0.044	0.046	0.046	0.041	0.044	0.046	NA	NA	NA
0.600	0.043	0.046	0.049	0.049	0.044	0.047	0.048	NA	NA	NA
0.500	0.045	0.047	0.050	0.050	0.045	0.048	0.049	NA	NA	NA
0.400	0.046	0.047	0.050	0.051	0.047	0.049	0.049	NA	NA	NA
0.300	0.045	0.047	0.050	0.051	0.047	0.049	0.049	NA	NA	NA
0.200	0.045	0.047	0.050	0.051	0.047	0.048	0.049	NA	NA	NA

Table 12. Power of max-Chow Test for Break at End.

$T = 3{,}200$, Nominal Size = 5%

Potential break correctly specified at end										
Actual break	1.000	0.900	0.800	0.700	0.600	0.500	0.400	0.300	0.200	0.100
0.900	1.000	1.000	1.000	1.000	0.940	0.539	0.196	NA	NA	NA
0.800	0.955	1.000	1.000	1.000	0.972	0.647	0.258	NA	NA	NA
0.700	0.338	0.706	0.982	1.000	0.979	0.694	0.291	NA	NA	NA
0.600	0.095	0.190	0.416	0.793	0.983	0.715	0.307	NA	NA	NA
0.500	0.055	0.075	0.113	0.216	0.434	0.725	0.314	NA	NA	NA
0.400	0.047	0.057	0.062	0.078	0.117	0.224	0.319	NA	NA	NA
0.300	0.045	0.053	0.054	0.057	0.062	0.086	0.159	NA	NA	NA
0.200	0.046	0.052	0.050	0.054	0.053	0.061	0.076	NA	NA	NA

0.790, 0.526, 0.265, and 0.189 in the same experiment. If we make the comparison with $T = 3{,}200$, Tables 10 and 11 indicate that the distinction is moot since each rejects the null 100% of the time for large break parameters a and b.

When the potential break locations are at the beginning of the sample the recursive Predictive test clearly dominates the Chow test so long as the actual break is not too near the end of the sample. That is, in the second panel of Tables 4, 7, and 10 the rows associated with $b = 0.8$ and 0.9 are uniformly larger than those associated with Tables 5, 8, and 11. Once again the Chow test is not as robust as the recursive Predictive test to large deviations of the potential and actual break locations. As soon as we consider the max-Chow results in Tables 6, 9, and 12 however, we find that it is usually the case that the max-Chow dominates the recursive Predictive test.

CONCLUSION

In this chapter we provide a set of refined asymptotics for standard Chow and Predictive tests for structural change in the intercept from a linear regression. By treating the locations of the actual and potential break point locations as general integer-valued functions we are able to establish the behavior of these tests under the null of no breaks and under a collection of alternatives that allow for breaks to occur "local to the end" of the sample. In particular we establish that the tests are asymptotically chi-square under the null and can be consistent against a range of alternatives. In doing so we also establish the rates at which the tests diverge against these alternatives. Monte Carlo simulations verify the asymptotic results and suggest that the tests can be useful in large samples.

NOTES

1. The Predictive test is more generally equivalent to a test of efficiency in the literature on out-of-sample testing. Given our simple model, this simplifies to a test of zero mean prediction error.
2. In the proof of Theorem 2.3 we are able to establish that the fixed Predictive test and the Chow test are asymptotically equivalent in probability under the null.
3. Note that the third panel of Tables 2 and 3 are identical by construction.

REFERENCES

Andrews, D. W. K. (1993). Tests for parameter instability and structural change with unknown change point. *Econometrica*, *61*, 821–856.

Andrews, D. W. K. (2003). End-of-sample instability tests. *Econometrica*, *71*, 1661–1694.
Andrews, D. W. K., & Kim, J.-Y (2006). End-of-sample cointegration breakdown tests. *Journal of Business and Economic Statistics*, *24*, 379–394.
Bai, J. (1997). Estimating multiple breaks one at a time. *Econometric Theory*, *13*, 315–352.
Chow, G. C. (1960). Tests of equality between sets of coefficients in two linear regressions. *Econometrica*, *28*, 591–605.
Clements, M. P., & Hendry, D. F. (1996). Intercept corrections and structural change. *Journal of Applied Econometrics*, *11*, 475–494.
Dufour, J., Ghysels, E., & Hall, A. (1994). Generalized predictive tests and structural change analysis in econometrics. *International Economic Review*, *35*, 199–229.
Ghysels, E., & Hall, A. (1990). A test for structural stability of Euler conditions parameters estimated via the generalized method of moments estimator. *International Economic Review*, *31*, 355–364.
Hall, P., & Heyde, C. C. (1980). *Martingale limit theory and its applications*. New York: Academic Press.
Hansen, B. E. (1992). Convergence to stochastic integrals for dependent heterogeneous processes. *Econometric Theory*, *8*, 489–500.
Kramer, W., Ploberger, W., & Alt, R. (1988). Testing for structural change in dynamic models. *Econometrica*, *56*, 1355–1369.
Ploberger, W., Kramer, W., & Kontrus, K. (1989). A new test for structural stability in the linear regression model. *Journal of Econometrics*, *40*, 307–318.
Sowell, F. (1996). Optimal tests for parameter instability in the generalized method of moments framework. *Econometrica*, *64*, 1085–1107.
Stock, J. H., & Watson, M. W. (1996). Evidence on structural stability in macroeconomic time series relations. *Journal of Business and Economic Statistics*, *14*, 11–30.
West, K. (1996). Asymptotic inference about predictive ability. *Econometrica*, *64*, 1067–1084.
West, K. D., & McCracken, M. W. (1998). Regression-based tests of predictive ability. *International Economic Review*, *39*, 817–840.
Wooldridge, J. M., & White, H. (1998). Central limit theorems for dependent, heterogeneous processes with trending moments. In H. White (Ed.), *Topics in Econometric Theory: The Selected Works of Halbert White*. Cheltenham: Edward Elgar.

APPENDIX

Theorem 1: Maintain Assumptions 1 and 2 and let z_1 and z_2 denote independent standard normal variates. (i) If $\lim R/T = \lambda \in (0,1)$, then $W \to_d [(1-\lambda)^{1/2} z_1 - \lambda^{1/2} z_2]^2$, (ii) if $\lim R/T = 0$, then $W \to_d z_1^2$, and (iii) if $\lim R/T = 1$, then $W \to_d z_2^2$.

Proof of Theorem 1: It is straightforward to show that $T^{-1}\sum_{t=1}^{T} \hat{v}_{2,t+1}^2 \to_p \sigma^2$ and hence we proceed immediately to the numerator. Note that for $M(T) = -JB_1(T)J' + B_2(T)$ and $F_1 = (T^{-1}\sum_{t=1}^{T} z_t z_t' - (T^{-1}\sum_{t=1}^{T} z_t)(T^{-1}\sum_{t=1}^{T} z_t'))^{-1}$

$$\sum_{t=1}^{T} (\hat{v}_{1,t+1}^2 - \hat{v}_{2,t+1}^2) = TH_2'(T)M(T)H_2(T)$$

$$= \left[1 - \left(\frac{PR}{T^2}\right)\left(P^{-1}\sum_{t=R}^{T} z_t - \sum_{t=R}^{T} z_t - R^{-1}\sum_{t=1}^{R-1} z_t\right)' F_1 \left(P^{-1}\sum_{t=R}^{T} z_t - R^{-1}\sum_{t=1}^{R-1} z_t\right)\right]^{-1}$$

$$\times \left\{ \left[\left(\frac{PR}{T^2}\right)^{1/2} \left(P^{-1}\sum_{t=R}^{T} z_t - R^{-1}\sum_{t=1}^{R-1} z_t\right)' F_1 \left(\left(T^{-1/2}\sum_{t=1}^{T} u_{t+1} z_t\right) \right.\right.\right.$$

$$\left.\left. - \left(T^{-1/2}\sum_{t=1}^{T} u_{t+1}\right)\left(T^{-1}\sum_{t=1}^{T} z_t\right)\right)\right] + \left(\frac{P}{R}\right)^{1/2}\left[T^{-1/2}\sum_{t=1}^{T} u_{t+1}\right]$$

$$\left. - \left(\frac{T}{R}\right)^{1/2}\left[P^{-1/2}\sum_{t=R}^{T} u_{t+1}\right]\right\}^2$$

$$= \left[1 - \left(\frac{PR}{T^2}\right)\left(P^{-1}\sum_{t=R}^{T} z_t - R^{-1}\sum_{t=1}^{R-1} z_t\right)' F_1 \left(P^{-1}\sum_{t=R}^{T} z_t - R^{-1}\sum_{t=1}^{R-1} z_t\right)\right]^{-1}$$

$$\times \left\{ \left[\left(\frac{PR}{T^2}\right)^{1/2} \left(P^{-1}\sum_{t=R}^{T} z_t - R^{-1}\sum_{t-1}^{R-1} z_t\right)' F_1 \left(\left(T^{-1/2}\sum_{t=1}^{T} u_{t+1} z_t\right) \right.\right.\right.$$

$$\left.\left. - \left(T^{-1/2}\sum_{t=1}^{T} u_{t+1}\right)\left(T^{-1}\sum_{t=1}^{T} z_t\right)\right)\right]$$

$$\left. + \left[\left(\frac{P}{T}\right)^{1/2}\left[R^{-1/2}\sum_{t=1}^{R} u_{t+1}\right] - \left(\frac{R}{T}\right)^{1/2}\left[P^{-1/2}\sum_{t=R}^{T} u_{t+1}\right]\right]\right\}^2$$

Assumption 2 suffices for each of F_1, $T^{-1/2}\sum_{t=1}^{T} u_{t+1} z_t$, $T^{-1/2}\sum_{t=1}^{T} u_{t+1}$, and $T^{-1}\sum_{t=1}^{T} z_t$ to be $O_p(1)$. Moreover we obtain $P - \sum_{t=R}^{T} z_t - R^{-1}$

$\sum_{t=1}^{R-1} z_t = o_p(1)$ since under the null, z_s is covariance stationary. By continuity we then obtain

$$\sum_{t=1}^{T}(\hat{v}_{1,t+1}^2 - \hat{v}_{2,t+1}^2) = \left\{ \left(\frac{P}{T}\right)^{1/2} \left[R^{-1/2} \sum_{t=1}^{R} u_{t+1} \right] - \left(\frac{R}{T}\right)^{1/2} \left[P^{-1/2} \sum_{t=R}^{T} u_{t+1} \right] \right\}^2 + o_p \quad (1)$$

Given Assumption 2, Theorem 3.2 of Hall and Heyde (1980) implies that each of $R^{-1/2}\sum_{t=1}^{R} u_{t+1}$ and $P^{-1/2}\sum_{t=R}^{T} u_{t+1}$ is asymptotically normal with zero mean and variance σ^2. Since the forecast errors are uncorrelated we also know that these two terms are asymptotically independent. Choosing $\lim R/T$ appropriately we obtain cases (i)–(iii) in the Theorem.

Corollary 1: Maintain Assumptions 1 and 2 and let z_1 and z_2 denote independent standard normal variates. If $\lim R_1/T = 0$ and $\lim R_2/T = 1$, then $W_{\max} = \max[W_1, W_2] \rightarrow_d \max[z_1^2, z_2^2]$.

Proof of Corollary 1: Proof follows directly from the fact that $R_1^{-1/2}\sum_{t=1}^{R_1} u_{t+1}$ and $P_2^{-1/2}\sum_{t=R_2}^{T} u_{t+1}$ are each asymptotically normal and uncorrelated.

Theorem 2: Maintain Assumptions 1 and 2′. (i) If $R \geqslant T_B$, then $W = O_p(PT_B^2/RT)$, and (ii) if $R \leqslant T_B$, then $W = O_p(RP_B^2/RT)$.

Proof of Theorem 2: It is straightforward to show that $T^{-1}\sum_{t=1}^{T} \hat{v}_{2,t+1}^2 \rightarrow_p \omega > 0$ and hence we proceed immediately to the numerator. Note that for

$$F_1 = \left(T^{-1} \sum_{t=1}^{T} z_t z_t' - \left(T^{-1} \sum_{t=1}^{T} z_t \right) \left(T^{-1} \sum_{t=1}^{T} z_t' \right) \right)^{-1}$$

$M(T) = -JB_1(T)J' + B_2(T)$ and

$$F_2 = \left(\frac{T^2}{RP}\right) \left[1 - \left(\frac{PR}{T^2}\right) \left(P^{-1} \sum_{t=R}^{T} z_t - R^{-1} \sum_{t=1}^{R-1} z_t \right)' \times F_1 \left(P^{-1} \sum_{t=R}^{T} z_t - R^{-1} \sum_{t=1}^{R-1} z_t \right) \right]^{-1}$$

$$\sum_{t=1}^{T}(\hat{v}_{1,t+1}^2 - \hat{v}_{2,t+1}^2)$$
$$= TH_2'(T)M(T)H_2(T) + 2TH_2'(T)(-JB_1(T)J'B_3^{-1}(T) + B_2(T)D_{23}(T))\beta_3^*$$
$$+T\beta_3^{*\prime}\begin{pmatrix} 0 & 0 \\ 0 & \left(\left(\frac{\min(P,P_B)}{T}\right)F_2^{1/2} - \left(\frac{P}{T}\right)\left(\frac{P_B}{T}\right)F_2^{1/2}\left(P^{-1}\sum_{t=R}^{T}x_{1,t}'\right)B_1(T)\left(P_B^{-1}\sum_{t=T_B}^{T}x_{1,t}\right)\right)^2 \end{pmatrix}\beta_3^*.$$

That $TH_2'(T)M(T)H_2(T) = O_p(\text{Eq.}(1))$ follows from Theorem 1. To show that the second term $2TH_2'(T)(-JB_1(T)J'B_3^{-1}(T) + B_2(T)D_{23}(T))\beta_3^*$ is of lower order than the third term follows from similar arguments to those used to derive the order of the third term. We, therefore, proceed to deriving the order of the third term.

Recall that under the alternative hypothesis $\beta_{3,2}^* \neq 0$. If we substitute in the definition of F_2 and compute the term $(P^{-1}\sum_{t=R}^{T}x_{1,t}')B_1(T)(P_B^{-1}\sum_{t=T_B}^{T}x_{1,t})$ the third term above can be rewritten as

$$T(\beta_{3,2}^*)^2\left(\left(\frac{\min(P,P_B)}{T}\right)F_2^{1/2} - \left(\frac{P}{T}\right)\left(\frac{P_B}{T}\right)F_2^{1/2}\left(P^{-1}\sum_{t=R}^{T}x_{1,t}'\right)\right.$$
$$\left.\times B_1(T)\left(P_B^{-1}\sum_{t=T_B}^{T}x_{1,t}\right)\right)^2 = (\beta_{3,2}^*)^2\left[1 - \left(\frac{PR}{T^2}\right)\left(P^{-1}\sum_{t=R}^{T}z_t - R^{-1}\sum_{t=1}^{R-1}z_t\right)'\right.$$
$$\times F_1\left(P^{-1}\sum_{t=R}^{T}z_t - R^{-1}\sum_{t=1}^{R-1}z_t\right)\Bigg]^{-1}\left(\frac{T^3}{RP}\right)\left[\frac{\min(P,P_B)}{T} - \left(\frac{PP_B}{T^2}\right)\right.$$
$$\left. - \left(\frac{PP_BRT_B}{T^4}\right)\left(P^{-1}\sum_{t=R}^{T}z_t - R^{-1}\sum_{t=1}^{R-1}z_t\right)'F_1\left(P_B^{-1}\sum_{t=T_B}^{T}z_t - T_B^{-1}\sum_{t=1}^{T_B-1}z_t\right)\right]^2.$$

If we now consider the two cases the above term can be rewritten as

$$T(\beta_{3,2}^*)^2\left(\left(\frac{\min(P,P_B)}{T}\right)F_2^{1/2} - \left(\frac{P}{T}\right)\left(\frac{P_B}{T}\right)F_2^{1/2}\left(P^{-1}\sum_{t=R}^{T}x_{1,t}'\right)B_1(T)\left(P_B^{-1}\sum_{t=T_B}^{T}x_{1,t}\right)\right)^2$$

$$
= \begin{cases}
\left(\dfrac{PT_B^2}{RT}\right)(\beta_{3,2}^*)^2 \\
\times \left[\dfrac{\left(1-(P_BR/T^2)(P^{-1}\sum_{t=R}^{T} z_t - R^{-1}\sum_{t=1}^{R-1} z_t)' F_1 (P_B^{-1}\sum_{t=T_B}^{T} z_t - T_B^{-1}\sum_{t=1}^{T_B-1} z_t)\right)^2}{1-(PR/T^2)(P^{-1}\sum_{t=R}^{T} z_t - R^{-1}\sum_{t=1}^{R-1} z_t)' F_1 (P^{-1}\sum_{t=R}^{T} z_t - R^{-1}\sum_{t=1}^{R-1} z_t)}\right] & P \le P_B \\
\left(\dfrac{P_B^2R}{PT}\right)(\beta_{3,2}^*)^2 \\
\times \left[\dfrac{(1-(PT_B/T^2)(P^{-1}\sum_{t=R}^{T} z_t - R^{-1}\sum_{t=1}^{R-1} z_t)' F_1 (P_B^{-1}\sum_{t=T_B}^{T} z_t - T_B^{-1}\sum_{t=1}^{T_B-1} z_t))^2}{1-(PR/T^2)(P^{-1}\sum_{t=R}^{T} z_t - R^{-1}\sum_{t=1}^{R-1} z_t)' F_1 (P^{-1}\sum_{t=R}^{T} z_t - R^{-1}\sum_{t=1}^{R-1} z_t)}\right] & P_B \le P.
\end{cases}
$$

Assumption 2′ suffices for each of F_1, $P^{-1}\sum_{t=R}^{T} z_t$, $R^{-1}\sum_{t=1}^{R-1} z_t$, $P_B^{-1}\sum_{t=T_B}^{T} z_t$, and $T_B^{-1}\sum_{t=1}^{T_B-1} z_t$ to be $O_p(1)$. By continuity we then know that each of

$$
\frac{(1-(P_BR/T^2)(P^{-1}\sum_{t=R}^{T} z_t - R^{-1}\sum_{t=1}^{R-1} z_t)' F_1 (P_B^{-1}\sum_{t=T_B}^{T} z_t - T_B^{-1}\sum_{t=1}^{T_B-1} z_t))^2}{1-(PR/T^2)(P^{-1}\sum_{t=R}^{T} z_t - R^{-1}\sum_{t=1}^{R-1} z_t)' F_1 (P^{-1}\sum_{t=R}^{T} z_t - R^{-1}\sum_{t=1}^{R-1} z_t)}
$$

and

$$
\frac{(1-(PT_B/T^2)(P^{-1}\sum_{t=R}^{T} z_t - R^{-1}\sum_{t=1}^{R-1} z_t)' F_1 (P_B^{-1}\sum_{t=T_B}^{T} z_t - T_B^{-1}\sum_{t=1}^{T_B-1} z_t))^2}{1-(PR/T^2)(P^{-1}\sum_{t=R}^{T} z_t - R^{-1}\sum_{t=1}^{R-1} z_t)' F_1 (P^{-1}\sum_{t=R}^{T} z_t - R^{-1}\sum_{t=1}^{R-1} z_t)}
$$

is $O_p(1)$ with strictly positive probability limits. This implies that the order of magnitude of the statistic is determined by the lead terms in the right-hand side of the previous equality and we have the desired result.

Corollary 2: Maintain Assumptions 1 and 2′ let $\lim R_1/T = 0$ and $\lim R_2/T = 1$. (i) If $R_1 \geqslant T_B$ then $W_{\max} = O_p(P_1T_B^2/R_1T)$, (ii) if $R_2 \leqslant T_B$ then $W_{\max} = O_p(R_2P_B^2/P_2T)$, (iii) if $R_2 > T_B$ and $R_1 < T_B$ then $W_{\max} = O_p(\max[R_1P_B^2/P_1T, P_2T_B^2/R_2T])$.

Proof of Corollary 2:

(i) From Theorem 2 we know that $W_1 = O_p(P_1T_B^2/R_1T)$ while $W_2 = O_p(P_2T_B^2/R_2T)$. The result is immediate since $R_2 > R_1$ implies $P_1/R_1 > P_2/R_2$.

(ii) From Theorem 2 w know that $W_1 = O_p(R_1 P_B^2 / P_1 T)$ while $W_2 = O_p(R_2 P_B^2 / P_2 T)$. The result is immediate since $R_2 > R_1$ implies $R_1/P_1 < R_2/P_2$.
(iii) The result is trivial given continuity of the max[.,.] function.

Theorem 3: Maintain Assumptions 1 and 2 and let $\lim R/T = 0$. When the rolling scheme is used make the additional assumption that $\lim P/R^{3/2} = 0$. For the rolling and fixed schemes, $V \rightarrow_d$ chi-square (1).

Proof of Theorem 3: Consider the fixed scheme with $\lim R/T = 0$. It is straightforward to show that $P^{-1}\sum_{t=R}^{T}\hat{u}_{1,t+1}^2 \rightarrow_p \sigma^2$ and hence we proceed immediately to the numerator. Since $(P^{-1/2}\sum_{t=R}^{T}\hat{u}_{1,t+1})^2/(1+P/R) = ((R/T)^{1/2}P^{-1/2}\sum_{t=R}^{T}\hat{u}_{1,t+1})^2$ we turn our attention to the limiting distribution of $(R/T)^{1/2}P^{-1/2}\sum_{t=R}^{T}\hat{u}_{1,t+1}$. Adding and subtracting appropriate terms we obtain

$$
\begin{aligned}
&(R/T)^{1/2}P^{-1/2}\sum_{t=R}^{T}\hat{u}_{1,t+1} \\
&= (R/T)^{1/2}P^{-1/2}\sum_{t=R}^{T}u_{t+1} - (R/T)^{1/2}P^{-1/2}\sum_{t=R}^{T}x_{1,t}'B_1(R)H_1(R) \\
&= (R/T)^{1/2}P^{-1/2}\sum_{t=R}^{T}u_{t+1} - (RP/T)^{1/2}\left(P^{-1}\sum_{t=R}^{T}x_{1,t}'\right)B_1(R)H_1(R) \\
&= (R/T)^{1/2}P^{-1/2}\sum_{t=R}^{T}u_{t+1} - (P/T)^{1/2}\left(R^{-1/2}\sum_{t=1}^{R-1}u_{t+1}\right) \\
&\quad + (P/T)^{1/2}\left(P^{-1}\sum_{t=R}^{T}z_t - R^{-1}\sum_{t=1}^{R}z_t\right)' F_{1,R}\left(\left(R^{-1/2}\sum_{t=1}^{R}u_{t+1}\right)\left(R^{-1}\sum_{t=1}^{R}z_t\right)\right. \\
&\quad \left. - \left(R^{-1/2}\sum_{t=1}^{R}u_{t+1}z_t\right)\right)
\end{aligned}
$$

where $F_{1,R} = (R^{-1}\sum_{t=1}^{R-1}z_t z_t' - (R^{-1}\sum_{t=1}^{R-1}z_t)(R^{-1}\sum_{t=1}^{R-1}z_t'))^{-1}$. Assumption 2 suffices for each of $F_{1,R}$, $R^{-1/2}\sum_{t=1}^{R-1}u_{t+1}z_t$, $R^{-1/2}\sum_{t=1}^{R-1}u_{t+1}$ and $R^{-1}\sum_{t=1}^{R-1}z_t$ to be $O_p(1)$. Moreover, we obtain $P^{-1}\sum_{t=R}^{T}z_t - R^{-1}\sum_{t=1}^{R-1}z_t = o_p(1)$ since under the null, z_t is covariance stationary. By continuity we then obtain

$$\left(\left(\frac{R}{T}\right)^{1/2} P^{-1/2}\sum_{t=R}^{T}\hat{u}_{1,t+1}\right)^2 = \left\{\left(\frac{P}{T}\right)^{1/2}\left[R^{-1/2}\sum_{t=1}^{R-1}u_{t+1}\right] - \left(\frac{R}{T}\right)^{1/2}\left[P^{-1/2}\sum_{t=R}^{T}u_{t+1}\right]\right\}^2 + o_p \tag{1}$$

The result then follows from Theorem 1 since this expansion is identical to that for the Chow test. We, therefore, obtain the additional result that the fixed Predictive test and the Chow are asymptotically equivalent in probability.

Consider the rolling scheme with $\lim R/T = 0$. It is straightforward to show that $P^{-1}\sum_{t=R}^{T}\hat{u}_{1,t+1}^2 \to_p \sigma^2$ and hence we proceed immediately to the numerator. Since $(P^{-1/2}\sum_{t=R}^{T}\hat{u}_{1,t+1})^2/(2R/3P) = ((3/2)^{1/2}R^{-1/2}\sum_{t=R}^{T}\hat{u}_{1,t+1})^2$ we turn our attention to the limiting distribution of $R^{-1/2}\sum_{t=R}^{T}\hat{u}_{1,t+1}$. Adding and subtracting appropriate terms we obtain

$$R^{-1/2}\sum_{t=R}^{T}\hat{u}_{1,t+1} = R^{-1/2}\sum_{t=R}^{T}u_{t+1} - R^{-1/2}\sum_{t=R}^{T}x'_{1,t}B_1(t)H_1(t) \tag{A.1}$$

Consider the second right-hand side term in (1). Note that under the null $(Ex_{1,t}x'_{1,t})^{-1} = B_1$ and $Ex_{1,t} = F$ since z_t is covariance stationary. Using the identities $B_1(t) = [B_1(t) - B_1] + B_1$ and $x_{1,t} = [x_{1,t} - F] + F$ we obtain

$$\begin{aligned}
R^{-1}&\sum_{t=R}^{T}x'_{1,t}B_1(t)\left(R^{-1/2}\sum_{s=t-R+1}^{t}h_{1,s+1}\right)\\
&= R^{-3/2}FB_1\sum_{t=R}^{T}\left(\sum\nolimits_{s=t-R+1}^{t}h_{1,s+1}\right) + R^{-1}\sum_{t=R}^{T}(x_{1,t}-F)'\\
&\quad\times B_1\left(R^{-1/2}\sum_{s=t-R+1}^{t}h_{1,s+1}\right) + R^{-1}\sum_{t=R}^{T}F'[B_1(t)-B_1]\left(R^{-1/2}\sum_{s=t-R+1}^{t}h_{1,s+1}\right)\\
&\quad + R^{-1}\sum_{t=R}^{T}(x_{1,t}-F)'(B_1(t)-B_1)\left(R^{-1/2}\sum_{s=t-R+1}^{t}h_{1,s+1}\right)
\end{aligned} \tag{A.2}$$

We will now show that the latter three right-hand side terms in Eq. (2) are $o_p(1)$. For both the second and third terms note that by taking absolute values we obtain

$$\left| R^{-1} \sum_{t=R}^{T} F'[B_1(t) - B_1] \left(R^{-1/2} \sum_{s=t-R+1}^{t} h_{1,s+1} \right) \right|$$
$$\leq \left(\frac{P}{R^{3/2}} \right) k^2 (|F|)(\sup_t R^{1/2} |B_1(t) - B_1|) \left(\sup_t \left| R^{-1/2} \sum_{s=t-R+1}^{t} h_{1,s+1} \right| \right)$$
$$\left| R^{-1} \sum_{t=R}^{T} (x_{1,t} - F)' B_1 \left(R^{-1/2} \sum_{s=t-R+1}^{t} h_{1,s+1} \right) \right|$$
$$\leq \left(\frac{P}{R^{3/2}} \right) k^2 \left(P^{-1} \sum_{t=R}^{T} |x_{1,t} - F| \right) (|B_1|) \left(\sup_t \left| R^{-1/2} \sum_{s=t-R+1}^{t} h_{1,s+1} \right| \right)$$

Assumption 2 is sufficient to show that each of $P^{-1}\sum_{t=R}^{T}|x_{1,t} - F|$, $\sup_t R^{1/2}|B_1(t) - B_1|$, and $\sup_t|R^{-1/2}\sum_{s=t-R+1}^{t} h_{1,s+1}|$ is $O_p(1)$. Since k^2 is finite, the result follows since $P/R^{3/2}$ is $o(1)$. For the fourth term in Eq. (2) note that

$$R^{-1} \sum_{t=R}^{T} (x_{1,t} - F)'(B_1(t) - B_1) \left(R^{-1/2} \sum_{s=t-R+1}^{t} h_{1,s+1} \right)$$
$$= \left(\frac{P}{R^2} \right)^{1/2} \left[\sum_{t=R}^{T} \left(R^{-1/2} \sum_{s=t-R+1}^{y} h'_{1,s+1} \right) \otimes (P^{-1/2}(x_{1,t} - F)) \right] \mathrm{vec}(B)$$

That $\sum_{t=R}^{T}(R^{-1/2}\sum_{s=t-R+1}^{t} h'_{1,s+1}) \otimes (P^{-1/2}(x_{1,t} - F))$ is $O_p(1)$ follows from Assumption 2 and Theorem 3.1 of Hansen (1992). The result follows since P/R^2 is $o(1)$ and B is finite.

Now return to the expression in Eq. (1). Note that since $x_{1,t}$ contains 1 in the first element, $FB_1 = (-1, 0, 0, \ldots, 0)$. Since this also implies that the first element of $h_{1,s+1}$ is u_{s+1}, we obtain

$$R^{-1/2} \sum_{t=R}^{T} \hat{u}_{1,t+1} = R^{-1/2} \sum_{t=R}^{T} u_{t+1} - R^{-3/2} \sum_{t=R}^{T} \left(\sum_{s=t-R+1}^{t} u_{s+1} \right) + o_p(1)$$

Now decompose $\sum_{t=R}^{T}(\sum_{s=t-R+1}^{t} u_{s+1})$ into the terms A_1, A_2, and A_3 as in the technical appendix in West (1996) but divide by $R^{3/2}$:

$$R^{-3/2} A_1 = R^{-3/2}[u_1 + 2u_2 + \cdots + Ru_R]$$

$$R^{-3/2} A_2 = R^{-1/2}[u_{R+1} + u_{R+2} + \cdots + u_P]$$

$$R^{-3/2} A_3 = R^{-3/2}[(R-1)u_{P+1} + (R-2)u_{P+2} + \cdots + u_T]$$

The term $R^{-3/2}A_2$ is precisely the first (*P-R*) terms in$R^{-1/2}\sum_{t=R}^{T}u_{t+1}$. Hence $R^{-1/2}\sum_{t=R}^{T}u_{t+1} - R^{-3/2}\sum_{t=R}^{T}(\sum_{s=t-R+1}^{t}u_{s+1})$ equals $R^{-3/2}A_1 + [R^{-1/2}\sum_{t=P+1}^{T}u_{t+1} - R^{-3/2}A_3]$. Algebra then reveals that

$$R^{-1/2}\sum_{t=R}^{T}\hat{u}_{1,t+1} = -R^{-3/2}\sum_{j=1}^{R}ju_j + R^{-3/2}\sum_{j=1}^{R-1}ju_{P+j} + o_p(1)$$

To show that the limiting variance of the sum is (2/3), first note that each of the two terms is uncorrelated with the other. Taking expectations and rearranging terms we find that $\mathrm{Var}(-R^{-3/2}\sum_{j=1}^{R}ju_j) = \sigma^2R^{-1}\sum_{s=1}^{R}(s/R)^2 = \mathrm{Var}(R^{-3/2}\sum_{j=1}^{R-1}ju_{P+j}) + o(1)$. If we add the two components we obtain $\mathrm{Var}(-R^{-3/2}\sum_{j=1}^{R}ju_j + R^{-3/2}\sum_{j-1}^{R-1}ju_{P+j}) = 2\sigma^2R^{-1}\sum_{s=1}^{R}(s/R)^2 + o(1)$. Using an argument akin to that in West (1996) we know that $R^{-1}\sum_{s=1}^{R}(s/R)^2 \to \int_0^1 j^2dj = 1/3$. From this we obtain the desired result.

We must then show that $-R^{-3/2}\sum_{j=1}^{R}ju_j + R^{-3/2}\sum_{j=1}^{R-1}ju_{P+j}$ is asymptotically normal. Define the sequence $Z_{t,T} = -tR^{-3/2}u_{t+1}$ for $1 \leqslant t \leqslant R$, $Z_{t,T} = 0$ for $R+1 \leqslant t \leqslant P$ and $Z_{t,T} = R^{-3/2}(T-t)u_{t+1}$ for $P+1 \leqslant t \leqslant T+1$. Define $\Omega_T \equiv \mathrm{Var}(\sum_{t=1}^{T}Z_{t,T})$ and $X_{T,t} \equiv \Omega_T^{-1/2}Z_{t,T}$. Then Theorem 3.1 of Wooldridge and White (1998) implies that $\sum_{t=1}^{T}X_{T,t}$ is limiting standard normal. Since $\lim \Omega_T = \Omega = (2/3)$ is p.d. we conclude that $R^{-1/2}\sum_{t=R}^{T}\hat{u}_{t+1}$ is limiting normal with asymptotic variance equal to (2/3).

Theorem 4: Maintain Assumptions 1 and 2′ and consider the recursive scheme. (i) If $R \leqslant T_B$ then $V_R = O_p(T_B^2(\sum_{t=T_B}^{T}t^{-1})^2/P)$, and (ii) If $R \geqslant T_B$, then $V_R = O_p(T_B^2(\sum_{t=R}^{T}t^{-1})^2/P)$.

Proof of Theorem 4: It is straightforward to show that $P^{-1}\sum_{t=R}^{T}\hat{u}_{1,t+1}^2 \to_p \omega > 0$ and hence we proceed immediately to the numerator. Adding and subtracting terms we obtain the expansion

$$\begin{aligned} P^{-1/2}\sum_{t=R}^{T}\hat{u}_{1,t+1} &= P^{-1/2}\sum_{t=R}^{T}u_{t+1} - P^{-1/2}\sum_{t=R}^{T}x'_{1,t}B_1(t)H_1(t) \\ &\quad - P^{-1/2}\sum_{t=R}^{T}x'_{3,t}(JB_1(t)J'B_3^{-1}(t)\beta_3^* \\ &= P^{-1/2}\sum_{t=R}^{T}u_{t+1} - P^{-1/2}\sum_{t=R}^{T}x'_{1,t}B_1(t)H_1(t) \end{aligned}$$

$$- P^{-1/2}\sum_{t=R}^{T}(x_{3,t} - Ex_{3,t})'(JB_1(t)J'B_3^{-1}(t) - I)\beta_3^*$$
$$- P^{-1/2}\sum_{t=R}^{T}(Ex_{3,t})'(JB_1(t)J'B_3^{-1}(t) - I)\beta_3^*.$$

That the first three right-hand side terms above are bounded in probability follows from arguments like those in Theorem 3. We now show that the fourth term is the appropriate order.

Recall that $x_{3,t}$ contains the term d_{t,T_B} in the final position. This term takes the value 1 if $t > T_B$ and 0 otherwise. We must therefore consider separately the cases in which $R > T_B$ or $R \leqslant T_B$. Doing so we find that $Ex_{3,t} = (Ex'_{1,t}, d_{t,T_B})'$ while

$$JB_1(t)J'B_3^{-1}(t) - I = \begin{cases} \begin{pmatrix} 0 & 0 \\ 0 & -1 \end{pmatrix} & T_B > t \\ \begin{pmatrix} 0 & B_1(t)\left(t^{-1}\sum\limits_{s=T_B}^{t} x_{1,s}\right) \\ 0 & -1 \end{pmatrix} & T_B \leq t \end{cases}$$

Algebra then reveals that

$$P^{-1/2}\sum_{t=R}^{T}(Ex_{3,t})'(JB_1(t)J'B_3^{-1}(t) - I)\beta_3^*$$
$$= \begin{cases} \left(T_B P^{-1/2}\sum\limits_{t=T_B}^{T} t^{-1}\right)\beta_{3,2}^* \\ \quad - P^{-1/2}\sum\limits_{t=T_B}^{T}(1 - T_B/t)\left(t^{-1}\sum\limits_{s=1}^{t} z_s - Ez_t\right)' & R < T_B \\ \quad \times F_{1,t}\left(t^{-1}\sum\limits_{s=1}^{t} z_s - (t - T_B)^{-1}\sum\limits_{s=T_B}^{t} z_s\right)\beta_{3,2}^* \\ \left(T_B P^{-1/2}\sum\limits_{t=R}^{T} t^{-1}\right)\beta_{3,2}^* \\ \quad - P^{-1/2}\sum\limits_{t=R}^{T}(1 - T_B/t)\left(t^{-1}\sum\limits_{t=R}^{T} z_s - Ez_t\right)' & R \geq T_B \\ \quad \times F_{1,t}\left(t^{-1}\sum\limits_{s=1}^{t} z_s - (t - T_B)^{-1}\sum\limits_{s=T_B}^{t} z_s\right)\beta_{3,2}^* \end{cases}$$

where $F_{1,t} = (t^{-1}\sum_{s=1}^{t} z_s z_s' - (t^{-1}\sum_{s=1}^{t} z_s)(t^{-1}\sum_{s=1}^{t} z_s'))^{-1}$. For each of the possible two right-hand side terms the lead term has the highest order. Cases (i) and (ii) follow from squaring those terms.

Theorem 5: Maintain Assumptions 1 and 2′ and consider the rolling scheme. (a) Let $0 \leqslant \lim R/T \leqslant 1/2$: (i) if $T_B \leqslant R$, then $V_L = O_p(\max[P/R, T_B^4/R^3])$, (ii) if $T_B > \max[R, P]$, then $V_L = O_p(\max[P/R, P_B^2(2R - P_B)^2/R^3])$, and (iii) if $R < T_B < P$, then $V_L = O_p(\max[P/R, R])$. (b) Let $1/2 < \lim R/T \leqslant 1$: (i) if $T_B \geqslant R$, then $V_L = O_p(P_B^2(2R - P_B)^2/PR^2)$, (ii) if $T_B < \min[R, P]$, then $V_L = O_p(T_B^4/PR^2)$, and (iii) if $P < T_B < R$, then $V_L = O_p(P(2T_B - P)^2/R^2)$.

Proof of Theorem 5: In either case (a) or (b) it is straightforward to show that $P^{-1}\sum_{t=R}^{T}\hat{u}_{1,t+1}^2 \to_p \omega > 0$ and hence we proceed immediately to the numerator.

(a) It suffices to show that $(R^{-1/2}\sum_{t=R}^{T}\hat{u}_{1,t+1})^2$ satisfies cases (i)–(iii). Adding and subtracting terms we obtain the expansion

$$
\begin{aligned}
R^{-1/2}\sum_{t=R}^{T}\hat{u}_{1,t+1} &= R^{-1/2}\sum_{t=R}^{T}u_{t+1} - R^{-1/2}\sum_{t=R}^{T}x_{1,t}'B_1(t)H_1(t) \\
&\quad - R^{-1/2}\sum_{t=R}^{T}x_{3,t}'(JB_1(t)J'B_3^{-1}(t) - I)\beta_3^* \\
&= (P/R)^{1/2}P^{-1/2}\sum_{t=R}^{T}u_{t+1} - (P/R)^{1/2}P^{-1/2}\sum_{t=R}^{T}x_{1,t}'B_1(t)H_1(t) \\
&\quad - (P/R)^{1/2}P^{-1/2}\sum_{t=R}^{T}(x_{3,t} - Ex_{3,t})'(JB_1(t)J'B_3^{-1}(t) - I)\beta_3^* \\
&\quad - R^{-1/2}\sum_{t=R}^{T}(Ex_{3,t})'(JB_1(t)J'B_3^{-1}(t) - I)\beta_3^* \qquad \text{(A.3)}
\end{aligned}
$$

Arguments like those in Theorem 3 suffice to show that $P^{-1/2}\sum_{t=R}^{T}(x_{3,t} - Ex_{3,t})'(JB_1(t)J'B_3^{-1}(t) - I)\beta_3^*$, $P^{-1/2}\sum_{t=R}^{T}x_{1,t}'B_1(t)H_1(t)$, and $P^{-1/2}\sum_{t=R}^{T}u_{t+1}$ are $O_p(1)$. It is clear then that the first three right-hand side terms in Eq. (3) are $O_p((P/R)^{1/2})$.

For the final term note that

$$
\begin{aligned}
&-P^{-1/2}\sum_{t=R}^{T}(Ex_{3,t})'(JB_1(t)J'B_3^{-1}(t) - I)\beta_3^* \\
&= -P^{-1/2}\sum_{t=R}^{T}\left[Ex_{1,t}'B_1(t)R^{-1}\sum_{s=t-R+1}^{t}x_{1,s}1(s > T_B) - 1(t > T_B)\right]\beta_{3,2}^*
\end{aligned}
$$

$$
= \begin{cases}
-\left(R^{-1/2}\sum_{t=R}^{R+T_B}\left(\frac{t-T_B}{R}-1\right)\right)\beta_{3,2}^{*} - R^{-1/2}\sum_{t=R}^{R+T_B}\left(R^{-1}\sum_{s=t-R+1}^{t} z_s - Ez_t\right)' \\
\times F_{1,t}\left((t-T_B)R^{-2}\sum_{s=t-R+1}^{t} z_s - R^{-1}\sum_{s=T_B}^{t} z_s\right)\beta_{3,2}^{*} & T_B \le R \\
-\left(R^{-1/2}\sum_{t=T_B}^{T}\left(\frac{t-T_B}{R}-1\right)\right)\beta_{3,2}^{*} - R^{-1/2}\sum_{t=T_B}^{T}\left(R^{-1}\sum_{s=t-R+1}^{t} z_s - Ez_t\right)' \\
\times F_{1,t}\left((t-T_B)R^{-2}\sum_{s=t-R+1}^{t} z_s - R^{-1}\sum_{s=T_B}^{t} z_s\right)\beta_{3,2}^{*} & T_B > \max[R,P] \\
-\left(R^{-1/2}\sum_{t=T_B}^{T_B+R}\left(\frac{t-T_B}{R}-1\right)\right)\beta_{3,2}^{*} - R^{-1/2}\sum_{t=T_B}^{T_B+R}\left(R^{-1}\sum_{s=t-R+1}^{t} z_s - Ez_t\right)' \\
\times F_{1,t}\left((t-T_B)R^{-2}\sum_{s=t-R+1}^{t} z_s - R^{-1}\sum_{s=T_B}^{t} z_s\right)\beta_{3,2}^{*} & R < T_B < P
\end{cases}
$$

where $F_{1,t} = (R^{-1}\sum_{s=t-R+1}^{t} z_s z_s' - (R^{-1}\sum_{s=t-R+1}^{t} z_s)(R^{-1}\sum_{s=t-R+1}^{t} z_s'))^{-1}$. For each of the three cases the lead term has the highest order. Consider the first for which $T_B \leqslant R$. Carrying through the summation we find that $-(R^{-1/2}\sum_{t=R}^{R+T_B}(((t-T_B)/R)-1) = (T_B^2 - T_B)/2R^{3/2}$. For the latter two we find that $-(R^{-1/2}\sum_{t=T_B}^{T}(t-T_B)/(R-1)) = (P_B(2R-P_B)-P_B)/2R^{3/2}$ and $-(R^{-1/2}\sum_{t=T_B}^{T_{B+R}}(t-T_B)/(R)-1) = (R^2-R)/2R^{3/2}$. Squaring these terms and keeping in mind that the square of the first three terms in Eq. (3) are $O_p(P/R)$ provides the desired result.

(b) Consider the case in which $1/2 \leqslant \lim R/T \leqslant 1$. Since $2/3 \le (1-(P/R)^2/3)^{-1} \le 1$ it suffices to show that $(P^{-1/2}\sum_{t=R}^{T}\hat{u}_{1,t+1})^2$ satisfies cases (i)–(iii). Adding and subtracting terms we obtain the expansion

$$
P^{-1/2}\sum_{t=R}^{T}\hat{u}_{1,t+1} = P^{-1/2}\sum_{t=R}^{T}u_{t+1} - P^{-1/2}\sum_{t=R}^{T}x_{1,t}'B_1(t)H_1(t) - P^{-1/2}\sum_{t=R}^{T}x_{3,t}'(JB_1(t)J'B_3^{-1}(t)-I)\beta_3^{*}
$$

$$= P^{-1/2}\sum_{t=R}^{T} u_{t+1} - P^{-1/2}\sum_{t=R}^{T} x'_{1,t}B_1(t)H_1(t)$$
$$- P^{-1/2}\sum_{t=R}^{T}(x_{3,t} - Ex_{3,t})'(JB_1(t)J'B_3^{-1}(t) - I)\beta_3^*$$
$$- P^{-1/2}\sum_{t=R}^{T}(Ex_{3,t})'(JB_1(t)J'B_3^{-1}(t) - I)\beta_3^*.$$

Arguments like those in Theorem 3 suffice to show that $P^{-1/2}\sum_{t=R}^{T}(x_{3,t} - Ex_{3,t})'(JB_1(t)J'B_3^{-1}(t) - I)\beta_3^*$, $P^{-1/2}\sum_{t=R}^{T}x'_{1,t}B_1(t)H_1(t)$, $P^{-1/2}\sum_{t=R}^{T}u_{t+1}$ and are $O_p(1)$. We must then show that the latter term satisfies (i)–(iii).

For the final term note that

$$- P^{-1/2}\sum_{t=R}^{T}(Ex_{3,t})'(JB_1(t)J'B_3^{-1}(t) - I)\beta_3^*$$

$$= - P^{-1/2}\sum_{t=R}^{T}\left[Ex'_{1,t}B_1(t)R^{-1}\sum_{s=t-R+1}^{t} x_{1,s}1(s>T_B) - 1(t>T_B)\right]\beta_{33}^*$$

$$= \begin{cases} -\left(P^{-1/2}\sum_{t=T_B}^{T}\left(\frac{t-T_B}{R}-1\right)\right)\beta_{3,2}^* - P^{-1/2}\sum_{t=T_B}^{T}\left(R^{-1}\sum_{s=t-R+1}^{t} z_s - Ez_t\right)' & \\ \times F_{1,t}\left((t-T_B)R^{-2}\sum_{s=t-R+1}^{t} z_s - R^{-1}\sum_{s=T_B}^{t} z_s\right)\beta_{3,2}^* & T_B \geq R \\ -\left(P^{-1/2}\sum_{t=R}^{R+T_B}\left(\frac{t-T_B}{R}-1\right)\right)\beta_{3,2}^* - P^{-1/2}\sum_{t=R}^{R+T_B}\left(R^{-1}\sum_{s=t-R+1}^{t} z_s - Ez_t\right)' & \\ \times F_{1,t}\left((t-T_B)R^{-2}\sum_{s=t-R+1}^{t} z_s - R^{-1}\sum_{s=T_B}^{t} z_s\right)\beta_{3,2}^* & T_B < \min[R,P] \\ -\left(P^{-1/2}\sum_{t=R}^{T}\left(\frac{t-T_B}{R}-1\right)\right)\beta_{3,2}^* - P^{-1/2}\sum_{t=R}^{T}\left(R^{-1}\sum_{s=t-R+1}^{t} z_s - Ez_t\right)' & \\ \times F_{1,t}\left((t-T_B)R^{-2}\sum_{s=t-R+1}^{t} z_s - R^{-1}\sum_{s=T_B}^{t} z_s\right)\beta_{3,2}^* & P < T_B < R \end{cases}$$

where $F_{1,t} = (R^{-1}\sum_{s=t-R+1}^{t} z_s z_s' - (R^{-1}\sum_{s=t-R+1}^{t} z_s)(R^{-1}\sum_{s=t-R+1}^{t} z_s'))^{-1}$. For each of the three cases the lead term has the highest order. Consider the first for which $T_B \geq R$. Carrying through the summation we find that $-(P^{-1/2}\sum_{t=T_B}^{T}(((t-T_B)/R)-1) = (P_B(2R-P_B)-P_B)/2R^{1/2}$. For the latter two are find that $-(P^{-1/2}\sum_{t=R}^{R+T_B}(((t-T_B)/R)-1) = (T_B^2-T_B)/2RP^{1/2}$ and $-(P^{-1/2}\sum_{t=R}^{T}(((t-T_B)/R)-1) = (P(2T_B-P)-P)/2RP^{1/2}$. Squaring these terms provides the desired result.

Theorem 6: Maintain Assumptions 1 and 2′ and consider the fixed scheme. (i) If $R \geqslant T_B$, then $V_F = \mathrm{O_p}(PT_B^2/RT)$, and (ii) if $R \leqslant T_B$, then $V_F = \mathrm{O}_p(RP_B^2/PT)$.

Proof of Theorem 6: It is straightforward to show that $P^{-1}\sum_{t=R}^{T}\hat{u}_{1,t+1}^2 \to_p \omega > 0$ and hence we proceed immediately to the numerator. Adding and subtracting terms we obtain the expansion

$$\left(\frac{R}{T}\right)^{1/2} P^{-1/2}\sum_{t=R}^{T}\hat{u}_{1,t+1} = \left(\frac{R}{T}\right)^{1/2} P^{-1/2}\sum_{t=R}^{T} u_{t+1} - \left(\frac{R}{T}\right)^{1/2} \times P^{-1/2}\sum_{t=R}^{T} x_{1,t}' B_1(t) H_1(t) \left(\frac{R}{T}\right)^{1/2} P^{-1/2} \times \sum_{t=R}^{T} x_{3,t}'(JB_1(t)J'B_3^{-1}(t) - I)\beta_3^*.$$

That the first three terms are $O_p(1)$ follows arguments like those in Theorem 3. We must therefore show that the third term is the appropriate order.

Since the fixed scheme is being used the third term can be rewritten as

$$\left(\frac{R}{T}\right)^{1/2} P^{-1/2}\sum_{t=R}^{T} x_{3,t}'(JB_1(t)J'B_3^{-1}(t) - I)\beta_3^* = \left(\frac{RP}{T}\right)^{1/2} \left(P^{-1}\sum_{t=R}^{T} x_{3,t}'\right)(JB_1(R)J'B_3^{-1}(R) - I)\beta_3^*$$

We must consider separately the cases in which $R > T_B$ or $R \leqslant T_B$. Doing so we find that

$$\left(\frac{RP}{T}\right)^{1/2} \left(P^{-1}\sum_{t=R}^{T} x'_{3,t}\right)(JB_1(R)J'B_3^{-1}(R) - I)\beta_3^*$$

$$= \begin{cases} -\left(\frac{R}{PT}\right)^{1/2} P_B & T_B > R \\ -\left(\frac{P}{RT}\right)^{1/2} T_B + \left(\frac{RP}{T}\right)^{1/2} (1 - T_B/R) \left(R^{-1} \sum_{t=1}^{R-1} z_t - P^{-1} \sum_{t=R}^{T} z_t \right)' & \\ F_1 \left(R^{-1} \sum_{t=1}^{R-1} z_t - (R - T_B)^{-1} \sum_{t=T_B}^{R-1} z_t \right) & T_B \leq R \end{cases}$$

where $F_1 = (R^{-1}\sum_{s=1}^{R} z_s z_s' - (R^{-1}\sum_{s=1}^{R} z_s)(R^{-1}\sum_{s=1}^{R} z_s'))^{-1}$. The lead term in each of the above cases is the higher order term. Cases (i) and (ii) follow from squaring these terms.

STEIN-RULE ESTIMATION AND GENERALIZED SHRINKAGE METHODS FOR FORECASTING USING MANY PREDICTORS

Eric Hillebrand and Tae-Hwy Lee

ABSTRACT

We examine the Stein-rule shrinkage estimator for possible improvements in estimation and forecasting when there are many predictors in a linear time series model. We consider the Stein-rule estimator of Hill and Judge (1987) that shrinks the unrestricted unbiased ordinary least squares (OLS) estimator toward a restricted biased principal component (PC) estimator. Since the Stein-rule estimator combines the OLS and PC estimators, it is a model-averaging estimator and produces a combined forecast. The conditions under which the improvement can be achieved depend on several unknown parameters that determine the degree of the Stein-rule shrinkage. We conduct Monte Carlo simulations to examine these parameter regions. The overall picture that emerges is that the Stein-rule shrinkage estimator can dominate both OLS and principal components estimators within an intermediate range of the signal-to-noise ratio. If the signal-to-noise ratio is low, the PC estimator is superior.

30th Anniversary Edition
Advances in Econometrics, Volume 30, 171–196

ISSN: 0731-9053/doi:10.1108/S0731-9053(2012)0000030011

If the signal-to-noise ratio is high, the OLS estimator is superior. In out-of-sample forecasting with AR(1) predictors, the Stein-rule shrinkage estimator can dominate both OLS and PC estimators when the predictors exhibit low persistence.

Keywords: Stein-rule; shrinkage; risk; variance-bias tradeoff; OLS; principal components

JEL classification: C1; C2; C5

INTRODUCTION

Recent contributions to the forecasting literature consider many predictors in data-rich environments and principal components (PCs), such as Bai (2003), Bai and Ng (2006, 2008), Bair, Hastie, Paul, and Tibshirani (2006), Hillebrand, Huang, Lee, and Li (2011), Huang and Lee (2010), Inoue and Kilian (2008), Stock and Watson (2002, 2006, 2011), among others. In particular, Stock and Watson (2011) note that many forecasting models in this environment can be written in a unified framework called the shrinkage representation. Although the notion of the generalized shrinkage representation can be found in much earlier publications (e.g., Judge and Bock, 1978), interest in shrinkage has been revived in the recent literature on out-of-sample (OOS) forecasting.

The issue of forecasting using many predictors was discussed earlier in econometrics and statistics under the subject heading of ill-conditioned data or multicolinearity. In particular, Hill and Judge (1987) studied "improved prediction in the presence of multicolinearity." They examined possible improvements in estimation and forecasting when there are many predictors in a linear regression model. The Stein-rule estimator proposed in their paper shrinks the unrestricted unbiased ordinary least squares (OLS) estimator toward a restricted biased principal component (PC) estimator.

Improvements are usually measured employing a risk function of the squared forecast error loss. While the asymptotic risk functions for the OLS and PC estimators are rather easily obtained, the risk of the Stein-rule is complicated as it depends on several unknown parameters and data-characteristics. It is not easy to understand conditions and situations under which improvements can be achieved. We conduct Monte Carlo simulations to shed light on the issue, both in-sample estimation and out-of-sample forecasting. In general, a key feature is that the desired improvement through Stein-rule shrinkage depends on the signal-to-noise ratio, which is affected by multiple determinants. The Stein-rule shrinkage estimator can dominate both OLS and PC estimators within an intermediate range of the signal-to-noise ratio. If the signal-to-noise ratio is low, the PC estimator tends to be superior. If the signal-to-noise ratio is high, the OLS estimator tends to be superior. In out-of-sample forecasting with AR(1) predictors, the Stein-rule shrinkage estimator can dominate both OLS and PC estimators when the predictors have low persistence.

Hill and Fomby (1992) examined the out-of-sample performance of a variety of biased estimation procedures such as ridge regression, principal component regression, and several Stein-like estimators. Their setup of evaluation was out-of-sample prediction in the sense that the out-of-sample data are different from the data used for parameter estimation, but not out-of-sample prediction in the context of the recent time series forecasting literature.

As the Stein-rule estimator of Hill and Judge (1987) combines OLS and PC estimators, it can be shown that it is a model-averaging estimator and thus produces a combined forecast. In fact, Hansen (2011) shows that the Stein-type shrinkage estimator is a Mallow-type combined estimator. Other papers have studied the relation between Stein-like shrinkage and forecast combinations. Fomby and Samanta (1991) use the Stein-rule for directly combining forecasts. Clark and McCracken (2009) examine the properties of combined forecasts of two nested models and note that their combined forecast is a Stein-type shrinkage forecast. Hence, the shrinkage principle provides insights not only into how to solve the issues of estimation in the presence of multicolinearity and forecasting using many predictors, but also how forecast combinations in the sense of Bates and Granger (1969) yield improvements.

The paper is organized as follows. Section "Shrinkage representation" presents the shrinkage representation for forecasting using principal components. In Section "Principal component model" we consider the OLS and PC estimators and their asymptotic risk of the squared error loss. In Section "Stein-rule estimator", the Stein-rule shrinkage estimator that combines the OLS and PC estimators is presented. Sections "In-sample performance of Stein-rule shrinkage estimator" and "Out-of-sample performance of Stein-rule shrinkage estimator" present Monte Carlo analysis for in-sample and out-of-sample performance of these three estimators – OLS, PC, and the Stein-rule estimators. Finally, Section "Concluding remarks" provides some concluding remarks.

SHRINKAGE REPRESENTATION

This section uses Stock and Watson's (2011) notation. Let the time series under study be denoted by y_t and let P_{it}, $i = 1, \ldots, K$, be a set of K orthonormal predictors such that $P'P/T = I_K$. These predictors can be thought of as the principal components of a possibly large dataset X_{t-1}. The statistical model is

$$y_t = \delta' P_{t-1} + \varepsilon_t \tag{1}$$

where $\delta \in \mathbb{R}^K$ is a parameter vector and ε_t is some error with mean zero and variance σ^2. Both y_t and P_t are assumed to have sample mean zero. Let $\tilde{y}_{T+1|T}$ be the forecast of y at time $T+1$ given information through time T. The theorems in Stock and Watson (2011) show that an array of forecasting methods, namely Normal Bayes, Bayesian Model Averaging, Empirical Bayes, and Bagging, have a *shrinkage representation*

$$\tilde{y}_{T+1|T} = \sum_{i=1}^{K} \psi(\kappa t_i)\hat{\delta}_i P_{iT} + o_P(1) \tag{2}$$

where $\hat{\delta}_i = T^{-1}\sum_{t=1}^{T} P_{i,t-1} y_t$ is the OLS estimator of δ_i, $t_i = \sqrt{T}\hat{\delta}_i/\hat{\sigma}$ is the t-statistic for $\hat{\delta}_i$, $\hat{\sigma}^2 = \sum_{t=1}^{T}(y_t - \hat{\delta}' P_{t-1})^2/(T-K)$ is the consistent estimator of σ^2, ψ is a function that is specific to a forecasting method, and κ is a constant that is specific to a forecasting method. For example, the shrinkage representation of the OLS estimator is $\psi(\kappa t_i) = 1$ for all i. A pretest estimator has shrinkage representation $\psi(\kappa t_i) = \mathbb{1}_{\{|t_i| > t_c\}}$ for some critical value t_c. The principal components estimator that retains the first K_1 principal components and discards the others has shrinkage representation $\psi(\kappa t_i) = 1$ for $i \in \{1, \ldots, K_1\}$ and $\psi(\kappa t_i) = 0$ else. See also Judge and Bock (1978, p. 231) and Hill and Fomby (1992, p. 6) for a general representation of a family of minimax shrinkage estimators.

PRINCIPAL COMPONENT MODEL

This section follows Hill and Judge (1987, 1990) and Hill and Fomby (1992), with adapted notation. Let the model in terms of the original predictor X be

$$y = X\beta + \varepsilon \tag{3}$$

where y is a $T \times 1$ time series, X is a $T \times K$ matrix of K predictors, β is a $K \times 1$ parameter vector, and ε is a $T \times 1$ error time series with the conditional mean zero $\mathbb{E}(\varepsilon|X) = 0$ and conditional variance $\mathbb{E}(\varepsilon\varepsilon'|X) = \sigma^2 I_K$. Note that we do not assume normality of ε in this section, while we generate it from the normal distribution in our simulation study in Sections "In-sample performance of Stein-rule shrinkage estimator" and "Out-of-sample performance of Stein-rule shrinkage estimator". Our interest is to forecast y when the number K of predictors in X is large. The location vector β is unknown and the objective is to estimate it by $\beta(y, X)$. We consider three estimators for β in this paper: (i) the OLS estimator denoted $\hat{\beta}$, (ii) the principal component (PC) estimator denoted $\hat{\beta}^*$, and (iii) the Stein-like combined estimator of $\hat{\beta}$ and $\hat{\beta}^*$, which is to be denoted as $\tilde{\beta}$ in the next section. In this section we examine the sampling properties of $\hat{\beta}$ and $\hat{\beta}^*$ in terms of the asymptotic risk under the weighted squared error loss. In Sections "In-sample performance of Stein-rule shrinkage estimator" and "Out-of-sample performance of Stein-rule shrinkage estimator" we compare them with the Stein-like combined estimator $\tilde{\beta}$.

The sampling performance of an estimator $\beta(y, X)$ is evaluated by its risk function, the expected weighted squared error loss with weight Q,

$$\text{Risk}(\beta, \beta(y, X), Q) = \mathbb{E}[(\beta(y, X) - \beta)' Q (\beta(y, X) - \beta)] \tag{4}$$

As we will examine the performance of the Stein-like estimator in dynamic models for forecasting with weakly dependent time series, the predictor matrix X is treated as stochastic. Hence, the expectation in Eq. (4) is taken over the joint probability law of (y, X). In this section we compute the weighted quadratic risk with a weight $Q = X'X$, which gives the squared conditional prediction error risk. In Sections "In-sample performance of Stein-rule shrinkage estimator" and "Out-of-sample performance of Stein-rule shrinkage estimator" we also consider a weight $Q = I_K$. The asymptotic risks of $\hat{\beta}$ and $\hat{\beta}^*$ are computed below based on the asymptotic covariances of $\hat{\beta}$ and $\hat{\beta}^*$.

The OLS estimator

$$\hat{\beta} = (X'X)^{-1} X'y \tag{5}$$

conditional on X, has the asymptotic sampling property

$$\sqrt{T}(\hat{\beta} - \beta)\Big|_X \xrightarrow{d} \mathcal{N}(0, \sigma^2 (X'X)^{-1}) \tag{6}$$

The asymptotic quadratic risk weighted with $Q = X'X$ of the OLS estimator $\hat{\beta}$ is

$$\begin{aligned}\text{Risk}(\beta, \hat{\beta}, X'X) &= \mathbb{E}\{(\hat{\beta}-\beta)'X'X(\hat{\beta}-\beta)\}\\ &= \text{tr } \mathbb{E}\{X'X(\hat{\beta}-\beta)(\hat{\beta}-\beta)'\}\\ &= \text{tr } \mathbb{E}\{X'X\mathbb{E}[(\hat{\beta}-\beta)(\hat{\beta}-\beta)'|X]\}\\ &= \text{tr } \mathbb{E}\{X'X\sigma^2(X'X)^{-1}\}\\ &= \text{tr}(\sigma^2 I_K)\\ &= K\sigma^2\end{aligned} \tag{7}$$

Since the bias $\mathbb{E}(\hat{\beta}-\beta|X)=0$ conditional on X, the risk contains only a variance component.

Turning to the PC estimator $\hat{\beta}^*$, let V be the $K \times K$ matrix of eigenvectors of $X'X = TV\Lambda V'$, where Λ is the diagonal matrix of eigenvalues in descending order. Then, $V'V = I_K$ and

$$y = X\beta + \varepsilon = XV\Lambda^{-\frac{1}{2}}\Lambda^{\frac{1}{2}}V'\beta + \varepsilon = P\delta + \varepsilon, \qquad P = XV\Lambda^{-\frac{1}{2}},\ \delta = \Lambda^{\frac{1}{2}}V'\beta \tag{8}$$

This is the principal components regression model; P contains the principal components of X, and $\hat{\delta} = (P'P)^{-1}P'y = T^{-1}P'y$ can be estimated either from the principal components or as $\hat{\delta} = \Lambda^{\frac{1}{2}}V'\hat{\beta}$ from the OLS estimator of β. So far, the principal components model is equivalent to the original model. When X has a large degree of colinearity, the eigenvalues in Λ vary greatly in magnitude, and some are close to zero. Then, the number of components is decomposed into $K = K_1 + K_2$, where K_1 is the number of eigenvalues that are relatively large and K_2 is the number of eigenvalues that are relatively close to zero. The K_2 principal components that correspond to the small eigenvalues are discarded; the remaining K_1 principal components are kept. The model becomes

$$y = P\delta + \varepsilon = (P_1\ P_2)\begin{pmatrix}\delta_1\\ \delta_2\end{pmatrix} + \varepsilon = P_1\delta_1 + P_2\delta_2 + \varepsilon \tag{9}$$

$$= X(V_1V_2)\Lambda^{-\frac{1}{2}}\Lambda^{\frac{1}{2}}(V_1V_2)'\beta + \varepsilon = XV_1\Lambda_1^{-\frac{1}{2}}\Lambda_1^{\frac{1}{2}}V_1'\beta + XV_2\Lambda_2^{-\frac{1}{2}}\Lambda_2^{\frac{1}{2}}V_2'\beta + \varepsilon \tag{10}$$

where Λ_1 and Λ_2 are the $K_1 \times K_1$ and $K_2 \times K_2$ diagonal matrices, respectively, that contain the corresponding eigenvalues, and $P_2\delta_2 = XV_2\Lambda_2^{-\frac{1}{2}}\Lambda_2^{\frac{1}{2}}V_2'\beta$ is deleted. Therefore, principal components regression with deleted components is equivalent to OLS estimation with the restriction

$$\delta_2 = R\beta = \Lambda_2^{\frac{1}{2}}V_2'\beta = 0, \tag{11}$$

where $R = \Lambda_2^{\frac{1}{2}} V_2'$ imposes K_2 linear restrictions on β. Note that $R = \Lambda_2^{\frac{1}{2}} V_2'$ is stochastic depending on X, and the risk of the restricted estimator is the expected loss with expectation taken over (y, X).

The principal components estimator of δ with K_2 deleted components, corresponding to the restrictions $\delta_2 = 0$, is

$$\hat{\delta}_1 = (P_1' P_1)^{-1} P_1' y = T^{-1} P_1' y. \tag{12}$$

The asymptotic distribution conditional on X is

$$\sqrt{T}(\hat{\delta}_1 - \delta_1)\Big|_X \xrightarrow{d} \mathcal{N}(0, \sigma^2 I_{K_1}). \tag{13}$$

The estimator $\hat{\delta}_1$ and setting $\delta_2 = 0$ result in the fit

$$y = P_1 \hat{\delta}_1 + \hat{\varepsilon} = X V_1 \Lambda_1^{-\frac{1}{2}} \hat{\delta}_1 + \hat{\varepsilon}, \tag{14}$$

and the principal components estimator of β is therefore

$$\hat{\beta}^* = V_1 \Lambda_1^{-\frac{1}{2}} \hat{\delta}_1. \tag{15}$$

This is a special case of the restricted least squares (RLS) estimators explored by Mittelhammer (1985). Fomby, Hill, and Johnson (1978) present an optimality property of $\hat{\beta}^*$ that the trace of the asymptotic covariance matrix of $\hat{\beta}^*$ obtained by deleting K_2 principal components associated with the smallest eigenvalues is at least as small as that for any other RLS estimator with $J \leq K_2$ restrictions. This optimality is in terms of the asymptotic quadratic risk weighted with $Q = I_K$, that is, Risk $(\beta, \hat{\beta}^*, I_K)$.

For forecasting, it is interesting to examine the asymptotic quadratic risk weighted with $Q = X'X$ of the PC estimator $\hat{\beta}^*$.

$$\begin{aligned}
\text{Risk}(\beta, \hat{\beta}^*, X'X) &= \mathbb{E}(\hat{\beta}^* - \beta)' X'X (\hat{\beta}^* - \beta) \\
&= \mathbb{E}(V_1 \Lambda_1^{-\frac{1}{2}} \hat{\delta}_1 - \beta)' (T V \Lambda V') (V_1 \Lambda_1^{-\frac{1}{2}} \hat{\delta}_1 - \beta) \\
&= T \mathbb{E}(V_1 \Lambda_1^{-\frac{1}{2}} \hat{\delta}_1 - \beta)' (V \Lambda^{1/2} \Lambda^{1/2} V') (V_1 \Lambda_1^{-\frac{1}{2}} \hat{\delta}_1 - \beta) \\
&= T \mathbb{E}(\hat{\delta}_1' \Lambda_1^{-\frac{1}{2}} V_1' V \Lambda^{1/2} - \beta' V \Lambda^{1/2}) (\Lambda^{1/2} V' V_1 \Lambda_1^{-\frac{1}{2}} \hat{\delta}_1 - \Lambda^{1/2} V' \beta) \\
&= T \mathbb{E}(\hat{\delta}_1' [I_{K_1} \;\; \mathbf{0}] - \delta') \left(\begin{bmatrix} I_{K_1} \\ \mathbf{0} \end{bmatrix} \hat{\delta}_1 - \delta \right)
\end{aligned}$$

$$\begin{aligned} &= T\mathbb{E}(\hat{\delta}_1 - \delta_1)'(\hat{\delta}_1 - \delta_1) + T\delta_2'\delta_2 \\ &= T \operatorname{tr} \mathbb{E}(\hat{\delta}_1 - \delta_1)(\hat{\delta}_1 - \delta_1)' + T\delta_2'\delta_2 \\ &= T \operatorname{tr} (T^{-1}\sigma^2 I_{K_1}) + T\delta_2'\delta_2 \\ &= K_1\sigma^2 + T\delta_2'\delta_2 \end{aligned} \tag{16}$$

where the first term corresponds to the variance term that declines as K_1 decreases and the second term corresponds to the bias term. The second to the last equality follows from Eq. (13).

To compare the asymptotic risks of the OLS $\hat{\beta}$ estimator and the PC estimator $\hat{\beta}^*$, look at the risk difference

$$\text{Risk}(\beta, \hat{\beta}, X'X) - \text{Risk}(\beta, \hat{\beta}^*, X'X) = K\sigma^2 - (K_1\sigma^2 + T\delta_2'\delta_2) = K_2\sigma^2 - T\delta_2'\delta_2,$$

which is positive when $\delta_2'\delta_2$ is small. This is the case if the restriction in Eq. (11) is reasonable. In that case the OLS estimator $\hat{\beta}$ is dominated by the PC estimator $\hat{\beta}^*$.

STEIN-RULE ESTIMATOR

Hill and Judge (1987, 1990) propose a Stein-rule estimator $\tilde{\beta}$ that shrinks the standard OLS estimator $\hat{\beta}$ toward the principal components estimator $\hat{\beta}^*$:

$$\begin{aligned} \tilde{\beta} &= \hat{\beta}^* + \left(1 - \frac{a\hat{\sigma}^2(T-K)}{\hat{\beta}' R'(R(X'X)^{-1}R')^{-1} R\hat{\beta}}\right)(\hat{\beta} - \hat{\beta}^*) \\ &= \hat{\beta}^* + \psi(\hat{\beta} - \hat{\beta}^*) \\ &= \psi\hat{\beta} + (1-\psi)\hat{\beta}^* \end{aligned} \tag{17}$$

where a is a constant, R is defined from Eq. (11), and the Stein coefficient ψ is the shrinkage from the OLS estimator $\hat{\beta}$ to the PC estimator $\hat{\beta}^*$. Using $R = \Lambda_2^{\frac{1}{2}} V_2'$ and $\hat{\beta} = V\Lambda^{-\frac{1}{2}}\hat{\delta}$, we obtain

$$\tilde{\beta} = V_1\Lambda_1^{-\frac{1}{2}}\hat{\delta}_1 + \left(1 - \frac{1}{T}\frac{a\hat{\sigma}^2(T-K)}{\hat{\delta}'\Lambda^{-\frac{1}{2}}V'V_2\Lambda_2^{\frac{1}{2}}(\Lambda_2^{\frac{1}{2}}V_2'V\Lambda^{-1}V'V_2\Lambda_2^{\frac{1}{2}})^{-1}\Lambda_2^{\frac{1}{2}}V_2'V\Lambda^{-\frac{1}{2}}\hat{\delta}}\right)(V\Lambda^{-\frac{1}{2}}\hat{\delta} - V_1\Lambda_1^{-\frac{1}{2}}\hat{\delta}_1) \tag{18}$$

Using that

$$V'V_2 = \begin{bmatrix} \mathbf{0} \\ I_{K_2} \end{bmatrix}, \quad V_2'V = [\mathbf{0} \; I_{K_2}], \quad \text{and} \quad (X'X)^{-1} = T^{-1}V\Lambda^{-1}V'$$

where I_{K_2} is the $K_2 \times K_2$ identity matrix, we obtain that

$$\begin{aligned} \tilde{\beta} &= \left(1 - \frac{1}{T}\frac{a\hat{\sigma}^2(T-K)}{\hat{\delta}_2'\hat{\delta}_2}\right)(V\Lambda^{-\frac{1}{2}}\hat{\delta} - V_1\Lambda_1^{-\frac{1}{2}}\hat{\delta}_1) + V_1\Lambda_1^{-\frac{1}{2}}\hat{\delta}_1, \\ &= V_1\Lambda_1^{-\frac{1}{2}}\hat{\delta}_1 + \psi(V\Lambda^{-\frac{1}{2}}\hat{\delta} - V_1\Lambda_1^{-\frac{1}{2}}\hat{\delta}_1). \end{aligned} \tag{19}$$

Further rearrangement yields the expression

$$\tilde{\beta} = V_1\Lambda_1^{-\frac{1}{2}}\hat{\delta}_1 + \psi V_2\Lambda_2^{-\frac{1}{2}}\hat{\delta}_2, \tag{20}$$

for the Stein-rule estimator, from which its shrinkage representation can now be read. Since the individual t-statistics of the principal components are given by $t_i = \sqrt{T}\hat{\delta}_{2,i}/\hat{\sigma}$, the coefficient of the K_2 terms in $\hat{\delta}$ corresponding to the discarded principal components can be written as

$$1 - \frac{1}{T}\frac{a\hat{\sigma}^2(T-K)}{\hat{\delta}_2'\hat{\delta}_2} = 1 - \frac{a(T-K)}{\sum_{K_1+1}^{K} t_i^2} = 1 - \frac{a(T-K)}{K_2 F_{K_2,T-K}}, \tag{21}$$

where $F_{K_2,T-K} = \sum_{K_1+1}^{K} t_i^2/K_2$ is the test statistic for $H_0 : \delta_2 = 0$, the restriction of Eq. (11). Note that

$$F_{K_2,T-K} = \frac{\sum_{K_1+1}^{K} t_i^2}{K_2} = \frac{T\hat{\delta}_2'\hat{\delta}_2/K_2}{\hat{\sigma}^2} = \frac{\text{signal from } K_2 \text{ discarded variables}}{\text{noise}}.$$

The Stein coefficient function ψ in the shrinkage representation of Section "shrinkage representation" is given by

$$\psi_i = \begin{cases} 1, & i \in \{1, \ldots, K_1\}, \\ \psi, & i \in \{K_1 + 1, \ldots, K\}. \end{cases} \tag{22}$$

The asymptotic quadratic risk weighted with $Q = X'X$ for the Stein estimator $\tilde{\beta}$

$$\text{Risk}\,(\beta, \tilde{\beta}, X'X) = \mathbb{E}[(\tilde{\beta} - \beta)'\,X'X(\tilde{\beta} - \beta)] \tag{23}$$

can be calculated here but it is rather complicated as it depends on parameters such as β, σ, a, and on data characteristics such as T, K, K_1, X (with X determining Λ, V). Hence, we use Monte Carlo analysis in the next two sections to examine the risk of $\tilde{\beta}$ in comparison with those of $\hat{\beta}$ and $\hat{\beta}^*$.

IN-SAMPLE PERFORMANCE OF STEIN-RULE SHRINKAGE ESTIMATOR

We conduct Monte Carlo analysis to compare the risk of $\tilde{\beta}$ with those of $\hat{\beta}$ and $\hat{\beta}^*$. The risk of the Stein-rule estimator depends on β, σ, a, T, K, K_1, X. For the risk comparisons we fix $T = 200$ and $K = 50$ while we vary β, σ, a, K_1, and X.

Simulation Design

The elementary model to be studied is the linear equation

$$y = X\beta + \varepsilon, \tag{24}$$

where y is a $T \times 1$ vector, X is a $T \times K$ matrix of regressors, ε is a $T \times 1$ random vector drawn from $\mathcal{N}(0, \sigma^2)$ distribution, $\beta \in \mathbb{R}^{K \times 1}$, and $\sigma \in \mathbb{R}_+$.

We compare the performance of the Stein-rule estimator in-sample with the standard OLS estimator and the principal components estimator and employ the following simulation design. We draw a matrix X_0 of $\mathcal{N}(0, 1)$ random variables of dimensions $T \times K$, $T = 200$, and $K = 50$. We aim to impose different eigenvalue structures on the regressor matrix X in the spirit of Hill and Judge (1987). To this end, we singular-value decompose X_0 into

$$X_0 = U\Lambda_0^{\frac{1}{2}}V'$$

and discard the diagonal matrix $\Lambda_0^{\frac{1}{2}}$. The regressor matrix X is then constructed as

$$X = U\Lambda^{\frac{1}{2}}V', \tag{25}$$

where Λ is constructed according to three different scenarios.

- The singular values are constant.

$$\Lambda^{\frac{1}{2}} = \mathrm{diag}(2, \ldots, 2). \tag{26}$$

- The singular values are linearly decreasing from 5 to 1.
- The singular values are exponentially decreasing from 5 to 1.

$$\operatorname{diag}(\Lambda^{\frac{1}{2}}) = 1 + 4e^{-0.10k}, \quad k = 1, \ldots, K. \tag{27}$$

In the data-generating process, we consider different scenarios for the variance σ^2 of the error process. In particular, we set

$$\sigma \in \{1, 3, 5, 7\}. \tag{28}$$

The data-generating parameter vector β is set to

$$\beta = \left[\frac{L}{K}\right]_{k\in\{1,\ldots,K\}} \tag{29}$$

such that its direction in parameter space is $(1/K)_k$ and its length is L. We consider different scenarios for the length L of the vector, in particular,

$$L \in \{0, 1, 2, 3\}. \tag{30}$$

Table 1 lists the population-R^2 for the different resulting scenarios, where

$$\text{population - } R^2 = \frac{\beta'\mathbb{E}(X'X)\beta}{\beta'\mathbb{E}(X'X)\beta + T\sigma^2}.$$

Our simulation design considers only a limited region of the space of simulation design parameters (T, K, L, σ, Λ). Estimating a response surface for a larger region could give some more indication on the range of data sets where gains from Stein-rule estimation can be expected. This is left for future research.

The performance of the estimators is measured in terms of their risk. The general risk function considered is

$$\text{Risk}\,(\beta, \beta(y, X), Q) = \mathbb{E}[(\beta\,(y, X) - \beta)'Q\,(\beta\,(y, X) - \beta)],$$

as shown in Eq. (4). We study the particular case where $Q = I$, which results in the standard mean squared error considered in James and Stein (1961) and Judge and Bock (1978) and the second case where $Q = X'X$ as considered in Hill and Fomby (1992), Hill and Judge (1987, 1990), Judge and Bock (1978). This risk measure can be interpreted as the square of the distance $(X\beta - X\hat{\beta})$ of the fitted value from the signal part of y.

There are a few estimator-specific settings to consider as well, in particular the number K_1 of principal components for the principal components estimator and the value of the parameter a in the Stein-rule shrinkage

Table 1. Population R^2 for the Different Simulation Scenarios.

Eigenvalues	Constant				Linear				Exponential			
L/σ^2	1	3	5	7	1	3	5	7	1	3	5	7
0	0	0	0	0	0	0	0	0	0	0	0	0
1	0.020	0.007	0.004	0.003	0.049	0.017	0.010	0.007	0.019	0.007	0.004	0.003
2	0.074	0.026	0.016	0.011	0.172	0.065	0.040	0.029	0.073	0.026	0.016	0.011
3	0.153	0.057	0.035	0.025	0.319	0.135	0.086	0.063	0.151	0.056	0.034	0.025

estimator. We consider $K_1 \in \{1, 5, 10, 20\}$ and $a \in (0, 1)$. The two numbers interact through the bounds for a given in Hill and Judge (1987, p. 87) and Judge and Bock (1978, p. 193):

$$0 \leq a \leq \frac{2(K - K_1 - 2)}{T - K + 2}. \tag{31}$$

Here, for $K_1 \in \{1, 5, 10, 20\}$, we obtain

$$0 \leq a \leq 0.62, 0.57, 0.50, 0.37,$$

so that we expect the region for a in which the Stein-rule shrinkage estimator performs better than OLS and PC estimators to move toward the origin as the number of components increases.

Choosing the Number of Principal Components

Selecting the number of principal components is a problem that has spawned a large literature (see, e.g., Anderson, 2003; Bai & Ng, 2002; Hallin & Liska, 2007; Onatski, 2009). In this paper, we restrict ourselves to studying the behavior of the Stein-rule estimator for a set of number of components, including the one-factor model, few components ($K_1 = 5$), a moderate number ($K_1 = 10$), and many factors ($K_1 = 20$). Recall that the number of regressors is $K = 50$.

Figs. 1–3 show the risk of the three estimators, Stein-rule shrinkage, OLS, and PC, as functions of the parameter a of the Stein-rule shrinkage estimator. Since the OLS and PC estimators do not depend on this parameter, they are constants in the graphs. The risk of the OLS estimator is depicted by a dotted line; the risk of the PC estimator is shown as a dashed line. The risk of the Stein-rule shrinkage estimator is shown as connected dots. The left panel of four plots in each figure shows the MSE risk ($Q = I$);

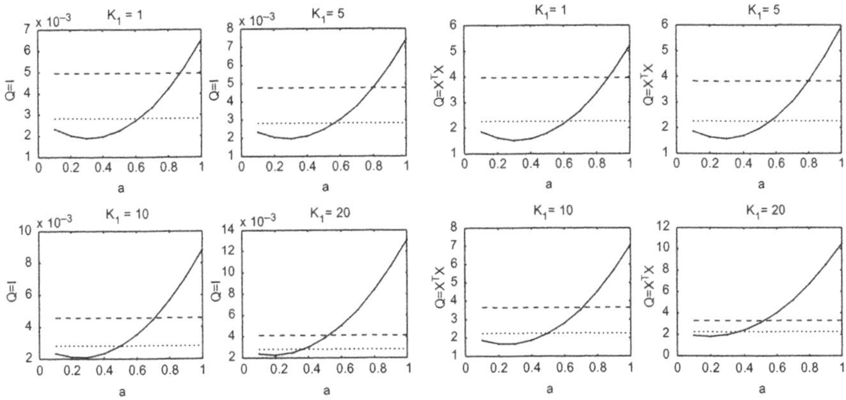

Fig. 1. Risk($\beta(y, X)$) = $\mathbb{E}[(\beta(y, X) - \beta)' Q(\beta(y, X) - \beta)]$ as Function of a. Left Panel: $Q = I$ (MSE), Right Panel: $Q = X'X$. The Data-Generating Singular Values are Constant and Equal to Two. Other Parameters Are Set to $L = 1$, $\sigma = 3$. The Connected Dots' Line Shows the Performance of the Stein-like Estimator. For Comparison, the Performance of the Standard OLS Estimator is Shown in Dots. The Performance of the Principal Components Estimator is Plotted with Dashes.

the right panel of four plots shows the risk for $Q = X'X$. The four plots show the different scenarios for the number $K_1 \in \{1, 5, 10, 20\}$ of principal components. Each figure shows a different singular value scenario, Fig. 1 shows the case of constant singular values equal to two; Fig. 2 shows the case of linearly decreasing singular values, and Fig. 3 shows the case of exponentially decreasing singular values.

The graphs show that the risk of the Stein-rule follows a parabola in a, which indicates that there is an optimal a, at least in the simulation scenarios considered. Unlike in the case of the original James and Stein (1961) estimator, this optimal a is not analytically known at this point. The minimum of the parabola is moving inward toward the origin as the number K_1 of components increases, as expected. For the scenarios where the singular values are constant and where they are linearly decreasing, the OLS estimator performs generally better than the PC estimator. For exponentially decreasing singular values, the PC estimator often performs better than the OLS estimator. The Stein-rule shrinkage estimator has a greater relative advantage over PC and OLS estimators for small number of components ($K_1 = 1, 5$). For larger K_1, the performance of the Stein-rule shrinkage estimator approaches that of the relatively better estimator

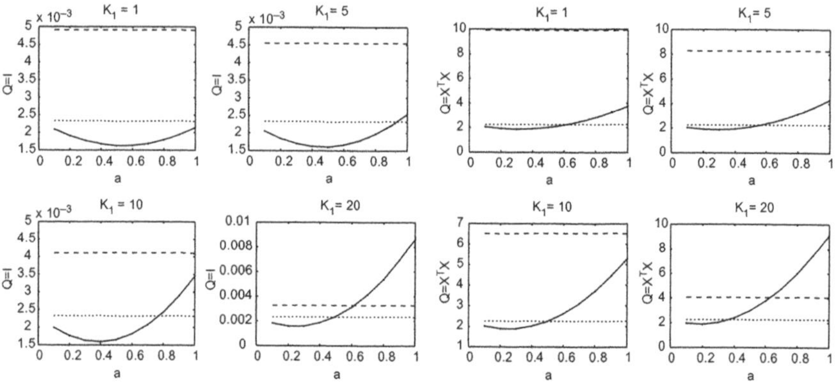

Fig. 2. $\text{Risk}(\beta(y, X)) = \mathbb{E}[(\beta(y, X) - \beta)' Q(\beta(y, X) - \beta)]$ as Function of a. Left Panel: $Q = I$ (MSE), Right Panel: $Q = X'X$. The Data-Generating Singular Values are Linearly decreasing from 5 to 1. Other Parameters are Set to $L = 1$, $\sigma = 3$. The Connected Dots' Line Shows the Performance of the Stein-like Estimator. For Comparison, the Performance of the Standard OLS Estimator is shown in Dots. The Performance of the Principal Components Estimator is Plotted with Dashes.

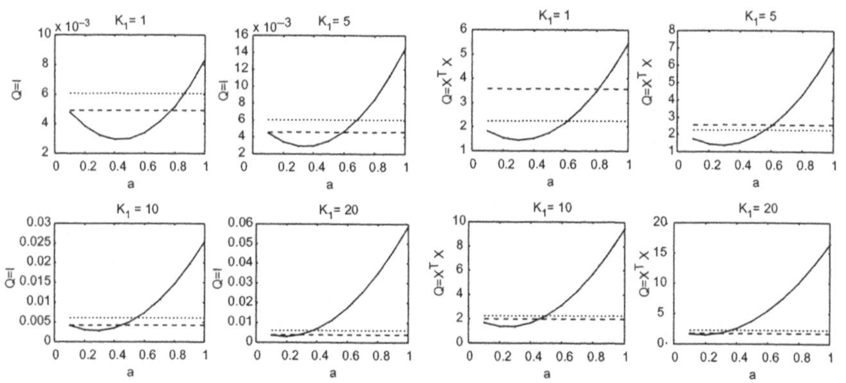

Fig. 3. $\text{Risk}(\beta(y, X)) = \mathbb{E}[(\beta(y, X) - \beta)' Q(\beta(y, X) - \beta)]$ as Function of a. Left Panel: $Q = I$ (MSE), Right Panel: $Q = X'X$. The Data-Generating Singular Values are Exponentially Decreasing from 5 to 1 at a Rate of 0.10. Other Parameters are Set to $L = 1$, $\sigma = 3$. The Connected Dots' Line Shows the Performance of the Stein-like Estimator. For Comparison, the Performance of the Standard OLS Estimator is Shown in Dots. The Performance of the Principal Components Estimator is Plotted with Dashes.

among OLS and PC. Note that the singular-value scenarios considered in this paper do not include values close to zero as in Hill and Judge (1987). We found that for most scenarios of this nature, where a strong degree of multicolinearity is present, the principal components estimator performs better than the Stein-rule shrinkage estimator.

Different Variance Scenarios

Fig. 4–6 report the performance of the estimators for different noise levels

$$\sigma \in \{1, 3, 5, 7\}.$$

The organization of the graphs is the same as described in Section "Choosing the number of principal components". Again, the risk of the Stein-rule shrinkage estimator describes a parabola in a, indicating the existence of an optimal parameter value. For low values of variance, OLS performs better than principal components, and as the noise level increases, the PC estimator outperforms OLS. The Stein-rule shrinkage estimator can outperform both OLS and PC estimators within an intermediate noise range.

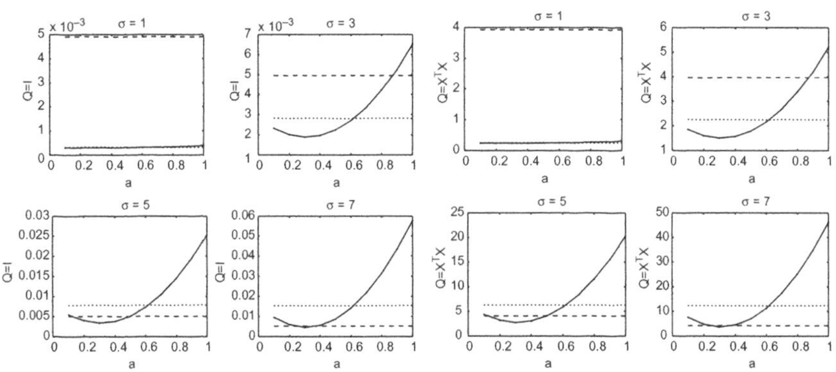

Fig. 4. Risk($\beta(y, X)$) = $\mathbb{E}[(\beta(y, X) - \beta)' Q(\beta(y, X) - \beta)]$ as Function of a. Left Panel: $Q = I$ (MSE), Right Panel: $Q = X'X$. The Data-Generating Singular Values are Constant and Equal to Two. Other Parameters are Set to $L = 1$, $K_1 = 1$. The Connected Dots' Line Shows the Performance of the Stein-like Estimator. For Comparison, the Performance of the Standard OLS Estimator is Shown in Dots. The Performance of the Principal Components Estimator is Plotted with Dashes.

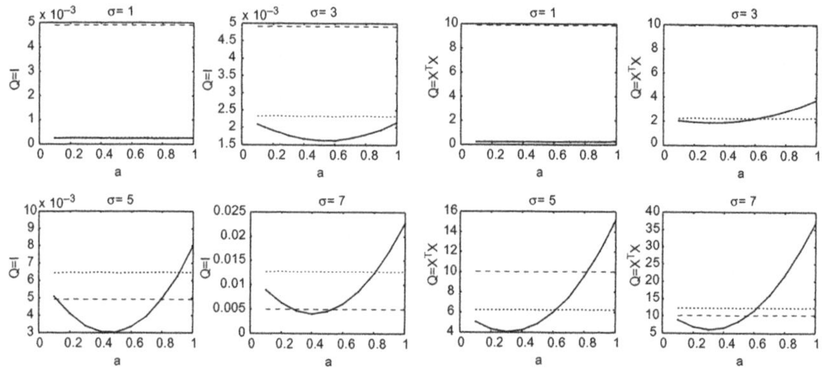

Fig. 5. Risk($\beta(y,X)$) = $\mathbb{E}[(\beta(y,X)-\beta)'Q(\beta(y,X)-\beta)]$ as Function of a. Left Panel: $Q = I$ (MSE), Right Panel: $Q = X'X$. The Data-Generating Singular Values are Linearly Decreasing from 5 to 1. Other Parameters Are Set to $L = 1$, $K_1 = 1$. The Connected Dots' Line Shows the Performance of the Stein-like Estimator. For Comparison, the Performance of the Standard OLS Estimator is Shown in Dots. The Performance of the Principal Components Estimator is Plotted with Dashes.

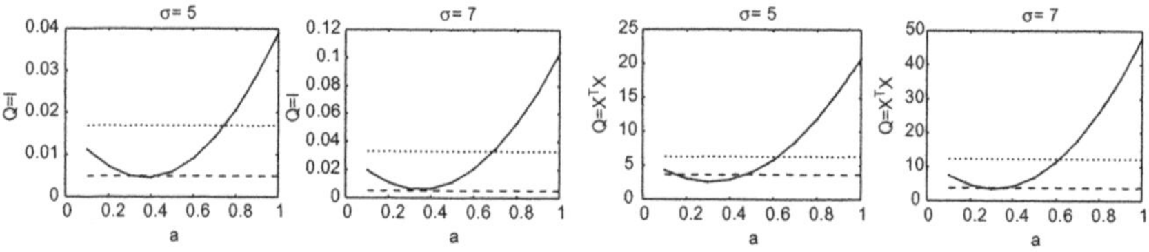

Fig. 6. Risk($\beta(y,X)$) = $\mathbb{E}[(\beta(y,X) - \beta)'Q(\beta(y,X) - \beta)]$ as Function of a. Left Panel: $Q = I$ (MSE), Right Panel: $Q = X'X$. The Data-Generating Singular Values are Exponentially Decreasing from 5 to 1 at a Rate of 0.10. Other Parameters are Set to $L = 1$, $K_1 = 1$. The Connected Dots' Line Shows the Performance of the Stein-like Estimator. For Comparison, the Performance of the Standard OLS Estimator is Shown in Dots. The Performance of the Principal Components Estimator is Plotted with Dashes.

Different Lengths of the Parameter Vector β

Fig. 7–9 display the performance of the estimators for different lengths L of the parameter vector $\beta = L/K$. The four plots of each panel show the risks of the estimators for

$$L \in \{0, 1, 2, 3\}.$$

The organization of the graphs is the same as described in Section "Choosing the number of principal components". If $L = 0$, that is, there is no signal in y, the PC estimator outperforms both OLS and the Stein-rule shrinkage estimators. For large values of L, OLS performs better than the other estimators. On an intermediate range, the Stein-rule shrinkage estimator can outperform both other estimators.

Recall from Eq. (11) that $\delta_2 = \Lambda_2^{\frac{1}{2}} V_2' \beta$ where $\beta = \left[\frac{L}{K}\right]$. Hence, the length L for β determines the length of δ_2. Because $T\delta_2'\delta_2$ is the second term in the asymptotic risk of the PC estimator corresponding to the bias due to the omission of the K_2 principal components, as shown in Eq. (16), a large value of L increases the risk of the PC estimator compared to the risk of the OLS estimator.

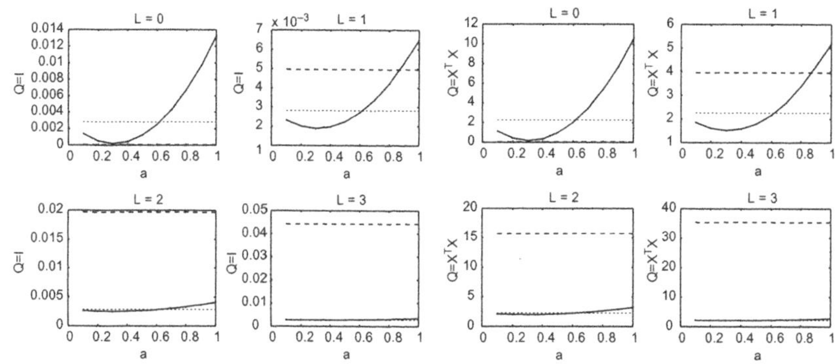

Fig. 7. Risk($\beta(y, X)$) = $\mathbb{E}[(\beta(y, X) - \beta)' Q(\beta(y, X) - \beta)]$ as Function of a. Left Panel: $Q = I$ (MSE), Right Panel: $Q = X'X$. The Data-Generating Singular Values are Constant and Equal to Two. Other Parameters are Set to $K_1 = 1$, $\sigma = 3$. The Connected Dots' Line Shows the Performance of the Stein-like Estimator. For Comparison, the Performance of the Standard OLS Estimator is Shown in Dots. The Performance of the Principal Components Estimator is Plotted with Dashes.

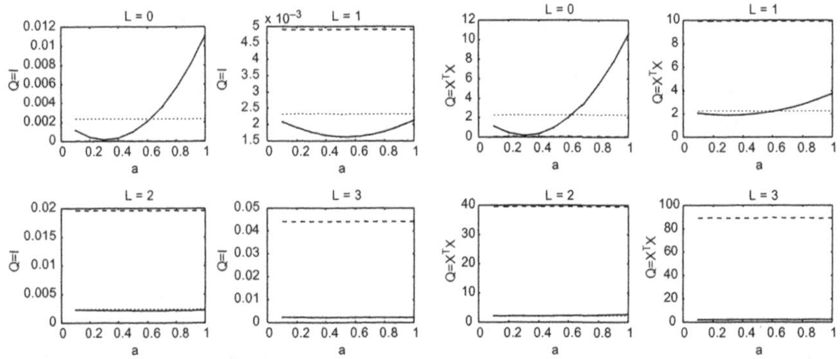

Fig. 8. Risk($\beta(y, X)$) = $\mathbb{E}[(\beta(y, X) - \beta)' Q(\beta(y, X) - \beta)]$ as Function of a. Left Panel: $Q = I$ (MSE), Right Panel: $Q = X'X$. The Data-Generating Singular Values are Linearly Decreasing from 5 to 1. Other Parameters are Set to $K_1 = 1$, $\sigma = 3$. The Connected Dots' Line Shows the Performance of the Stein-like Estimator. For Comparison, the Performance of the Standard OLS Estimator is Shown in Dots. The Performance of the Principal Components Estimator is Plotted with Dashes.

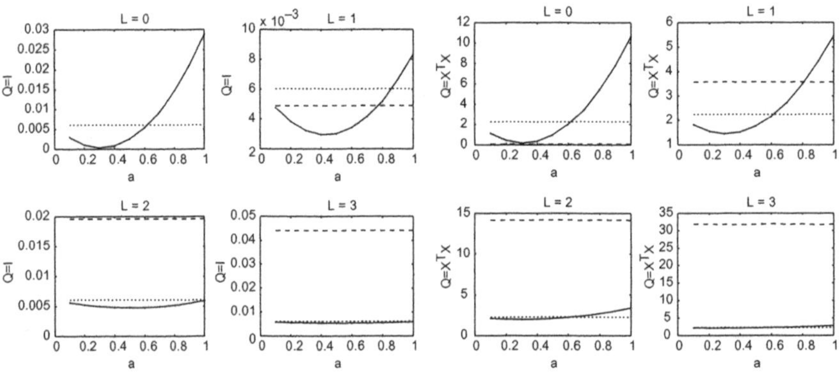

Fig. 9. Risk($\beta(y, X)$) = $\mathbb{E}[(\beta(y, X) - \beta)' Q(\beta(y, X) - \beta)]$ as Function of a. Left Panel: $Q = I$ (MSE), Right Panel: $Q = X'X$. The Data-Generating Singular Values are Exponentially Decreasing from 5 to 1 at a Rate of 0.10. Other Parameters are Set to $K_1 = 1$, $\sigma = 3$. The Connected Dots' Line Shows the Performance of the Stein-like Estimator. For Comparison, the Performance of the Standard OLS Estimator is Shown in Dots. The Performance of the Principal Components Estimator is Plotted with Dashes.

For the Stein-rule estimator, a large value of L increases $T\delta_2'\delta_2$, which in turn will increase the F statistic defined in Eq. (21)

$$F_{K_2,T-K} = \frac{T\hat{\delta}_2'\hat{\delta}_2/K_2}{\hat{\sigma}^2}$$

and hence increases the Stein-rule coefficient ψ and thus reduces the shrinkage from the OLS estimator $\hat{\beta}$ to the PC estimator $\hat{\beta}^*$.

OUT-OF-SAMPLE PERFORMANCE OF STEIN-RULE SHRINKAGE ESTIMATOR

Simulation Design

We assess the out-of-sample (OOS) performance of the Stein-rule shrinkage estimator in two different simulation setups. One is exactly the same as described in Section "Simulation design", only that the forecast performance on $T_2 = 100$ out-of-sample observations is evaluated. The two risk functions considered for the OOS comparison are the mean squared forecast error (MSFE)

$$\text{MSFE}(\beta(y, X)) = \mathbb{E}[(\hat{y} - y)'(\hat{y} - y)], \tag{32}$$

where $\hat{y} = X\beta(y, X)$, and the squared signal-to-prediction distance as considered in (4) with $Q = X'X$,

$$\text{Risk}(\beta, \beta(y, X), X'X) = \mathbb{E}[(\beta(y, X) - \beta)'X'X(\beta(y, X) - \beta)]. \tag{33}$$

The second simulation environment that we study has AR(1) time series in the columns of the regressor matrix X. That is,

$$X = [\{x_{1,t}\}\{x_{2,t}\}, \ldots, \{x_{K,t}\}]_{t\in\{1,\ldots,T\}}, \tag{34}$$

and the individual columns follow

$$x_{k,t} = \phi x_{k,t-1} + \sigma_{X,k}\xi_{t,k}, \quad k = 1, \ldots, K, \xi_{t,k} \sim \mathcal{N}(0, 1). \tag{35}$$

The standard deviations of the AR(1) processes in the columns of X are chosen to correspond to the exponentially decaying sequence employed in Eq. (27):

$$\sqrt{\text{Var} x_{k,t}} = \frac{\sigma_{X,k}}{\sqrt{1 - \phi^2}} = 1 + 4e^{-0.10k}, \quad k = 1, \ldots, K. \tag{36}$$

Thus, $\sigma_{X,k} = \sigma_{X,k}(\phi) = (1 + 4e^{-0.10k})\sqrt{1 - \phi^2}$. Varying ϕ replaces the variance dimension considered in the in-sample study. The standard deviation of the noise in y is set to $\sigma = 3$ and $T_2 = 100$ out-of-sample observations are evaluated.

Note that principal components are linear combinations of the columns of X,

$$P_1 = XV_1\Lambda_1^{-\frac{1}{2}}, \tag{37}$$

and therefore the individual components are, with some coefficients $w_{k,j}$ determined by V_1 and Λ_1,

$$P_{j,t} = \sum_{k=1}^{K} w_{k,j}\, x_{k,t} = \phi \sum_{k=1}^{K} w_{k,j}\, x_{k,t-1} + \sum_{k=1}^{K} w_{k,j}\, \sigma_{X,k}\, \xi_{t,k} = \phi P_{j,t-1} + \eta_{j,t},$$

where $\eta_{j,t} = \sum_{k=1}^{K} w_{k,j}\sigma_{X,k}\xi_{t,k}$. As long as the AR(1) parameter ϕ is the same across all columns of X, the principal components will themselves be AR(1) processes with the same decorrelation length as the individual columns. If different ϕ_k are chosen across the columns, the principal components will be linear combinations of AR(1) processes with different persistence parameters, which can lead to long memory behavior of the components, as described in Granger (1980).

Choosing the Number of Principal Components

Figs. 10 and 11 display the out-of-sample performance of the estimators for different numbers of principal components

$$K_1 \in \{1, 5, 10, 20\}.$$

The organization of the graphs is similar to the one described in Section "Choosing the number of principal components". Instead of different singular-value scenarios, two different simulation designs are considered. Fig. 10 shows the case where the regressor matrix X is drawn from independent $\mathcal{N}(0, 1)$ distributions; Fig. 11 shows the case where the regressors are AR(1) time series. Unlike in the in-sample study, here the relative performance of PC and OLS estimators changes with the number K_1 of principal components. For small numbers, OLS performs better than PC, and for large K_1, PC performs better than OLS. The Stein-rule shrinkage

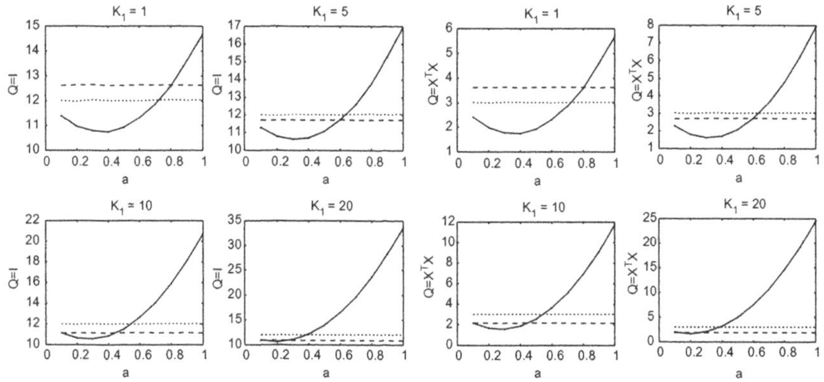

Fig. 10. Left Panel: $\text{MSFE}(\beta(y, X)) = \mathbb{E}[(\hat{y} - y)'(\hat{y} - y)]$, Where $\hat{y}$ is a T_2-Vector of Forecasts of y, as Function of a. Right Panel: $\text{Risk}(\beta, \beta(y, X), X'X) = \mathbb{E}[(\beta(y, X) - \beta)'X'X(\beta(y, X) - \beta)]$ for the T_2-Period Forecast Sample. The Data-Generating Eigenvalues are Exponentially Decreasing from 5 to 1 at a Rate of 0.10. Other Parameters are Set to $L = 1$, $\sigma = 3$. The Connected Dots' Line Shows the Performance of the Stein-like Estimator. For Comparison, the Performance of the Standard OLS Estimator is Shown in Dots. The Performance of the Principal Components Estimator is Plotted with Dashes.

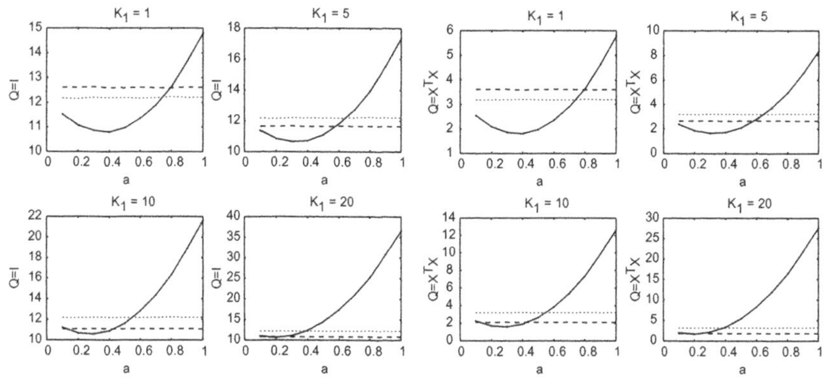

Fig. 11. Left Panel: $\text{MSFE}(\beta(y, X)) = \mathbb{E}[(\hat{y} - y)'(\hat{y} - y)]$, Where $\hat{y}$ Is a T_2-Vector of Forecasts of y, as Function of a. Right Panel: $\text{Risk}(\beta, \beta(y, X), X'X) = \mathbb{E}[(\beta(y, X) - \beta)'X'X(\beta(y, X) - \beta)]$ for the T_2-Period Forecast Sample. The Columns of the Regressor Matrix X are AR(1) Processes. Other Parameters are Set to $L = 1$, $\sigma = 3$. The Connected Dots' Line Shows the Performance of the Stein-like Estimator. For Comparison, the Performance of the Standard OLS Estimator is Shown in Dots. The Performance of the Principal Components Estimator is Plotted with Dashes.

estimator dominates for up to 10 components. There is no obvious difference between the i.i.d. and the AR(1) simulation scenarios.

Different Variance Scenarios

Fig. 12 shows the performance of the estimator when X is drawn from an $\mathcal{N}(0, 1)$ distribution. Similar to the in-sample study, OLS performs best for low noise levels and PC performs best for high noise levels. The Stein-rule shrinkage estimator can outperform both in an intermediate noise range.

Fig. 13 shows the performance for the estimators when the columns of X follow AR(1) dynamics. The four plots in each panel show the situation for different values $\phi \in \{0.30, 0.50, 0.90, 0.99\}$ of the AR-parameter. The standard deviation of the error in the AR model is then set through $\sigma_{X,k}(\phi) = (1 + 4e^{-0.10k})\sqrt{1 - \phi^2}$ such that the standard deviation of the column follows Eq. (36). The figure shows that the Stein-rule shrinkage estimator outperforms OLS and PC estimators in low-persistence scenarios ($\phi = 0.30, 0.50$), whereas in high-persistence scenarios ($\phi = 0.90, 0.99$) the

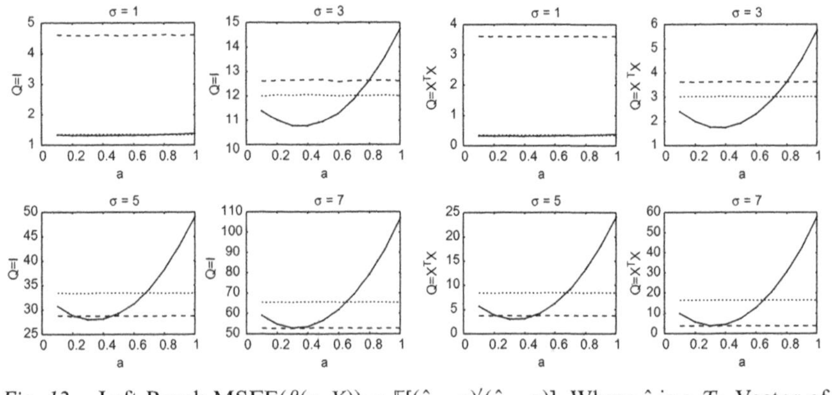

Fig. 12. Left Panel: $\text{MSFE}(\beta(y, X)) = \mathbb{E}[(\hat{y} - y)'(\hat{y} - y)]$, Where $\hat{y}$ is a T_2-Vector of Forecasts of y, as Function of a. Right Panel: $\text{Risk}(\beta, \beta(y, X), X'X) = \mathbb{E}[(\beta(y, X) - \beta)'X'X(\beta(y, X) - \beta)]$ for the T_2-Period Forecast Sample. The Data-Generating Eigenvalues are Exponentially Decreasing from 5 to 1 at a Rate of 0.10. Other Parameters are Set to $K_1 = 1$, $L = 1$. The Connected Dots' Line Shows the Performance of the Stein-like Estimator. For Comparison, the Performance of the Standard OLS Estimator is Shown in Dots. The Performance of the Principal Components Estimator is Plotted with Dashes.

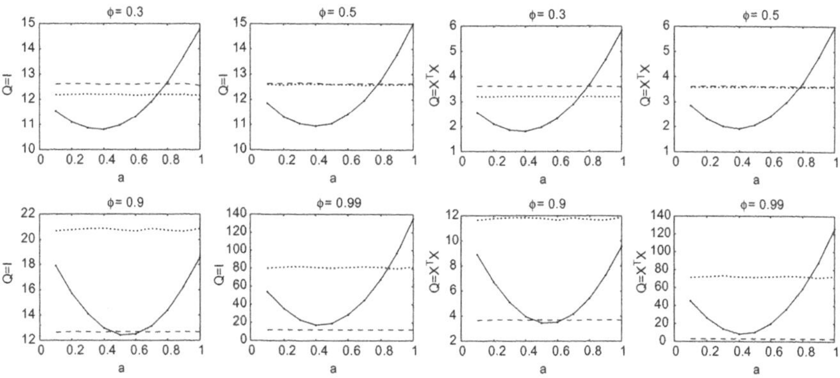

Fig. 13. Left Panel: $\text{MSFE}(\beta(y, X)) = \mathbb{E}[(\hat{y} - y)'(\hat{y} - y)]$, Where $\hat{y}$ is a T_2-Vector of Forecasts of y, as Function of a. Right Panel: $\text{Risk}(\beta, \beta(y, X), X'X) = \mathbb{E}[(\beta(y, X) - \beta)'X'X(\beta(y, X) - \beta)]$ for the T_2-Period Forecast Sample. The Columns of the Regressor Matrix X are AR(1) Processes. Other Parameters are Set to $K_1 = 1$, $L = 1$. The Connected Dots' Line Shows the Performance of the Stein-like Estimator. For Comparison, the Performance of the Standard OLS Estimator is Shown in Dots. The Performance of the Principal Components Estimator is Plotted with Dashes.

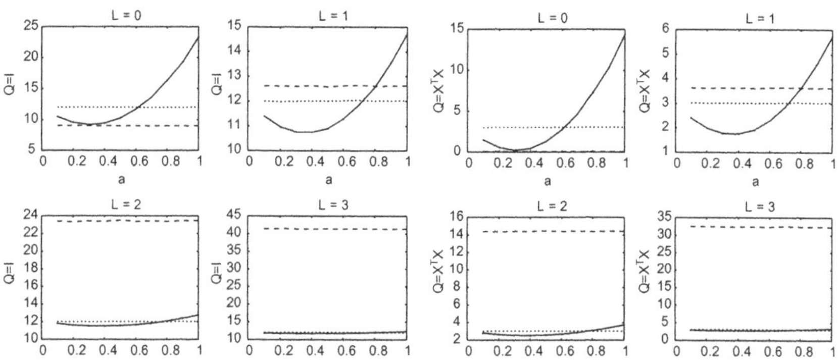

Fig. 14. Left Panel: $\text{MSFE}(\beta(y, X)) = \mathbb{E}[(\hat{y} - y)'(\hat{y} - y)]$, Where $\hat{y}$ Is a T_2-Vector of Forecasts of y, as Function of a. Right Panel: $\text{Risk}(\beta, \beta(y, X), X'X) = \mathbb{E}[(\beta(y, X) - \beta)'X'X(\beta(y, X) - \beta)]$ for the T_2-Period Forecast Sample. The Data-Generating Eigenvalues are Exponentially Decreasing from 5 to 1 at a Rate of 0.10. Other Parameters are Set to $K_1 = 1$, $\sigma = 3$. The Connected Dots' Line Shows the Performance of the Stein-like Estimator. For Comparison, the Performance of the Standard OLS Estimator is Shown in Dots. The Performance of the Principal Components Estimator is Plotted with Dashes.

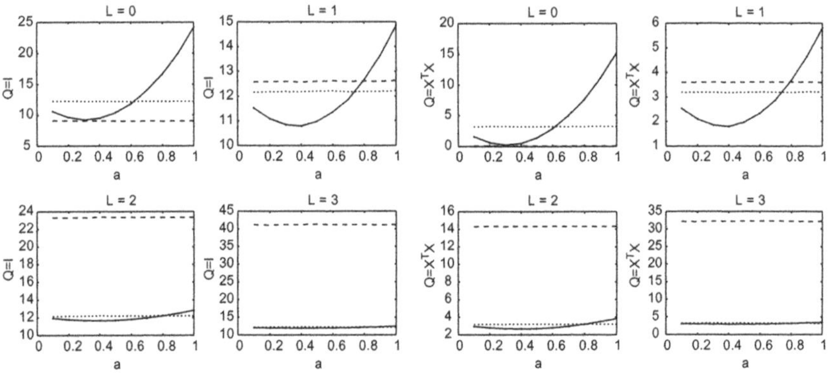

Fig. 15. Left Panel: $\text{MSFE}(\beta(y, X)) = \mathbb{E}[(\hat{y} - y)'(\hat{y} - y)]$, Where $\hat{y}$ is a T_2-Vector of Forecasts of y, as Function of a. Right Panel: $\text{Risk}(\beta, \beta(y, X), X'X) = \mathbb{E}[(\beta(y, X) - \beta)'X'X(\beta(y, X) - \beta)]$ for the T_2-Period Forecast Sample. The Columns of the Regressor Matrix X are AR(1) Processes. Other Parameters are Set to $K_1 = 1, \sigma = 3$. The Connected Dots' Line Shows the Performance of the Stein-like Estimator. For Comparison, the Performance of the Standard OLS Estimator is Shown in Dots. The Performance of the Principal Components Estimator is Plotted with Dashes.

PC estimator outperforms both Stein-rule and OLS. The relative performance of OLS and PC estimators also changes with persistence: In low persistence scenarios, OLS performs better than PC, and vice versa for high persistence.

Different Lengths of the Parameter Vector β

Figs. 14 and 15 show the performance of the estimators for different lengths $L \in 0, 1, 2, 3$ of the parameter vector. As in the in-sample case, when $L = 0$, PC performs best. For $L = 1$, OLS performs better than PC, but both are dominated by the Stein-rule shrinkage estimator. For higher values of L, OLS performs best among all three estimators. This holds true for both simulation environments, i.i.d. regressors and AR(1) regressors.

CONCLUDING REMARKS

In this paper, we have shown that the Stein-rule shrinkage estimator that shrinks the OLS estimator toward the PC estimator, as proposed in Hill and

Judge (1987, 1990), can be represented as a shrinkage estimator for a forecasting model as proposed in Stock and Watson (2011). We examined the performance of the estimator in a variety of simulation environments, both in-sample and out-of-sample. The overall picture that emerges is that the Stein-rule shrinkage estimator can dominate both OLS and principal components estimators within an intermediate range of the signal-to-noise ratio. If the noise level is high (high variance of noise terms) or if the signal is low (short parameter vector), the principal components estimator is superior. If the noise level is low (low variance of noise terms) or if the signal is high (long parameter vector), OLS is superior. In out-of-sample simulations with AR(1) regressors, the Stein-rule shrinkage estimator can dominate both OLS and principal components estimators in low-persistence situations.

ACKNOWLEDGMENTS

We would like to thank the participants of the Advances in Econometrics conference in Baton Rouge in March 2012 for their helpful comments, in particular Carter Hill, Tom Fomby, Stan Johnson, Mike McCracken, and Lee Adkins. We would also like to thank the organizers of the conference, in particular Dek Terrell, for a job superbly done and for many useful comments on this paper. The usual disclaimer applies.

REFERENCES

Anderson, T. W. (2003). *An introduction to multivariate statistical analysis* (3rd ed.). Hoboken, NJ: Wiley.

Bai, J. (2003). Inferential theory for factor models of large dimensions. *Econometrica*, *71*(1), 135–171.

Bai, J., & Ng, S. (2002). Determining the number of factors in approximate factor models. *Econometrica*, *70*(1), 191–221.

Bai, J., & Ng, S. (2006). Confidence intervals for diffusion index forecasts and inference for factor-augmented regressions. *Econometrica*, *74*(4), 1133–1150.

Bai, J., & Ng, S. (2008). Forecasting economic time series using targeted predictors. *Journal of Econometrics*, *146*, 304–317.

Bair, E., Hastie, T., Paul, D., & Tibshirani, R. (2006). Prediction by supervised principal components. *Journal of the American Statistical Association*, *101*(473), 119–137.

Bates, J. M., & Granger, C. W. J. (1969). The combination of forecasts. *Operations Research Quarterly*, *20*, 451–468.

Clark, T. E., & McCracken, M. W. (2009). Combining forecasts from nested models. *Oxford Bulletin of Economics and Statistics*, *71*(3), 303–329.

Fomby, T. B., Hill, R. C., & Johnson, S. R. (1978). An optimal property of principal components in the context of restricted least squares. *Journal of the American Statistical Association, 73*(361), 191–193.

Fomby, T. B., & Samanta, S. K. (1991). Application of Stein Rules to combination forecasting. *Journal of Business and Economic Statistics, 9*(4), 391–407.

Granger, C. W. J. (1980). Long memory relationships and the aggregation of dynamic models. *Journal of Econometrics, 14*, 227–238.

Hallin, M., & Liska, R. (2007). Determining the number of factors in the general dynamic factor model. *Journal of the American Statistical Association, 102*(478), 603–617.

Hansen, B. (2011). *Efficient shrinkage in parametric models.* Manuscript, University of Wisconsin, Madison.

Hill, R. C., & Fomby, T. B. (1992). The effect of extrapolation on minimax Stein-Rule prediction. In W. Griffiths, H. Lütkepohl & M. E. Bock (Eds.), *Readings in econometric theory and practice, A Volume in Honor of George G. Judge* (pp. 3–32). Amsterdam: North-Holland.

Hill, R. C., & Judge, G. G. (1987). Improved prediction in the presence of multicolinearity. *Journal of Econometrics, 35*, 83–100.

Hill, R. C., & Judge, G. G. (1990). Improved estimation under colinearity and squared error loss. *Journal of Multivariate Analysis, 32*, 296–312.

Hillebrand, E., Huang, H., Lee, T.-H., & Li, C. (2011). *Using the yield curve in forecasting output growth and inflation.* Manuscript, Aarhus University and UC Riverside.

Huang, H., & Lee, T.-H. (2010). To combine forecasts or to combine information? *Econometric Reviews, 29*, 534–570.

Inoue, A., & Kilian, L. (2008). How useful is bagging in forecasting economic time series? A case study of U.S. CPI inflation. *Journal of the American Statistical Association, 103*(482), 511–522.

James, W., & Stein, C. M. (1961). Estimation with quadratic loss. *Proceedings of the Fourth Berkeley Symposium on Mathematical Statistics and Probability, 1*, 361–379.

Judge, G. G., & Bock, M. E. (1978). *The statistical implications of pre-test and stein-rule estimators in econometrics.* Amsterdam: North-Holland.

Mittelhammer, R. C. (1985). Quadratic risk domination of restricted least squares estimators via Stein-Rules auxiliary constraints. *Journal of Econometrics, 29*, 289–304.

Onatski, A. (2009). Testing hypotheses about the number of factors in large factor models. *Econometrica, 77*(5), 1447–1479.

Stock, J. H., & Watson, M. W. (2002). Forecasting Using Principal Components from a Large Number of Predictors. *Journal of the American Statistical Association, 97*, 1167–1179.

Stock, J. H., & Watson, M. W. (2006). Forecasting with many predictors. In G. Elliott, C. W. J. Granger & A. Timmermann (Eds.), *Handbook of Economic Forecasting* (Vol. 1, pp. 515–554). Amsterdam: North Holland.

Stock, J. H., & Watson, M. W. (2011). *Generalized shrinkage methods for forecasting using many predictors.* Manuscript, Harvard University and Princeton University.

PART III
INSTRUMENTAL VARIABLES

ON THE ESTIMATION AND TESTING OF FIXED EFFECTS PANEL DATA MODELS WITH WEAK INSTRUMENTS

Badi H. Baltagi, Chihwa Kao and Long Liu

ABSTRACT

This chapter studies the asymptotic properties of within-groups k-*class estimators in a panel data model with weak instruments. Weak instruments are characterized by the coefficients of the instruments in the reduced form equation shrinking to zero at a rate proportional to* $\sqrt{n}T^{\delta}$, *where* n *is the dimension of the cross-section and* T *is the dimension of the time series. Joint limits as* $(n,T) \to \infty$ *show that this within-group* k-*class estimator is consistent if* $0 \leq \delta < \frac{1}{2}$ *and inconsistent if* $\frac{1}{2} \leq \delta < \infty$.

Keywords: Weak instrument; panel data; fixed effects; Pitman drift local-to-zero

INTRODUCTION

This chapter contributes to the literature on weak instrumental variable (IV) for panel data models with fixed effects. The problem of weak instruments have attracted considerable attention in recent years; see Stock, Wright, and Yogo

30th Anniversary Edition
Advances in Econometrics, Volume 30, 199–235

ISSN: 0731-9053/doi:10.1108/S0731-9053(2012)0000030012

(2002) for an excellent survey. Weak instruments are characterized by the coefficients of the instruments in the reduced form equation shrinking to zero at a rate proportional to the square root of the sample size. In case of weak instruments, the usual asymptotic normal approximations of the two-stage least squares (2SLS) estimator can be quite poor, even if the number of observations is large. Staiger and Stock (1997) use *weak-instrument asymptotics* to show that the 2SLS estimator is inconsistent (i.e., converges to a random variable) and has a nonstandard limiting distribution. This is a serious problem as inference, test of hypotheses and confidence intervals in the case of weak instruments becomes unreliable and misleading.

Bai and Ng (2010) show that for panel data models in which all regressors are endogenous but share exogenous common factors, valid instruments can be constructed from the endogenous regressors that are themselves invalid instruments in a conventional sense. This requires both dimensions of the panel n and T to be large. More recently, Cai, Fang, and Li (2012) argue that there may be benefits to using panel data when the available instruments are cross-sectionally weak. They consider the within-group 2SLS (W2SLS) estimator in a panel context where the degree of weakness of the instruments depends upon the number of cross-sectional observations n only. For large n, and fixed T, they show that the bias of W2SLS is of order $1/T$ as $n \to \infty$. They argue that leting the degree of weakness of the instruments depend on n only is an "analytical device" and with T fixed, "it is natural to relate the degree of weakness to n only." However, from Staiger and Stock (1997), the degree of weakness of the instruments depends upon the total number of observations nT and how n and T tend to infinity is crucial for the asymptotics of weak instruments in panel data.[1] This chapter extends the results presented in Cai et al. (2012) to the case where the weak instruments are modeled as "Pitman drift" local-to-zero sense, and the degree of weakness of the instruments is allowed to depend upon both n and T, but with different impact. To be specific, we let the degree of weakness of the instruments depend upon $\sqrt{n}T^{\delta}$ where $\delta \geq 0$. When $\delta = 0$, it reduces to the weak instrument case in Cai et al. (2012). When $\delta = 1/2$, it reduces to the weak instrument case in Staiger and Stock (1997). The basic argument is that with enough time periods observed, panel data may provide enough information to yield consistent estimation. In fact, it is well known that for cross-sectional data, when the concentration parameter stays constant as the sample size grows, the signal of the model is too weak compared to the corresponding noise. Hence the model is weakly identified, and 2SLS converges to a random variable. However, in the panel data set-up, if the time-series dimension is large, the weak signal can be strengthened by the repeating regression across the time-series dimension.

This argument is similar in spirit to the argument of establishing consistency for the panel spurious regression; see, for example, Phillips and Moon (1999) and Kao (1999).

Cai et al. (2012) also considered the case where the degree of weakness of the instruments depends upon n^{δ}, where $\delta \geq 0$. For a fixed T, when $0 < \delta < \frac{1}{2}$, the correlation between the instruments and endogenous variables converges to zero more slowly than the square root of the sample size, as $n \to \infty$. This corresponds to the *nearly weak instruments* case of Hahn and Kuersteiner (2002) and Hahn, Hausman and Kuersteiner (2004). For $\delta = 1/2$, this is the *weak instruments* case, and for $\delta > \frac{1}{2}$, this is the *nearly non-identified* case because the correlation converges to zero faster than the square root of the sample size, as $n \to \infty$. For cross-section or time-series models, Hahn and Kuersteiner (2002) showed that 2SLS for the *nearly weak instruments* case is consistent and its limiting distribution is normal. However, for the *weak instruments* case as well as the *nearly non-identified* case, 2SLS is inconsistent and its limiting distribution is not normal. Cai et al. (2012) similarly showed that for panel data models with fixed T, the bias of W2SLS estimator with *weak* or *nearly non-identified* instruments is of order $1/T$ as $n \to \infty$. They argue that as $T \to \infty$, W2SLS is consistent and asymptotically normal. They also consider a mixed case where some IVs are weak and others are nearly weak and show that as $n \to \infty$, with T fixed, the W2SLS estimator of the weak instruments is biased of order $1/T$, while the W2SLS estimator of the nearly weak instruments is consistent. We generalize the Cai et al. (2012) panel data results by studying the asymptotic properties of the general within-group k-class estimator, which includes W2SLS and within-group Limited Information Maximum Likelihood (LIML) as special cases. We allow the degree of weakness of the instruments to depend upon $\sqrt{n}T^{\delta}$ where $\delta \geq 0$. We study the asymptotics using joint limits in n and T, rather than fixing T and letting $n \to \infty$. We show that for the simple case of one right-hand-side endogenous variable and no included exogenous variables, W2SLS is consistent if $0 \leq \delta < \frac{1}{2}$ and inconsistent if $\frac{1}{2} \leq \delta < \infty$. Next, we generalize these results to the within-group k-class estimator with included exogenous regressors applied to fixed effects panel data. We show using joint limits that this within-group k-class estimator is consistent if $0 \leq \delta < \frac{1}{2}$ and inconsistent if $\frac{1}{2} \leq \delta < \infty$. We characterize these conditions for three special cases of the within-group k-class estimator including W2SLS, within-group LIML, and within-group bias-adjusted 2SLS (B2SLS). We also generalize the test for weak instruments proposed by Cragg and Donald (1993) and Stock and Yogo (2005) to the case of fixed effects panel data as well as test of hypothesis that is robust to weak instruments in the fixed effects panel data set-up. We study the asymptotic properties of these tests as both $(n,T) \to \infty$.[2]

The rest of the chapter is organized as follows. Section "Model and Assumptions" introduces the fixed effects panel data model with weak instruments. Section "Estimation" discusses the within-group k-class estimator. Section "Testing for Weak Instruments" generalizes the test for weak instruments proposed by Cragg and Donald (1993) and Stock and Yogo (2005) to the case of fixed effects panel data. Section "Robust Inference with Weak Instruments" considers the problem of hypothesis testing whose size is robust to weak instruments in the fixed effects panel data set-up. Section "Monte Carlo Simulation" provides Monte Carlo results, while the last section concludes. All the proofs are relegated to the appendix. All the limits in the chapter are taken as $(n,T) \to \infty$ jointly, except when otherwise noted.

MODEL AND ASSUMPTIONS

Consider the following panel IV regression model with endogenous regressors

$$y_t = Y_t\beta + X_t\gamma + \mu + u_t \quad (1)$$

and

$$Y_t = Z_t\Pi + X_t\Gamma + \alpha + V_t \quad (2)$$

for $t=1,2,\ldots,T$, where y_t is a $n \times 1$ vector and Y_t is a $n \times L$ matrix of endogenous variables, X_t is a $n \times K_1$ matrix of K_1 exogenous regressors, Z_t is a $n \times K_2$ matrix of K_2 instruments, and β, γ, Π, and Γ are unknown parameters. μ and α denote the individual effects which are of dimensions $n \times 1$ and $n \times L$, respectively. The remainder disturbances $(u_t, V'_t)'$ are of dimensions $n \times 1$ and $n \times L$ respectively. These disturbances $(u_t, V_t)'$ are assumed to be i.i.d. $N(0,\Sigma)$ across $t = 1, 2, \ldots, T$, with the elements of Σ denoted by σ_{uu}, Σ_{Vu}, and Σ_{VV}. Let $Z^* = [X, Z]$, $Y^* = [y, Y]$ and let $\Phi = EZ^{*\prime}_{it}Z^*_{it}$, partitioned so that $EX_{it}X'_{it} = \Phi_{XX}$, $EX_{it}Z'_{it} = \Phi_{XZ}$ and $EZ_{it}Z'_{it} = \Phi_{ZZ}$. It is assumed throughout that $EZ^*_{it}(u_{it}, V'_{it}) = 0$ for all i and t. This i.i.d. assumption for the errors can be relaxed to allow for weak dependence across the time-series and cross-section dimensions at the expense of more complicated notation. This will be taken up in a future extension of this chapter. Eq. (1) is the structural equation and β is the parameter of interest. The reduced form Eq. (2) relates the endogenous regressors to the instruments. In matrix form, Eqs. (1) and (2) can be rewritten as

$$y = Y\beta + X\gamma + \mu \otimes \iota_T + u \quad (3)$$

and

$$Y = Z\Pi + X\Gamma + \alpha \otimes \iota_T + V \tag{4}$$

where $y = (y'_1, y'_2, \ldots, y'_T)'$ is a $nT \times 1$ vector, ι_T is a vector of ones of dimension T, and Y, X, Z, u, and V are similarly defined.

To wipe out the individual effects, we premultiply Eqs. (3) and (4) by the within transformation $Q = I_n \otimes E_T$, where $E_T = I_T - \bar{J}_T$, $\bar{J}_T = J_T/T$; J_T is a matrix of ones of dimension T and I_n is an identity matrix of dimension n. This yields

$$\tilde{y} = \tilde{Y}\beta + \tilde{X}\gamma + \tilde{u} \tag{5}$$

and

$$\tilde{Y} = \tilde{Z}\Pi + \tilde{X}\Gamma + \tilde{V}, \tag{6}$$

where $\tilde{y} = Qy$, and $\tilde{Y}$, $\tilde{X}$, $\tilde{Z}$, $\tilde{u}$, and $\tilde{V}$ are similarly defined. This wipes out possible correlation between these individual effects and the regressors. It also wipes out time-invariant variables that may cause omission bias if not included in the model. We model weak instruments by focussing on Π being local to zero, which is analogous to the local-to-unity panel unit root literature as in Moon, Perron, and Phillips (2007).

Assumption 1. *Let* $\Pi = \frac{C}{\sqrt{n}T^{\delta}}$, *where* C *is a* $K_2 \times L$ *constant matrix and* $\delta \geq 0$.

Assumption 1 controls the relative magnitude of the instrument strength, as measured by δ. When $\delta = 1/2$, it is the standard weak instrument case introduced by Staiger and Stock (1997). When $\delta = 0$, it reduces to the weak instrument case in Cai et al. (2012).

Following Staiger and Stock (1997), we assume:

Assumption 2. *The following joint limits hold, as* $(n, T) \to \infty$:

1. $\left(\frac{1}{Tn}\sum_{t=1}^{T} u_t' u_t, \frac{1}{Tn}\sum_{t=1}^{T} V_t' u_t, \frac{1}{Tn}\sum_{t=1}^{T} V_t' V_t\right) \xrightarrow{p} (\sigma_{uu}, \Sigma_{Vu}, \Sigma_{VV})$;

2. $\frac{1}{Tn}\sum_{t=1}^{T}\left(\tilde{X}_t, \tilde{Z}_t\right)'\left(\tilde{X}_t, \tilde{Z}_t\right) \xrightarrow{p} \Phi \equiv \begin{pmatrix} \Phi_{XX} & \Phi_{XZ} \\ \Phi_{ZX} & \Phi_{ZZ} \end{pmatrix}$;

3. $\left(\frac{1}{\sqrt{Tn}}\sum_{t=1}^{T} X_t' u_t, \frac{1}{\sqrt{Tn}}\sum_{t=1}^{T} Z_t' u_t, \frac{1}{\sqrt{Tn}}\sum_{t=1}^{T} X_t' V_t, \frac{1}{\sqrt{Tn}}\sum_{t=1}^{T} Z_t' V_t\right) \xrightarrow{d} (\Psi_{Xu}, \Psi_{Zu}, \Psi_{XV}, \Psi_{ZV})$, *where* $\Psi = (\Psi'_{Xu}, \Psi'_{Zu}, vec(\Psi_{XV})', vec(\Psi_{ZV})')'$ *is distributed* $N(0, \Sigma \otimes \Phi)$.

Note that Assumption 2 implies that $\left(\frac{1}{Tn}\sum_{t=1}^{T}\tilde{u}_t'\tilde{u}_t, \frac{1}{Tn}\sum_{t=1}^{T}\tilde{V}_t'\ \tilde{u}_t, \frac{1}{Tn}\sum_{t=1}^{T}\tilde{V}_t'\ \tilde{V}_t\right) \xrightarrow{p} (\sigma_{uu}, \Sigma_{Vu}, \Sigma_{VV})$ and $\left(\frac{1}{\sqrt{Tn}}\sum_{t=1}^{T}\tilde{X}_t'\tilde{u}_t, \frac{1}{\sqrt{Tn}}\sum_{t=1}^{T}\tilde{Z}_t'\tilde{u}_t, \frac{1}{\sqrt{Tn}}\sum_{t=1}^{T}\tilde{X}_t'\tilde{V}_t, \frac{1}{\sqrt{Tn}}\sum_{t=1}^{T}\tilde{Z}_t'\tilde{V}_t\right) \xrightarrow{p} (\Psi_{Xu}, \Psi_{Zu}, \Psi_{XV}, \Psi_{ZV})$ since $\frac{1}{T}\sum_{t=1}^{T}u_t \xrightarrow{p} 0$ and $\frac{1}{T}\sum_{t=1}^{T}V_t \xrightarrow{p} 0$. Following Staiger and Stock (1997), we define

$$\lambda = \Omega^{1/2} C \Sigma_{VV}^{-1/2}, \tag{7}$$

where $\Omega = \Phi_{ZZ} - \Phi_{ZX}\Phi_{XX}^{-1}\Phi_{XZ}$. Also define

$$z_u = \Omega^{-1/2'}(\Psi_{Zu} - \Phi_{ZX}\Phi_{XX}^{-1}\Psi_{Xu})\sigma_{uu}^{-1/2} \tag{8}$$

and

$$z_V = \Omega^{-1/2'}(\Psi_{ZV} - \Phi_{ZX}\Phi_{XX}^{-1}\Psi_{XV})\Sigma_{VV}^{-1/2}. \tag{9}$$

The random variable $[z'_u, vec(z_V)']'$ is distributed $N(0, \bar{\Sigma} \otimes I_{K_2})$, where

$$\bar{\Sigma} = \begin{pmatrix} 1 & \rho' \\ \rho & I_L \end{pmatrix} \tag{10}$$

with $\rho = \Sigma_{VV}^{-1/2'}\Sigma_{Vu}\sigma_{uu}^{-1/2}$ and I_L is an identity matrix of dimension L.

ESTIMATION

Most of the theorems in this section are developed for the within-group k-class estimator. However, we start by deriving the asymptotic properties of W2SLS for the simple case of one right-hand-side endogenous regressor and no included exogenous regressors.

A special case: 2SLS when L = *1 and* $K_1 = 0$

Let $P_{\tilde{Z}} = \tilde{Z}(\tilde{Z}'\tilde{Z})^{-1}\tilde{Z}'$ be the projection matrix on the space spanned by the columns of $\tilde{Z}$. The W2SLS estimator is defined as

$$\hat{\beta}_{W2SLS} = \frac{\tilde{Y}' P_{\tilde{Z}} \tilde{y}}{\tilde{Y}' P_{\tilde{Z}} \tilde{Y}}$$

Theorem 1. *As* $(n, T) \to \infty$, *for* $0 \le \delta < \frac{1}{2}$

$$T^{\frac{1}{2}-\delta}(\hat{\beta}_{W2SLS} - \beta) \xrightarrow{d} N(0, \sigma_{uu}(C'\Phi_{ZZ}C)^{-1})$$

and for $\frac{1}{2} \le \delta < \infty$

$$\hat{\beta}_{W2SLS} - \beta = O_p(1).$$

The results in Theorem 1 imply that $\hat{\beta}_{W2SLS}$ is consistent only if $0 \le \delta < \frac{1}{2}$ and inconsistent if $\frac{1}{2} \le \delta < \infty$. The strength of the instruments is measured by the following concentration matrix $\Lambda_{Tn} = \Sigma_{VV}^{-1/2'}\Pi'\tilde{Z}'\tilde{Z}\Pi\Sigma_{VV}^{-1/2}$. Using Assumptions 1 and 2, we have

$$\begin{aligned}\Lambda_{Tn} &= \frac{1}{T^{2\delta-1}}\Sigma_{VV}^{-1/2'}C'\frac{\tilde{Z}'\tilde{Z}}{Tn}C\Sigma_{VV}^{-1/2} \\ &= \frac{1}{T^{2\delta-1}}\Sigma_{VV}^{-1/2'}C'\Phi_{ZZ}C\Sigma_{VV}^{-1/2} + o_p(1) \\ &= O_p(T^{1-2\delta}).\end{aligned}$$

Note that $T^{1-2\delta}$ can be interpreted as the rate at which Λ_{Tn} grows as T increases. Clearly, for the consistency of W2SLS, one needs $\Lambda_{Tn} \to \infty$ as $T^{1-2\delta} \to \infty$ which holds if $0 \le \delta < \frac{1}{2}$. We also note from Theorem 1 that the limiting distribution near the point of non-identification, i.e., $\delta = \frac{1}{2}$, is discontinuous.[3]

Within-Group k*-Class Panel Data Estimators*

We now generalize the results to the within-group k-class estimator with included regressors.[4] Let $P_{\tilde{X}} = \tilde{X}(\tilde{X}'\tilde{X})^{-1}\tilde{X}'$ be the projection matrix on the space spanned by the columns of $\tilde{X}$ and $M_{\tilde{X}} = I - P_{\tilde{X}}$. Premultiplying Eqs. (3) and (4) by $M_{\tilde{X}}$, we get

$$\tilde{y}^{\perp} = \tilde{Y}^{\perp}\beta + \tilde{u}^{\perp}$$

and

$$\tilde{Y}^{\perp} = \tilde{Z}^{\perp}\Pi + V^{\perp},$$

where the superscript "$\perp$" denotes the residuals from the projection on $\tilde{X}$, such as $\tilde{y}^{\perp} = M_{\tilde{X}}\tilde{y}$, $\tilde{Z}^{\perp} = M_{\tilde{X}}\tilde{Z}$, and $\tilde{Y}^{\perp} = M_{\tilde{X}}\tilde{Y}$. The within-group k-class estimator of β is given by

$$\hat{\beta}(k) = [\tilde{Y}^{\perp'}(I - kM_{\tilde{Z}^{\perp}})\tilde{Y}^{\perp}]^{-1}[\tilde{Y}^{\perp'}(I - kM_{\tilde{Z}^{\perp}})\tilde{y}^{\perp}]$$

for some choice of k. Note that the W2SLS estimator is a special case of the within-group k-class estimator when $k=1$. Theorem 2 derives the asymptotic properties of this within-group k-class panel data estimator.

Theorem 2. *Under Assumptions 1 and 2. As* $(n, T) \to \infty$ *we have*

1. *For* $0 \le \delta < \frac{1}{2}$, *joint with* $\kappa_{Tn} = T^{1/2+\delta}n(k-1) \xrightarrow{d} \kappa$,

$$T^{1/2-\delta}(\hat{\beta}(k) - \beta) + \theta_{Tn} \xrightarrow{d} N\left(0, \sigma_{uu}(\Sigma_{VV}^{1/2'}\lambda'\lambda\Sigma_{VV}^{1/2})^{-1}\right),$$

where $\theta_{Tn} = \left[\frac{1}{T^{1-2\delta}}\tilde{Y}^{\perp'}(I - kM_{\tilde{Z}^{\perp}})\tilde{Y}^{\perp}\right]^{-1}\kappa_{Tn}\Sigma_{Vu}$.

2. *For* $\delta = \frac{1}{2}$, *joint with* $\kappa_{Tn} = Tn(k-1) \xrightarrow{d} \kappa$,

$$\hat{\beta}(k) - \beta \xrightarrow{d} \sigma_{uu}^{1/2}\Sigma_{VV}^{-1/2}\Delta_1(\kappa),$$

where $\Delta_1(\kappa) = [(\lambda + z_V)'(\lambda + z_V) - \kappa I_{Tn}]^{-1}[(\lambda + z_V)'z_u - \kappa\rho]$.

3. *For* $\frac{1}{2} < \delta < \infty$, *joint with* $\kappa_{Tn} = Tn(k-1) \xrightarrow{d} \kappa$,

$$\hat{\beta}(k) - \beta \xrightarrow{d} \sigma_{uu}^{1/2}\Sigma_{VV}^{-1/2}\Delta_2(\kappa),$$

where $\Delta_2(\kappa) = [z'_V z_V - \kappa I_{Tn}]^{-1}[z'_V z_u - \kappa\rho]$.

Similar to the results of Theorem 1 for $\hat{\beta}_{W2SLS}$, Theorem 2 shows that $\hat{\beta}(k)$ is consistent if $0 \le \delta < \frac{1}{2}$ and inconsistent if $\frac{1}{2} \le \delta < \infty$. Similarly, using Assumptions 1 and 2 and Lemma 2, the strength of the instruments is measured by the following concentration matrix:

$$\begin{aligned} \Lambda_{Tn} &= \Sigma_{VV}^{-1/2'}\Pi'\tilde{Z}^{\perp'}\tilde{Z}^{\perp}\Pi\Sigma_{VV}^{-1/2} \\ &= \frac{1}{T^{2\delta-1}}\Sigma_{VV}^{-1/2'}C'\frac{\tilde{Z}^{\perp'}\tilde{Z}^{\perp}}{Tn}C\Sigma_{VV}^{-1/2} \\ &= \frac{1}{T^{2\delta-1}}\Sigma_{VV}^{-1/2'}C'\Omega C\Sigma_{VV}^{-1/2} + o_p(1) \\ &= O_p(T^{1-2\delta}), \end{aligned}$$

Note that $T^{1-2\delta}$ can be interpreted as the rate at which Λ_{Tn} grows as T increases. Clearly, for consistency of the within-group k-class estimator, one needs $\Lambda_{Tn} \to \infty$ as $T^{1-2\delta} \to \infty$ which holds if $0 \le \delta < \frac{1}{2}$.

For the W2SLS estimator with $k=1$, it follows that $T^{1/2+\delta}n(k-1)=0$ and $Tn(k-1)=0$. Therefore, the W2SLS estimator satisfies the conditions of κ_{Tn} for the three cases considered in Theorem 2.

The within-group k-class estimator also includes the within-group bias-adjusted 2SLS (B2SLS) described in Donald and Newey (2001) for the cross-section or time-series regression case. This is a special case of the k-class estimator with $k=nT/(nT-K_2+2)$. Rothenberg (1984) showed that B2SLS is unbiased to the second order for the fixed-instrument, normal error model. For this special case, $T^{1/2+\delta}n(k-1)=(K_2-2)/T^{1/2-\delta}=o_p(1)$ and $Tn(k-1)=K_2-2=O_p(1)$. Hence, the within-group B2SLS estimator satisfies the conditions of κ_{Tn} for the three cases considered in Theorem 2.

For the within-group LIML estimator in panel data, we obtain the following results:

Theorem 3. *Under Assumptions 1 and 2, with $\bar{\Sigma}$ is defined in Eq. (10), we have*

1. *For* $0 \le \delta < \frac{1}{2}$, $T^{2\delta}n(\hat{k}_{LIML}-1) \xrightarrow{p} 0$.

2. *For* $\delta=\frac{1}{2}$, $Tn(\hat{k}_{LIML}-1) \xrightarrow{d} \kappa^*_{LIML}$, *where* κ^*_{LIML} *is the smallest root of the determinantal equation,* $\left|\Xi_2-\kappa\bar{\Sigma}\right|=0$, *where* $\Xi_2=\begin{pmatrix} z_u'z_u & z_u'(\lambda+z_V) \\ (\lambda+z_V)'z_u & (\lambda+z_V)'(\lambda+z_V) \end{pmatrix}$.

3. *For* $\frac{1}{2}<\delta<\infty$, $Tn(\hat{k}_{LIML}-1) \xrightarrow{d} \kappa^*_{LIML}$, *where* κ^*_{LIML} *is the smallest root of the determinantal equation,* $\left|\Xi_3-\kappa\bar{\Sigma}\right|=0$, *where* $\Xi_3=\begin{pmatrix} z_u'z_u & z_u'z_V \\ z_V'z_u & z_V'z_V \end{pmatrix}$.

Wald Test Under Weak Identification

Next, we consider testing the q linear restrictions $R\beta=r$, where R is $q\times L$. The standard formula for the Wald statistic, based on the within-group k-class estimator, is given by

$$W(k)=\left[R\hat{\beta}(k)-r\right]'\left\{\hat{\sigma}_{uu}(k)R\left[\tilde{Y}^{\perp\prime}(I-kM_{\tilde{Z}^{\perp}})\tilde{Y}^{\perp}\right]^{-1}R'\right\}^{-1}\left[R\hat{\beta}(k)-r\right],$$

where $\hat{\sigma}_{uu}(k)=\hat{\tilde{u}}(k)'\hat{\tilde{u}}(k)/(Tn-K_1-L)$, and $\hat{\tilde{u}}(k)=\tilde{y}-\tilde{Y}\hat{\beta}(k)-\tilde{X}\hat{\gamma}(k)=\tilde{y}^{\perp}-\tilde{Y}^{\perp}\hat{\beta}(k)$.

Theorem 4. *Under Assumptions 1 and 2. As $(n, T) \to \infty$ we have*

1. *For $0 \leq \delta < \frac{1}{2}$, joint with $\kappa_{Tn} = T^{1/2+\delta} n(k-1) \xrightarrow{d} \kappa$,*

$$W(k) \xrightarrow{d} \chi^2(q, \Lambda)$$

a noncentral chi-squared distribution with q degrees of freedom and noncentrality parameter $\Lambda = \theta' R' [\sigma_{uu} R (\Sigma_{VV}^{1/2'} \lambda' \lambda \Sigma_{VV}^{1/2})^{-1} R']^{-1} R\theta$, where $\theta = \kappa (\Sigma_{VV}^{1/2'} \lambda' \lambda \Sigma_{VV}^{1/2})^{-1} \Sigma_{Vu}$.

2. *For $\delta = \frac{1}{2}$, joint with $\kappa_{Tn} = Tn(k-1) \xrightarrow{d} \kappa$,*

$$W(k) \xrightarrow{d} \Delta_1'(\kappa) \Sigma_{VV}^{-1/2} R' \Big\{ S(\Delta_1(\kappa))$$
$$R \Big\{ \Sigma_{VV}^{1/2'} [(\lambda + z_V)'(\lambda + z_V) - \kappa I_{Tn}] \Sigma_{VV}^{1/2} \Big\}^{-1} R' \Big\}^{-1} R \Sigma_{VV}^{-1/2} \Delta_1(\kappa).$$

3. *For $\frac{1}{2} < \delta < \infty$, joint with $\kappa_{Tn} = Tn(k-1) \xrightarrow{d} \kappa$,*

$$W(k) \xrightarrow{d} \Delta_2'(\kappa) \Sigma_{VV}^{-1/2} R' \Big\{ S(\Delta_2(\kappa)) R \Big\{ \Sigma_{VV}^{1/2'} [z'_V z_V - \kappa I_{Tn}] \Sigma_{VV}^{1/2} \Big\}^{-1} R' \Big\}^{-1} R \Sigma_{VV}^{-1/2} \Delta_2(\kappa).$$

Note that for $0 \leq \delta < \frac{1}{2}$, if $\kappa = 0$, then $\theta = 0$ and $\Lambda = 0$. Hence $W(k) \xrightarrow{d} \chi^2(q)$ a central chi-squared distribution with q degrees of freedom.

TESTING FOR WEAK INSTRUMENTS

Following Stock and Yogo (2005), we focus in this section on testing the null hypothesis that the set of instruments is weak against the alternative that they are strong. In this case, the instruments are defined to be strong if W2SLS inference is reliable for any linear combination of the coefficients. From the results in Theorems 2 and 4, weak instruments can produce biased IV estimators and test of hypotheses with large size distortions, e.g., when $\frac{1}{2} \leq \delta < \infty$. The Stock and Yogo (2005) test is based on the partial identification test statistic proposed by Cragg and Donald (1993). For our case, this statistic is $g_{\min}$, the smallest eigenvalue of the matrix analog of the F statistic from the first-stage regression of W2SLS, i.e.,

$$g_{\min} = \text{mineval} G_{Tn},$$

where

$$G_{Tn} = \frac{\hat{\Sigma}_{VV}^{-1/2'} \tilde{Y}^{\perp'} P_{\tilde{Z}^{\perp}} \tilde{Y}^{\perp} \hat{\Sigma}_{VV}^{-1/2}}{K_2}$$

with $\hat{\Sigma}_{VV} = \tilde{Y}' M_{\tilde{Z}} \tilde{Y}/(Tn - K_1 - K_2)$. A small $g_{\min}$ indicates that the instruments are weak; see Stock and Yogo (2005). Let $W(K, \Omega, \Upsilon)$ denote the Wishart distribution with K denoting the degrees of freedom, Ω denoting the covariance matrix, and Υ denoting the noncentrality matrix, we have the following result:

Theorem 5. *Under Assumptions 1 and 2, we have*

1. *For* $0 \le \delta < \frac{1}{2}$, $\frac{1}{T^{1-2\delta}} K_2 G_{Tn} \xrightarrow{p} \lambda'\lambda$.
2. *For* $\delta = \frac{1}{2}$, $K_2 G_{Tn} \xrightarrow{d} (\lambda + z_V)'(\lambda + z_V) \sim W(K_2, I_L, \lambda'\lambda)$.
3. *For* $\frac{1}{2} < \delta < \infty$, $K_2 G_{Tn} \xrightarrow{d} z'_V z_V \sim W(K_2, I_L, 0)$.

Note that for $0 \le \delta < \frac{1}{2}$, $G_{Tn} \cong T^{1-2\delta}\lambda'\lambda/K_2 \to \infty$. For $\delta = \frac{1}{2}$, $E((\lambda + z_V)'(\lambda + z_V)) = K_2 I_L + \lambda'\lambda$, hence $G_{Tn} \cong I_L + \lambda'\lambda/K_2$. For $\frac{1}{2} < \delta < \infty$, $E(z'_V z_V) = K_2 I_L$, hence $G_{Tn} \cong I_L$. Therefore, as pointed out in Stock et al. (2002), $tr(G_{Tn})/L$ can be thought of as a measure of the strength of the instruments. It is clear that $g_{\min} \to \infty$ if $0 \le \delta < \frac{1}{2}$, $g_{\min} \xrightarrow{d}$ mineval $\frac{(\lambda+z_V)'(\lambda+z_V)}{K_2}$ if $\delta = \frac{1}{2}$, and $g_{\min} \xrightarrow{d}$ mineval $\frac{z'_V z_V}{K_2}$ if $\frac{1}{2} < \delta < \infty$. Next we discuss how to use $g_{\min}$ to detect the presence of weak instruments.

When $\delta = \frac{1}{2}$, $K_2 G_{Tn} \xrightarrow{d} (\lambda + z_V)'(\lambda + z_V)$ which has a noncentral Wishart distribution with noncentrality matrix $\lambda'\lambda$. This noncentrality matrix is the limit of the concentration matrix

$$\Lambda_{Tn} = \Sigma_{VV}^{-1/2'} \Pi' \tilde{Z}' \tilde{Z} \Pi \Sigma_{VV}^{-1/2} \xrightarrow{p} \lambda'\lambda.$$

On the other hand, when $0 \le \delta < \frac{1}{2}$, $K_2 G_{Tn} \to \infty$, because $\Lambda_{Tn} \to \infty$. Also note that $z'_V z_V$ has a Wishart distribution (i.e., $\lambda'\lambda = 0$). This corresponds to

$$\Lambda_{Tn} \xrightarrow{p} 0$$

when $\frac{1}{2} < \delta < \infty$. Let $\delta_{\min}$ be the smallest eigenvalue of $\lambda'\lambda$. Following Stock and Yogo (2005), we propose using the conservative critical value x which satisfies the relationship

$$P(g_{\min} \leq x) \leq P(\chi^2(K_2, \delta_{\min}) \geq vx),$$

where $\chi^2(v, \delta_{\min})$ denotes the noncentral chi-squared random variable with v degrees of freedom and non-centrality parameter $\delta_{\min}$.[5] Stock and Yogo (2005) focus on the worst-behaved linear combination and it is in this sense that this test is conservative. We refer the reader to their tables for critical values.

ROBUST INFERENCE WITH WEAK INSTRUMENTS

The above results indicate that for $\delta \geq \frac{1}{2}$, the within-group k-class estimator is inconsistent. In this section, we discuss hypothesis testing whose size is robust to the weak instruments in the panel data set-up. Following the survey by Stock et al. (2002), we will discuss the AR test of Anderson and Rubin (1949), the Lagrange multiplier (LM) test of Kleibergen (2002) and Moreira (2009), and the conditional likelihood ratio (CLR) test of Moreira (2003) but applied to the *fixed effects* panel data model. For simplicity, we only consider the case of one right-hand-side endogenous variable, i.e., $L = 1$.[6]

For convenience, we assume that $\tilde{X}_t$ and $\tilde{Z}_t$ are non-stochastic such that $\tilde{Z}'_t\tilde{X}_t = 0$. The reduced form equations corresponding to the structural Eqs. (5) and (6) are as follows:

$$\tilde{y} = \tilde{Z}\Pi\beta + \tilde{X}\gamma^* + \tilde{u}^*$$

and

$$\tilde{Y} = \tilde{Z}\Pi + \tilde{X}\Gamma + \tilde{V}$$

where $\gamma^* = \gamma + \Gamma\beta$ and $\tilde{u}^* = \tilde{u} + \tilde{V}\beta$. The reduced form errors are assumed to be homoskedastic with covariance matrix

$$\Sigma^* = \begin{bmatrix} \sigma^*_{uu} & \Sigma^{*\prime}_{Vu} \\ \Sigma^*_{Vu} & \Sigma_{VV} \end{bmatrix}.$$

The concentration parameter can be rewritten as $\Sigma_{VV}^{-1/2'}\Pi'\tilde{Z}'\tilde{Z}\Pi\Sigma_{VV}^{-1/2}$. Consider the null hypothesis

$$H_0 : \beta = \beta_0$$

Define

$$\mathbb{S} = \frac{(\tilde{Z}'\tilde{Z})^{-1/2}\tilde{Z}'\tilde{Y}^* b_0}{(b'_0\Sigma^{*-1}b_0)^{1/2}}$$

and

$$\mathbb{T} = \frac{(\tilde{Z}'\tilde{Z})^{-1/2}\tilde{Z}'\tilde{Y}^{*}\Sigma^{*-1}a_0}{(a'_0\Sigma^{*-1}a_0)^{1/2}}$$

where $b_0 = [1, -\beta_0]'$, $a_0 = [\beta_0, 1]'$, and $\tilde{Y}^{*} = [\tilde{y}, \tilde{Y}]$. Define

$$\Theta = \begin{bmatrix} \mathbb{S}' \\ \mathbb{T}' \end{bmatrix} [\mathbb{S}, \mathbb{T}] = \begin{bmatrix} \mathbb{S}'\mathbb{S} & \mathbb{T}'\mathbb{S} \\ \mathbb{S}'\mathbb{T} & \mathbb{T}'\mathbb{T} \end{bmatrix} = \begin{bmatrix} \Theta_{\mathbb{S}} & \Theta_{\mathbb{ST}} \\ \Theta_{\mathbb{ST}} & \Theta_{\mathbb{T}} \end{bmatrix}.$$

Three test statistics that are functions of Θ are the LM statistics of Kleibergen (2002) and Moreira (2009), the Anderson and Rubin (1949) (AR) statistic, and the Moreira (2003) CLR statistic. We now define the AR, LM, and CLR test statistics as follows:

$$\mathrm{AR} = \frac{\Theta_{\mathbb{S}}}{K_2}$$

$$\mathrm{LM} = \frac{\Theta_{\mathbb{ST}}^2}{\Theta_{\mathbb{T}}}$$

and

$$\mathrm{CLR} = \frac{1}{2}\left(\Theta_{\mathbb{S}} - \Theta_{\mathbb{T}} + \sqrt{(\Theta_{\mathbb{S}} - \Theta_{\mathbb{T}})^2 + 4\Theta_{\mathbb{ST}}^2}\right);$$

see Andrews and Stock (2006, 2007). Define

$$\mu_{\pi} = (\tilde{Z}'\tilde{Z})^{1/2}\Pi,$$

$$c_{\beta} = \frac{\beta - \beta_0}{(b'_0\Sigma^{*-1}b_0)^{1/2}},$$

and

$$d_{\beta} = \frac{a'\Sigma^{*-1}a_0}{(a'_0\Sigma^{*-1}a_0)^{1/2}}.$$

Note that $\tilde{Y}^{*}$ can be written as

$$\tilde{Y}^{*} = \tilde{Z}\Pi a + \tilde{X}\eta + [\tilde{u}^{*}, \tilde{V}].$$

where $a = (\beta, 1)'$, and $\eta = [\gamma^*, \Gamma]'$. That is, $\tilde{Y}$ is multivariate normal with mean matrix $\tilde{Z}\Pi a + \tilde{X}\eta$. Then $\mathbb{S}$ is $K_2 \times 1$ multivariate normal with mean

$$E[\mathbb{S}] = E\left[\frac{(\tilde{Z}'\tilde{Z})^{-1/2}\tilde{Z}'\tilde{Y}b_0}{(b'_0\Sigma^{*-1}b_0)^{1/2}}\right] = \frac{(\tilde{Z}'\tilde{Z})^{-1/2}\tilde{Z}'(\tilde{Z}\Pi a + \tilde{X}\eta)b_0}{(b'_0\Sigma^{*-1}b_0)^{1/2}} = \frac{(\tilde{Z}'\tilde{Z})^{1/2}\Pi(\beta - \beta_0)}{(b'_0\Sigma^{*-1}b_0)^{1/2}} = c_\beta\mu_\pi$$

and

$$Var[\mathbb{S}] = I_{K_2}$$

using

$$\tilde{Z}'\tilde{X} = 0$$

and

$$a'b_0 = (\beta, 1)\begin{pmatrix} 1 \\ -\beta_0 \end{pmatrix} = \beta - \beta_0.$$

Similarly, $\mathbb{T}$ is $K_2 \times 1$ multivariate normal with mean

$$E[\mathbb{T}] = d_\beta\mu_\pi$$

and variance

$$Var[\mathbb{T}] = I_{K_2}.$$

It is also easy to show that $\mathbb{S}$ and $\mathbb{T}$ are independent using $b'_0 a_0 = 0$. Under the null, $H_0 : \beta = \beta_0$,

$$\mathbb{S} \sim N(0, I_{K_2})$$

which does not depend on Π since $c_\beta = 0$. However, the distribution of $\mathbb{T}$ depends on Π under the null.

Assume $\frac{\tilde{Z}'\tilde{Z}}{Tn} \to D_Z$. The asymptotic distributions of $\mathbb{S}$ and $\mathbb{T}$ are given in the following theorem:

Theorem 6. *Suppose Assumptions 1 and 2 hold. We have*

$$\mathbb{S} - c_\beta\mu_\pi \xrightarrow{d} \mathbb{S}^* \sim N(0, I_{K_2})$$

and

$$\mathbb{T} - d_\beta\mu_\pi \xrightarrow{d} \mathbb{T}^* \sim N(0, I_{K_2}),$$

where $\mathbb{S}$ *and* $\mathbb{T}$ *are independent with*

1. *If* $0 \leq \delta < \frac{1}{2}$, $\mu_\pi \to \infty$,
2. *If* $\delta = \frac{1}{2}$, $\mu_\pi = O(1)$, *and*
3. *If* $\frac{1}{2} < \delta < \infty$, $\mu_\pi = o(1)$.

Hence, under the null

$$\mathbb{S} \xrightarrow{d} N\left(0, I_{K_2}\right)$$

and

$$AR = \frac{\Theta_{\mathbb{S}}}{K_2} \xrightarrow{d} \frac{\chi^2_{K_2}}{K_2}$$

for all values of δ. Note that

$$LM = \frac{\Theta^2_{\mathbb{ST}}}{\Theta_{\mathbb{T}}} = \frac{(\mathbb{S}'\mathbb{T})^2}{\mathbb{T}'\mathbb{T}} = \mathbb{S}' P_{\mathbb{T}} \mathbb{S},$$

where $P_{\mathbb{T}} = \mathbb{T}(\mathbb{T}'\mathbb{T})^{-1}\mathbb{T}'$ is a symmetric idempotent matrix with $rank(P_{\mathbb{T}}) = K_2$. By Proposition B.3.1 in Lütkepohl (2005), we have

$$LM \xrightarrow{d} \chi^2_{K_2}$$

for all values of δ. Because Σ^* is unknown, it must be replaced by a consistent estimator, $\hat{\Sigma}^*$. Critical values for the CLR statistic can be found in Andrews, Moreira, and Stock (2006).

CLR, LM, and AR tests have good size properties under all values of δ, i.e., strong and weak IVs. However, they may have different power properties. Deriving the asymptotic power envelops and power upper bounds is an interesting question which we leave for future research.

MONTE CARLO SIMULATION

In this section, we report some Monte Carlo results that examine the finite sample properties of the cross-section 2SLS and the panel data W2SLS estimator when $L = 1$. Following Staiger and Stock (1997), instruments Z_{1t}, Z_{2t}, Z_{3t}, and Z_{4t} are assumed to be standard normal variables and X_t is the constant 1; the errors $(u_t, V_t)'$ are generated from an i.i.d. bivariate normal distribution with $\Sigma = \begin{pmatrix} 1 & 0.99 \\ 0.99 & 1 \end{pmatrix}$. The true value of β, γ, and Γ is set as 0, 1, and 1 respectively. We set $C=0.5$. Individual fixed effects μ and α are generated from independent standard normal distributions. To summarize, the data generating process (DGP) is given by

Table 1. Bias, IQR, and RMSE of W2SLS and LIML Estimators.

			Bias		IQR		RMSE	
n	T	δ	W2SLS	LIML	W2SLS	LIML	W2SLS	LIML
50	1	0	0.473	0.447	0.647	1.692	0.673	1.331
50	10	0	0.259	−0.014	0.293	0.485	0.338	0.359
50	10	0.2	0.490	0.010	0.299	0.779	0.537	0.577
50	10	0.5	0.798	0.329	0.262	0.913	0.821	0.752
50	10	0.8	0.933	0.735	0.194	1.057	0.944	1.074
50	20	0	0.142	0.004	0.225	0.297	0.219	0.220
50	20	0.2	0.367	0.012	0.275	0.581	0.419	0.430
50	20	0.5	0.769	0.300	0.263	0.822	0.793	0.679
50	20	0.8	0.943	0.776	0.180	0.825	0.952	0.988
50	50	0	0.059	0.010	0.173	0.189	0.142	0.140
50	50	0.2	0.253	0.021	0.275	0.416	0.325	0.309
50	50	0.5	0.758	0.299	0.260	0.795	0.782	0.661
50	50	0.8	0.954	0.858	0.147	0.693	0.960	1.000
50	100	0	0.037	0.008	0.121	0.132	0.097	0.098
50	100	0.2	0.179	0.020	0.231	0.332	0.248	0.246
50	100	0.5	0.758	0.327	0.242	0.796	0.779	0.674
50	100	0.8	0.969	0.928	0.131	0.502	0.974	1.000
100	1	0	0.447	0.375	0.621	1.593	0.641	1.238
100	10	0	0.246	−0.027	0.287	0.488	0.325	0.362
100	10	0.2	0.477	0.004	0.292	0.777	0.523	0.576
100	10	0.5	0.784	0.303	0.287	0.887	0.813	0.724
100	10	0.8	0.929	0.723	0.214	1.156	0.943	1.121
100	20	0	0.139	0.007	0.241	0.324	0.227	0.240
100	20	0.2	0.361	0.014	0.271	0.605	0.413	0.449
100	20	0.5	0.767	0.317	0.271	0.936	0.793	0.762
100	20	0.8	0.941	0.798	0.183	0.885	0.951	1.033
100	50	0	0.052	−0.005	0.161	0.186	0.130	0.138
100	50	0.2	0.224	−0.010	0.267	0.430	0.299	0.319
100	50	0.5	0.755	0.266	0.260	0.915	0.780	0.728
100	50	0.8	0.964	0.857	0.157	0.684	0.971	0.995
100	100	0	0.031	0.000	0.126	0.135	0.098	0.100
100	100	0.2	0.170	−0.001	0.237	0.349	0.244	0.258
100	100	0.5	0.760	0.303	0.246	0.953	0.782	0.768
100	100	0.8	0.973	0.916	0.142	0.511	0.978	0.991

$$y_t = 1 + \mu + u_t \tag{11}$$

and

$$Y_t = 1 + \frac{0.5}{\sqrt{n}T^{\delta}}(Z_{1t} + Z_{2t} + Z_{3t} + Z_{4t}) + \alpha + V_t. \tag{12}$$

for $t = 1,2,\ldots,T$. The cross-section sample size n takes the values (50,100), while the time-series sample size T takes the values (1,10,20,50,100). δ takes the values (0,0.2,0.5,0.8) for the panel data case, i.e., for $T > 1$ and $\delta = 0$ for the cross-sectional case, i.e., $T = 1$. For each experiment, we perform 1,000.

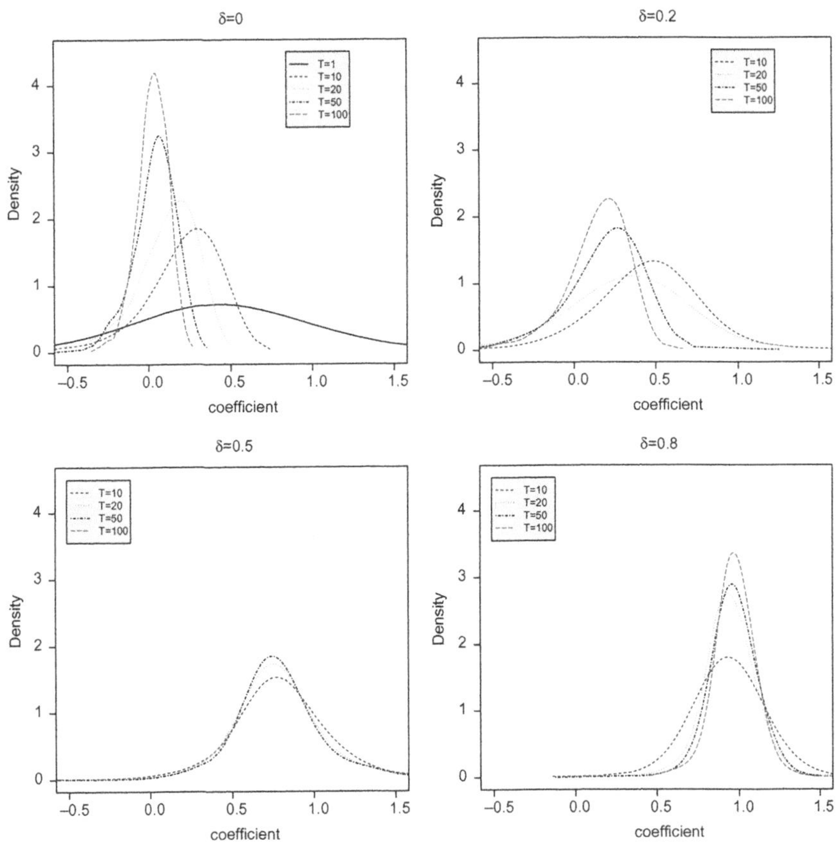

Fig. 1. Density of the W2SLS Estimator (n = 100).

replications. For each replication we estimate the model using W2SLS and LIML estimators of β. Table 1 reports the root mean squared error (RMSE) of these estimators for various values of n, T, and δ. Following Kelejian and Prucha (1999), RMSE is defined as $[bias^2 + (IQR/1.35)^2]^{1/2}$, where $bias$ is the difference between the median and the true parameter value and IQR is the interquantile range. That is $IQR = c_1 - c_2$, where c_1 and c_2 are the 0.75 and 0.25 quantiles, respectively. As explained in Kelejian and Prucha (1999),

Table 2. Size of t-Test.

n	T	δ	W2SLS	LIML
50	1	0	0.153	0.098
50	10	0	0.347	0.091
50	10	0.2	0.587	0.127
50	10	0.5	0.851	0.208
50	10	0.8	0.946	0.393
50	20	0	0.191	0.075
50	20	0.2	0.461	0.108
50	20	0.5	0.862	0.213
50	20	0.8	0.972	0.473
50	50	0	0.118	0.062
50	50	0.2	0.331	0.087
50	50	0.5	0.835	0.214
50	50	0.8	0.967	0.534
50	100	0	0.098	0.055
50	100	0.2	0.252	0.091
50	100	0.5	0.862	0.201
50	100	0.8	0.988	0.634
100	1	0	0.132	0.080
100	10	0	0.309	0.084
100	10	0.2	0.565	0.115
100	10	0.5	0.850	0.188
100	10	0.8	0.935	0.384
100	20	0	0.211	0.081
100	20	0.2	0.444	0.116
100	20	0.5	0.847	0.215
100	20	0.8	0.963	0.468
100	50	0	0.102	0.049
100	50	0.2	0.289	0.085
100	50	0.5	0.849	0.161
100	50	0.8	0.974	0.527
100	100	0	0.080	0.054
100	100	0.2	0.232	0.077
100	100	0.5	0.879	0.198
100	100	0.8	0.974	0.609

these characteristics are closely related to the standard measures of bias and RMSE but, unlike these measures, are assured to exist. We can see that LIML has a smaller bias than the W2SLS estimator; however, W2SLS has a smaller IQR and RMSE than the LIML estimator. Fig. 1 shows the density function of W2SLS estimator for $n = 100$. As we can see in the graph, when $\delta = 0$ or 0.2, the distribution tends to center at zero as T increases. When $\delta = 0.5$ or 0.8, the distribution does not change much as T increases. Table 2

Table 3. Size of Robust Tests.

n	T	δ	AR	LM	CLR
50	1	0	0.065	0.065	0.069
50	10	0	0.055	0.057	0.056
50	10	0.2	0.055	0.055	0.055
50	10	0.5	0.055	0.055	0.051
50	10	0.8	0.055	0.050	0.053
50	20	0	0.050	0.048	0.046
50	20	0.2	0.050	0.050	0.052
50	20	0.5	0.050	0.058	0.059
50	20	0.8	0.050	0.062	0.057
50	50	0	0.058	0.053	0.053
50	50	0.2	0.058	0.053	0.052
50	50	0.5	0.058	0.057	0.057
50	50	0.8	0.058	0.056	0.056
50	100	0	0.050	0.042	0.042
50	100	0.2	0.050	0.042	0.042
50	100	0.5	0.050	0.045	0.047
50	100	0.8	0.050	0.054	0.048
100	1	0	0.054	0.050	0.050
100	10	0	0.040	0.053	0.053
100	10	0.2	0.040	0.053	0.054
100	10	0.5	0.040	0.048	0.047
100	10	0.8	0.040	0.046	0.046
100	20	0	0.050	0.045	0.045
100	20	0.2	0.050	0.046	0.046
100	20	0.5	0.050	0.042	0.041
100	20	0.8	0.050	0.044	0.043
100	50	0	0.043	0.047	0.047
100	50	0.2	0.043	0.048	0.048
100	50	0.5	0.043	0.049	0.050
100	50	0.8	0.043	0.041	0.054
100	100	0	0.044	0.056	0.056
100	100	0.2	0.044	0.058	0.058
100	100	0.5	0.044	0.064	0.061
100	100	0.8	0.044	0.057	0.055

reports the size of the t-test for $\beta = 0$. Results from Table 2 confirm that the t-tests using the W2SLS and LIML estimators are not robust with respect to weak instruments. Table 3 reports the results of the AR, LM, and CLR tests. Table 3 indicates that the robust tests are indeed robust to the weak instruments in this panel data design.

CONCLUSION

Following the extensive literature on weak instruments surveyed by Stock et al. (2002), this chapter proposes k-class IV estimators and test statistics and studies their behavior when the available instruments are weak in a fixed effects panel data model. It is important to note that Cai et al. (2012) studied this fixed effects panel data model, but they let the degree of weakness of the instruments depend upon n^{δ}, where $\delta \geq 0$, and studied the asymptotic properties of W2SLS and pivotal statistics for fixed T and $n \to \infty$. In contrast, our study let the degree of weakness of the instruments depend upon $\sqrt{n}T^{\delta}$ and studies the asymptotic properties of k-class IV estimators and pivotal test statistics as both $(n,T) \to \infty$. Both papers argue that there are benefits to panel data in reducing the bias of W2SLS and k-class IV estimators in case of weak instruments. Monte Carlo results confirm these asymptotic results.

ACKNOWLEDGMENTS

We would like to thank the editor Dek Terrell and the participants of the 30th Anniversary *Advances in Econometrics* Conference held at *Louisiana State University*, Baton Rouge, Louisiana, March 23–25, 2012 for their comments and suggestions. We dedicate this chapter in honor of Carter Hill and Thomas Fomby for their enormous contributions as series editors of the *Advances in Econometrics* and for their valuable contributions to econometrics.

NOTES

1. How n and T tend to infinity was emphasized by Phillips and Moon (1999) for panel unit root testing.

2. Cai et al. (2012) also consider some asymptotically pivotal tests in the case of fixed effects panel data and study their asymptotic properties for fixed T and $n \to \infty$.

3. This is the Hahn and Kuersteiner (2002) result for a cross-sectional IV regression.

4. See Stock et al. (2002) for an important summary of the advantages and disadvantages of k-class estimators.

5. Stock and Yogo (2005) observe that the limiting distribution of $g_{\min}$ will depend upon all of the eigenvalues of $\lambda'\lambda$.

6. It is important to note that Cai et al. (2012) also considered the Anderson and Rubin (1949), the Kleibergen (2002), and the Moreira (2003) CLR and studied their asymptotic properties for fixed T and $n \to \infty$.

REFERENCES

Anderson, T. W., & Rubin, H. (1949). Estimators of the parameters of a single equation in a complete set of stochastic equations. *Annals of Mathematical Statistics*, *21*, 570–582.

Andrews, D. W. K., Moreira, M., & Stock, J. H. (2006). Optimal two-sided invariant similar tests for instrumental variable regression. *Econometrica*, *74*, 715–752.

Andrews, D. W. K., & Stock, J. H. (2006). Testing with many weak instruments. *Journal of Econometrics*, *138*, 24–46.

Andrews, D. W. K., & Stock, J. H. (2007). Inference with weak instruments. In R. Blundell, W. K. Newey & T. Persson (Eds.), *Advances in economics and econometrics: Theory and applications, Ninth world congress of the econometric society*. Cambridge, UK: Cambridge University Press.

Bai, J., & Ng, S. (2010). Instrumental variable estimation in a data rich environment. *Econometric Theory*, *26*(6), 1577–1606.

Cai, Z., Fang, Y., & Li, H. (2012). Weak instrumental variables models for longitudinal data. *Econometric Reviews*, *31*, 361–389.

Cragg, J. G., & Donald, S. G. (1993). Testing identifiability and specification in instrumental variable models. *Econometric Theory*, *9*, 222–240.

Donald, S. G., & Newey, W. K. (2001). Choosing the number of instruments. *Econometrica*, *69*, 1161–1191.

Hahn, J., & Kuersteiner, G. (2002). Discontinuities of weak instrument limiting distributions. *Economics Letters*, *75*, 325–331.

Hahn, J., Hausman, J., & Kuersteiner, G. (2004). Estimation with weak instruments: Accuracy of higher-order bias and MSE approximations. *The Econometrics Journal*, *7*(1), 272–306.

Kao, C. (1999). Spurious regression and residual-based tests for cointegration in panel data. *Journal of Econometrics*, *90*, 1–44.

Kelejian, H. H., & Prucha, I. R. (1999). A generalized moments estimator for the autoregressive parameter in spatial model. *International Economic Review*, *40*, 509–533.

Kleibergen, F. (2002). Pivotal statistics for testing structural parameters in instrumental variable regression. *Econometrica*, *70*, 1781–1803.

Lütkepohl, H. (2005). *New introduction to multiple time series analysis*. Berlin: Springer.

Moon, H. R., Perron, B., & Phillips, P. C. B. (2007). Incidental trends and the power of panel unit root tests. *Journal of Econometrics*, *141*, 416–459.

Moreira, M. J. (2003). A conditional likelihood ratio test for structural models. *Econometrica, 71*, 1027–1048.

Moreira, M. J. (2009). Tests with correct size when instruments can be arbitrarily weak. *Journal of Econometrics, 152*, 131–140.

Phillips, P. C. B., & Moon, H. R. (1999). Linear regression limit theory for nonstationary panel data. *Econometrica, 67*, 1057–1111.

Rothenberg, T. J. (1984). Approximating the distribution of econometric estimators and test statistics. In Z. Griliches & M. D. Intriligator (Eds.), *Handbook of econometrics* (Vol. 2, 881–935). Amsterdam: North-Holland.

Staiger, D., & Stock, J. H. (1997). Instrumental variables regression with weak instruments. *Econometrica, 65*, 557–587.

Stock, J. H., Wright, H. J., & Yogo, M. (2002). A survey of weak instruments and weak identification in generalized method of moments. *Journal of Business and Economic Statistics, 20*, 518–529.

Stock, J. H., & Yogo, M. (2005). Testing for weak instruments in linear IV regression. In D. W. K. Andrews & J. H. Stock (Eds.), *Identification and inference for econometric models: Essays in honor of Thomas Rothenberg* (Ch. 5, pp. 80–108). Cambridge, UK: Cambridge University Press.

APPENDIX

To prove Theorem 1, we need the following Lemma.

Lemma 1. *As* $(n, T) \to \infty$, *for* $0 \le \delta < \frac{1}{2}$

$$\frac{1}{T^{1-\delta}n^{\frac{1}{2}}}\tilde{Z}'\tilde{Y} = \Phi_{ZZ}C + O_p\left(\frac{1}{T^{\frac{1}{2}-\delta}}\right),$$

for $\delta = \frac{1}{2}$

$$\frac{1}{\sqrt{Tn}}\tilde{Z}'\tilde{Y} \xrightarrow{d} \Phi_{ZZ}C + \Psi_{ZV},$$

and for $\frac{1}{2} < \delta < \infty$.

$$\frac{1}{\sqrt{Tn}}\tilde{Z}'\tilde{Y} \xrightarrow{d} \Psi_{ZV}$$

Proof. First we note that for $0 \le \delta \le \frac{1}{2}$

$$\begin{aligned}
\frac{1}{T^{1-\delta}n^{\frac{1}{2}}}\tilde{Z}'\tilde{Y} &= \frac{1}{T^{1-\delta}n}\sum_{t=1}^{T}\tilde{Z}'_t\tilde{Y}_t \\
&= \frac{1}{T^{1-\delta}n^{\frac{1}{2}}}\sum_{t=1}^{T}\tilde{Z}'_t\tilde{Z}_t\Pi + \frac{1}{T^{1-\delta}n^{\frac{1}{2}}}\sum_{t=1}^{T}\tilde{Z}'_t\tilde{V}_t \\
&= \frac{1}{Tn}\sum_{t=1}^{T}\tilde{Z}'_t\tilde{Z}_t\left(T^{\delta}n^{\frac{1}{2}}\Pi\right) + \frac{1}{T^{1-\delta}n^{\frac{1}{2}}}\sum_{t=1}^{T}\tilde{Z}'_t\tilde{V}_t \\
&= \frac{1}{Tn}\sum_{t=1}^{T}\tilde{Z}'_t\tilde{Z}_tC + \frac{1}{T^{\frac{1}{2}-\delta}}\frac{1}{\sqrt{Tn}}\sum_{t=1}^{T}\tilde{Z}'_t\tilde{V}_t \\
&= \Phi_{ZZ}C + O_p\left(\frac{1}{T^{\frac{1}{2}-\delta}}\right)
\end{aligned}$$

as $(n, T) \to \infty$ since $\frac{1}{\sqrt{Tn}}\sum_{t=1}^{T}\tilde{Z}'_t\tilde{V}_t = O_p(1)$.
For $\delta = \frac{1}{2}$

$$\frac{1}{\sqrt{Tn}}\tilde{Z}'\tilde{Y} = \frac{1}{Tn}\sum_{t=1}^{T}\tilde{Z}'_t\tilde{Z}_t\left(\sqrt{Tn}\Pi\right) + \frac{1}{\sqrt{Tn}}\sum_{t=1}^{T}\tilde{Z}'_t\tilde{V}_t \xrightarrow{d} \Phi_{ZZ}C + \Psi_{ZV}.$$

Finally, for $\frac{1}{2} < \delta < \infty$

$$\begin{aligned}
\frac{1}{\sqrt{Tn}}\tilde{Z}'\tilde{Y} &= \frac{1}{\sqrt{Tn}}\sum_{t=1}^{T}\tilde{Z}'_t\tilde{Y}_t \\
&= \frac{1}{\sqrt{Tn}}\sum_{t=1}^{T}\tilde{Z}'_t\tilde{Z}_t\Pi + \frac{1}{\sqrt{Tn}}\sum_{t=1}^{T}\tilde{Z}'_t\tilde{V}_t \\
&= \left(\frac{1}{Tn}\sum_{t=1}^{T}\tilde{Z}'_t\tilde{Z}_t\right)\left(\frac{C}{T^{\delta-1/2}}\right) + \frac{1}{\sqrt{Tn}}\sum_{t=1}^{T}\tilde{Z}'_t\tilde{V}_t \xrightarrow{d} \Psi_{ZV} + O_p\left(\frac{1}{T^{\delta-1/2}}\right)
\end{aligned}$$

using $\frac{1}{Tn}\sum_{t=1}^{T}\tilde{Z}'_t\tilde{Z}_t = O_p(1)$ and $\frac{1}{\sqrt{Tn}}\sum_{t=1}^{T}\tilde{Z}'_t\tilde{V}_t \xrightarrow{d} \Psi_{ZV}$. This proves the lemma.

Proof of Theorem 1

Proof. Consider

$$\hat{\beta}_{W2SLS} - \beta = \frac{\tilde{Y}'P_{\tilde{Z}}\tilde{u}}{\tilde{Y}'P_{\tilde{Z}}\tilde{Y}} = \frac{\tilde{Y}'\tilde{Z}(\tilde{Z}'\tilde{Z})^{-1}\tilde{Z}'\tilde{u}}{\tilde{Y}'\tilde{Z}(\tilde{Z}'\tilde{Z})^{-1}\tilde{Z}'\tilde{Y}}.$$

For $0 < \delta < \frac{1}{2}$

$$\begin{aligned}
T^{\frac{1}{2}-\delta}(\hat{\beta}_{W2SLS} - \beta) &= \frac{\left(\frac{1}{T^{1-\delta}\sqrt{n}}\tilde{Y}'\tilde{Z}\right)\left(\frac{1}{nT}\tilde{Z}'\tilde{Z}\right)^{-1}\left(\frac{1}{\sqrt{Tn}}\tilde{Z}'\tilde{u}\right)}{\left(\frac{1}{T^{1-\delta}\sqrt{n}}\tilde{Y}'\tilde{Z}\right)\left(\frac{1}{nT}\tilde{Z}'\tilde{Z}\right)^{-1}\left(\frac{1}{T^{1-\delta}\sqrt{n}}\tilde{Z}'\tilde{Y}\right)} \xrightarrow{d} \frac{\Psi_{Zu}}{\Phi_{ZZ}C} \\
&= N(0, \sigma_{uu}(C'\Phi_{ZZ}C)^{-1}),
\end{aligned}$$

for $\delta = \frac{1}{2}$

$$\begin{aligned}
\hat{\beta}_{W2SLS} - \beta &= \frac{\left(\frac{1}{\sqrt{Tn}}\tilde{Y}'\tilde{Z}\right)\left(\frac{1}{nT}\tilde{Z}'\tilde{Z}\right)^{-1}\left(\frac{1}{\sqrt{Tn}}\tilde{Z}'\tilde{u}\right)}{\left(\frac{1}{\sqrt{Tn}}\tilde{Y}'\tilde{Z}\right)\left(\frac{1}{nT}\tilde{Z}'\tilde{Z}\right)^{-1}\left(\frac{1}{\sqrt{Tn}}\tilde{Z}'\tilde{Y}\right)} \\
&= \frac{(\Phi_{ZZ}C + \Psi_{ZV})'\Phi_{ZZ}^{-1}\Psi_{Zu}}{(\Phi_{ZZ}C + \Psi_{ZV})'\Phi_{ZZ}^{-1}(\Phi_{ZZ}C + \Psi_{ZV})} = O_p(1)
\end{aligned}$$

and for $\frac{1}{2} < \delta < \infty$

$$\hat{\beta}_{W2SLS} - \beta = \frac{\tilde{Y}'P_{\tilde{Z}}\tilde{u}}{\tilde{Y}'P_{\tilde{Z}}\tilde{Y}} = \frac{\left(\frac{1}{\sqrt{Tn}}\tilde{Y}'\tilde{Z}\right)\left(\frac{1}{nT}\tilde{Z}'\tilde{Z}\right)^{-1}\left(\frac{1}{\sqrt{Tn}}\tilde{Z}'\tilde{u}\right)}{\left(\frac{1}{\sqrt{Tn}}\tilde{Y}'\tilde{Z}\right)\left(\frac{1}{nT}\tilde{Z}'\tilde{Z}\right)^{-1}\left(\frac{1}{\sqrt{Tn}}\tilde{Z}'\tilde{Y}\right)} = \frac{\Psi'_{ZV}\Phi_{ZZ}^{-1}\Psi_{Zu}}{\Psi'_{ZV}\Phi_{ZZ}^{-1}\Psi_{ZV}} = O_p(1).$$

This proves the theorem.

To prove Theorem 2, we need the following Lemma.

Lemma 2. *Under Assumptions 1 and 2, as* $(n, T) \to \infty$:

1. $\frac{1}{Tn}\tilde{u}^{\perp'}\tilde{u}^{\perp} \xrightarrow{p} \sigma_{uu}$, $\frac{1}{Tn}\tilde{Y}^{\perp'}\tilde{u}^{\perp} \xrightarrow{p} \Sigma_{Vu}$, $\frac{1}{Tn}\tilde{Y}^{\perp'}\tilde{Y}^{\perp} \xrightarrow{p} \Sigma_{VV}$, *and* $\frac{1}{Tn}\tilde{Z}^{\perp'}\tilde{Z}^{\perp} \xrightarrow{p} \Phi_{ZZ} - \Phi_{Z\tilde{X}}\Phi_{\tilde{X}\tilde{X}}^{-1}\Phi_{\tilde{X}Z} = \Omega$
2. $\frac{1}{\sqrt{Tn}}\tilde{Z}^{\perp'}\tilde{u}^{\perp} \xrightarrow{d} \Psi_{Zu} - \Phi_{Z\tilde{X}}\Phi_{\tilde{X}\tilde{X}}^{-1}\Psi_{\tilde{X}u}$, $P_{\tilde{Z}^{\perp}}^{1/2}\tilde{u}^{\perp} \xrightarrow{d} \sigma_{uu}^{1/2} z_u$, *and* $\tilde{u}^{\perp'} P_{\tilde{Z}^{\perp}}\tilde{u}^{\perp} \xrightarrow{d} \sigma_{uu} z'_u z_u$
3. *For* $0 \le \delta \le \frac{1}{2}$, $\frac{1}{T^{1-\delta}n^{\frac{1}{2}}}\tilde{Z}^{\perp'}\tilde{Y}^{\perp} \xrightarrow{p} \Omega C$, $\frac{1}{T^{1/2-\delta}}P_{\tilde{Z}^{\perp}}^{1/2}\tilde{Y}^{\perp} \xrightarrow{p} \lambda\Sigma_{VV}^{1/2}$, $\frac{1}{T^{1/2-\delta}}\tilde{Y}^{\perp'}P_{\tilde{Z}^{\perp}}\tilde{u}^{\perp} \xrightarrow{d} \sigma_{uu}^{1/2}\Sigma_{VV}^{1/2'}\lambda' z_u$, *and* $\frac{1}{T^{1-2\delta}}\tilde{Y}^{\perp'}P_{\tilde{Z}^{\perp}}\tilde{Y}^{\perp} \xrightarrow{p} \Sigma_{VV}^{1/2'}\lambda'\lambda\Sigma_{VV}^{1/2}$; *for* $\delta = \frac{1}{2}$, $\frac{1}{\sqrt{Tn}}\tilde{Z}^{\perp'}\tilde{Y}^{\perp} \xrightarrow{d} \Omega C + \Omega^{1/2} z_V \Sigma_{VV}^{1/2}$, $P_{\tilde{Z}^{\perp}}^{1/2}\tilde{Y}^{\perp} \xrightarrow{d} (\lambda + z_V)\Sigma_{VV}^{1/2}$, $\tilde{Y}^{\perp'}P_{\tilde{Z}^{\perp}}\tilde{u}^{\perp} \xrightarrow{d} \sigma_{uu}^{1/2}\Sigma_{VV}^{1/2'}(\lambda + z_V)' z_u$, *and* $\tilde{Y}^{\perp'}P_{\tilde{Z}^{\perp}}\tilde{Y}^{\perp} \xrightarrow{d} \Sigma_{VV}^{1/2'}(\lambda + z_V)'(\lambda + z_V)\Sigma_{VV}^{1/2}$; *for* $\frac{1}{2} < \delta < \infty$, $\frac{1}{\sqrt{Tn}}\tilde{Z}^{\perp'}\tilde{Y}^{\perp} \xrightarrow{d} \Omega^{1/2} z_V \Sigma_{VV}^{1/2}$, $P_{\tilde{Z}^{\perp}}^{1/2}\tilde{Y}^{\perp} \xrightarrow{d} z_V\Sigma_{VV}^{1/2}$, $\tilde{Y}^{\perp'}P_{\tilde{Z}^{\perp}}\tilde{u}^{\perp} \xrightarrow{d} \sigma_{uu}^{1/2}\Sigma_{VV}^{1/2'} z'_V z_u$, *and* $\tilde{Y}^{\perp'}P_{\tilde{Z}^{\perp}}\tilde{Y}^{\perp} \xrightarrow{d} \Sigma_{VV}^{1/2'} z'_V z_V \Sigma_{VV}^{1/2}$.

Proof. Consider part 1. First we note that

$$\frac{1}{Tn}\tilde{u}^{\perp'}\tilde{u}^{\perp} = \frac{1}{Tn}\tilde{u}' M_{\tilde{X}}\tilde{u} = \frac{1}{Tn}\tilde{u}'\tilde{u} - \left(\frac{1}{Tn}\tilde{u}'\tilde{X}\right)\left(\frac{1}{Tn}\tilde{X}'\tilde{X}\right)^{-1}\left(\frac{1}{Tn}\tilde{X}'\tilde{u}\right) \xrightarrow{p} \sigma_{uu}$$

using $\frac{1}{Tn}\tilde{u}'\tilde{u} \xrightarrow{p} \sigma_{uu}$, $\frac{1}{Tn}\tilde{X}'\tilde{X} = O_p(1)$, and $\frac{1}{Tn}\tilde{X}'\tilde{u} = o_p(1)$. Next:

$$\begin{aligned}
\frac{1}{Tn}\tilde{Y}^{\perp'}\tilde{u}^{\perp} &= \frac{1}{Tn}\tilde{Y}' M_{\tilde{X}}\tilde{u} \\
&= \frac{1}{Tn}\Pi'\tilde{Z}' M_{\tilde{X}}\tilde{u} + \frac{1}{Tn}\tilde{V}' M_{\tilde{X}}\tilde{u} \\
&= \Pi'\left[\frac{1}{Tn}\tilde{Z}'\tilde{u} - \left(\frac{1}{Tn}\tilde{Z}'\tilde{X}\right)\left(\frac{1}{Tn}\tilde{X}'\tilde{X}\right)^{-1}\left(\frac{1}{Tn}\tilde{X}'\tilde{u}\right)\right] \\
&\quad + \left[\frac{1}{Tn}\tilde{V}'\tilde{u} - \left(\frac{1}{Tn}\tilde{V}'\tilde{X}\right)\left(\frac{1}{Tn}\tilde{X}'\tilde{X}\right)^{-1}\left(\frac{1}{Tn}\tilde{X}'\tilde{u}\right)\right] \xrightarrow{p} \Sigma_{Vu}
\end{aligned}$$

using $\Pi = o(1)$, $\frac{1}{Tn}\tilde{Z}'\tilde{u} = o_p(1)$, $\frac{1}{Tn}\tilde{Z}'\tilde{X} = O_p(1)$, $\frac{1}{Tn}\tilde{X}'\tilde{X} = O_p(1)$, $\frac{1}{Tn}\tilde{X}'\tilde{u} = o_p(1)$, $\frac{1}{Tn}\tilde{V}'\tilde{X} = o_p(1)$, and $\frac{1}{Tn}\tilde{V}'\tilde{u} \xrightarrow{p} \Sigma_{Vu}$. Similarly:

$$\begin{aligned}
\frac{1}{Tn}\tilde{Y}^{\perp'}\tilde{Y}^{\perp} &= \frac{1}{Tn}\tilde{Y}' M_{\tilde{X}}\tilde{Y} \\
&= \frac{1}{Tn}\Pi'\tilde{Z}' M_{\tilde{X}}\tilde{Z}\Pi + \frac{1}{Tn}\tilde{V}' M_{\tilde{X}}\tilde{Z}\Pi + \frac{1}{Tn}\Pi'\tilde{Z}' M_{\tilde{X}}\tilde{V} + \frac{1}{Tn}\tilde{V}' M_{\tilde{X}}\tilde{V} \\
&= \left[\frac{1}{Tn}\tilde{V}'\tilde{u} - \left(\frac{1}{Tn}\tilde{V}'\tilde{X}\right)\left(\frac{1}{Tn}\tilde{X}'\tilde{X}\right)^{-1}\left(\frac{1}{Tn}\tilde{X}'\tilde{V}\right)\right] + o_p(1) \xrightarrow{p} \Sigma_{VV}
\end{aligned}$$

since $\Pi = o(1)$, $\frac{1}{Tn}\tilde{X}'\tilde{X} \xrightarrow{p} \Phi_{\tilde{X}\tilde{X}}$, $\frac{1}{Tn}\tilde{V}'\tilde{X} \xrightarrow{p} 0$, and $\frac{1}{Tn}\tilde{V}'\tilde{u} \xrightarrow{p} \Sigma_{Vu}$. Finally:

$$\frac{1}{Tn}\tilde{Z}^{\perp'}\tilde{Z}^{\perp} = \frac{1}{Tn}\tilde{Z}'M_{\tilde{X}}\tilde{Z} = \frac{1}{Tn}\tilde{Z}'\tilde{Z} - \left(\frac{1}{Tn}\tilde{Z}'\tilde{X}\right)\left(\frac{1}{Tn}\tilde{X}'\tilde{X}\right)^{-1}\left(\frac{1}{Tn}\tilde{X}'\tilde{Z}\right)$$
$$\xrightarrow{p} \Phi_{ZZ} - \Phi_{Z\tilde{X}}\Phi_{\tilde{X}\tilde{X}}^{-1}\Phi_{\tilde{X}Z} = \Omega$$

as $(n, T) \to \infty$ proving part (1).

Consider part 2. Note that

$$\frac{1}{\sqrt{Tn}}\tilde{Z}^{\perp'}\tilde{u}^{\perp} = \frac{1}{\sqrt{Tn}}\tilde{Z}'M_{\tilde{X}}\tilde{u}$$
$$= \frac{1}{\sqrt{Tn}}\tilde{Z}'\tilde{u} - \left(\frac{1}{Tn}\tilde{Z}'\tilde{X}\right)\left(\frac{1}{Tn}\tilde{X}'\tilde{X}\right)^{-1}\left(\frac{1}{\sqrt{Tn}}\tilde{X}'\tilde{u}\right)$$
$$\xrightarrow{d} \Psi_{Zu} - \Phi_{Z\tilde{X}}\Phi_{\tilde{X}\tilde{X}}^{-1}\Psi_{\tilde{X}u}$$

since $\frac{1}{\sqrt{Tn}}\tilde{X}'\tilde{u} \xrightarrow{d} \Psi_{\tilde{X}u}$ and $\frac{1}{\sqrt{Tn}}\tilde{Z}'\tilde{u} \xrightarrow{d} \Psi_{Zu}$. Therefore:

$$P_{\tilde{Z}^{\perp}}^{1/2}\tilde{u}^{\perp} = \left(\frac{1}{Tn}\tilde{Z}^{\perp'}\tilde{Z}^{\perp}\right)^{-1/2}\left(\frac{1}{\sqrt{Tn}}\tilde{Z}^{\perp'}\tilde{u}^{\perp}\right) \xrightarrow{d} \Omega^{-1/2}(\Psi_{Zu} - \Phi_{Z\tilde{X}}\Phi_{\tilde{X}\tilde{X}}^{-1}\Psi_{\tilde{X}u}) = \sigma_{uu}^{1/2}z_u$$

and

$$\tilde{u}^{\perp'}P_{\tilde{Z}^{\perp}}\tilde{u}^{\perp} = \left(\tilde{u}^{\perp'}P_{\tilde{Z}^{\perp}}^{1/2}\right)\left(P_{\tilde{Z}^{\perp}}^{1/2}\tilde{u}^{\perp}\right) \xrightarrow{d} \sigma_{uu}z'_u z_u.$$

Consider part 3. For $0 \leq \delta < \frac{1}{2}$, we have

$$\frac{1}{T^{1-\delta}n^{\frac{1}{2}}}\tilde{Z}^{\perp'}\tilde{Y}^{\perp} = \frac{1}{T^{1-\delta}n^{\frac{1}{2}}}\tilde{Z}'M_{\tilde{X}}\tilde{Z}\Pi + \frac{1}{T^{1-\delta}n^{\frac{1}{2}}}\tilde{Z}'M_{\tilde{X}}\tilde{V}$$
$$= \left[\left(\frac{1}{Tn}\tilde{Z}'\tilde{Z}\right) - \left(\frac{1}{Tn}\tilde{Z}'\tilde{X}\right)\left(\frac{1}{Tn}\tilde{X}'\tilde{X}\right)^{-1}\left(\frac{1}{Tn}\tilde{X}'\tilde{Z}\right)\right](T^{\delta}n^{\frac{1}{2}}\Pi)$$
$$+ \frac{1}{T^{\frac{1}{2}-\delta}}\left[\left(\frac{1}{\sqrt{Tn}}\tilde{Z}'\tilde{V}\right) - \left(\frac{1}{Tn}\tilde{Z}'\tilde{X}\right)\left(\frac{1}{Tn}\tilde{X}'\tilde{X}\right)^{-1}\left(\frac{1}{\sqrt{Tn}}\tilde{X}'\tilde{V}\right)\right]$$
$$= \Omega C + O_p\left(\frac{1}{T^{\frac{1}{2}-\delta}}\right)$$

with $\Omega = \Phi_{ZZ} - \Phi_{Z\tilde{X}}\Phi_{\tilde{X}\tilde{X}}^{-1}\Phi_{\tilde{X}Z}$ because $\frac{1}{\sqrt{Tn}}\tilde{Z}'\tilde{V} \xrightarrow{d} \Psi_{ZV}$, $\frac{1}{Tn}\tilde{Z}'\tilde{X} \xrightarrow{p} \Phi_{Z\tilde{X}}$, $\frac{1}{Tn}\tilde{X}'\tilde{X} \xrightarrow{p} \Phi_{\tilde{X}\tilde{X}}$, and $\frac{1}{\sqrt{Tn}}\tilde{X}'\tilde{V} \xrightarrow{d} \Psi_{\tilde{X}V}$. Together with Lemmas 1 and 2, we obtain

$$\frac{1}{T^{\frac{1}{2}-\delta}} P_{\tilde{Z}^{\perp}}^{1/2} \tilde{Y}^{\perp} = \left(\frac{1}{Tn} \tilde{Z}^{\perp'} \tilde{Z}^{\perp}\right)^{-1/2} \left(\frac{1}{T^{1-\delta} n^{\frac{1}{2}}} \tilde{Z}^{\perp'} \tilde{Y}^{\perp}\right) \xrightarrow{p} \Omega^{-1/2} \Omega C = \Omega^{1/2} C = \lambda \Sigma_{VV}^{1/2},$$

$$\frac{1}{\sqrt{Tn}} \tilde{Y}^{\perp'} P_{\tilde{Z}^{\perp}} \tilde{u}^{\perp} = \left(\frac{1}{\sqrt{Tn}} \tilde{Y}^{\perp'} P_{\tilde{Z}^{\perp}}^{1/2}\right) \left(P_{\tilde{Z}^{\perp}}^{1/2} \tilde{u}^{\perp}\right) \xrightarrow{d} \sigma_{uu}^{1/2} \Sigma_{VV}^{1/2'} \lambda' z_u$$

and

$$\frac{1}{T^{1-2\delta}} \tilde{Y}^{\perp'} P_{\tilde{Z}^{\perp}} \tilde{Y}^{\perp} = \left(\frac{1}{T^{1/2-\delta}} \tilde{Y}^{\perp'} P_{\tilde{Z}^{\perp}}^{1/2}\right) \left(\frac{1}{T^{1/2-\delta}} P_{\tilde{Z}^{\perp}}^{1/2} \tilde{Y}^{\perp}\right) \xrightarrow{p} \Sigma_{VV}^{1/2'} \lambda' \lambda \Sigma_{VV}^{1/2}.$$

Next, for $\delta = \frac{1}{2}$, we have

$$\begin{aligned}
\frac{1}{\sqrt{Tn}} \tilde{Z}^{\perp'} \tilde{Y}^{\perp} &= \frac{1}{\sqrt{Tn}} \tilde{Z}' M_{\tilde{X}} \tilde{Z} \Pi + \frac{1}{\sqrt{Tn}} \tilde{Z}' M_{\tilde{X}} \tilde{V} \\
&= \left[\left(\frac{1}{Tn} \tilde{Z}' \tilde{Z}\right) - \left(\frac{1}{Tn} \tilde{Z}' \tilde{X}\right) \left(\frac{1}{Tn} \tilde{X}' \tilde{X}\right)^{-1} \left(\frac{1}{Tn} \tilde{X}' \tilde{Z}\right)\right] \left(\sqrt{Tn} \Pi\right) \\
&\quad + \left[\left(\frac{1}{\sqrt{Tn}} \tilde{Z}' \tilde{V}\right) - \left(\frac{1}{Tn} \tilde{Z}' \tilde{X}\right) \left(\frac{1}{Tn} \tilde{X}' \tilde{X}\right)^{-1} \left(\frac{1}{\sqrt{Tn}} \tilde{X}' \tilde{V}\right)\right] \\
&\xrightarrow{d} (\Phi_{ZZ} - \Phi_{Z\tilde{X}} \Phi_{\tilde{X}\tilde{X}}^{-1} \Phi_{\tilde{X}Z}) C + (\Psi_{ZV} - \Phi_{Z\tilde{X}} \Phi_{\tilde{X}\tilde{X}}^{-1} \Psi_{\tilde{X}V}) \\
&= \Omega C + \Omega^{1/2} z_V \Sigma_{VV}^{1/2}
\end{aligned}$$

by the definition of Ω and z_V. Together with Lemmas 1 and 2, we have

$$\begin{aligned}
P_{\tilde{Z}^{\perp}}^{1/2} \tilde{Y}^{\perp} &= \left(\frac{1}{Tn} \tilde{Z}^{\perp'} \tilde{Z}^{\perp}\right)^{-1/2} \left(\frac{1}{\sqrt{Tn}} \tilde{Z}^{\perp'} \tilde{Y}^{\perp}\right) \xrightarrow{d} \Omega^{-1/2} [\Omega C + \Omega^{1/2} z_V \Sigma_{VV}^{1/2}] \\
&= \Omega^{1/2} C + z_V \Sigma_{VV}^{1/2} = (\lambda + z_V) \Sigma_{VV}^{1/2} \\
\tilde{Y}^{\perp'} P_{\tilde{Z}^{\perp}} \tilde{u}^{\perp} &= \left(\tilde{Y}^{\perp'} P_{\tilde{Z}^{\perp}}^{1/2}\right) \left(P_{\tilde{Z}^{\perp}}^{1/2} \tilde{u}^{\perp}\right) \xrightarrow{d} \sigma_{uu}^{1/2} \Sigma_{VV}^{1/2'} (\lambda + z_V)' z_u
\end{aligned}$$

and

$$\tilde{Y}^{\perp'} P_{\tilde{Z}^{\perp}} \tilde{Y}^{\perp} = \left(\tilde{Y}^{\perp'} P_{\tilde{Z}^{\perp}}^{1/2}\right) \left(P_{\tilde{Z}^{\perp}}^{1/2} \tilde{Y}^{\perp}\right) \xrightarrow{d} \Sigma_{VV}^{1/2'} (\lambda + z_V)' (\lambda + z_V) \Sigma_{VV}^{1/2}.$$

Finally, for $\frac{1}{2} < \delta < \infty$, we have

$$\frac{1}{\sqrt{Tn}}\tilde{Z}^{\perp'}\tilde{Y}^{\perp} = \frac{1}{\sqrt{Tn}}\tilde{Z}'M_{\tilde{X}}\tilde{Z}\Pi + \frac{1}{\sqrt{Tn}}\tilde{Z}'M_{\tilde{X}}\tilde{V}$$
$$= \frac{1}{T^{\delta-1/2}}\left[\left(\frac{1}{Tn}\tilde{Z}'\tilde{Z}\right) - \left(\frac{1}{Tn}\tilde{Z}'\tilde{X}\right)\left(\frac{1}{Tn}\tilde{X}'\tilde{X}\right)^{-1}\left(\frac{1}{Tn}\tilde{X}'\tilde{Z}\right)\right](\sqrt{n}T^{\delta}\Pi)$$
$$+ \left[\left(\frac{1}{\sqrt{Tn}}\tilde{Z}'\tilde{V}\right) - \left(\frac{1}{Tn}\tilde{Z}'\tilde{X}\right)\left(\frac{1}{Tn}\tilde{X}'\tilde{X}\right)^{-1}\left(\frac{1}{\sqrt{Tn}}\tilde{X}'\tilde{V}\right)\right]$$
$$= O_p\left(\frac{1}{T^{\delta-1/2}}\right) + (\Psi_{ZV} - \Phi_{Z\tilde{X}}\Phi_{\tilde{X}\tilde{X}}^{-1}\Psi_{\tilde{X}V}) = \Omega^{1/2}z_V\Sigma_{VV}^{1/2} + O_p\left(\frac{1}{T^{\delta-1/2}}\right).$$

Using Lemmas 1 and 2, we have

$$P_{\tilde{Z}^{\perp}}^{1/2}\tilde{Y}^{\perp} = \left(\frac{1}{Tn}\tilde{Z}^{\perp'}\tilde{Z}^{\perp}\right)^{-1/2}\left(\frac{1}{\sqrt{Tn}}\tilde{Z}^{\perp'}\tilde{Y}^{\perp}\right) \xrightarrow{d} \Omega^{-1/2}\Omega^{1/2}z_V\Sigma_{VV}^{1/2} = z_V\Sigma_{VV}^{1/2},$$

$$\tilde{Y}^{\perp'}P_{\tilde{Z}^{\perp}}\tilde{u}^{\perp} = \left(\tilde{Y}^{\perp'}P_{\tilde{Z}^{\perp}}^{1/2}\right)\left(P_{\tilde{Z}^{\perp}}^{1/2}\tilde{u}^{\perp}\right) \xrightarrow{d} \sigma_{uu}^{1/2}\Sigma_{VV}^{1/2}z'_V z_u$$

and

$$\tilde{Y}^{\perp'}P_{\tilde{Z}^{\perp}}\tilde{Y}^{\perp} = \left(\tilde{Y}^{\perp'}P_{\tilde{Z}^{\perp}}^{1/2}\right)\left(P_{\tilde{Z}^{\perp}}^{1/2}\tilde{Y}^{\perp}\right) \xrightarrow{d} \Sigma_{VV}^{1/2}z'_V z_V\Sigma_{VV}^{1/2}.$$

This proves part 3.

Proof of Theorem 2

Proof. We write

$$\hat{\beta}(k) = \beta + \left[\tilde{Y}^{\perp'}\left(I - kM_{\tilde{Z}^{\perp}}\right)\tilde{Y}^{\perp}\right]^{-1}\left[\tilde{Y}^{\perp'}\left(I - kM_{\tilde{Z}^{\perp}}\right)\tilde{u}^{\perp}\right].$$

First for $0 \leq \delta < \frac{1}{2}$, let $\kappa_{Tn} = T^{1/2+\delta}n(k-1)$. Using $k = \left(1 + \frac{\kappa_{Tn}}{T^{1/2+\delta}n}\right)$ and $M_{\tilde{Z}^{\perp}} = I - P_{\tilde{Z}^{\perp}}$, by Lemmas 2.1 and 2.3, we have

$$\frac{1}{T^{1-2\delta}}\tilde{Y}^{\perp\prime}\left(I-kM_{\tilde{Z}^{\perp}}\right)\tilde{Y}^{\perp}=\frac{1}{T^{1-2\delta}}\tilde{Y}^{\perp\prime}\left[I-\left(1+\frac{\kappa_{Tn}}{T^{1/2+\delta}n}\right)\left(I-P_{\tilde{Z}^{\perp}}\right)\right]\tilde{Y}^{\perp}$$
$$=\left(1+\frac{\kappa_{Tn}}{\sqrt{Tn}}\right)\left(\frac{1}{T^{1-2\delta}}\tilde{Y}^{\perp\prime}P_{\tilde{Z}^{\perp}}\tilde{Y}^{\perp}\right)-\frac{\kappa_{Tn}}{T^{1/2-\delta}}\left(\frac{1}{Tn}\tilde{Y}^{\perp\prime}\tilde{Y}^{\perp}\right)\xrightarrow{p}\Sigma_{VV}^{1/2\prime}\lambda'\lambda\Sigma_{VV}^{1/2}$$

and

$$\frac{1}{T^{1/2-\delta}}\tilde{Y}^{\perp\prime}\left(I-kM_{\tilde{Z}^{\perp}}\right)\tilde{u}^{\perp}=\frac{1}{T^{1/2-\delta}}\tilde{Y}^{\perp\prime}\left[I-\left(1+\frac{\kappa_{Tn}}{T^{1/2+\delta}n}\right)\left(I-P_{\tilde{Z}^{\perp}}\right)\right]\tilde{u}^{\perp}$$
$$=\left(1+\frac{\kappa_{Tn}}{\sqrt{Tn}}\right)\left(\frac{1}{T^{1/2-\delta}}\tilde{Y}^{\perp\prime}P_{\tilde{Z}^{\perp}}\tilde{u}^{\perp}\right)-\kappa_{Tn}\left(\frac{1}{Tn}\tilde{Y}^{\perp\prime}\tilde{u}^{\perp}\right)\xrightarrow{d}\sigma_{uu}^{1/2}\Sigma_{VV}^{1/2\prime}\lambda' z_u-\kappa\Sigma_{Vu}$$

uniformly in k. Therefore:

$$T^{1/2-\delta}\left(\hat{\beta}(k)-\beta\right)+\left[\frac{1}{T^{1-2\delta}}\tilde{Y}^{\perp\prime}\left(I-kM_{\tilde{Z}^{\perp}}\right)\tilde{Y}^{\perp}\right]^{-1}\kappa_{Tn}\Sigma_{Vu}$$
$$=\left[\frac{1}{T^{1-2\delta}}\tilde{Y}^{\perp\prime}\left(I-kM_{\tilde{Z}^{\perp}}\right)\tilde{Y}^{\perp}\right]^{-1}\left[\frac{1}{T^{1/2-\delta}}\tilde{Y}^{\perp\prime}\left(I-kM_{\tilde{Z}^{\perp}}\right)\tilde{u}^{\perp}-\kappa_{Tn}\left(\frac{1}{Tn}\tilde{Y}^{\perp\prime}\tilde{u}^{\perp}\right)\right.$$
$$\left.+\kappa_{Tn}\Sigma_{Vu}\right]\xrightarrow{d}\left(\Sigma_{VV}^{1/2\prime}\lambda'\lambda\Sigma_{VV}^{1/2}\right)^{-1}\left(\sigma_{uu}^{1/2}\Sigma_{VV}^{1/2\prime}\lambda' z_u\right)\sim N\left(0,\sigma_{uu}\left(\Sigma_{VV}^{1/2\prime}\lambda'\lambda\Sigma_{VV}^{1/2}\right)^{-1}\right)$$

as $(n,T)\rightarrow\infty$ joint with $\kappa_{Tn}\xrightarrow{d}\kappa$.

Consider $\delta=\frac{1}{2}$ and let $\kappa_{Tn}=Tn(k-1)$. Using $k=\left(1+\frac{\kappa_{Tn}}{Tn}\right)$ and $M_{\tilde{Z}^{\perp}}=I-P_{\tilde{Z}^{\perp}}$, by Lemmas 2.1 and 2.3, we have

$$\tilde{Y}^{\perp\prime}\left(I-kM_{\tilde{Z}^{\perp}}\right)\tilde{Y}^{\perp}=\tilde{Y}^{\perp\prime}\left[I-\left(1+\frac{\kappa_{Tn}}{Tn}\right)(I-P_{\tilde{Z}^{\perp}})\right]\tilde{Y}^{\perp}$$
$$=\left(1+\frac{\kappa_{Tn}}{Tn}\right)\left(\tilde{Y}^{\perp\prime}P_{\tilde{Z}^{\perp}}\tilde{Y}^{\perp}\right)-\kappa_{Tn}\left(\frac{1}{Tn}\tilde{Y}^{\perp\prime}\tilde{Y}^{\perp}\right)$$
$$\xrightarrow{d}\Sigma_{VV}^{1/2\prime}(\lambda+z_V)'(\lambda+z_V)\Sigma_{VV}^{1/2}-\kappa\Sigma_{VV}$$
$$=\Sigma_{VV}^{1/2\prime}[(\lambda+z_V)'(\lambda+z_V)-\kappa I_{Tn}]\Sigma_{VV}^{1/2}$$

and

$$\tilde{Y}^{\perp\prime}\left(I-kM_{\tilde{Z}^{\perp}}\right)\tilde{u}^{\perp}=\tilde{Y}^{\perp\prime}\left[I-\left(1+\frac{\kappa_{Tn}}{T^{1/2+\delta}n}\right)\left(I-P_{\tilde{Z}^{\perp}}\right)\right]\tilde{u}^{\perp}$$
$$=\left(1+\frac{\kappa_{Tn}}{\sqrt{Tn}}\right)\left(\tilde{Y}^{\perp\prime}P_{\tilde{Z}^{\perp}}\tilde{u}^{\perp}\right)-\kappa_{Tn}\left(\frac{1}{Tn}\tilde{Y}^{\perp\prime}\tilde{u}^{\perp}\right)$$
$$\xrightarrow{d}\sigma_{uu}^{1/2}\Sigma_{VV}^{1/2\prime}(\lambda+z_V)'z_u-\kappa\Sigma_{Vu}=\sigma_{uu}^{1/2}\Sigma_{VV}^{1/2\prime}[(\lambda+z_V)'z_u-\kappa\rho]$$

uniformly in κ. Therefore:

$$\hat{\beta}(k)-\beta \xrightarrow{d} \{\Sigma_{VV}^{1/2'}[(\lambda+z_V)'(\lambda+z_V)-\kappa I_{Tn}]\Sigma_{VV}^{1/2}\}^{-1}\{\sigma_{uu}^{1/2}\Sigma_{VV}^{1/2'}[(\lambda+z_V)'z_u-\kappa\rho]\}$$
$$=\sigma_{uu}^{1/2}\Sigma_{VV}^{-1/2}\Delta_1(\kappa),$$

as $(n,T)\rightarrow\infty$ joint with $\kappa_{Tn}\xrightarrow{d}\kappa$, where $\Delta_1(\kappa)=[(\lambda+z_V)'(\lambda+z_V)-\kappa I_{Tn}]^{-1}[(\lambda+z_V)'z_u-\kappa\rho]$.

Consider $\frac{1}{2}<\delta<\infty$ and let $\kappa_{Tn}=Tn(k-1)$. Using $k=\left(1+\frac{\kappa_{Tn}}{Tn}\right)$ and $M_{\tilde{Z}^\perp}=I-P_{\tilde{Z}^\perp}$, by Lemmas 2.1 and 2.3, we have

$$\begin{aligned}\tilde{Y}^{\perp'}\left(I-kM_{\tilde{Z}^\perp}\right)\tilde{Y}^\perp &= \tilde{Y}^{\perp'}\left[I-\left(1+\frac{\kappa_{Tn}}{Tn}\right)\left(I-P_{\tilde{Z}^\perp}\right)\right]\tilde{Y}^\perp\\ &= \left(1+\frac{\kappa_{Tn}}{Tn}\right)\left(\tilde{Y}^{\perp'}P_{\tilde{Z}^\perp}\tilde{Y}^\perp\right)-\kappa_{Tn}\left(\frac{1}{Tn}\tilde{Y}^{\perp'}\tilde{Y}^\perp\right)\\ &\xrightarrow{d} \Sigma_{VV}^{1/2'}z'_V z_V\Sigma_{VV}^{1/2}-\kappa\Sigma_{VV}\\ &= \Sigma_{VV}^{1/2'}[z'_V z_V-\kappa I_{Tn}]\Sigma_{VV}^{1/2}\end{aligned}$$

and

$$\begin{aligned}\tilde{Y}^{\perp'}\left(I-kM_{\tilde{Z}^\perp}\right)\tilde{u}^\perp &= \tilde{Y}^{\perp'}\left[I-\left(1+\frac{\kappa_{Tn}}{Tn}\right)\left(I-P_{\tilde{Z}^\perp}\right)\right]\tilde{u}^\perp\\ &= \left(1+\frac{\kappa_{Tn}}{Tn}\right)\left(\tilde{Y}^{\perp'}P_{\tilde{Z}^\perp}\tilde{u}^\perp\right)-\kappa_{Tn}\left(\frac{1}{Tn}\tilde{Y}^{\perp'}\tilde{u}^\perp\right)\\ &\xrightarrow{d} \sigma_{uu}^{1/2}\Sigma_{VV}^{1/2'}z'_V z_u-\kappa\Sigma_{Vu}\\ &= \sigma_{uu}^{1/2}\Sigma_{VV}^{1/2'}[z'_V z_u-\kappa\rho].\end{aligned}$$

Therefore,

$$\hat{\beta}(k)-\beta\xrightarrow{d}\{\Sigma_{VV}^{1/2'}[z'_V z_V-\kappa I_{Tn}]\Sigma_{VV}^{1/2}\}^{-1}\{\sigma_{uu}^{1/2}\Sigma_{VV}^{1/2'}[z'_V z_u-\kappa\rho]\}=\sigma_{uu}^{1/2}\Sigma_{VV}^{-1/2}\Delta_2(\kappa),$$

as $(n,T)\rightarrow\infty$ joint with $\kappa_{Tn}\xrightarrow{d}\kappa$, where $\Delta_2(\kappa)=[z'_V z_V-\kappa I_{Tn}]^{-1}\,[z'_V z_u-\kappa\rho]$.

Proof of Theorem 3

Proof. Let us denote J, partitioned conformably with $\tilde{Y}^{*\perp}$, to be $J_{11}=I_T$, $J_{21}=-\beta\otimes\imath_T$, $J_{11}=0\otimes\imath'_T$ and $J_{11}=I_{Tn}$. Because $\tilde{y}^\perp=\tilde{Y}^\perp\beta+\tilde{u}^\perp$, hence $\tilde{Y}^{*\perp}J=[\tilde{u}^\perp,\tilde{Y}^\perp]$. By Lemma 2.1, we have $\frac{1}{Tn}J'\tilde{Y}^{*\perp'}\tilde{Y}^{*\perp}$

$J = \begin{pmatrix} \frac{1}{Tn}\tilde{u}^{\perp'}\tilde{u}^{\perp} & \frac{1}{Tn}\tilde{u}^{\perp'}\tilde{Y}^{\perp} \\ \frac{1}{Tn}\tilde{Y}^{\perp'}\tilde{u}^{\perp} & \frac{1}{Tn}\tilde{Y}^{\perp'}\tilde{Y}^{\perp} \end{pmatrix} \xrightarrow{p} \begin{pmatrix} \sigma_{uu} & \Sigma_{uV} \\ \Sigma_{Vu} & \Sigma_{VV} \end{pmatrix} = \Upsilon'\bar{\Sigma}\Upsilon$ where $\Upsilon = diag(\sigma_{uu}^{1/2}, \Sigma_{VV}^{1/2})$ and $\bar{\Sigma}$ is defined in Eq. (10). If $\frac{1}{Tn}J'\tilde{Y}^{*\perp'}P_{\tilde{Z}^{*\perp}}\tilde{Y}^{*\perp}J = o_p(1)$ which will be shown below, then we have $\frac{1}{Tn}J'\tilde{Y}^{*\perp'}M_{\tilde{Z}^{*\perp}}\tilde{Y}^{*\perp}J = \frac{1}{Tn}J'\tilde{Y}^{*\perp'}\tilde{Y}^{*\perp}J - \frac{1}{Tn}J'\tilde{Y}^{*\perp'}P_{\tilde{Z}^{*\perp}}\tilde{Y}^{*\perp}J \xrightarrow{p} \Upsilon'\bar{\Sigma}\Upsilon$.

For $0 \le \delta < \frac{1}{2}$, note that for any nonsingular $(n+1)T \times (n+1)T$ matrix J, the roots of $\left|\tilde{Y}^{*\perp'}\left[I_{Tn} - \hat{k}_{LIML}M_{\tilde{Z}^{*\perp}}\right]\tilde{Y}^{*\perp}\right| = 0$ are the same as the roots of

$$\begin{aligned} &\left|\frac{1}{T^{1-2\delta}}J'\tilde{Y}^{*\perp'}\left[I_{Tn} - \hat{k}_{LIML}M_{\tilde{Z}^{*\perp}}\right]\tilde{Y}^{*\perp}J\right| \\ &= \left|\frac{1}{T^{1-2\delta}}J'\tilde{Y}^{*\perp'}\left[P_{\tilde{Z}^{*\perp}} - (\hat{k}_{LIML}-1)M_{\tilde{Z}^{*\perp}}\right]\tilde{Y}^{*\perp}J\right| \\ &= \left|\left(\frac{1}{T^{1-2\delta}}J'\tilde{Y}^{*\perp'}P_{\tilde{Z}^{*\perp}}\tilde{Y}^{*\perp}J\right) - \hat{\kappa}_{LIML}\left(\frac{1}{Tn}J'\tilde{Y}^{*\perp'}M_{\tilde{Z}^{*\perp}}\tilde{Y}^{*\perp}J\right)\right| = 0 \end{aligned}$$

where $\hat{\kappa}_{LIML} = T^{2\delta}n(\hat{k}_{LIML} - 1)$. By Lemmas 2.2 and 2.3, we have

$$\begin{aligned} &\frac{1}{T^{1-2\delta}}J'\tilde{Y}^{*\perp'}P_{\tilde{Z}^{*\perp}}\tilde{Y}^{*\perp}J \\ &= \begin{pmatrix} \frac{1}{T^{1-2\delta}}\left(\tilde{u}^{\perp'}P_{\tilde{Z}^{*\perp}}\tilde{u}^{\perp}\right) & \frac{1}{T^{1/2-\delta}}\left(\frac{1}{T^{1/2-\delta}}\tilde{u}^{\perp'}P_{\tilde{Z}^{*\perp}}\tilde{Y}^{\perp}\right) \\ \frac{1}{T^{1/2-\delta}}\left(\frac{1}{T^{1/2-\delta}}\tilde{Y}^{\perp'}P_{\tilde{Z}^{*\perp}}\tilde{u}^{\perp}\right) & \left(\frac{1}{T^{1-2\delta}}\tilde{Y}^{\perp'}P_{\tilde{Z}^{*\perp}}\tilde{Y}^{\perp}\right) \end{pmatrix} \\ &\xrightarrow{d} \begin{pmatrix} 0 & 0 \\ 0 & \Sigma_{VV}^{1/2'}\lambda'\lambda\Sigma_{VV}^{1/2} \end{pmatrix} \\ &= \Upsilon'\Xi_1\Upsilon \end{aligned}$$

where $\Xi_1 = \begin{pmatrix} 0 & 0 \\ 0 & \lambda'\lambda \end{pmatrix}$. The solutions to $\left|\left(\frac{1}{T^{1-2\delta}}J'\tilde{Y}^{*\perp'}P_{\tilde{Z}^{*\perp}}\tilde{Y}^{*\perp}J\right) - \hat{\kappa}_{LIML}\left(\frac{1}{Tn}J'\tilde{Y}^{*\perp'}M_{\tilde{Z}^{*\perp}}\tilde{Y}^{*\perp}J\right)\right| = 0$ therefore converge to those of $\left|\Xi_1 - \kappa\bar{\Sigma}\right| = 0$, among them the smallest root is zero. Thus $\hat{\kappa}_{LIML} = T^{2\delta}n(\hat{k}_{LIML} - 1) \xrightarrow{p} 0$.

For $\delta = \frac{1}{2}$, the roots of $\left|\tilde{Y}^{*\perp'}\left[I_{Tn} - \hat{k}_{LIML}M_{\tilde{Z}^{*\perp}}\right]\tilde{Y}^{*\perp}\right| = 0$ are the same as the roots of

$$\left|J'\tilde{Y}^{*\perp'}\left[P_{\tilde{Z}^{*\perp}}-(\hat{k}_{LIML}-1)M_{\tilde{Z}^{*\perp}}\right]\tilde{Y}^{*\perp}J\right|$$
$$=\left|\left(J'\tilde{Y}^{*\perp'}P_{\tilde{Z}^{*\perp}}\tilde{Y}^{*\perp}J\right)-\hat{\kappa}_{LIML}\left(\frac{1}{Tn}J'\tilde{Y}^{*\perp'}M_{\tilde{Z}^{*\perp}}\tilde{Y}^{*\perp}J\right)\right|=0,$$

where $\hat{\kappa}_{LIML}=Tn(\hat{k}_{LIML}-1)$. By Lemmas 2.2 and 2.3, we have

$$J'\tilde{Y}^{*\perp'}P_{\tilde{Z}^{*\perp}}\tilde{Y}^{*\perp}J=\begin{pmatrix}\tilde{u}^{\perp'}P_{\tilde{Z}^{*\perp}}\tilde{u}^{\perp} & \tilde{u}^{\perp'}P_{\tilde{Z}^{*\perp}}\tilde{Y}^{\perp}\\ \tilde{Y}^{\perp'}P_{\tilde{Z}^{*\perp}}\tilde{u}^{\perp} & \tilde{Y}^{\perp'}P_{\tilde{Z}^{*\perp}}\tilde{Y}^{\perp}\end{pmatrix}$$
$$\xrightarrow{d}\begin{pmatrix}\sigma_{uu}z'_u z_u & \sigma_{uu}^{1/2}z'_u(\lambda+z_V)\Sigma_{VV}^{1/2}\\ \sigma_{uu}^{1/2}\Sigma_{VV}^{1/2'}(\lambda+z_V)'z_u & \Sigma_{VV}^{1/2'}(\lambda+z_V)'(\lambda+z_V)\Sigma_{VV}^{1/2}\end{pmatrix}$$
$$=\Upsilon'\Xi_2\Upsilon$$

where $\Xi_2=\begin{pmatrix}z'_u z_u & z'_u(\lambda+z_V)\\ (\lambda+z_V)'z_u & (\lambda+z_V)'(\lambda+z_V)\end{pmatrix}$. The solutions to $\left|\left(J'\tilde{Y}^{*\perp'}P_{\tilde{Z}^{*\perp}}\tilde{Y}^{*\perp}J\right)-\hat{\kappa}_{LIML}\left(\frac{1}{Tn}J'\tilde{Y}^{*\perp'}M_{\tilde{Z}^{*\perp}}\tilde{Y}^{*\perp}J\right)\right|=0$ therefore converge to those of $|\Xi_2-\kappa\bar{\Sigma}|=0$. Thus $\hat{\kappa}_{LIML}=Tn(\hat{k}_{LIML}-1)\xrightarrow{d}\kappa^*_{LIML}$, where κ^*_{LIML} is the smallest root of $|\Xi_2-\kappa\bar{\Sigma}|=0$.

For $\frac{1}{2}<\delta<\infty$, the roots of $\left|\tilde{Y}^{*\perp'}\left[I_{Tn}-\hat{k}_{LIML}M_{\tilde{Z}^{*\perp}}\right]\tilde{Y}^{*\perp}\right|=0$ are the same as the roots of

$$\left|J'\tilde{Y}^{*\perp'}\left[P_{\tilde{Z}^{*\perp}}-(\hat{k}_{LIML}-1)M_{\tilde{Z}^{*\perp}}\right]\tilde{Y}^{*\perp}J\right|$$
$$=\left|\left(J'\tilde{Y}^{*\perp'}P_{\tilde{Z}^{*\perp}}\tilde{Y}^{*\perp}J\right)-\hat{\kappa}_{LIML}\left(\frac{1}{Tn}J'\tilde{Y}^{*\perp'}M_{\tilde{Z}^{*\perp}}\tilde{Y}^{*\perp}J\right)\right|=0$$

where $\hat{\kappa}_{LIML}=Tn(\hat{k}_{LIML}-1)$. By Lemmas 2.2 and 2.3, we have

$$J'\tilde{Y}^{*\perp'}P_{\tilde{Z}^{*\perp}}\tilde{Y}^{*\perp}J=\begin{pmatrix}\tilde{u}^{\perp'}P_{\tilde{Z}^{*\perp}}\tilde{u}^{\perp} & \tilde{u}^{\perp'}P_{\tilde{Z}^{*\perp}}\tilde{Y}^{\perp}\\ \tilde{Y}^{\perp'}P_{\tilde{Z}^{*\perp}}\tilde{u}^{\perp} & \tilde{Y}^{\perp'}P_{\tilde{Z}^{*\perp}}\tilde{Y}^{\perp}\end{pmatrix}$$
$$\xrightarrow{d}\begin{pmatrix}\sigma_{uu}z'_u z_u & \sigma_{uu}^{1/2}z'_u z_V\Sigma_{VV}^{1/2}\\ \sigma_{uu}^{1/2}\Sigma_{VV}^{1/2'}z'_V z_u & \Sigma_{VV}^{1/2'}z'_V z_V\Sigma_{VV}^{1/2}\end{pmatrix}$$
$$=\Upsilon'\Xi_3\Upsilon,$$

where $\Xi_3 = \begin{pmatrix} z'_u z_u & z'_u z_V \\ z'_V z_u & z'_V z_V \end{pmatrix}$. The solutions to $\left|(J'\tilde{Y}^{*\perp'} P_{\tilde{Z}^{*\perp}} \tilde{Y}^{*\perp} J) - \hat{\kappa}_{LIML} \left(\frac{1}{Tn} J'\tilde{Y}^{*\perp'} M_{\tilde{Z}^{*\perp}} \tilde{Y}^{*\perp} J\right)\right| = 0$ therefore converge to those of $\left|\Xi_3 - \kappa\bar{\Sigma}\right| = 0$.

Thus $\hat{\kappa}_{LIML} = Tn(\hat{k}_{LIML} - 1) \xrightarrow{d} \kappa^*_{LIML}$, where κ^*_{LIML} is the smallest root of $\left|\Xi_3 - \kappa\bar{\Sigma}\right| = 0$.

Proof of Theorem 4

Proof. Using $R\hat{\beta}(k) - r = R[\hat{\beta}(k) - \beta]$, we have

$$W(k) = [R\hat{\beta}(k) - r]'\{\hat{\sigma}_{uu}(k)R[\tilde{Y}^{\perp'}(I - kM_{\tilde{Z}^{\perp}})\tilde{Y}^{\perp}]^{-1}R'\}^{-1}[R\hat{\beta}(k) - r]$$
$$= \{[\hat{\beta}(k) - \beta]'R'\}\{\hat{\sigma}_{uu}(k)R[\tilde{Y}^{\perp'}(I - kM_{\tilde{Z}^{\perp}})\tilde{Y}^{\perp}]^{-1}R'\}^{-1}\{R[\hat{\beta}(k) - \beta]\}$$

and note that $\hat{\tilde{u}}(k) = \tilde{y}^{\perp} - \tilde{Y}^{\perp}\hat{\beta}(k) = \tilde{u}^{\perp} - \tilde{Y}^{\perp}[\hat{\beta}(k) - \beta]$, so

$$\begin{aligned}
\hat{\sigma}_{uu}(k) &= \frac{1}{Tn - K_1 - L}\hat{\tilde{u}}(k)'\hat{\tilde{u}}(k) \\
&= \frac{1}{Tn - K_1 - L}\{\tilde{u}^{\perp} - \tilde{Y}^{\perp}[\hat{\beta}(k) - \beta]\}'\left\{\tilde{u}^{\perp} - \tilde{Y}^{\perp}[\hat{\beta}(k) - \beta]\right\} \\
&= \frac{Tn}{Tn - K_1 - L}\left\{\left(\frac{1}{Tn}\tilde{u}^{\perp'}\tilde{u}^{\perp}\right) - 2[\hat{\beta}(k) - \beta]'\left(\frac{1}{Tn}\tilde{Y}^{\perp'}\tilde{u}^{\perp}\right)\right. \\
&\quad \left. + [\hat{\beta}(k) - \beta]'\left(\frac{1}{Tn}\tilde{Y}^{\perp'}\tilde{Y}^{\perp}\right)[\hat{\beta}(k) - \beta]\right\}.
\end{aligned}$$

For $0 \leq \delta < \frac{1}{2}$, Theorem 2 implies $\hat{\beta}(k) - \beta \xrightarrow{p} 0$. By Lemma 2.1,

$$\hat{\sigma}_{uu}(k) \xrightarrow{p} \sigma_{uu}$$

as $(n, T) \to \infty$. Because $\theta_{Tn} = \left[\frac{1}{T^{1-2\delta}}\tilde{Y}^{\perp'}\left(I - kM_{\tilde{Z}^{\perp}}\right)\tilde{Y}^{\perp}\right]^{-1}\kappa_{Tn}\Sigma_{Vu} \xrightarrow{p} \kappa(\Sigma_{VV}^{1/2'}\lambda'\lambda\Sigma_{VV}^{1/2})^{-1}\Sigma_{Vu} = \theta$, then

$$RT^{1/2-\delta}(\hat{\beta}(k) - \beta) \sim N(R\theta, \sigma_{uu}R(\Sigma_{VV}^{1/2'}\lambda'\lambda\Sigma_{VV}^{1/2})^{-1}R')$$

Recall that $\frac{1}{T^{1-2\delta}}\tilde{Y}^{\perp'}\left(I - kM_{\tilde{Z}^{\perp}}\right)\tilde{Y}^{\perp} \xrightarrow{p} \Sigma_{VV}^{1/2'}\lambda'\lambda\Sigma_{VV}^{1/2}$ by the proof of Theorem 2. By Proposition B.7 in Lütkepohl (2005), we have

$$W(k) = \left\{T^{1/2-\delta}[\hat{\beta}(k) - \beta]'R'\right\}\left\{\hat{\sigma}_{uu}(k)R\left[\frac{1}{T^{1-2\delta}}\tilde{Y}^{\perp'}\left(I - kM_{\tilde{Z}^{\perp}}\right)\tilde{Y}^{\perp}\right]^{-1}R'\right\}^{-1}$$
$$\times\{RT^{1/2-\delta}[\hat{\beta}(k) - \beta]\}$$
$$= \{T^{1/2-\delta}[\hat{\beta}(k) - \beta]'R'\}\{\sigma_{uu}R(\Sigma_{VV}^{1/2'}\lambda'\lambda\Sigma_{VV}^{1/2})^{-1}R'\}^{-1}\{RT^{1/2-\delta}[\hat{\beta}(k) - \beta]\}$$
$$+ o_p(1) \xrightarrow{d} \chi^2_*(\Phi, \Lambda),$$

which is a noncentral chi-squared distribution with Φ degrees of freedom and noncentrality parameter $\Lambda = \theta' R'[\sigma_{uu}R(\Sigma_{VV}^{1/2'}\lambda'\lambda\Sigma_{VV}^{1/2})^{-1}R']^{-1}R\theta$. Note that if $\kappa = 0$, $\theta = 0$, and $\Lambda = 0$, hence $W(k) \xrightarrow{d} \chi^2(\Phi)$, a central chi-squared distribution with Φ degrees of freedom.

For $\delta = \frac{1}{2}$, by Lemma 2.1

$$\hat{\sigma}_{uu}(k) \xrightarrow{d} \sigma_{uu} - 2\sigma_{uu}^{1/2}[(\lambda + z_V)'z_u - \kappa\rho]'[(\lambda + z_V)'(\lambda + z_V) - \kappa I_{Tn}]^{-1}\Sigma_{VV}^{-1/2}\Sigma_{Vu}$$
$$+ \sigma_{uu}[(\lambda + z_V)'z_u - \kappa\rho]'[(\lambda + z_V)'(\lambda + z_V) - \kappa I_{Tn}]^{-1}[(\lambda + z_V)'(\lambda + z_V) - \kappa I_{Tn}]^{-1}$$
$$[(\lambda + z_V)'z_u - \kappa\rho] = \sigma_{uu}S(\Delta_1(\kappa)),$$

where $S(b) = 1 - 2\rho' b + b'b$ and $\Delta_1(k) = [(\lambda + z_V)'(\lambda + z_V) - \kappa I_{Tn}]^{-1}[(\lambda + z_V)'z_u - \kappa\rho]$. Therefore

$$W(k) = \{[\hat{\beta}(k) - \beta]'R'\}\left\{\hat{\sigma}_{uu}(k)R\left[\tilde{Y}^{\perp'}\left(I - kM_{\tilde{Z}^{\perp}}\right)\tilde{Y}^{\perp}\right]^{-1}R'\right\}^{-1}\{R[\hat{\beta}(k) - \beta]\}$$
$$\xrightarrow{d}[\sigma_{uu}^{1/2}\Sigma_{VV}^{-1/2}\Delta_1(\kappa)]'R'\{\sigma_{uu}S(\Delta_1(\kappa))R\{\Sigma_{VV}^{1/2'}[(\lambda + z_V)'(\lambda + z_V)$$
$$- \kappa I_{Tn}]\Sigma_{VV}^{1/2}\}^{-1}R'\}^{-1}R[\sigma_{uu}^{1/2}\Sigma_{VV}^{-1/2}\Delta_1(\kappa)]$$
$$= \Delta_1'(\kappa)\Sigma_{VV}^{-1/2}R'\{S(\Delta_1(\kappa))R\{\Sigma_{VV}^{1/2'}[(\lambda + z_V)'(\lambda + z_V) - \kappa I_{Tn}]$$
$$\times \Sigma_{VV}^{1/2}\}^{-1}R'\}^{-1}R\Sigma_{VV}^{-1/2}\Delta_1(\kappa)$$

as $(n, T) \to \infty$ because $\tilde{Y}^{\perp'}\left(I - kM_{\tilde{Z}^{\perp}}\right)\tilde{Y}^{\perp} \xrightarrow{d} \Sigma_{VV}^{1/2'}[(\lambda + z_V)'(\lambda + z_V) - \kappa I_{Tn}]\Sigma_{VV}^{1/2}$.

For $0 \leq \delta < \frac{1}{2}$, by Lemma 2.1

$$\hat{\sigma}_{uu}(k) \xrightarrow{d} \sigma_{uu} - 2\sigma_{uu}^{1/2}[z'_V z_u - \kappa\rho]'[z'_V z_V - \kappa I_{Tn}]^{-1}\Sigma_{VV}^{-1/2}\Sigma_{Vu} + \sigma_{uu}\left[z'_V z_u - \kappa\rho\right]'$$
$$\left[z'_V z_V - \kappa I_{Tn}\right]^{-1}\left[z'_V z_V - \kappa I_{Tn}\right]^{-1}\left[z'_V z_u - \kappa\rho\right] = \sigma_{uu}S(\Delta_2(\kappa)),$$

where $S(b) = 1 - 2\rho' b + b' b$ and $\Delta_2(\kappa) = [z'_V z_V - \kappa I_{Tn}]^{-1}[z'_V z_u - \kappa\rho]$. Therefore:

$$W(k) = \{[\hat{\beta}(k) - \beta]' R'\}\left\{\hat{\sigma}_{uu}(k) R\left[\tilde{Y}^{\perp'}\left(I - kM_{\tilde{Z}^{\perp}}\right)\tilde{Y}^{\perp}\right]^{-1} R'\right\}^{-1}\{R[\hat{\beta}(k) - \beta]\}$$

$$\xrightarrow{d} [\sigma_{uu}^{1/2}\Sigma_{VV}^{-1/2}\Delta_2(\kappa)]' R' \{\sigma_{uu} S(\Delta_2(\kappa)) R \{\Sigma_{VV}^{1/2'}[z'_V z_V - \kappa I_{Tn}]\Sigma_{VV}^{1/2}\}^{-1} R'\}^{-1}$$
$$\times R[\sigma_{uu}^{1/2}\Sigma_{VV}^{-1/2}\Delta_2(\kappa)]$$
$$= \Delta'_2(\kappa)\Sigma_{VV}^{-1/2} R' \{S(\Delta_2(\kappa)) R \{\Sigma_{VV}^{1/2'}[z'_V z_V - \kappa I_{Tn}]\Sigma_{VV}^{1/2}\}^{-1} R'\}^{-1}$$
$$\times R\Sigma_{VV}^{-1/2}\Delta_2(\kappa)$$

as $(n, T) \to \infty$ because $\tilde{Y}^{\perp'}\left(I - kM_{\tilde{Z}^{\perp}}\right)\tilde{Y}^{\perp} \xrightarrow{d} \Sigma_{VV}^{1/2'}[z'_V z_V - \kappa I_{Tn}]\Sigma_{VV}^{1/2}$.

Proof of Theorem 5

Proof. Note that $M_{\tilde{Z}^*}\tilde{Y} = M_{\tilde{Z}^*}\tilde{V}$, we have

$$\hat{\Sigma}_{VV} = \frac{1}{Tn - K_1 - K_2}\tilde{V}' M_{\tilde{Z}^*}\tilde{V}$$
$$= \frac{Tn}{Tn - K_1 - K_2}\left[\left(\frac{1}{Tn}\tilde{V}'\tilde{V}\right) - \frac{1}{Tn}\left(\frac{1}{\sqrt{Tn}}\tilde{V}'\tilde{Z}^*\right)\left(\frac{1}{Tn}\tilde{Z}^{*'}\tilde{Z}^*\right)^{-1}\right.$$
$$\left.\times\left(\frac{1}{\sqrt{Tn}}\tilde{Z}^{*'}\tilde{V}\right)\right] \xrightarrow{p} \Sigma_{VV}$$

because $\frac{1}{\sqrt{Tn}}\tilde{V}'\tilde{Z}^* = \left(\frac{1}{\sqrt{Tn}}\sum_{t=1}^{T}\tilde{V}'_t\tilde{X}_t, \frac{1}{\sqrt{Tn}}\sum_{t=1}^{T}\tilde{V}'_t\tilde{Z}_t\right) \xrightarrow{d} (\Psi'_{\tilde{X}V}, \Psi'_{ZV})$ and $\frac{1}{Tn}\sum_{t=1}^{T}\tilde{Z}_t^{*'}\tilde{Z}_t^* \xrightarrow{p} \Phi$.

For $0 \leq \delta < \frac{1}{2}$, by Lemma 2.3,

$$\frac{K_2}{T^{1-2\delta}}G_{Tn} = \hat{\Sigma}_{VV}^{-1/2'}\left(\frac{1}{T^{1-2\delta}}\tilde{Y}^{\perp'}P_{\tilde{Z}^{\perp}}\tilde{Y}^{\perp}\right)\hat{\Sigma}_{VV}^{-1/2}$$
$$\xrightarrow{p} \Sigma_{VV}^{-1/2'}(\Sigma_{VV}^{1/2'}\lambda'\lambda\Sigma_{VV}^{1/2})\Sigma_{VV}^{-1/2} = \lambda'\lambda.$$

For $\delta = \frac{1}{2}$, by Lemma 2.3,

$$K_2 G_{Tn} = \hat{\Sigma}_{VV}^{-1/2'} \left(\tilde{Y}^{\perp'} P_{\tilde{Z}^{\perp}} \tilde{Y}^{\perp} \right) \hat{\Sigma}_{VV}^{-1/2} \xrightarrow{d} \Sigma_{VV}^{-1/2'} [\Sigma_{VV}^{1/2'} (\lambda + z_V)'(\lambda + z_V) \Sigma_{VV}^{1/2}] \Sigma_{VV}^{-1/2}$$
$$= (\lambda + z_V)'(\lambda + z_V),$$

which is a noncentral Wishart distribution with K_2 degrees of freedom and noncentrality parameter $\lambda'\lambda$.

For $\frac{1}{2} < \delta < \infty$, by Lemma 2.3,

$$K_2 G_{Tn} = \hat{\Sigma}_{VV}^{-1/2'} \left(\tilde{Y}^{\perp'} P_{\tilde{Z}^{\perp}} \tilde{Y}^{\perp} \right) \hat{\Sigma}_{VV}^{-1/2} \xrightarrow{d} \Sigma_{VV}^{-1/2'} (\Sigma_{VV}^{1/2'} z'_V z_V \Sigma_{VV}^{1/2}) \Sigma_{VV}^{-1/2} = z'_V z_V,$$

which is a Wishart distribution with K_2 degrees of freedom.

Proof of Theorem 6

Consider part 1. When $0 \leq \delta < \frac{1}{2}$, we note that

$$\mu_\pi = \left(\tilde{Z}'\tilde{Z}\right)^{1/2} \pi = \left(\frac{\tilde{Z}'\tilde{Z}}{Tn}\right)^{1/2} (Tn)^{1/2} \frac{C}{\sqrt{n} T^\delta} = \left(\frac{\tilde{Z}'\tilde{Z}}{Tn}\right)^{1/2} T^{1/2} \frac{C}{T^\delta} = O\left(T^{\frac{1}{2}-\delta}\right).$$

It follows that

$$S - c_\beta \mu_\pi \xrightarrow{d} N(0, I_k)$$

and

$$T - d_\beta \mu_\beta \xrightarrow{d} N(0, I_k).$$

Consider part 2. When $\delta = \frac{1}{2}$, we have

$$\mu_\pi \to D_Z C = O_p(1).$$

Then

$$S \xrightarrow{d} N(c_\beta D_Z C, I_k)$$

and

$$T \xrightarrow{d} N(d_\beta D_Z C, I_k).$$

Consider part 3. When $\frac{1}{2} < \delta < \infty$

$$\mu_\pi = o_p(1).$$

Then

$$S \xrightarrow{d} N(0, I_k)$$

and

$$T \xrightarrow{d} N(0, I_k).$$

This proves the theorem.

A RISK SUPERIOR SEMIPARAMETRIC ESTIMATOR FOR OVERIDENTIFIED LINEAR MODELS

George G. Judge and Ron C. Mittelhammer

ABSTRACT

In the context of competing IV econometric models and estimators, we demonstrate a semiparametric Stein-like estimator (SSLE) that, under quadratic loss, has superior risk performance. The method eliminates the need for pretesting to decide whether covariate endogeneity is present and makes use of a pretest estimator choice between IV and non-IV methods unnecessary. A sampling study is used to illustrate finite sample performance over a range of sampling designs, including its performance relative to pretest estimators. An important applied problem from the literature is analyzed to indicate possible applied implications and the relation of SSLE to other modern IV estimators.

Keywords: Combining estimators; semiparametric estimation and inference; quadratic loss; temparal structure

JEL classifications: C13; C14; C36

30th Anniversary Edition
Advances in Econometrics, Volume 30, 237–255

ISSN: 0731-9053/doi:10.1108/S0731-9053(2012)0000030013

INTRODUCTION

Carter Hill and Thomas B. Fomby have been important contributors to the area of econometrics over the last four decades. Although they have contributed to and left their footprints on a range of econometric areas, in this chapter we go back to the future and focus, in a simultaneous equations context, on specification-pretesting and instrumental variables. Instrumental Variable (IV) estimators have played an important role in econometrics since the introduction in the 1940s of the simultaneous equations-endogenous data process and least squares bias. Modern finite sample IV methods have recently received renewed emphasis, in general and for estimating causal-treatment effects in contexts where controlled-designed experiments are not possible-feasible. Although IV estimators are consistent, an often heard complaint is that their precision is low in sample sizes often found in practice. Angrist and Krueger (2001) speak to this problem and present a useful survey of the history and uses of instrumental variable methods. Since in overidentified models, over fitting can cause small sample bias, jackknife IV (JIVE) estimators have been proposed to address this problem. These are well summarized in a paper by Ackerberg and Devereux (2009). In another approach to efficiency gains with IV estimators, Hansen, McDonald, and Newey (2009) use moment conditions that are nonlinear in the equation error noise.

These approaches focus on a particular econometric model and seek how to improve estimator and inference performance. In this chapter, with the same objective of efficiency gain, we recognize that economic systems and processes are seldom in equilibrium and there are usually conflicting ways to describe the indirect noisy observations. In order to accommodate the underlying unknown data sampling process, we demonstrate an estimator that is a combination of the IV–2SLS and classical linear model (least squares – LS) estimators that is superior under quadratic loss. This means that the often used minimum mean squared error criterion is built in as a part of the estimation process.

THE ECONOMETRIC MODELS AND THE COMBINING IDEA

Suppose the observed data result from a sampling process of the form

$$y_i = \mathbf{X}_{i.}\mathbf{B} + \varepsilon_i \tag{1}$$

where i indexes observations, y_i is the dependent-endogenous variable, $\mathbf{X}_{i.}$ represents a $1 \times k$ vector of covariates, ε_i is an unobservable error with $E[\varepsilon_i] = 0$, and $\mathbf{B}$ is a $k \times 1$ vector of unknown parameters. In addition, in a simultaneous equation context, assume a $1 \times m$ vector of instrumental variables, $\mathbf{Z}_{i.}$, that are independent of ε_i and the data sampling process is *iid* $[\mathbf{X}_{i.}, \mathbf{Z}_{i.}, \varepsilon_i]$, for $i = 1, ..., n$.

In this context the LS estimator is given by

$$\mathbf{B}_{LS} = (\mathbf{X}'\mathbf{X})^{-1}\mathbf{X}'\mathbf{y} = \mathbf{B} + (\mathbf{X}'\mathbf{X})^{-1}\mathbf{X}'\boldsymbol{\varepsilon} \tag{2}$$

which may be biased and inconsistent, the instrumental variable (IV) estimator is defined by

$$\mathbf{B}_{IV} = (\mathbf{Z}'\mathbf{X})^{-1}\mathbf{Z}'\mathbf{y} \tag{3}$$

and the 2SLS or generalized method of moments estimator is represented by

$$\mathbf{B}_{2SLS} = (\mathbf{X}'\mathbf{P}_{\mathbf{Z}}\mathbf{X})^{-1}\mathbf{X}'\mathbf{P}_{\mathbf{Z}}\mathbf{y} \tag{4}$$

where $\mathbf{P}_{\mathbf{Z}} = \mathbf{Z}(\mathbf{Z}\mathbf{Z})^{-1}\mathbf{Z}'$. Under appropriate regularity conditions, the estimators (3) and/or (4) are consistent, asymptotically normal, and efficient estimators of $\mathbf{B}$.

The Idea of Combining Estimators

Given estimators that exhibit distinct and dissimilar sampling properties, the objective of this chapter is to investigate the statistical-econometric implications of choosing a combination of competing estimation problems. The goal is to allow change that is to take place over time and to develop natural adaptive semiparametric estimation and inference methods that have superior risk performance. To do this we focus on an idea, developed in Judge and Mittelhammer (2012a, 2012b), of defining a combination of the 2SLS and LS estimators that, in the case where some regressors are not orthogonal to the noise term, is superior to either estimator separately under quadratic loss.

The estimator is defined as a combination of the 2SLS and LS estimators and is given by

$$\mathbf{B}_{Conv} = \alpha\mathbf{B}_{2SLS} + (1 - \alpha)\mathbf{B}_{LS} \quad \text{for } \alpha \in [0, 1] \tag{5}$$

The MSE of this combination is defined by

$$\begin{aligned}\mathrm{MSE}(\mathbf{B}_{Conv}) &= E(||\alpha(\mathbf{B}_{2SLS}-\mathbf{B})+(1-\alpha)(\mathbf{B}_{LS}-\mathbf{B})||^2)\\ &= \alpha^2\mathrm{MSE}(\mathbf{B}_{2SLS})+(1-\alpha)^2\mathrm{MSE}(\mathbf{B}_{LS})\\ &\quad +2\alpha(1-\alpha)E[(\mathbf{B}_{2SLS}-\mathbf{B})'(\mathbf{B}_{LS}-\mathbf{B})]\end{aligned} \tag{6}$$

The estimation problem is then one of choosing the value of α so as to minimize the MSE of the combination of the two estimators. When proceeding semiparametrically, it is possible to define large sample approximations to all of the components of the MSE calculation, and thereby derive an approximation to the optimal combination of the LS and 2SLS estimators. The details of such an approach are developed in Judge and Mittelhammer (2012a).

For known values of the MSE components, the optimal choice for α is straightforwardly shown to be

$$\alpha_o = 1 - \frac{tr(Cov(\mathbf{B}_{2SLS})) - tr(\Sigma_\times)}{E(\gamma_n)'E(\gamma_n) + tr(Cov(\mathbf{B}_{LS})) + tr(Cov(\mathbf{B}_{2SLS})) - 2tr(\Sigma_\times)} \tag{7}$$

where $\hat{\gamma}_n = (\mathbf{X}'\mathbf{X})^{-1}\mathbf{X}'\hat{\mathbf{e}}$ and $\Sigma_\times = E(\mathbf{B}_{2SLS}-\mathbf{B})(\mathbf{B}_{LS}-\mathbf{B})'$. An estimated minimum MSE combined estimator is defined by Eq. (5) with an estimated value of α_o inserted. Judge and Mittelhammer (2012a) develop consistent estimators for each for the components of Eq. (7). They also show that an alternative consistent estimator of the denominator term in Eq. (7) is given by $(\mathbf{B}_{2SLS}-\mathbf{B}_{LS})'(\mathbf{B}_{2SLS}-\mathbf{B}_{LS})$. An estimate of Eq. (7) can then be expressed alternatively as[1]

$$\hat{\alpha}_o = 1 - \frac{tr(C\hat{o}v(\mathbf{B}_{2SLS})) - tr(\hat{\Sigma}_\times)}{(\mathbf{B}_{2SLS}-\mathbf{B}_{LS})'(\mathbf{B}_{2SLS}-\mathbf{B}_{LS})} \tag{8}$$

Using Eq. (8), the estimated optimal estimator may be expressed in "Stein-like" form (James & Stein, 1961) as

$$\mathbf{B}_{Conv} = \mathbf{B}_{2SLS} - \frac{\hat{\tau}_o}{||\mathbf{B}_{2SLS}-\mathbf{B}_{LS}||^2}(\mathbf{B}_{2SLS}-\mathbf{B}_{LS}) \tag{9}$$

where $\hat{\tau}_o \equiv tr(C\hat{o}v(\mathbf{B}_{2SLS})) - tr(\hat{\Sigma}_\times)$.

Asymptotically, the estimated minimum MSE estimator, when centered and scaled by root n, has a limit distribution demonstrating that it is quadratic risk superior to either the 2SLS or LS estimators, by definition of α_o which is consistently estimated by Eq. (8). For inference purposes, the limiting distribution can be estimated by replacing α_o and ε with their consistently estimated counterparts, and estimating the limit matrices with appropriate moments based on observed **X** and **Z**. First order asymptotic properties of the combined estimator are demonstrated in Judge and Mittelhammer (2012a). In the next section a sampling study is used to illustrate finite sample performance over a range of designs and model specification – pretest situations.

SAMPLING EXPERIMENT

In this section we illustrate the finite sample properties of the combined estimator (5). We first compare the combined estimator to the 2SLS estimator that would be typically used by applied econometricians in this data sampling context. We then consider using a pretest estimator to choose between the 2SLS and the LS estimator, based on Hausman's (1978) test for endogeneity, as an alternative to the combined estimator and compare the performance of the combined estimator to a pretest estimator derived from an application of the test. The combined estimator is defined using Eq. (7) with all unknown vector and matrix components replaced by their respective consistent component estimators.

Design of the Sampling Experiment

We implement an MC design that is similar to Hahn and Hausman (2002, 2004), and is also closely related to a design implemented by Staiger and Stock (1997). The model contains K instrumental variables that are used as a basis for estimating the model given by

$$\mathrm{Y}_i = \mathrm{X}_i\mathrm{B} + \varepsilon_i, \quad i = 1, \ldots, n \tag{10}$$

where

$$\begin{bmatrix} \varepsilon \\ X \\ \hline Z_K \end{bmatrix} \sim iid\ N\left([\mathbf{0}], \left[\begin{array}{cc|c} 1 & \rho_{\varepsilon,X} & \mathbf{0}'_K \\ \rho_{\varepsilon,X} & 1 & \rho_{Z,X}\mathbf{1}'_K \\ \hline \mathbf{0}_K & \rho_{Z,X}\mathbf{1}_K & \boldsymbol{I}_K \end{array}\right]\right) \tag{11}$$

$\mathbf{0}_K$ is a $K \times 1$ vector of zeros, $\mathbf{1}_K$ is a $K \times 1$ vector of ones, and $B = 1$. In this sampling design, the regressor X is correlated with the noise term ε at level $\rho_{\varepsilon,X}$, and each of the K instrumental variables in $\mathbf{Z}_K$ are correlated with the endogenous X at level $\rho_{Z,X}$. Regarding strength of instruments, the design implies that the theoretical R^2 in a regression of X on $\mathbf{Z}_K$ is given simply by $R^2 = K\rho^2_{Z,X}$, which bounds the maximal value of $\rho^2_{Z,X}$ at K^{-1}. We set $K=3$.

We examine five degrees of nonorthogonality between X and ε, ranging from strong to weak, as $\rho_{\varepsilon,X} = .90,\ .75,\ .50,\ .25,$ and .1. Regarding the strength of the instruments, we utilize four levels as $R^2_{X,Z} = .75,\ .5,\ .25,$ and .1, where the latter two levels represent cases of notably weak instruments.[2] The model is estimated using sample sizes of $n=25$, 50, 100, 200, and 500, and employs M = 10,000 repeated samples to calculate MSE measures of fit defined by $\hat{\text{MSE}}(B) = M^{-1}\sum_{i=1}^{M}(\hat{B}_i - B)^2$. The numerical results of the sampling experiments are presented in Tables A1–A4.

Sampling Performance Comparison: $\mathbf{B}_{2SLS}$ *versus* $\mathbf{B}_{Conv}$

Graphical representations of the MSE of $\mathbf{B}_{Conv}$, relative to the MSE of $\mathbf{B}_{2SLS}$, are presented in Figs. 1–4. In this sampling comparison, when the instruments are relatively weak ($R^2_{X,Z} = .25$ or .10), $\mathbf{B}_{Conv}$ is dramatically superior to $\mathbf{B}_{2SLS}$ for small and moderate size samples, for all but the highest levels of correlation between X and ε. Moreover, regardless of the sample size, the combined estimator competes very favorably with the asymptotically optimal 2SLS across all weak instrument scenarios.

For scenarios in which the instruments are stronger ($R^2_{X,Z} = .50$ or .75), $\mathbf{B}_{Conv}$ still remains notably superior to $\mathbf{B}_{2SLS}$ for the smaller sample sizes across all levels of nonorthogonality. As the sample size increases,

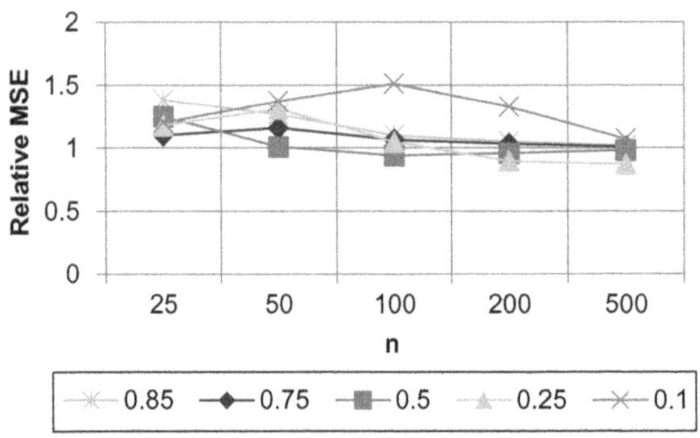

Fig. 1. $\text{MSE}(\mathbf{B}_{2SLS})/\text{MSE}(\mathbf{B}_{Conv})$ for $R^2_{X,Z} = .75$ and $\rho_{X,\varepsilon} = .85,\ .75,\ .5,\ .25,\ .1.$

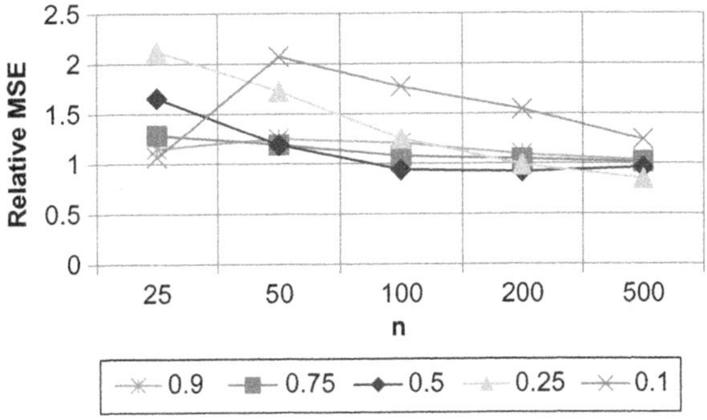

Fig. 2. MSE($\mathbf{B}_{2SLS}$)/MSE($\mathbf{B}_{Conv}$) for $R^2_{X,Z} = .50$ and $\rho_{X,\varepsilon} = .90,\ .75,\ .5,\ .25,\ .1$.

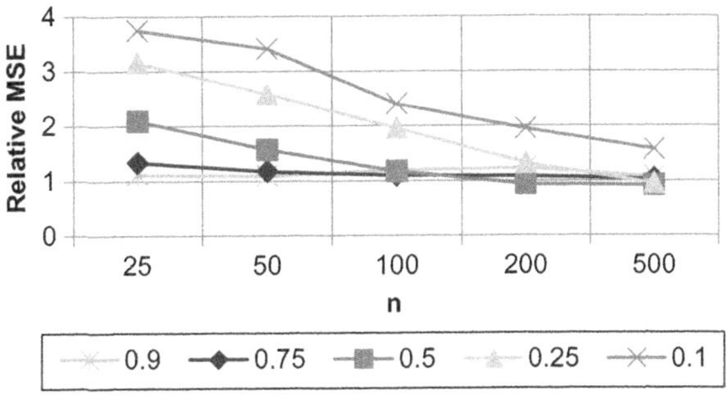

Fig. 3. MSE($\mathbf{B}_{2SLS}$)/MSE($\mathbf{B}_{Conv}$) for $R^2_{X,Z} = .25$ and $\rho_{X,\varepsilon} = .90,\ .75,\ .5,\ .25,\ .1$.

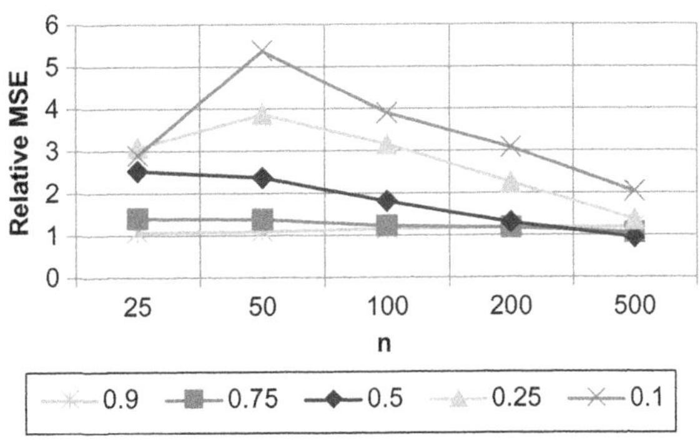

Fig. 4. MSE($\mathbf{B}_{2SLS}$)/MSE($\mathbf{B}_{Conv}$) for $R^2_{X,Z} = .10$ and $\rho_{X,\varepsilon} = .90,\ .75,\ .5,\ .25,\ .1$.

superiority wanes, especially for the intermediate levels of nonorthogonality. However, again, $\mathbf{B}_{Conv}$ remains mean square error competitive with $\mathbf{B}_{2SLS}$ regardless of sample size, and is never notably inferior to the 2SLS estimator. Overall, across the 100 MC scenarios, $\mathbf{B}_{Conv}$ is superior in 87 of the cases, and where it is not, the MSEs of the two estimators are in any case small in value, indicating that both estimators are effective for estimating **B**. The consistency of both estimators is evident in the strong downward trend in the MSEs as the sample size increases (see the tables in the appendix).

While the details are not shown here,[3] the source of the MSE levels is somewhat different for the combined and 2SLS estimators. In particular, the combined estimator has a tendency to be somewhat more biased but not as variable as the 2SLS estimator. This is as one would expect from combining the more biased but notably less variable LS estimator with the asymptotically unbiased but substantially more variable 2SLS estimator. *Moreover*, it is evident that the variability of the 2SLS estimator is very substantial in the case of weaker instruments, both for moderate and strong nonorthogonality, and it is only for large sample sizes that this variability dissipates significantly. The combined estimator substantially dominates 2SLS in precision in this case, and remains dominant even when the instruments are stronger. The bias component is most pronounced in the case of substantial nonorthogonality and weak instruments. Overall, the decomposition supports the notion that the combined estimator produces a clear stabilizing effect on the estimation of the parameters, combining the substantially lower variability of the LS estimator with the lower bias of the 2SLS estimator, producing an estimator with superior MSE performance in most cases, and MSE competitive performance in all cases.

Sampling Performance Comparison: $\mathbf{B}_{Conv}$ *versus* $\mathbf{B}_{\text{Pretest}}$

The Hausman pretest estimator calculated in this section is the original and familiar "contrast version" of the test based on the usual χ^2 statistic calculated as

$$H_v = (\mathbf{B}_{2SLS} - \mathbf{B}_{LS})'[Cov(\mathbf{B}_{2SLS}) - Cov(\mathbf{B}_{LS})]^{+}(\mathbf{B}_{2SLS} - \mathbf{B}_{LS}) \sim ChiSquare(v) \tag{12}$$

where "+" denotes a generalized inverse of the covariance matrix difference, and the degrees of freedom v is equal to the number of potentially endogenous RHS variables in the model specification. Based on the data sampling process described in section "Design of the Sampling Experiment," $v = 1$ in the current application. The Hausman pretest estimator is defined as

$$\mathbf{B}_{H_v} = \left\{ \begin{array}{c} \mathbf{B}_{LS} \\ \mathbf{B}_{2SLS} \end{array} \right\} \text{ if } H_v \left\{ \begin{array}{c} < \\ \geq \end{array} \right\} \chi^2(1-\alpha, v) \tag{13}$$

where $\chi^2(1-\alpha, v)$ is the $(1-\alpha)$th percentile of the χ^2 distribution with v degrees of freedom. See Chmelarova and Hill (2010) for further discussion, and some variations on the implementation of the test.

A difficulty in implementing the pretest estimator is the issue of the level of type I error to choose in applying the test. In the current application, we choose the typical value of $\alpha = .05$ that practitioners might use in testing the hypothesis of endogeneity. The MSE results of the sampling experiments are presented in Tables A1–A4, and graphical representations of the MSE of $\mathbf{B}_{Conv}$, relative to the MSE of $\mathbf{B}_{Pretest}$, are presented in Figs. 5–8.

The relative MSE results indicate that the combined estimator is once again the MSE superior estimator in 59 of 60 scenarios when endogeneity is appreciable ($\rho_{X,\varepsilon} \geq .5$), regardless of the strength of instruments or the sample size, the one minor deviation being the case of smallest sample size $n = 25$, and $\rho_{X,\varepsilon} = .5$. For low levels of endogeneity ($\rho_{X,\varepsilon} \leq .25$) it is not surprising that the pretest estimator exhibits MSE superiority for smaller sample sizes since the Hausman test is defined to protect against type I error at .95 probability, and thus is frequently "choosing" the OLS estimator in a sampling context in which the OLS estimator exhibits favorable MSE behavior. However, as the sample size increases, the $\mathbf{B}_{Conv}$ estimator closes

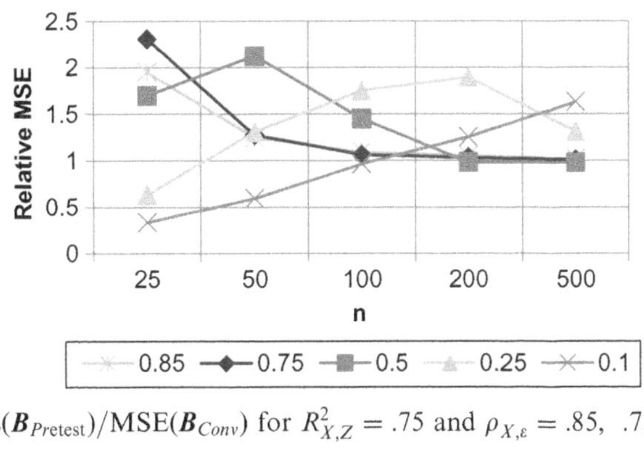

Fig. 5. MSE($\boldsymbol{B}_{Pretest}$)/MSE($\boldsymbol{B}_{Conv}$) for $R^2_{X,Z} = .75$ and $\rho_{X,\varepsilon} = .85,\ .75,\ .5,\ .25,\ .1$.

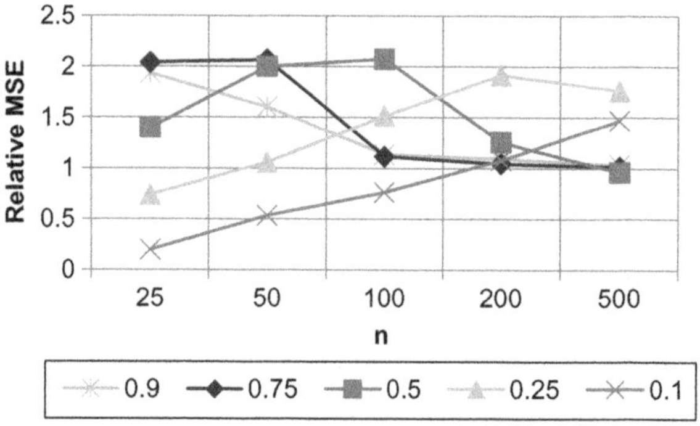

Fig. 6. MSE($\boldsymbol{B}_{Pretest}$)/MSE($\boldsymbol{B}_{Conv}$) for $R^2_{X,Z} = .50$ and $\rho_{X,\varepsilon} = .90, .75, .5, .25, .1$.

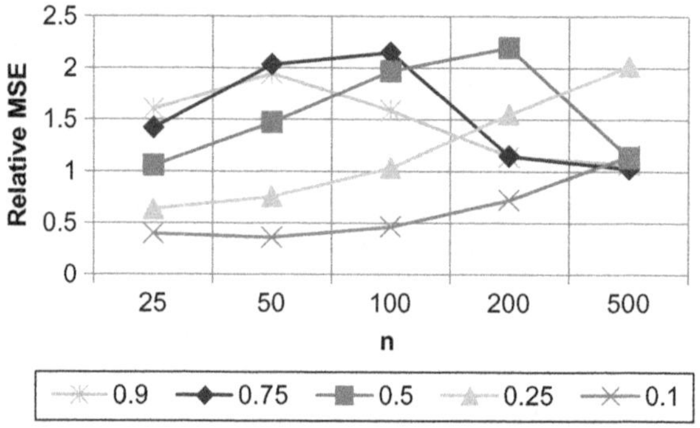

Fig. 7. MSE($\boldsymbol{B}_{Pretest}$)/MSE($\boldsymbol{B}_{Conv}$) for $R^2_{X,Z} = .25$ and $\rho_{X,\varepsilon} = .90, .75, .5, .25, .1$.

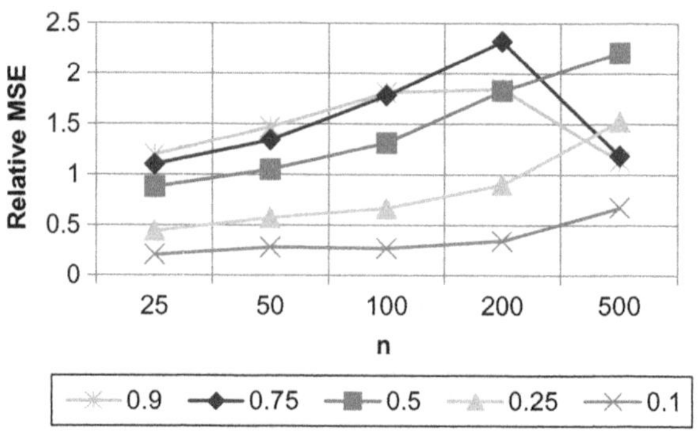

Fig. 8. MSE($\boldsymbol{B}_{Pretest}$)/MSE($\boldsymbol{B}_{Conv}$) for $R^2_{X,Z} = .10$ and $\rho_{X,\varepsilon} = .90, .75, .5, .25, .1$.

the MSE gap, and superiority is achieved in all cases except when endogeneity is very minor ($\rho_{X,\varepsilon} = .1$) and the instruments are the weakest ($R^2_{X,Z} = .10$). The latter case is a situation where OLS exhibits its greatest MSE superiority over 2SLS, and is also where the combined estimator exhibits it greatest MSE gains over 2SLS.

The MSE decomposition results for the two estimators[4] again suggest differing sources of the superior performance of the estimators. However, unlike the comparison to the 2SLS estimator, it is not the case that the combined estimator always tends to be more biased, and the pretest estimator always tends to be more variable. It depends on the characteristics of the data generating process as to which is the more biased estimator. However, in all cases where the bias is appreciable, it is the combined estimator that has the lower bias. This is in fact demonstrably the case for the weak instrument scenarios. When the pretest estimator exhibits bias superiority, the bias for both estimators is effectively minimal.

AN EMPIRICAL APPLICATION OF THE COMBINING METHODOLOGY

In this section we illustrate the empirical application of the combining estimation methodology by applying it to data originally analyzed by Angrist and Krueger (1991), and a returns to schooling model that was estimated by Ackerberg and Devereux (2009, p. 359), using LS, 2SLS, and JIVE methods. In their model, quarter of birth is interacted with year of birth to form 30 external (to the structural equation) instrumental variables for use in estimating the model with IV techniques. The data is based on an extract from the 1980 U.S. Census consisting of $n = 329{,}509$ observations on men who were born between 1930 and 1939. Variables in the dataset used in this analysis include the natural log of weekly wages, the year of birth (1930–1939), quarter of birth (1–4), and education level (highest grade completed, ranging from 0 to 20).

Following Ackerberg and Devereux (2009), the simultaneous equation specification of the model uses the log of wages as the dependent variable, with explanatory variables consisting of education level and indicators for the year of birth. The coefficient of primary interest is the education level parameter, which measures the returns to years of schooling. Indicator variables for the quarter of birth were used in the construction of the instrumental variables because of their role as exogenous explanatory

Table 1. Results for the Returns to Schooling Model.

Estimator	$\mathbf{B}_{LS}$	$\mathbf{B}_{2SLS}$	$\mathbf{B}_{JIVE}$	$\mathbf{B}_{Conv}$
Treatment effect	0.071	0.089	0.096	0.081
Standard error	0.0003	0.016	0.022	0.009
$\hat{\alpha}_o$	0.556			
Hausman $\chi^2(1)$	1.254			
(*p-value*)	(0.263)			

variables in the determination of education level. Specifically, interactions between indicators for years of birth and quarter of birth provide 30 external instrumental variables for use in 2SLS estimation (Ackerberg & Devereux, 2009, p. 359).

Regarding the strength of instruments in this case, the regression of years of schooling on the 30 external instruments resulted in an R^2 of .0025, suggesting a case of weak instruments, and representing a case that is notably weaker than any of the Monte Carlo experiments reported in section "Sampling Experiment." The F-test of the significance of the 30 external instruments in the first stage of two-stage least squares resulted in an F-value of 4.91, which upon using the rule of thumb value of 10 (Stock, Wright, & Yogo, 2002), corroborates that this is indeed a case of substantially weak instruments.

Estimates of the return to schooling parameter of the structural equation and the standard errors of the estimates, generated by LS, 2SLS, and JIVE, are reported in Table 1 (our results reproduce those reported in Ackerberg & Devereux, 2009, Table 3). We also report an estimate of $\mathbf{B}_{Conv}$, based on the estimate of the combining weight that is also provided in the table. Finally, we report the outcome of the Hausman test for choosing between the LS and 2SLS estimator.

The MSE-minimizing combined estimator recognizes the differing sampling behavior of the LS and 2SLS estimators and applies slightly more than 1/2 weight (0.556) to the 2SLS estimator. The resulting $\mathbf{B}_{Conv}$ estimate measures the value of more education, while also reflecting the effect of quarter of birth on how much education individuals received. By using a minimum MSE estimation objective and recognizing the uncertain nature of the returns to schooling model, the average impact of the treatment estimated by the combined estimate $\mathbf{B}_{Conv}$ is nearly a percentage point less than implied by the impact suggested by the single-model 2SLS criterion estimate. The $\mathbf{B}_{Conv}$ estimate indicates that the average effect of the treatment is one and one-half percentage points less than that implied by JIVE.

In addition to being a minimum MSE estimate of the returns to schooling, in terms of inference, the estimated precision of $\mathbf{B}_{Conv}$ is substantially greater than either $\mathbf{B}_{2SLS}$ or $\mathbf{B}_{JIVE}$, where the standard error of the MQL estimate is less than half that of JIVE and more than 40% less than 2SLS. The Hausman pretest results suggest that the $\mathbf{B}_{LS}$ estimator should be chosen at any conventional level of significance.

In summary, three established estimation methodologies provide different estimation and inference results concerning the rate of return to schooling. The combined estimator recognizes the possible uncertain nature of the data-model process and provides, in the case of this overidentified experimental economic model, a minimum MSE alternative to the single model-based estimators that are often employed in estimating treatment effects.

SUMMARY AND IMPLICATIONS

To recognize the disequilibrium nature of economic processes-data, and the uncertainty surrounding model-estimator choice, in this chapter we have specified a combined estimator, with quadratic loss as an arbitrator. Sampling experiments used to indicate small sample performance indicate the optimality of the combined estimator under quadratic loss, the nonoptimality in most situations of the pretest-model selection estimator under quadratic loss, and the applicability of the combined estimator for a dataset taken from the experimental economic literature. We have developed our minimum quadratic loss formulation in a simultaneous equation context. This may be readily extended to combine other estimator alternatives when there is heterogeneity in terms of estimator sampling behavior.

The debate on model selection, specification, and identification, which started decades ago, continues unabated – see for example Kurozumi and Tuvaandorj (2011). What we have done in this chapter is to extend the idea to estimation and recognize that given our knowledge of economic processes-data, there are many possible models, each with its own probability, and the testing–pretesting solution to model choice may have serious negative statistical implications.

NOTES

1. Judge and Mittelhammer (2012a) define a consistent estimator for the $\Sigma_\times$ term as

$$\hat{\Sigma}_\times = ((\mathbf{X}'\mathbf{Z})(\mathbf{Z}'\mathbf{Z})^{-1}(\mathbf{Z}'\mathbf{X}))^{-1}(\mathbf{X}'\mathbf{Z})(\mathbf{Z}'\mathbf{Z})^{-1} \times \left[\sum_{i=1}^{n}\left(\hat{e}_i\mathbf{Z}'_{i.} - n^{-1}\sum_{i=1}^{n}\mathbf{Z}'_{i.}\hat{e}_i\right)\left(\hat{e}_i\mathbf{X}'_{i.} - n^{-1}\sum_{i=1}^{n}\mathbf{X}'_{i.}\hat{e}_i\right)'\right](\mathbf{X}'\mathbf{X})^{-1}$$

where $\hat{e}_i$, for $i = 1,\ldots,n$, are 2SLS residuals.

2. Given the sampling design, the theoretical upper bound to $\rho_{\varepsilon,X}$ when $R^2 = .75$ is .866. We set $\rho_{\varepsilon,X} = .85$ for this value of R^2.

3. The details of the decomposition of the MSE into bias and variance effects is available from the authors upon request.

4. The details of the decomposition of the MSE into bias and variance effects is available from the authors upon request.

REFERENCES

Ackerberg, D., & Devereux, P. (2009). Improved JIVE estimators for over identified linear models without heteroscedasticity. *The Review of Economics and Statistics*, *91*, 351–362.

Angrist, J., & Krueger, A. (1991). Does compulsory school attendance affect schooling and earning? *Quarterly Journal of Economics*, *106*, 979–1014.

Angrist, J., & Krueger, A. (2001). Instrumental variables and the search for identification: From the supply and demand to natural experiments. *Journal of Economic Perspectives*, *15*, 6985.

Chmelarova, V., & Hill, C. (2010). The Hausman pretest estimator. *Economic Letters*, *108*, 96–99.

Hahn, J., & Hausman, J. (2002). Notes on bias in estimators for simultaneous equation models. *Economic Letters*, *75*, 237–241.

Hansen, C., McDonald, J., & Newey, W. (2009). Instrumental variables estimation with flexible distributions. *Journal of Business and Economic Statistics*, *28*, 1325.

Hausman, J. (1978). Specification tests in econometrics. *Econometrica*, *46*, 1251–1271.

James, W., & Stein, C. (1961). Estimation with quadratic loss. *Proceedings of the Fourth Berkeley Symposium on Mathematical Statistics and Probability*, University of California Press, Berkeley, CA (Vol. 1, pp. 361–380).

Judge, G., & Mittelhammer, R. (2012a). A minimum mean squared error semiparametric combining estimator. *Advances in Econometrics*, *29*, 55–86.

Judge, G., & Mittelhammer, R. (2012b). *An information theoretic approach to econometrics*. New York, NY: Cambridge University Press.

Kurozumi, E., & Tuvaandorj, P. (2011). Model selection criteria in multivariate models with multiple structural changes. *Journal of Econometrics*, *164*, 218–238.

Staiger, D., & Stock, J. H. (1997). Instrumental variables regression with weak instruments. *Econometrica*, *65*, 557–586.

Stock, J. H., Wright, J. H., & Yogo, M. (2002). A survey of weak instruments and weak identification in generalized method of moments. *Journal of Business and Economic Statistics*, *20*, 518–529.

APPENDIX

Table A1. MSE Results Estimators for $R^2_{X,Z} = .1$.

n	$\rho_{\varepsilon,X}$	MSE($\mathbf{B}_{LS}$)	MSE($\mathbf{B}_{2SLS}$)	MSE($\mathbf{B}_{Conv}$)	MSE($\mathbf{B}_{Pretest}$)
25	.9	0.818334	0.735036	0.681391	0.814032
50	.9	0.813223	0.577371	0.526454	0.773757
100	.9	0.811539	0.380366	0.329424	0.595976
200	.9	0.810818	0.174397	0.147032	0.271162
500	.9	0.810243	0.065362	0.054575	0.061843
25	.75	0.580706	0.739068	0.525661	0.579485
50	.75	0.571162	0.581236	0.420329	0.564672
100	.75	0.567065	0.360192	0.294206	0.524492
200	.75	0.564111	0.18509	0.157087	0.364506
500	.75	0.563267	0.064683	0.060798	0.072182
25	.5	0.280411	0.809048	0.319732	0.280225
50	.5	0.265533	0.59859	0.252691	0.265377
100	.5	0.258208	0.353225	0.195065	0.256271
200	.5	0.253085	0.170784	0.130674	0.239026
500	.5	0.251507	0.065626	0.069246	0.152956
25	.25	0.102873	0.721704	0.234274	0.10321
50	.25	0.082139	0.558641	0.144759	0.082975
100	.25	0.07237	0.358599	0.113819	0.075636
200	.25	0.066894	0.180951	0.080715	0.07249
500	.25	0.064413	0.065962	0.048241	0.073642
25	.1	0.053433	0.781225	0.269485	0.054079
50	.1	0.030766	0.617528	0.114944	0.031706
100	.1	0.020176	0.353346	0.090825	0.024062
200	.1	0.014873	0.190086	0.061937	0.020917
500	.1	0.012	0.066293	0.032697	0.02203

Table A2. MSE Results Estimators for $R^2_{X,Z} = .25$.

n	$\rho_{\varepsilon,X}$	MSE($\mathbf{B}_{LS}$)	MSE($\mathbf{B}_{2SLS}$)	MSE($\mathbf{B}_{Conv}$)	MSE($\mathbf{B}_{Pretest}$)
25	.9	0.818334	0.535372	0.481253	0.769533
50	.9	0.813223	0.281822	0.259549	0.5047
100	.9	0.811539	0.132833	0.111684	0.176964
200	.9	0.810818	0.066186	0.054057	0.061653
500	.9	0.810243	0.024807	0.023266	0.024807
25	.75	0.580706	0.539422	0.402886	0.57276
50	.75	0.571162	0.290649	0.247365	0.503928
100	.75	0.567065	0.138511	0.126254	0.271722
200	.75	0.564111	0.064884	0.059836	0.06871
500	.75	0.563267	0.024875	0.024223	0.024875
25	.5	0.280411	0.554553	0.265034	0.279616
50	.5	0.265533	0.279068	0.178129	0.262178
100	.5	0.258208	0.13835	0.118089	0.232977
200	.5	0.253085	0.063298	0.067811	0.14912
500	.5	0.251507	0.025011	0.027464	0.031152
25	.25	0.102873	0.51627	0.163605	0.103695
50	.25	0.082139	0.288602	0.111786	0.084229
100	.25	0.07237	0.150077	0.076808	0.079129
200	.25	0.066894	0.063519	0.048333	0.074924
500	.25	0.064413	0.025141	0.026396	0.053314
25	.1	0.053433	0.514065	0.137275	0.054372
50	.1	0.030766	0.311658	0.091415	0.032823
100	.1	0.020176	0.143364	0.059791	0.027631
200	.1	0.014873	0.063477	0.032719	0.023616
500	.1	0.012	0.0252	0.016082	0.018531

Table A3. MSE Results Estimators for $R^2_{X,Z} = .50$.

n	$\rho_{\varepsilon,X}$	MSE($\mathbf{B}_{LS}$)	MSE($\mathbf{B}_{2SLS}$)	MSE($\mathbf{B}_{Conv}$)	MSE($\mathbf{B}_{Pretest}$)
25	.9	0.818334	0.30982	0.269161	0.520759
50	.9	0.813223	0.137006	0.109629	0.175696
100	.9	0.811539	0.064156	0.053179	0.060217
200	.9	0.810818	0.031287	0.028649	0.031287
500	.9	0.810243	0.012154	0.011795	0.012154
25	.75	0.580706	0.330984	0.257133	0.525076
50	.75	0.571162	0.143275	0.120318	0.249153
100	.75	0.567065	0.063522	0.059064	0.065845
200	.75	0.564111	0.030922	0.029608	0.030922
500	.75	0.563267	0.012273	0.012085	0.012273
25	.5	0.280411	0.328984	0.198114	0.277023
50	.5	0.265533	0.142293	0.119426	0.239174
100	.5	0.258208	0.06432	0.068446	0.142173
200	.5	0.253085	0.030656	0.033312	0.041836
500	.5	0.251507	0.012337	0.012843	0.012348
25	.25	0.102873	0.300757	0.141549	0.104409
50	.25	0.082139	0.144395	0.083582	0.088281
100	.25	0.07237	0.06562	0.052425	0.079521
200	.25	0.066894	0.030654	0.031066	0.059547
500	.25	0.064413	0.012374	0.014618	0.025797
25	.1	0.053433	0.303157	0.28488	0.055905
50	.1	0.030766	0.13993	0.067533	0.035971
100	.1	0.020176	0.066665	0.037665	0.028697
200	.1	0.014873	0.030672	0.019884	0.021461
500	.1	0.012	0.012389	0.009974	0.014665

Table A4. MSE Results Estimators for $R^2_{X,Z} = .75$.

n	$\rho_{\varepsilon,X}$	MSE($\mathbf{B}_{LS}$)	MSE($\mathbf{B}_{2SLS}$)	MSE($\mathbf{B}_{Conv}$)	MSE($\mathbf{B}_{Pretest}$)
25	.85	0.734514	0.22563	0.163283	0.318419
50	.85	0.727611	0.090908	0.071808	0.089315
100	.85	0.72507	0.041328	0.037616	0.040996
200	.85	0.723581	0.020545	0.019565	0.020545
500	.85	0.722916	0.008063	0.007925	0.008063
25	.75	0.580706	0.184569	0.167765	0.387981
50	.75	0.571162	0.088608	0.076443	0.09756
100	.75	0.567065	0.041787	0.039235	0.041735
200	.75	0.564111	0.020397	0.019739	0.020397
500	.75	0.563267	0.008114	0.00802	0.008114
25	.5	0.280411	0.197983	0.158676	0.269532
50	.5	0.265533	0.087919	0.087163	0.185197
100	.5	0.258208	0.042273	0.04505	0.065377
200	.5	0.253085	0.020252	0.021149	0.020779
500	.5	0.251507	0.008167	0.008324	0.008167
25	.25	0.102873	0.197552	0.167668	0.105853
50	.25	0.082139	0.089495	0.068369	0.089608
100	.25	0.07237	0.042763	0.040814	0.071466
200	.25	0.066894	0.020237	0.022604	0.04284
500	.25	0.064413	0.008191	0.009424	0.012381
25	.1	0.053433	0.203873	0.16961	0.056553
50	.1	0.030766	0.088754	0.064681	0.038423
100	.1	0.020176	0.043053	0.0285	0.027477
200	.1	0.014873	0.020246	0.015246	0.019116
500	.1	0.012	0.008198	0.007652	0.012475

SPATIAL DEPENDENCE IN REGRESSORS AND ITS EFFECT ON PERFORMANCE OF LIKELIHOOD-BASED AND INSTRUMENTAL VARIABLE ESTIMATORS

R. Kelley Pace, James P. LeSage and Shuang Zhu

ABSTRACT

Most spatial econometrics work focuses on spatial dependence in the regressand or disturbances. However, Lesage and Pace (2009) as well as Pace and LeSage2009 showed that the bias in β *from applying OLS to a regressand generated from a spatial autoregressive process was exacerbated by spatial dependence in the regressor. Also, the marginal likelihood function or restricted maximum likelihood (REML) function includes a determinant term involving the regressors. Therefore, high dependence in the regressor may affect the likelihood through this term. In addition, Bowden and Turkington (1984) showed that regressor temporal autocorrelation had a non-monotonic effect on instrumental variable estimators.*

We provide empirical evidence that many common economic variables used as regressors (e.g., income, race, and employment) exhibit high levels of spatial dependence. Based on this observation, we conduct a

30th Anniversary Edition
Advances in Econometrics, Volume 30, 257–295

ISSN: 0731-9053/doi:10.1108/S0731-9053(2012)0000030014

Monte Carlo study of maximum likelihood (ML), REML and two instrumental variable specifications for spatial autoregressive (SAR) and spatial Durbin models (SDM) in the presence of spatially correlated regressors.

Findings indicate that as spatial dependence in the regressor rises, REML outperforms ML and that performance of the instrumental variable methods suffer. The combination of correlated regressors and the SDM specification provides a challenging environment for instrumental variable techniques.

We also examine estimates of marginal effects and show that these behave better than estimates of the underlying model parameters used to construct marginal effects estimates. Suggestions for improving design of Monte Carlo experiments are provided.

Keywords: Regressor autocorrelation; spatial Durbin model; REML; spatial autoregression; maximum likelihood; spatial econometrics

INTRODUCTION

In spatial settings, dependence of each observation on potentially every other observation impedes theoretical derivation of finite sample results. Consequently, most investigations of finite sample performance for spatial econometric methods rely on Monte Carlo experiments. These Monte Carlo experiments usually employ independent identically distributed (*iid*) random variables for the regressors. However, typical regressors used in applied practice often display a substantial amount of spatial dependence.

Spatial dependence in the regressors can affect estimators as diverse as maximum likelihood (ML), restricted or residual maximum likelihood (REML), and instrumental variables (IVs) (Anselin, 1988; Kelejian & Prucha, 1998; Lee, 2007). In this study, we document high levels of spatial dependence for typical economic and social regressors such as income, education, and race, and show that this dependence can result in substantial differences in performance across estimation methods.

In the case of REML, spatial dependence in the regressors may be accommodated by a determinant term that appears in the REML likelihood. For IV methods, since spatial dependence reduces information content of instruments, this exacerbates the weak instrument problem

(Bound, Jaeger, & Baker, 1995; Staiger & Stock, 1997). This occurs because spatial dependence can reduce the contributions of the instruments (partial fit) in first-stage regressions in two-stage least squares (2SLS) and increase the extent of dependence among first-stage disturbances. In addition, we show that estimation of the common spatial Durbin model (SDM) specification (Anselin, 1988) is particularly sensitive to traditional weak instrument considerations in the presence of spatial dependence in the regressors.

In section "Spatial Dependence in *y* and Regressors," we describe data generating processes (DGPs) for cases where the regressors exhibit spatial dependence, and give evidence that high levels of spatial dependence may be common in regressors that are frequently used in applied spatial regression work. Theoretical implications of dependence in the regressors are examined with respect to REML and IV procedures. Section "Some Empirical Examples" provides three empirical examples based on data sets varying in size that result in substantial differences in estimates produced by ML, REML and IV methods as a motivation for Monte Carlo experiments carried on in section "Monte Carlo Experiments." Recommendations for design of Monte Carlo experiments in spatial settings that more accurately reflect the presence of spatial dependence in the regressors appear in section "Suggestions for Spatial Monte Carlo Experiments."

SPATIAL DEPENDENCE IN Y AND REGRESSORS

In section "Spatial Models," we set forth the DGPs for the spatial autoregressive (SAR) and SDM, and provide empirical evidence documenting that spatial dependence may be quite common in regressors used by these models in section "Spatial Dependence in the Regressors." This motivates a theoretical examination of the role played by a DGP which includes a SAR process that governs the regressors.

We explore theoretical implications of dependence in the regressors for likelihood-based estimation methods, ML, and REML in section "Likelihood-Based Estimators." Section "Instrumental Variables" examines implications for 2SLS and best instruments approaches to estimating the dependence parameter in SAR and SDM. The focus here is on how these model specifications impact the instrument set. Finally, we discuss marginal effects estimates that are derived from the underlying model parameters in section "Marginal Effects and Bias Correction."

Spatial Models

Let the n by 1 dependent variable y follow a SAR DGP (often abbreviated as SAR) shown in Eq. (1).[1]

$$y = (I_n - \rho W)^{-1} X\beta + (I_n - \rho W)^{-1}\varepsilon \tag{1}$$

$$\varepsilon \sim N(0, \sigma_\varepsilon^2 I_n) \tag{2}$$

where X contains n observations on k exogenous regressors, ε is a n by 1 vector of normal *iid* disturbances with variance σ_ε^2, and β is a k by 1 vector of regression parameters. The spatial character of Eq. (1) is determined by the n by n spatial weight matrix W and the spatial scalar parameter ρ. When observation or region j is a neighbor to observation or region i, $W_{ij}>0$, otherwise $W_{ij}=0$. These values are exogenous. By convention, observations cannot serve as neighbors to themselves and therefore $W_{ii}=0$. For simplicity, assume that W has real eigenvalues such as occur when a matrix is similar to a real, symmetric matrix. As demonstrated by Ord (1975), row-normalized (known as row-stochastic matrices in linear algebra) weight matrices can still have all real eigenvalues, denoted by the n by 1 vector λ.[2] Also, since scaling a matrix by its principal eigenvalue can always yield a new matrix with a principal eigenvalue of 1, assume W has a principal eigenvalue of 1 and a minimum eigenvalue of $\lambda_{\min}$. Because the diagonal of W contains zeros, $tr(W)=0$. Therefore, the sum of the eigenvalues equals 0. Given a positive principal eigenvalue, $\lambda_{\min}<0$. Consequently, $\rho \in (\lambda_{\min}^{-1}, 1)$ yields a symmetric positive definite $(I_n-\rho W)$.

Given Eq. (1), the corresponding estimation equation is shown in Eq. (3).

$$y = X\beta + \rho W y + \varepsilon \tag{3}$$

Estimation of Eq. (3) by ordinary least-squares (OLS) is biased (Ord, 1975), so estimation requires ML or IV methods.

As with any model, the composition of the explanatory variable matrix X becomes important. The SAR model has two alternative specifications. The first is in Eq. (4), where ι_n is an n by 1 vector of ones and U is a n by p matrix of non-constant regressors. This specification contains the regressors X_{SAR} in Eq. (4) and the associated p by 1 parameter vector γ for the non-constant variables along with a scalar parameter α for the constant vector ι_n leading to $k=p+1$ regressors.

$$X_{\mathrm{SAR}} = \begin{bmatrix} \iota_n & U \end{bmatrix} \tag{4}$$

$$\beta_{SAR} = [\alpha\ \gamma']' \tag{5}$$

The second specification relies on regressors X_{SDM} in Eq. (6), which consists of the regressors (and associated p by 1 parameter vector γ), as well as their spatial lags (and associated p by 1 parameter vector θ) in addition to the constant vector and intercept parameter, which leads to $k = 2p + 1$ regressors. This second variant, labeled the SDM, nests the SAR specification (when $\theta = 0$), a spatial error model (SEM) specification (when $\theta = -\rho\gamma$), a specification labeled the spatial lag of X model (SLX) (when $\rho = 0$), and the conventional *iid* disturbance model (when $\rho = 0$ and $\theta = 0$). LeSage and Pace (2009) as well as Pace and LeSage (2009) show that the SDM model arises naturally when there is spatial dependence in the disturbances and omitted variables that are correlated with included variables.

$$X_{SDM} = \begin{bmatrix} \iota_n & U & WU \end{bmatrix} \tag{6}$$

$$\beta_{SDM} = [\alpha\ \gamma'\ \theta']' \tag{7}$$

Spatial Dependence in the Regressors

In section "Empirical Results on Regressors," we show how common variables often used in economic modeling exhibit substantial amounts of spatial autocorrelation. We employ this information in section "Theoretical Regressor Data Generating Process" to arrive at a DGP appropriate for the regressors used in the theoretical discussions in later sections and in the Monte Carlo experiments.

Empirical Results on Regressors

Many regressors found in economic models display spatial dependence. To obtain an idea of the magnitude of dependence, we produced ML estimates of dependence using a univariate SAR model for a number of commonly used regressors. Varying spatial scales were used, including county data from the contiguous US states, census tract data from New York State, and block group data from the Bronx. All variables were logged and missing values were excluded from the sample. The initial weight matrix used 30 ($o = 1, \ldots, 30$) nearest neighbors, where the $(o+1)$th nearest neighbor received 5 percent less weight than the oth order neighbor. The final weight matrix was this initial matrix plus its transpose reweighted to make it symmetric and doubly stochastic with a principal eigenvalue of 1.

Table 1. Estimates of Spatial Dependence for Selected Census Variables in Different Levels of Geography by Spatial Autoregressive Regression Model.

Variables	County	Census Tract	Block Group
Black	0.930	0.940	0.938
Age 60+	0.871	0.805	0.812
Age 16+	0.853	0.859	0.858
Bachelor's Degree	0.869	0.928	0.909
Civilian Employment	0.839	0.782	0.752
Female Employment	0.837	0.796	0.751
Median HH Income	0.953	0.937	0.918
Per Capita Income	0.916	0.940	0.943
Median Rent	0.950	0.938	0.886
Median House Price	0.962	0.938	0.823
Number of Observations	3,109	4,907	5,733

Table 1 documents that many common variables display a high degree of spatial dependence. These range from a high of 0.9620 for the autocorrelation in county house prices to 0.751 for autocorrelation in female employment across block groups in the Bronx.

Theoretical Regressor Data Generating Process

Based on the empirical evidence in section "Empirical Results on Regressors," we use an autoregressive DGP for the regressors in the theoretical developments described here, and Monte Carlo experiments described later. Specifically, we assume the regressors follow a univariate SAR model shown in Eqs. (8) and (9) with a DGP in Eq. (10).

$$U_j = \varphi W U_j + R_j, \quad j = 1, \ldots, p \tag{8}$$

$$R_j \sim N(0, I_n) \tag{9}$$

$$U_j = (I_n - \varphi W)^{-1} R_j \tag{10}$$

We assume, without loss of generality, that the variance of the underlying n by 1 random disturbance R_j for $j = 1, \ldots, p$ equals 1. In Eq. (8), R and U are n by p matrices and the scalar parameter φ lies in the open interval $(\lambda_{\min}^{-1}, 1)$. Therefore, given the previous assumptions on W, $(I_n - \varphi W)$ is a symmetric, positive definite matrix.

An important assumption made in the theoretical development is that no covariance exists between the random process underlying the regressors and the disturbances ε in the DGP for y in Eq. (1), that is

$$E(R_j\varepsilon') = 0, \quad j = 1, \ldots, p \tag{11}$$

Because no covariance exists between ε and R, this further implies no covariance between ε and U. This lack of covariance could be a result of independence between ε and R, when these are generated contemporaneously, or R could have been generated previously (with independence over time). In the Monte Carlo experiments, U and therefore X have been previously determined and remain fixed in repeated sampling.

Even when X is fixed, its characteristics affect estimation performance. For example, in other statistical areas, dependence across the columns (multicollinearity) can degrade estimation precision. Leverage points, if due to erroneous data, can cause estimators to break down (robustness). In a spatial context, dependence across the rows (spatial dependence) can also degrade estimation precision.

Although we have focused on an autoregressive DGP for the regressors, other types of regressors can exhibit substantial spatial dependence. For example, Cressie (1993) suggests using a quadratic surface in the location coordinates as a regressor. Economic models commonly include regional dummy variables that collectively have substantial spatial content. Many of our findings for the case of SAR regressors may apply to other types of spatial regressors.

Likelihood-Based Estimators

One approach to ML works with the concentrated log-likelihood function in Eq. (12) where κ is a scalar constant. The value of ρ that maximizes the concentrated log-likelihood function (Eq. (12)) equals the ML estimate, $\tilde{\rho}_{ML}$.

$$L_{ML}(\rho) = \kappa + \ln|I_n - \rho W| - \frac{n}{2}\ln(e(\rho)'e(\rho)) \tag{12}$$

$$e(\rho) = (I_n - \rho W)y - X\tilde{\beta} \tag{13}$$

$$\tilde{\beta} = (X'X)^{-1}X'(I_n - \rho W)y \tag{14}$$

Restricted or residual maximum likelihood (REML) as introduced by Patterson and Thompson (1971) results in a concentrated log-likelihood function that augments the usual log-likelihood with another determinant term, and multiplies the log of the sum-of-squared errors term by $(n-k)/2$ rather than $n/2$. In the spatial statistics literature, REML has been used in applications and Durban and Currie (2000) devised a REML type correction for SAR models that reduced bias in estimation of ρ. In addition, the form of REML is very similar to the marginal likelihood for the case of the normal distribution.

The REML concentrated log-likelihood in Eq. (15) has a determinant term that can be affected by spatial dependence in X.

$$\begin{aligned} L_{\mathrm{RE}}(\rho) = \kappa - \frac{1}{2}\ln|X'(I_n-\rho W)'(I_n-\rho W)X| \\ + \ln|I_n-\rho W| - \frac{n-k}{2}\ln(e(\rho)'e(\rho)) \end{aligned} \tag{15}$$

To obtain some insight into the log-determinant term involving X in Eq. (15), we will examine the interaction between X and the symmetric, real W with eigenvalues λ. Suppose X is univariate and equals U (SAR process) with symmetric W, and that U follows an autoregressive process as described by Eq. (10). In this case $X'(I_n-\rho W)'(I_n-\rho W)X$ equals $U'(I_n-\rho W)^2U$ which has a further expansion in Eq. (16). Let Λ represent the n by n matrix with the eigenvalues of W on the diagonal (and zeros otherwise) as shown in Eq. (17). The rotation of the n by 1 vector of unit independent normal variates R_1 by the orthonormal real eigenvectors of W, represented by E, still yields a n by 1 vector of *iid* unit independent normal variates, R_{rot} as shown in Eq. (18).

$$U'(I_n-\rho W)^2U = R_1'(I-\varphi W)^{-2}(I_n-\rho W)^2R_1 \tag{16}$$

$$\Lambda_{ii} = \lambda_i, \quad \Lambda_{ij} = 0 \text{ for } i \neq j \tag{17}$$

$$W = E\Lambda E', \qquad R_{\mathrm{rot}} = E'R_1 \tag{18}$$

$$U'(I_n-\rho W)^2U = R'_{\mathrm{rot}}(I_n-\varphi\Lambda)^{-2}(I_n-\rho\Lambda)^2R_{\mathrm{rot}} \tag{19}$$

Using the relation that $\mathrm{tr}(A) = E(R'_{\mathrm{rot}}AR_{\mathrm{rot}})$ for the n by 1 vector, R_{rot} composed of *iid* random normal unit random variates for an arbitrary n by n real matrix A produces (Eq. (20)). Since this is a scalar (the determinant of a scalar is that scalar), this would be an expression for the determinant term in the REML likelihood for cases where regressors exhibit dependence.

$$E(R'_{\text{rot}}(I_n - \varphi\Lambda)^{-2}(I_n - \rho\Lambda)^2 R_{\text{rot}}) = \text{tr}[(I_n - \varphi\Lambda)^{-2}(I_n - \rho\Lambda)^2] \tag{20}$$

$$= \sum_{i=1}^{n} \frac{1-\rho\lambda_i}{1-\varphi\lambda_i} \tag{21}$$

Because W typically has a number of eigenvalues near 1, the term (Eq. (21)) can be dominated by these. Therefore, when $0<\rho<\varphi$ and φ is near 1, the term in Eq. (21) will be large. This suggests that substantial spatial dependence in a regressor could lead to REML estimates that differ from ML in such cases.

Instrumental Variables

Instrumental variable techniques have been widely used in spatial econometrics (Anselin, 1988; Kelejian & Prucha, 1998; Lee 2007). In practice, due to the widespread availability of software and intuitive appeal, 2SLS has often been used in applications (Anselin, 1988; Byme, 2005; Land & Deane, 1992; Millimet & Rangaprasad, 2007; Richards & Padilla, 2009).

Given X, 2SLS uses the IV matrix Z that includes X and some instrumental variables V as shown in Eq. (22). The problematic element in estimating Eq. (3) is the simultaneity between y and Wy. 2SLS replaces Wy with its predicted value based on the IV matrix $P_Z Wy$ as in Eq. (24).

$$Z = \begin{bmatrix} X & V \end{bmatrix} \tag{22}$$

$$P_Z = Z(Z'Z)^{-1}Z' \tag{23}$$

$$y = X\beta + \rho(P_Z Wy) + \varepsilon \tag{24}$$

It remains to specify V. Kelejian and Prucha (1998) motivated possible instruments by considering the Taylor series expansion of $(I_n - \rho W)^{-1}$ in Eq. (25) and the expression for the expectation of Wy in the SAR model shown in Eq. (26).

$$(I_n - \rho W)^{-1} = I_n + \rho W + \rho^2 W^2 + \rho^3 W^3 + \ldots \tag{25}$$

$$E(Wy) = WX\beta + \rho W^2 X\beta + \rho^2 W^3 X\beta + \ldots \tag{26}$$

Their recommended instruments for the SAR model appear in Eq. (27) which involves the first two terms from $E(Wy)$. By analogy, the first two terms of $E(Wy)$ in the SDM model (not contained in the regressors) appear in Eq. (28).

$$V_{\text{SAR}} = [WU \quad W^2U] \tag{27}$$

$$V_{\text{SDM}} = \left[W^2U \quad W^3U \right] \tag{28}$$

Lee (2003) adopted a different IV approach, termed the best instruments, which also used (Eq. (26)) to motivate replacing Wy with $W(I_n - \rho W)^{-1}X\beta$. In this theoretical form (which uses the true parameters), the best instruments provide a bound to the performance of feasible IV approaches (Kelejian, Prucha, & Yuzefovich, 2004).

Examination of the finite sample properties of these techniques has largely been through use of Monte Carlo experiments (Kelejian et al., 2004, Klotz, 2004; Lee, 2007), and most of these examine cases with moderate to high levels of signal-to-noise. For example, Kelejian et al. (2004) excluded low signal-to-noise scenarios. In contrast, Lee (2007) examined a case where the R^2 was 0.04 and found that 2SLS performed poorly.

This suggests that 2SLS weak instrument problems noted in other contexts (Bound et al., 1995; Staiger & Stock, 1997) may carry over to spatial estimation. Factors affecting the weak instrument problem include goodness-of-fit and number of instruments. These problems can reduce performance even in relatively large sample sizes.

In addition to these well-known problems, the spatial dependence of regressors may affect the weak instrument problem. In fact, Bowden and Turkington (1984, pp. 77–81) found in a time series context that increasing the autocorrelation in the regressors had a non-monotonic effect. On one hand, this autocorrelation can increase correlation between the instrumented variable and its instruments. On the other hand, the autocorrelation increases correlation between instruments and included exogenous variables, which can lower the marginal contribution of the instruments.

The SDM may potentially further impact performance of IV, since including spatial lags of regressors already included in the model precludes use of these for identification of Wy. Specifically, the SDM by construction reduces the possible number of identifying instruments for Wy relative to the SAR model.

To gauge the sensitivity of two-stage estimation performance to spatial dependence in U, we use Monte Carlo simulation to produce a partial R^2 (with reference to V) for the first-stage regression of the endogenous variable Wy on Z. The DGP of y is from Eq. (3) where $X = U$ for the SAR model and $X = \left[U \; WU \right]$ for the SDM model. The Eqs. (8)–(10) define the univariate variable U. The first-stage regression is defined as in Eq. (29), where $V = WU$ is the instrument for the SAR model and $V = W^2U$ is the instrument for the SDM model. Sample size was 3,000, and for each set of results plotted, we fix R^2 and ρ for the main regression. The partial R^2 represents the marginal explanatory power of the instruments after filtering out included exogenous variables from the regressand Wy and instrument V of Eq. (29) as given by Eq. (31). This leads to estimated residuals $\hat{\varepsilon}$ in Eq. (32) and estimated partial R^2 in Eq. (33).

$$Wy = X\delta^{(1)} + V\delta^{(2)} + \varepsilon \tag{29}$$

$$M_X = I_n - X(X'X)^{-1}X' \tag{30}$$

$$M_X Wy = M_X V\delta^{(2)} + \varepsilon \tag{31}$$

$$\hat{\varepsilon} = M_X Wy - M_X V\hat{\delta}^{(2)} \tag{32}$$

$$\hat{R}_p^2 = 1 - \frac{\hat{\varepsilon}'\hat{\varepsilon}}{y'W'M_X Wy} \tag{33}$$

Fig. 1, shows that the marginal explanatory power of the IV is always lower for the SDM model than for the SAR model. The partial R^2 of the instrument is monotonically decreasing with spatial dependence in U for the SDM model, but not monotonic for the SAR model. However, for high levels of spatial dependence in U, the marginal information of instruments for the SAR model also diminishes very rapidly.

To further examine the weak instrument problem, we note that the SAR model in Eq. (3) along with the regressors given by Eq. (6) defines the SDM model as shown in Eq. (34). This has the associated DGP in Eq. (35). Recognizing that in Eq. (35) the term $(I_n - \rho W)^{-1}U\gamma$ also equals $U\gamma + (I_n - \rho W)^{-1}WU\rho\gamma$, and combining terms yields Eq. (36). Note that matrices commute with functions of themselves so that $W(I_n - \rho W)^{-1}$ equals $(I_n - \rho W)^{-1}W$.

$$y = \alpha + U\gamma + WU\theta + \rho Wy + \varepsilon \tag{34}$$

$$y = \alpha^* + (I_n - \rho W)^{-1}U\gamma + (I_n - \rho W)^{-1}WU\theta + (I_n - \rho W)^{-1}\varepsilon \tag{35}$$

$$y = \alpha^* + U\gamma + (I_n - \rho W)^{-1}WU(\rho\gamma + \theta) + (I_n - \rho W)^{-1}\varepsilon \tag{36}$$

An important part of the 2SLS method is the regression involving the endogenous variable Wy with the theoretical DGP for this shown in Eq. (37)

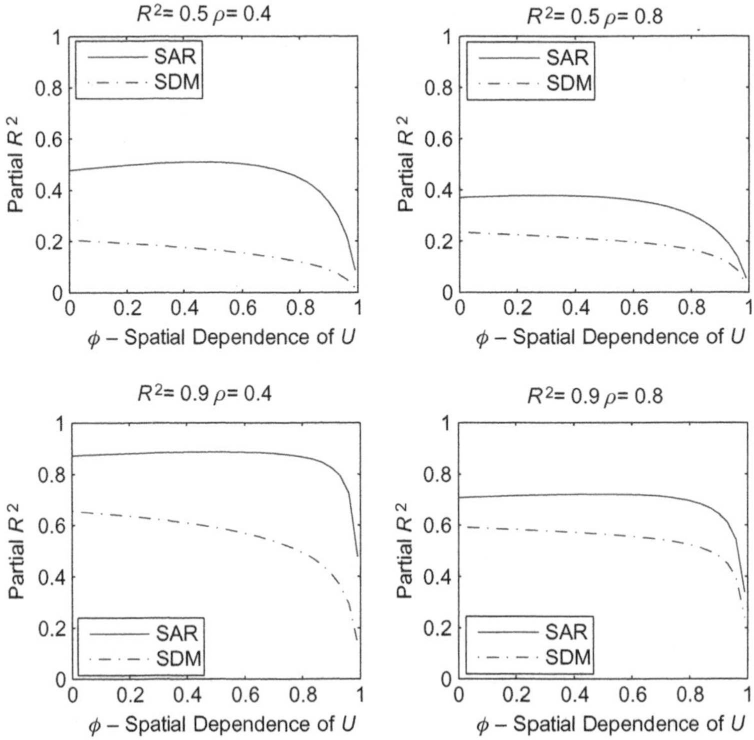

Fig. 1. Partial R^2 of Instruments from First-Stage Regressions versus Spatial Dependence of U.

for the case of dependence in the regressors. The IV matrix Z is shown in Eq. (38), with its expanded counterpart in Eq. (39), which represents the first-stage regression.

$$Wy = \alpha^{**} + WU\gamma + (I_n - \rho W)^{-1} W^2 U(\rho\gamma + \theta) + (I_n - \rho W)^{-1} W\varepsilon \tag{37}$$

$$Wy = Z\delta + \varepsilon \tag{38}$$

$$Wy = X_{\text{SDM}}\delta^{(1)} + V_{\text{SDM}}\delta^{(2)} + \varepsilon \tag{39}$$

Part of the weak instrument problem is the size of the marginal contribution from $V_{\text{SDM}}\delta^{(2)}$. To address this more directly we employ the

Frisch–Waugh–Lovell theorem to examine the regression (Eq. (39)), where both the regressand and regressors are multiplied by, in this case, $M_{X_{\mathrm{SDM}}}$.

$$M_{X_{\mathrm{SDM}}} = I_n - X_{\mathrm{SDM}}(X'_{\mathrm{SDM}} X_{\mathrm{SDM}})^{-1} X'_{\mathrm{SDM}} \tag{40}$$

$$M_{X_{\mathrm{SDM}}} Wy = M_{X_{\mathrm{SDM}}} V_{\mathrm{SDM}} \delta^{(2)} + \varepsilon \tag{41}$$

Consider the "true" signal of the IV in Eq. (41) (as opposed to the noise). Substituting the DGP Eq. (37) into Eq. (41) and noting that $M_{X_{\mathrm{SDM}}}$ annihilates WU and the constant term while the expectation eliminates the disturbance terms lead to Eq. (43), the marginal signal content of the IV. If the marginal signal has a small magnitude relative to the noise, this sets up conditions leading to the weak instrument problem.

$$E(M_{X_{\mathrm{SDM}}} V_{\mathrm{SDM}} \delta^{(2)}) = E(M_{X_{\mathrm{SDM}}} Wy) \tag{42}$$

$$= M_{X_{\mathrm{SDM}}} (I_n - \rho W)^{-1} W^2 U(\rho\gamma + \theta) \tag{43}$$

Note that the signal has a zero magnitude when the true DGP has parameters such that $(\theta = -\rho\gamma)$. In this case the 2SLS technique will not work as this violates conditions underlying the proof of consistency (instruments have explanatory power). This has been shown in detail by Lee, Liu, and Lin (2010) in Lemma 5.1. In practice, θ rarely exactly equals $-\rho\gamma$. However, it often has the opposite sign of γ and this suggests that weak instruments will frequently be a problem when estimating the SDM using 2SLS.

Although the discussion of weak instruments focuses on the problems of low signal-to-noise in the first-stage regression, in this spatial setting the first-stage regression may lose efficiency due to high dependence in the disturbances.

$$WX\beta + \rho W^2 X\beta + \rho^2 W^3 X\beta + \ldots + W(I_n - \rho W)^{-1}\varepsilon = Z\delta + \varepsilon \tag{44}$$

Applying the Frisch–Waugh–Lovell theorem again (where $M_Z = I_n - P_Z$) to the first-stage regression equation results in Eq. (45).

$$\varepsilon = \sum_{j=1}^{\infty} \rho^{j-1} M_Z W^j X\beta + M_Z \xi \tag{45}$$

$$\xi = W[(I_n - \rho W)^{-1}\varepsilon] = \sum_{j=1}^{\infty} \rho^{j-1} W^j \varepsilon \tag{46}$$

$$E(\xi\xi') = (I_n - \rho W)^{-2} W^2 \tag{47}$$

By construction, ξ in Eq. (46) has greater spatial dependence than the disturbances in the DGP, $(I_n - \rho W)^{-1}\varepsilon$. Even if M_Z could succeed in annihilating all terms involving the regressor on the LHS, ξ will exhibit spatial dependence and OLS can have lower efficiency in this case. However, M_Z will not succeed in annihilating all terms (just the low order terms such as $j = 1,2$). The terms $\rho^{j-1} M_Z W^j X\beta$ for larger j will potentially exhibit substantial high-order dependence involving W^j, which may further degrade estimation efficiency in the first-stage regression. Moreover, substantial spatial dependence in X will further complicate this situation.

We measure this dependence in first-stage residuals by estimating $\tilde{\phi}$ the degree of autocorrelation, ϕ, as specified in Eq. (48) by ML.

$$\hat{\varepsilon} = \iota_n \alpha + \phi W \hat{\varepsilon} + \eta \tag{48}$$

In summary, various settings can result in low signal in the first-stage regressions, and these regressions may not be well-estimated due to substantial dependence in the first-stage disturbances. Both of these effects can combine to create difficulties for 2SLS.

We have focused on the point estimates for ρ, but these issues may pertain to other quantities of interest such as confidence intervals and marginal effects (to be discussed below). Note that unless the standard error of the IV estimate of ρ is small, confidence intervals will often include the region $\rho > 1$ which further complicates the use of standard IV formulas in this setting.

Marginal Effects and Bias Correction

In many cases, the marginal effects are more important than the parameter estimates. Since the marginal effects are a non-linear function of the underlying parameter estimates, it is important to examine how well estimation procedures perform in producing valid estimates of the marginal effects. A number of authors have pointed out that the derivative of the regressand with respect to the regressor does not equal β in the SAR and SDM models. Unlike the case of *iid* models fitted by OLS, partial derivatives for the expected value of the regressand with respect to regressor j (marginal effects) for the SAR DGP equals Eq. (49)

$$\frac{\partial E(y)}{\partial X'_j} = (I_n - \rho W)^{-1} \beta_j \tag{49}$$

$$= S(\rho)_j \tag{50}$$

Since this is an n by n matrix and because there are $k-1$ non-constant variables, we have $(k-1)n^2$ partial derivatives which constitutes an overwhelming amount of information. LeSage and Pace (2009) suggest summarizing these partial derivatives using averages of all column or row sums to produce a scalar summary measure of the average total effect or impact. An average of the diagonal elements from $S(\rho)_j$ is proposed as a scalar summary for the average direct effect or impact, and the off-diagonal elements of $S(\rho)_j$ are cumulated and averaged to arrive at a scalar representing the average indirect effect or impact.

As shown in LeSage and Pace (2009) for a row or doubly stochastic W the average total effect is a simple non-linear function of ρ as in Eq. (51).

$$T(\rho)_j = (1-\rho)^{-1}\beta_j \tag{51}$$

SOME EMPIRICAL EXAMPLES

In section "Individual Model Estimates," we examine (SDM, SAR) estimates across the REML, ML, 2SLS and OLS procedures for a given weight matrix (30 nearest neighbors) and instrument set. In "Sensitivity to Different Weight Matrices and Instrument Sets," we focus only on the autoregressive parameter estimate across estimators and examine its sensitivity to different weight matrices (based on 4, 10, and 30 nearest neighbors) and to four different instrument sets.

Individual Model Estimates

Three empirical examples help motivate issues posed by correlated regressors for the different estimators. The examples include the Harrison and Rubinfeld (1978) model and 506 census tract data set that relates pollution to house prices with spatial information added by Gilley and Pace (1996), a model of voter participation in the 1980 presidential election (Pace & Barry, 1997) based on 3,107 county-level observations, and a relationship between 1990 census variables and housing price appreciation between 1990 and 2000 for a sample of 62,226 census tract observations.

We use 30 nearest neighbors and a symmetric, doubly stochastic weight matrix in all examples and estimate SAR and SDM for each data set using ML, REML, 2SLS, and OLS. The instruments include the regressors and the first two powers of W not contained in the regressors.

Table 2. Spatial Autoregressive Model Harrison–Rubinfeld Data.

Variables	REML	ML	2SLS	OLS
Intercept	3.0840	3.1263	3.1357	3.0585
CRIM	−0.0101	−0.0101	−0.0101	−0.0100
ZN	0.0005	0.0005	0.0005	0.0005
INDUS	−0.0010	−0.0010	−0.0010	−0.0011
CHAS	0.0169	0.0191	0.0196	0.0156
NOXSQ	−0.5227	−0.5260	−0.5267	−0.5207
RM2	0.0062	0.0062	0.0062	0.0062
AGE	0.0002	0.0002	0.0002	0.0002
DIS	−0.2511	−0.2496	−0.2493	−0.2521
RAD	0.0719	0.0724	0.0725	0.0716
TAX	−0.0003	−0.0003	−0.0003	−0.0003
PTRATIO	−0.0157	−0.0161	−0.0162	−0.0155
B	0.0003	0.0003	0.0003	0.0003
LSTAT	−34.1670	−34.2621	−34.2832	−34.1098
$\tilde{\rho}$	0.3880	0.3769	0.3744	0.3947

For the SAR model, estimates for the Harrison and Rubinfeld data appear in Table 2, where we see that all estimators (including OLS) arrive at approximately the same estimate for the model's dependence parameter. The partial R^2 equaled a surprisingly high 0.961, and the first-stage regression residuals had a measured spatial dependence parameter of 0.9594.

The SDM estimates appear in Table 3, where we see a drop in the partial R^2 to 0.672. Residuals in the first-stage regression still have a high autoregressive parameter estimate of 0.958, and estimates for these data diverge. The REML estimate of 0.821 greatly exceeds the ML estimate of 0.632, and the 2SLS estimate of 1.293 greatly exceeds the upper bound of 1 for the parameter space.

Table 4 contains the estimates from the alternative estimators for the SAR model of voting behavior. The partial R^2 was 0.514, while the residuals had a autoregressive parameter estimate above 0.99. In this case ML and REML estimates agree, but 2SLS gives a substantially lower estimate, and OLS gives a moderately higher estimate than ML.

Table 5 contains estimates for the SDM of voting behavior. The partial R^2 was 0.204, while the residuals had an autoregressive parameter estimate over 0.99. Again, ML and REML estimates are similar, but the 2SLS estimates of 0.9996 is very close to the parameter space boundary of 1, and the OLS estimate of 1.022 falls outside the parameter space.

Table 3. Spatial Durbin Model of Pollution Harrison–Rubinfeld Data.

Variables	REML	ML	2SLS	OLS
Intercept	2.6776	3.7965	−0.1226	0.6291
CRIM	−0.0106	−0.0107	−0.0105	−0.0105
ZN	0.0006	0.0006	0.0005	0.0005
INDUS	0.0000	−0.0001	0.0005	0.0004
CHAS	−0.0009	0.0018	−0.0074	−0.0057
NOXSQ	−0.6217	−0.6454	−0.5624	−0.5783
RM2	0.0067	0.0066	0.0070	0.0070
AGE	−0.0005	−0.0005	−0.0005	−0.0005
DIS	−0.3112	−0.3188	−0.2922	−0.2973
RAD	0.0486	0.0498	0.0454	0.0463
TAX	−0.0003	−0.0003	−0.0003	−0.0003
PTRATIO	−0.0149	−0.0150	−0.0147	−0.0147
B	0.0006	0.0006	0.0006	0.0006
LSTAT	−31.0642	−31.0827	−31.0181	−31.0305
Lag CRIM	−0.0092	−0.0147	0.0047	0.0010
Lag ZN	0.0009	0.0011	0.0004	0.0005
Lag INDUS	−0.0164	−0.0183	−0.0117	−0.0130
Lag CHAS	0.3114	0.3577	0.1955	0.2266
Lag NOXSQ	−0.0066	−0.1785	0.4235	0.3080
Lag RM2	−0.0328	−0.0342	−0.0293	−0.0302
Lag AGE	0.0054	0.0060	0.0040	0.0044
Lag DIS	0.2433	0.2023	0.3458	0.3183
Lag RAD	0.0046	0.0345	−0.0702	−0.0501
Lag TAX	0.0014	0.0014	0.0014	0.0014
Lag PTRATIO	−0.0096	−0.0215	0.0203	0.0123
Lag B	−0.0009	−0.0009	−0.0009	−0.0009
Lag LSTAT	−8.9874	−17.9298	13.3918	7.3837
$\tilde{\rho}$	0.8208	0.6320	1.2933	1.1664

Table 4. Spatial Autoregressive Model of Voting Behavior.

Variables	REML	ML	2SLS	OLS
Intercept	0.6120	0.6128	0.7285	0.5850
Voting Pop	−0.6113	−0.6118	−0.6855	−0.5941
Education	0.1595	0.1602	0.2722	0.1334
Home Ownership	0.4106	0.4107	0.4165	0.4093
Income	0.0123	0.0120	−0.0370	0.0237
$\tilde{\rho}$	0.6898	0.6885	0.4722	0.7403

Table 5. Spatial Durbin Model of Voting Behavior.

Variables	REML	ML	2SLS	OLS
Intercept	0.2442	0.2533	0.0563	0.0252
Voting Pop	−0.7254	−0.7253	−0.7283	−0.7287
Education	0.2075	0.2073	0.2117	0.2124
Home Ownership	0.4349	0.4349	0.4352	0.4353
Income	0.0373	0.0374	0.0354	0.0351
Lag Voting Pop	0.6384	0.6326	0.7580	0.7778
Lag Education	−0.1001	−0.0955	−0.1960	−0.2119
Lag Home Ownership	−0.3797	−0.3767	−0.4427	−0.4532
Lag Income	−0.1150	−0.1171	−0.0710	−0.0637
$\tilde{\rho}$	0.8670	0.8606	0.9996	1.0215

Table 6. Spatial Autoregressive Model of Housing Appreciation Rates.

Variables	REML	ML	2SLS	OLS
Intercept	0.0589	0.0590	0.0748	0.0515
House Age	0.0223	0.0223	0.0269	0.0201
Employ	−0.0164	−0.0164	−0.0507	−0.0002
Education	0.0436	0.0437	0.0769	0.0280
Age	−0.0335	−0.0335	−0.0601	−0.0209
Land Area	0.0171	0.0171	0.0232	0.0142
HHs	−0.0317	−0.0317	−0.0038	−0.0449
White HHs	0.0010	0.0010	0.0023	0.0004
ρ	0.8533	0.8532	0.6494	0.9495

Table 6 contains the SAR model estimates of year 2000 Census tract housing appreciation rates (the log of 2000 median prices minus the log of 1990 median house prices) regressed on socio economic variables from the 1990 Census. All variables are in log form. These estimates produce anticipated signs for most demographic and economic variables. The ML estimate for ρ and REML both equaled 0.853, and we see a substantially lower 2SLS estimate of 0.649. The OLS estimate took on a very high value of 0.950. The partial R^2 was 0.115, but the residuals in the first-stage regression still produced a high autoregressive parameter estimate of over 0.99.

Finally, Table 7 contains SDM estimates of housing appreciation rates. In this case, the 2SLS and OLS estimates were almost identical and substantially higher than REML and ML at 0.9879 and 0.9884. The partial R^2 value was low at 0.0343, and the ML autoregressive coefficient for the first-stage residuals was over 0.99.

Table 7. Spatial Durbin Model of Housing Appreciation Rates.

Variables	REML	ML	2SLS	OLS
Intercept	−0.0625	−0.0625	−0.0702	−0.0703
House Age	0.0319	0.0319	0.0320	0.0320
Employ	−0.0497	−0.0496	−0.0505	−0.0505
Education	−0.0180	−0.0180	−0.0185	−0.0185
Age	−0.0656	−0.0656	−0.0658	−0.0658
Land Area	0.0347	0.0347	0.0348	0.0348
HHs	−0.0158	−0.0158	−0.0147	−0.0147
White HHs	0.0106	0.0106	0.0104	0.0104
Lag House Age	−0.0239	−0.0239	−0.0305	−0.0305
Lag Employ	0.0005	0.0004	0.0428	0.0430
Lag Edu	0.1344	0.1346	0.0626	0.0622
Lag Age	0.0190	0.0189	0.0622	0.0624
Lag Area	−0.0286	−0.0286	−0.0341	−0.0342
Lag HHs	0.0459	0.0460	0.0180	0.0179
Lag White	−0.0101	−0.0101	−0.0113	−0.0113
$\tilde{\rho}$	0.8732	0.8729	0.9879	0.9884

In summary, the various estimators diverged greatly across different models (SAR, SDM) and across data sets. ML and REML were usually similar, except in the case of the Harrison and Rubinfeld SDM, where the REML estimate was substantially larger. The 2SLS estimates were sometimes similar to ML, but could be much lower or exceed a value of 1 which lies outside the parameter space. OLS showed the anticipated positive bias relative to ML. The partial R^2 values were often low, but in only one case were they under 0.05. However, the first-stage residuals exhibited very high levels of spatial dependence, and this may account for some of the poor performance of 2SLS. Undoubtedly, some or all of these examples have varying forms of misspecification. For example, if powers of W^jX for $j>0$ actually belong in the specification, these will not work well as instruments. Nevertheless, these examples show that the various estimation procedures can produce estimates that diverge greatly in applied practice.

Sensitivity to Different Weight Matrices and Instrument Sets

In this section we examine the sensitivity of the estimators to variants of W, and sensitivity of 2SLS to different instrument sets. For all examples we use $m(m = 4, 10, 30)$ nearest neighbors to form a symmetric, doubly stochastic weight matrix. We estimate a SAR and SDM for each data set using ML, REML, 2SLS, and OLS. We also look at four different instrument sets for 2SLS estimation of the SAR (shown in Eqs. (52)–(55)) and SDM (in Eqs. (57)–(60)). The n by 5 matrix S contains a bivariate quadratic polynomial (with 5 terms) in the two location coordinates as shown in Eq. (56).

$$V_{\text{SAR}}^{(1)} = [\,WX \quad W^2X\,] \tag{52}$$

$$V_{\text{SAR}}^{(2)} = [\,WX \quad W^2X \quad W^3X\,] \tag{53}$$

$$V_{\text{SAR}}^{(3)} = [\,WX \quad W^2X \quad S\,] \tag{54}$$

$$V_{\text{SAR}}^{(4)} = [\,WX \quad W^2X \quad W^3X \quad S\,] \tag{55}$$

$$S = \left[\text{lat} \quad \text{lon} \quad \text{lat}\cdot\text{lon} \quad \text{lat}^2 \quad \text{lon}^2\right] \tag{56}$$

$$V_{\text{SDM}}^{(1)} = [\,W^2X \quad W^3X\,] \tag{57}$$

$$V_{\text{SDM}}^{(2)} = [\,W^2X \quad W^3X \quad W^4X\,] \tag{58}$$

$$V_{\text{SDM}}^{(3)} = [\,W^2X \quad W^3X \quad S\,] \tag{59}$$

$$V_{\text{SDM}}^{(4)} = [\,W^2X \quad W^3X \quad W^4X \quad S\,] \tag{60}$$

For the Harrison and Rubinfeld data scenarios in Table 8, we see that the estimates of ρ for the SAR model reach their maximum across all estimators with a 10 nearest neighbor weight matrix. The largest $\tilde{\rho}$ for REML was 0.5156 while the largest estimated $\hat{\rho}$ for 2SLS was 0.5034. Although there was variation across the instruments, 2SLS performed similarly to REML and ML. For the first-stage regression, the partial R^2 levels were high, but so were the measured autocorrelation in the residuals, $\tilde{\phi}$.

For the Harrison and Rubinfeld data in Table 8, we see that the estimates of ρ for the SDM reach their maximum with a 30 nearest neighbor weight matrix for REML ($\tilde{\rho}$= 0.8208) and with a 10 nearest neighbor weight matrix for ML ($\tilde{\rho}$= 0.6614). This demonstrates that REML and ML can show substantial differences in applied work.

For the SDM, 2SLS also reaches its maximum estimate of $\rho(\hat{\rho}$= 1.5761) with a 30 nearest neighbor weight matrix, which is far outside the parameter space. The partial R^2 levels were lower for the SDM and varied substantially by instrument set. As before, the measured autocorrelation in the residuals,

Table 8. Harrison and Rubinfeld Data Results.

NN	SAR	SDM	R_p^2	$\tilde{\phi}$	REML	ML	2SLS	OLS
4	1.0000	0.0000	0.7273	0.7798	0.4655	0.4591	0.4218	0.5270
4	2.0000	0.0000	0.7706	0.7754	0.4655	0.4591	0.4499	0.5270
4	3.0000	0.0000	0.7403	0.7690	0.4655	0.4591	0.4330	0.5270
4	4.0000	0.0000	0.7854	0.7491	0.4655	0.4591	0.4622	0.5270
4	0.0000	1.0000	0.3201	0.7754	0.6019	0.5804	0.8802	0.8161
4	0.0000	2.0000	0.4058	0.7490	0.6019	0.5804	0.9382	0.8161
4	0.0000	3.0000	0.3639	0.7491	0.6019	0.5804	0.9066	0.8161
4	0.0000	4.0000	0.4574	0.7158	0.6019	0.5804	0.9592	0.8161
10	1.0000	0.0000	0.8088	0.9764	0.5156	0.5071	0.4555	0.5530
10	2.0000	0.0000	0.8405	0.9658	0.5156	0.5071	0.4871	0.5530
10	3.0000	0.0000	0.8230	0.9641	0.5156	0.5071	0.4704	0.5530
10	4.0000	0.0000	0.8574	0.9524	0.5156	0.5071	0.5034	0.5530
10	0.0000	1.0000	0.3846	0.9658	0.7099	0.6614	0.9013	0.9006
10	0.0000	2.0000	0.5124	0.9585	0.7099	0.6614	0.8891	0.9006
10	0.0000	3.0000	0.4497	0.9524	0.7099	0.6614	0.9611	0.9006
10	0.0000	4.0000	0.5775	0.9365	0.7099	0.6614	0.9380	0.9006
30	1.0000	0.0000	0.9610	0.9594	0.3880	0.3769	0.3744	0.3947
30	2.0000	0.0000	0.9780	0.9581	0.3880	0.3769	0.3831	0.3947
30	3.0000	0.0000	0.9646	0.9488	0.3880	0.3769	0.3783	0.3947
30	4.0000	0.0000	0.9790	0.9416	0.3880	0.3769	0.3842	0.3947
30	0.0000	1.0000	0.6716	0.9581	0.8208	0.6320	1.2933	1.1664
30	0.0000	2.0000	0.7987	0.7465	0.8208	0.6320	1.5334	1.1664
30	0.0000	3.0000	0.6865	0.9416	0.8208	0.6320	1.2956	1.1664
30	0.0000	4.0000	0.8068	0.6617	0.8208	0.6320	1.5761	1.1664

$\tilde{\phi}$, was high, although it was at its lowest for the four nearest neighbor scenarios.

For the election data scenarios in Table 9, the ML and REML estimates of ρ were almost identical, and reached their highest levels for both SDM and SAR models at 30 nearest neighbors. These results showed that the 2SLS estimate of ρ for the SAR model varied materially with the instrument set. For example, in the four nearest neighbor case, the 2SLS SAR estimate varied from 0.2319 (set 1) to 0.5083 (set 4). For comparison, the REML estimate was 0.4897 in this case. The quadratic term S is the difference between instruments sets 1 and 2 versus 3 and 4. Adding the quadratic expansion of latitude and longitude (some truly exogenous variables) caused the 2SLS estimates of ρ to double. Results also showed sensitivity to the instrument set in the case of four nearest neighbors, where $\hat{\rho}$ ranged from 0.8138 to 0.9764. For comparison, the REML estimate $\tilde{\rho}$ equaled 0.6001 for

Table 9. Election Data Results.

NN	SAR	SDM	R_p^2	$\tilde{\phi}$	REML	ML	2SLS	OLS
4	1	0	0.3530	0.8316	0.4897	0.4890	0.2319	0.6226
4	2	0	0.3671	0.8333	0.4897	0.4890	0.2538	0.6226
4	3	0	0.5456	0.7778	0.4897	0.4890	0.5066	0.6226
4	4	0	0.5530	0.7791	0.4897	0.4890	0.5083	0.6226
4	0	1	0.0521	0.8333	0.6001	0.5986	0.8138	0.8353
4	0	2	0.0702	0.8295	0.6001	0.5986	0.8920	0.8353
4	0	3	0.3304	0.7791	0.6001	0.5986	0.9800	0.8353
4	0	4	0.3361	0.7814	0.6001	0.5986	0.9764	0.8353
10	1	0	0.3903	0.9804	0.6027	0.6017	0.3318	0.6999
10	2	0	0.4153	0.9786	0.6027	0.6017	0.3731	0.6999
10	3	0	0.6071	0.9747	0.6027	0.6017	0.5772	0.6999
10	4	0	0.6119	0.9754	0.6027	0.6017	0.5760	0.6999
10	0	1	0.0816	0.9786	0.7378	0.7347	0.9837	0.9363
10	0	2	0.1309	0.9760	0.7378	0.7347	0.9684	0.9363
10	0	3	0.3904	0.9754	0.7378	0.7347	1.0003	0.9363
10	0	4	0.3999	0.9734	0.7378	0.7347	1.0019	0.9363
30	1	0	0.5144	0.9900	0.6898	0.6885	0.4722	0.7403
30	2	0	0.5675	0.9900	0.6898	0.6885	0.5327	0.7403
30	3	0	0.7038	0.9900	0.6898	0.6885	0.6194	0.7403
30	4	0	0.7177	0.9900	0.6898	0.6885	0.6250	0.7403
30	0	1	0.2035	0.9900	0.8670	0.8606	1.0565	1.0215
30	0	2	0.2890	0.9900	0.8670	0.8606	1.0642	1.0215
30	0	3	0.4802	0.9900	0.8670	0.8606	1.0087	1.0215
30	0	4	0.5241	0.9900	0.8670	0.8606	1.0379	1.0215

this case. For 30 nearest neighbors, all 2SLS estimates of ρ for the SDM were above 1.

As before, partial R^2 levels varied substantially across instrument sets and were lower for the SDM than for the SAR model. The measured autocorrelation in the first-stage residuals, $\tilde{\phi}$, was very high and typically increased with the number of neighbors.

The third illustrative model explains appreciation rates (difference between the log of the median house price in 2000 and the log of the median house price in 1990) over 62,226 census tracts as a function of 1990 census tract regressors (all in log form). Specifically, the regressors were median age of houses in 1990, employment, median years of educational attainment, median age of residents, land area of the tract, number of households, and number of white households.

The results appear in Table 10. In this example, the instrument set makes a difference. In particular, for the SAR specifications adding the quadratic

Table 10. Housing Appreciation Rates 1990–2000 Results (No 1990 Price).

NN	SAR	SDM	R_p^2	$\tilde{\phi}$	REML	ML	2SLS	OLS
4	1.0000	0.0000	0.0601	0.8530	0.5926	0.5925	0.4442	0.7977
4	2.0000	0.0000	0.0665	0.8523	0.5926	0.5925	0.5040	0.7977
4	3.0000	0.0000	0.3735	0.8010	0.5926	0.5925	0.9109	0.7977
4	4.0000	0.0000	0.3757	0.8004	0.5926	0.5925	0.9120	0.7977
4	0.0000	1.0000	0.0150	0.8523	0.6020	0.6019	1.0072	0.8214
4	0.0000	2.0000	0.0213	0.8519	0.6020	0.6019	0.9845	0.8214
4	0.0000	3.0000	0.3412	0.8004	0.6020	0.6019	1.0001	0.8214
4	0.0000	4.0000	0.3438	0.7999	0.6020	0.6019	0.9989	0.8214
10	1.0000	0.0000	0.0822	0.9764	0.7392	0.7391	0.5635	0.8950
10	2.0000	0.0000	0.0925	0.9766	0.7392	0.7391	0.6215	0.8950
10	3.0000	0.0000	0.4364	0.9681	0.7392	0.7391	0.9181	0.8950
10	4.0000	0.0000	0.4411	0.9680	0.7392	0.7391	0.9188	0.8950
10	0.0000	1.0000	0.0247	0.9766	0.7530	0.7528	1.0259	0.9254
10	0.0000	2.0000	0.0343	0.9765	0.7530	0.7528	0.9966	0.9254
10	0.0000	3.0000	0.3993	0.9680	0.7530	0.7528	0.9994	0.9254
10	0.0000	4.0000	0.4039	0.9676	0.7530	0.7528	1.0004	0.9254
30	1.0000	0.0000	0.1156	0.9900	0.8533	0.8532	0.6494	0.9495
30	2.0000	0.0000	0.1290	0.9900	0.8533	0.8532	0.6868	0.9495
30	3.0000	0.0000	0.5002	0.9900	0.8533	0.8532	0.9191	0.9495
30	4.0000	0.0000	0.5064	0.9900	0.8533	0.8532	0.9207	0.9495
30	0.0000	1.0000	0.0343	0.9900	0.8732	0.8729	0.9879	0.9884
30	0.0000	2.0000	0.0483	0.9900	0.8732	0.8729	0.9713	0.9884
30	0.0000	3.0000	0.4527	0.9900	0.8732	0.8729	0.9996	0.9884
30	0.0000	4.0000	0.4574	0.9900	0.8732	0.8729	0.9986	0.9884

surface to the instruments dramatically raises the 2SLS estimate of ρ and the partial R^2, R_p^2. For example, going from instrument set number 2 to 3 for the SAR with four nearest neighbors causes the 2SLS estimate to go from 0.504 to 0.911. In contrast, the ML estimate is 0.593. The partial R^2 levels can be low in this problem. For the SDM with the first instrument set $R_p^2 = 0.015$. The estimated spatial dependence in the first-stage residuals $\tilde{\phi}$ rises with the number of neighbors and varies between 0.8 and 0.99. Cases with 4 and 10 nearest neighbors both produce results where the 2SLS estimate of ρ exceeds 1.

In summary, the three data sets provide illustrations showing that estimates can diverge depending on the estimator, model specification, weight matrix, and instrument set. For the SDM and 30 nearest neighbors, the Harrison and Rubinfeld example showed that REML and ML could differ substantially. The election and housing appreciation example showed that instrument sets can matter a great deal for 2SLS, and this included the SAR model. All data sets produced 2SLS estimates of $\hat{\rho} > 1$. Partial R^2 levels varied substantially by instrument set, and were generally lower for the SDM specification. Estimated spatial dependence in the first-stage residuals was usually quite high.

Of course, we do not know which method produced the correct estimates based on these illustrations involving actual data sets, because the true DGP is not known. In the next section, we carry out Monte Carlo experiments to explore problems posed by spatially dependent regressors for the various estimation methods.

MONTE CARLO EXPERIMENTS

In this section, we discuss the design of the Monte Carlo experiments and associated experimental results. Specifically, in section "Results by Signal-to-Noise, Spatial Dependence in Regressors and Regressand," we examine a Monte Carlo experiment with different levels of spatial dependence in y, goodness-of-fit, and dependence in the regressors. In section "Performance by Sample Size," we take the most difficult case from the previous experiments and see how estimators perform as a function of sample size. In section "Marginal Effects," we examine performance of the various estimators when producing marginal effects estimates. Finally, in section "Sensitivity to W," we examine sensitivity of the results to different W.

All of the experiments use instruments based on WX and W^2X for SAR and W^2X and W^3X for SDM in the 2SLS procedures (see Eqs. (27) and (28)). There are 15 non-constant explanatory variables, so the number of additional instruments equals 30. We do not focus on the large number of instruments problem, but one would expect our experimental findings to be mitigated through use of fewer variables (which result in fewer additional instruments), and exacerbated with more variables (resulting in additional instruments).

Researchers often have prior beliefs about the level of spatial dependence in y and signal-to-noise levels for data used in their field. In the experiments we set σ_ε^2 to the level required to yield a given signal-to-noise ratio as defined in Eq. (62). To maintain succinct notation, we call this measure R^2.

$$R^2 = 1 - \frac{E(\varepsilon'(I_n - \rho W)^{-2}\varepsilon)}{\beta' X'(I_n - \rho W)^{-2} X\beta + E(\varepsilon'(I_n - \rho W)^{-2}\varepsilon)} \tag{61}$$

$$= 1 - \frac{\sigma_\varepsilon^2 \operatorname{tr}((I_n - \rho W)^{-2})}{\beta' X'(I_n - \rho W)^{-2} X\beta + \sigma_\varepsilon^2 \operatorname{tr}((I_n - \rho W)^{-2})} \tag{62}$$

The actual R^2 would vary across trials, but the average actual R^2 was reasonably close to the design value in Eq. (62). The form in Eq. (62) is simply a way of selecting levels of σ_ε^2 that correspond to levels of fit that researchers see in practice.

For each experiment we generated the n by p matrix R and derived U based on design values of φ. In other words, X_{SAR} or X_{SDM} are the same across all trials associated with the same level of φ. However, ε differs across trials.

In all the experiments, the intercept is set to 0 while the regression parameters γ associated with U in X_{SAR} (shown in Eq. (4)) and X_{SDM} (shown in Eq. (6)) are set to 1, and the regression parameters θ associated with WU in X_{SDM} are set to 0. Therefore, the DGP is always SAR. We did this to facilitate comparisons across the SAR and SDM results (especially when producing marginal effects estimates). This does have some ramifications. For example, as previously discussed at the end of section "Instrumental Variables," when $(\theta = -\rho\gamma)$ 2SLS is not identified. The choice of $\gamma = 1$ and $\theta = 0$ does allow identification. However, as ρ approaches 0, the 2SLS estimator could behave badly because it is approaching a situation where it is unidentified. Although 2SLS may have problems for low ρ, we will not focus on this as most researchers only use these techniques when they believe that ρ is moderate or large. In the Monte Carlo experiments, we use $\rho = 0.4$ to capture moderate dependence and $\rho = 0.8$ to capture high dependence.

The best IV estimator uses the true values of β and ρ in computing $(I_n - \rho W)^{-1}\ WX\beta$, which is obviously is the best case scenario for an IV technique.

Results by Signal-to-Noise, Spatial Dependence in Regressors and Regressand

In the following experiments, we only examine a sample size of 3,000 observations, normally considered a large sample size. We design the experiments to facilitate matching cases to researchers' knowledge of their data. Specifically, we simulate using Eqs. (1) and (10) with levels of ρ equal to 0.4 and 0.8, which reflect moderate and strong spatial dependence in y. Given the empirical levels of spatial dependence in the regressors shown in

Table 1, we set φ equal to 0, 0.5, and 0.95 to represent the *iid* case, the case of moderate spatial dependence in the regressors, and finally the case of high spatial dependence in the regressors. The disturbance variance σ_ε^2 was set to a level that assured an R^2 defined by Eq. (62) equal to 0.1 (low signal-to-noise), 0.5 (moderate signal-to-noise), or 0.9 (high signal-to-noise). Therefore, the experiment examined 18 cases that vary by spatial dependence in the regressors, signal-to-noise, and spatial dependence in y. Each case involves 1,000 trials.

In Tables 11 and 13 we record the mean estimate of ρ as well as the standard deviation of the estimated values of ρ from all the estimators for SAR and SDM. However, we discard all trials where the estimate of ρ for that estimator exceeds 0.99 to control the problems, which occurred due to sometimes wild estimates from the best and 2SLS IV estimators. In Tables 12 and 15, we report the root mean square error (RMSE) for SAR and SDM (less discarded trials). In Table 14, we report the percentage of estimates outside the (0, 1) range for SDM (now using all trials). In Tables 12–15 the subscripts RE, B, IV, ML stand for REML, the Best IV, 2SLS, and ML.

To simplify the presentation, we discuss all SAR results in section "SAR Results" and all SDM results in section "SDM Results."

SAR Results

Table 11 shows the usual IV results in the spatial literature in the presence of high or moderate signal-to-noise ($R^2 = 0.9, 0.5$) and no spatial dependence in the regressors ($\varphi = 0$). Namely, both the best IV specification and 2SLS show very low levels of bias in those cases. Also, the results confirm those of Lee (2007) for the low signal-to-noise cases where 2SLS shows greater bias. Note that the bias of ML monotonically increases with φ. Although REML displays some bias, it is lower than ML and the lowest of the estimators.

However, Tables 11 and 12 show the great importance of spatial dependence in the regressors. All of the worst cases in terms of bias and RMSE happen in the high spatial dependence of regressor ($\varphi = 0.95$) cases for the Best IV, 2SLS, and ML estimators. The performance of the IV estimators is non-monotonic with the highest relative RMSEs occurring for both the lowest and the highest levels of φ.

In terms of producing estimates outside of the interval (0, 1), the Best IV estimator with true parameters yields estimates of ρ of less than 0 in 3 out of 1,000 trials, and greater than 1 in 4 out of 1,000 trials when $R^2 = 0.1$, $\varphi = 0.95$, and $\rho = 0.4$ (case 15). 2SLS produces estimates greater than 1 in 9.6 percent of the trials for this case.

Table 11. Estimation of the SAR Model via Instrumental Variables and Maximum Likelihood.

Case	ρ	R^2	φ	$\tilde{\rho}_{ML}$	$\tilde{\rho}_{RE}$	$\tilde{\rho}_{B}$	$\tilde{\rho}_{IV}$	$\tilde{\rho}_{OLS}$
1	0.4000	0.9000	0.0000	0.4004	0.4003	0.4004	0.4022	0.4317
1	0.4000	0.9000	0.0000	0.0104	0.0104	0.0113	0.0112	0.0110
2	0.4000	0.9000	0.5000	0.3996	0.3999	0.4005	0.4017	0.4270
2	0.4000	0.9000	0.5000	0.0097	0.0097	0.0102	0.0102	0.0101
3	0.4000	0.9000	0.9500	0.3962	0.3995	0.4006	0.4035	0.4659
3	0.4000	0.9000	0.9500	0.0149	0.0149	0.0170	0.0172	0.0172
4	0.8000	0.9000	0.0000	0.8001	0.8001	0.8003	0.8015	0.8296
4	0.8000	0.9000	0.0000	0.0055	0.0055	0.0060	0.0062	0.0051
5	0.8000	0.9000	0.5000	0.7997	0.7997	0.8003	0.8016	0.8300
5	0.8000	0.9000	0.5000	0.0055	0.0055	0.0061	0.0060	0.0054
6	0.8000	0.9000	0.9500	0.7968	0.7979	0.8003	0.8032	0.8755
6	0.8000	0.9000	0.9500	0.0082	0.0082	0.0109	0.0108	0.0082
7	0.4000	0.5000	0.0000	0.4005	0.4001	0.4012	0.4164	0.5562
7	0.4000	0.5000	0.0000	0.0191	0.0191	0.0340	0.0337	0.0237
8	0.4000	0.5000	0.5000	0.3988	0.3997	0.4016	0.4133	0.5422
8	0.4000	0.5000	0.5000	0.0195	0.0194	0.0305	0.0299	0.0246
9	0.4000	0.5000	0.9500	0.3916	0.3992	0.4018	0.4264	0.6146
9	0.4000	0.5000	0.9500	0.0226	0.0225	0.0514	0.0516	0.0337
10	0.8000	0.5000	0.0000	0.7996	0.7994	0.8009	0.8117	0.9446
10	0.8000	0.5000	0.0000	0.0103	0.0103	0.0179	0.0178	0.0088
11	0.8000	0.5000	0.5000	0.7986	0.7986	0.8008	0.8115	0.9453
11	0.8000	0.5000	0.5000	0.0102	0.0102	0.0182	0.0171	0.0096
12	0.8000	0.5000	0.9500	0.7928	0.7952	0.8011	0.8255	1.0164
12	0.8000	0.5000	0.9500	0.0127	0.0124	0.0327	0.0300	0.0115
13	0.4000	0.1000	0.0000	0.4006	0.4001	0.4035	0.5122	0.6791
13	0.4000	0.1000	0.0000	0.0234	0.0233	0.1028	0.0850	0.0339
14	0.4000	0.1000	0.5000	0.3981	0.3994	0.4049	0.4944	0.6711
14	0.4000	0.1000	0.5000	0.0236	0.0235	0.0918	0.0809	0.0346
15	0.4000	0.1000	0.9500	0.3901	0.3991	0.4053	0.5642	0.6850
15	0.4000	0.1000	0.9500	0.0249	0.0246	0.1615	0.1289	0.0371
16	0.8000	0.1000	0.0000	0.7996	0.7993	0.8027	0.8762	1.0539
16	0.8000	0.1000	0.0000	0.0126	0.0126	0.0545	0.0387	0.0104
17	0.8000	0.1000	0.5000	0.7982	0.7982	0.8025	0.8756	1.0536
17	0.8000	0.1000	0.5000	0.0125	0.0125	0.0546	0.0393	0.0108
18	0.8000	0.1000	0.9500	0.7920	0.7948	0.8030	0.9278	1.0727
18	0.8000	0.1000	0.9500	0.0136	0.0134	0.1031	0.0573	0.0109

SDM Results

Tables 13–15 demonstrate the challenges posed by the SDM model for the ML, Best IV, and 2SLS estimators. Only REML displays very low bias across all the cases. Even ML shows increasing bias as dependence in the regressors (φ) rises.

Table 12. SAR Model RMSE from Instrumental Variables and Maximum Likelihood.

Case	ρ	R^2	φ	$\tilde{R}_{ML}$	$\tilde{R}_{RE}$	$\hat{R}_B$	$\hat{R}_{IV}$	$\hat{R}_{OLS}$
1	0.4000	0.9000	0.0000	0.0070	0.0071	0.0076	0.0079	0.0317
2	0.4000	0.9000	0.5000	0.0066	0.0065	0.0070	0.0072	0.0270
3	0.4000	0.9000	0.9500	0.0107	0.0101	0.0115	0.0119	0.0659
4	0.8000	0.9000	0.0000	0.0037	0.0037	0.0040	0.0044	0.0296
5	0.8000	0.9000	0.5000	0.0038	0.0038	0.0041	0.0042	0.0300
6	0.8000	0.9000	0.9500	0.0065	0.0060	0.0073	0.0079	0.0755
7	0.4000	0.5000	0.0000	0.0129	0.0128	0.0229	0.0266	0.1562
8	0.4000	0.5000	0.5000	0.0130	0.0130	0.0210	0.0239	0.1422
9	0.4000	0.5000	0.9500	0.0159	0.0152	0.0348	0.0413	0.2146
10	0.8000	0.5000	0.0000	0.0067	0.0068	0.0120	0.0152	0.1446
11	0.8000	0.5000	0.5000	0.0070	0.0069	0.0124	0.0152	0.1453
12	0.8000	0.5000	0.9500	0.0097	0.0087	0.0222	0.0290	0.2164
13	0.4000	0.1000	0.0000	0.0157	0.0154	0.0696	0.1138	0.2791
14	0.4000	0.1000	0.5000	0.0156	0.0152	0.0620	0.0985	0.2711
15	0.4000	0.1000	0.9500	0.0178	0.0166	0.1086	0.1690	0.2850
16	0.8000	0.1000	0.0000	0.0084	0.0085	0.0354	0.0762	0.2539
17	0.8000	0.1000	0.5000	0.0087	0.0086	0.0363	0.0756	0.2536
18	0.8000	0.1000	0.9500	0.0108	0.0095	0.0670	0.1279	0.2727

In the worst case (case 15) where $\varphi = 0.95$, $\rho = 0.4$, and $R^2 = 0.1$, ML gives a mean estimate of 0.3895 for a bias of 0.0105. Relative to REML, the RMSE for ML was 13.5 percent higher. In contrast, REML gives a mean estimate of 0.4001 for a bias of 0.0001. More dramatically, the Best IV gives estimates of ρ of 0.5664 or a bias of 0.1664, and 2SLS produces average estimates of 0.9423 for a bias of 0.5423. The RMSE for the Best IV (with true starting values) and 2SLS increases by factors of over 40 and 35 relative to REML.

Table 14 shows that for the Best IV as little as 43.5 percent of the trials fell in the (0,1) interval in the worst case (case 15). For ML and REML, all of the ρ estimates fell in this interval. For 2SLS in the worst case (case 18) when $\rho = 0.8$, $\varphi = 0.95$, and $R^2 = 0.1$ only 28.8 percent of the trials had ρ estimates within (0,1). Even when R^2 equaled 0.5, some of the cases resulted in ρ estimates outside of (0,1) for the Best IV and 2SLS estimators.

Performance by Sample Size

In section "SDM Results," the most difficult case (case 15) for the best instruments in estimating the SDM was for $\rho = 0.4$, $R^2 = 0.1$, and ($\varphi = 0.95$.

Table 13. Estimation of the SDM Model via Instrumental Variables and Maximum Likelihood.

Case	ρ	R^2	φ	$\tilde{\rho}_{ML}$	$\tilde{\rho}_{RE}$	$\hat{\rho}_{B}$	$\hat{\rho}_{IV}$	$\hat{\rho}_{OLS}$
1	0.4000	0.9000	0.0000	0.3960	0.3986	0.3949	0.4333	0.6329
1	0.4000	0.9000	0.0000	0.0211	0.0210	0.0598	0.0567	0.0302
2	0.4000	0.9000	0.5000	0.3940	0.3988	0.3942	0.4317	0.6262
2	0.4000	0.9000	0.5000	0.0218	0.0217	0.0586	0.0558	0.0305
3	0.4000	0.9000	0.9500	0.3885	0.3986	0.3983	0.5086	0.6690
3	0.4000	0.9000	0.9500	0.0228	0.0227	0.1037	0.0911	0.0330
4	0.8000	0.9000	0.0000	0.7980	0.7981	0.7989	0.8035	0.8925
4	0.8000	0.9000	0.0000	0.0087	0.0087	0.0123	0.0122	0.0077
5	0.8000	0.9000	0.5000	0.7972	0.7980	0.7990	0.8041	0.8971
5	0.8000	0.9000	0.5000	0.0093	0.0092	0.0126	0.0130	0.0082
6	0.8000	0.9000	0.9500	0.7926	0.7976	0.8010	0.8184	0.9882
6	0.8000	0.9000	0.9500	0.0111	0.0109	0.0242	0.0237	0.0103
7	0.4000	0.5000	0.0000	0.3954	0.3984	0.3847	0.6185	0.6936
7	0.4000	0.5000	0.0000	0.0226	0.0224	0.1852	0.1308	0.0339
8	0.4000	0.5000	0.5000	0.3931	0.3985	0.3827	0.6109	0.6905
8	0.4000	0.5000	0.5000	0.0224	0.0223	0.1797	0.1262	0.0344
9	0.4000	0.5000	0.9500	0.3890	0.3995	0.3953	0.7906	0.6936
9	0.4000	0.5000	0.9500	0.0232	0.0229	0.3130	0.1505	0.0354
10	0.8000	0.5000	0.0000	0.7966	0.7969	0.7966	0.8362	1.0289
10	0.8000	0.5000	0.0000	0.0119	0.0119	0.0371	0.0314	0.0111
11	0.8000	0.5000	0.5000	0.7952	0.7966	0.7969	0.8385	1.0318
11	0.8000	0.5000	0.5000	0.0118	0.0117	0.0383	0.0338	0.0109
12	0.8000	0.5000	0.9500	0.7907	0.7969	0.8027	0.9072	1.0678
12	0.8000	0.5000	0.9500	0.0126	0.0124	0.0732	0.0512	0.0107
13	0.4000	0.1000	0.0000	0.3960	0.3990	0.4098	0.8546	0.7023
13	0.4000	0.1000	0.0000	0.0230	0.0229	0.5872	0.1711	0.0348
14	0.4000	0.1000	0.5000	0.3934	0.3989	0.3783	0.8588	0.6994
14	0.4000	0.1000	0.5000	0.0227	0.0225	0.5758	0.1617	0.0345
15	0.4000	0.1000	0.9500	0.3895	0.4001	0.5664	0.9423	0.6965
15	0.4000	0.1000	0.9500	0.0231	0.0229	0.9076	0.1504	0.0349
16	0.8000	0.1000	0.0000	0.7968	0.7971	0.7897	0.9537	1.0725
16	0.8000	0.1000	0.0000	0.0123	0.0123	0.1177	0.0587	0.0106
17	0.8000	0.1000	0.5000	0.7952	0.7968	0.7918	0.9580	1.0731
17	0.8000	0.1000	0.5000	0.0126	0.0124	0.1206	0.0550	0.0105
18	0.8000	0.1000	0.9500	0.7907	0.7970	0.8217	1.0354	1.0809
18	0.8000	0.1000	0.9500	0.0128	0.0125	0.2174	0.0603	0.0110

Both the best instruments and 2SLS produce biased estimates for ρ when $n=$ 3,000. However, the theory establishes the consistency of these techniques. In this section, we examine how the bias and performance of the IV techniques varies with sample size. In this section, we report the usual

Table 14. Proportion of the SDM Model Estimates Outside (0, 1) for Instrumental Variables.

Case	ρ	R^2	φ	$\hat{\rho}_B < 0$	$\hat{\rho}_B > 1$	$\hat{\rho}_{IV} > 1$	$\hat{\rho}_{OLS} > 1$
1	0.4000	0.9000	0.0000	0.0000	0.0000	0.0000	0.0000
2	0.4000	0.9000	0.5000	0.0000	0.0000	0.0000	0.0000
3	0.4000	0.9000	0.9500	0.0010	0.0000	0.0000	0.0000
4	0.8000	0.9000	0.0000	0.0000	0.0000	0.0000	0.0000
5	0.8000	0.9000	0.5000	0.0000	0.0000	0.0000	0.0000
6	0.8000	0.9000	0.9500	0.0000	0.0000	0.0000	0.1110
7	0.4000	0.5000	0.0000	0.0490	0.0000	0.0000	0.0000
8	0.4000	0.5000	0.5000	0.0430	0.0000	0.0000	0.0000
9	0.4000	0.5000	0.9500	0.1500	0.0060	0.0510	0.0000
10	0.8000	0.5000	0.0000	0.0000	0.0000	0.0000	0.9930
11	0.8000	0.5000	0.5000	0.0000	0.0000	0.0000	0.9960
12	0.8000	0.5000	0.9500	0.0020	0.0000	0.0200	1.0000
13	0.4000	0.1000	0.0000	0.2740	0.1010	0.1720	0.0000
14	0.4000	0.1000	0.5000	0.2780	0.0870	0.1710	0.0000
15	0.4000	0.1000	0.9500	0.2840	0.2810	0.3400	0.0000
16	0.8000	0.1000	0.0000	0.0200	0.0040	0.2200	1.0000
17	0.8000	0.1000	0.5000	0.0160	0.0060	0.2210	1.0000
18	0.8000	0.1000	0.9500	0.0660	0.1450	0.7120	1.0000

arithmetic averages, RMSEs, and standard deviations of the various estimates (no medians). Table 16 shows that the best instruments show low bias after 500,000 observations while the 2SLS estimates take 2,500,000 observations to achieve low bias. In contrast, ML displayed low bias across all sample sizes (despite using an approximation to the log-determinant) while OLS displayed high (but relatively constant) bias across all sample sizes.

Because some of the discussions concerning the methods in the literature promote methods based on ease-of-computation, we examine the computational aspects for these large samples. In terms of computation, we used a Delaunay triangle routine to determine W based on contiguity. This took slightly over 1 minute on a 2005 vintage computer. We used the Barry and Pace (1999) log-determinant approximation in calculating ML (with some modifications described in LeSage & Pace (2009)). The reported times for ML combined the times for computing the log-determinant and estimating the model once. Naturally, in the simulation the log-determinant approximation was only run once and the marginal time required for estimating the model for another y approximately equals OLS. The Best IV

Table 15. SDM Model RMSE from Instrumental Variables and Maximum Likelihood.

Case	ρ	R^2	φ	$\tilde{R}_{\mathrm{ML}}$	$\tilde{R}_{\mathrm{RE}}$	$\hat{R}_{\mathrm{B}}$	$\hat{R}_{\mathrm{IV}}$	$\hat{R}_{\mathrm{OLS}}$
1	0.4000	0.9000	0.0000	0.0146	0.0142	0.0404	0.0469	0.2329
2	0.4000	0.9000	0.5000	0.0151	0.0146	0.0402	0.0455	0.2262
3	0.4000	0.9000	0.9500	0.0167	0.0151	0.0683	0.1115	0.2690
4	0.8000	0.9000	0.0000	0.0059	0.0059	0.0081	0.0086	0.0925
5	0.8000	0.9000	0.5000	0.0066	0.0064	0.0086	0.0091	0.0971
6	0.8000	0.9000	0.9500	0.0096	0.0075	0.0160	0.0216	0.1882
7	0.4000	0.5000	0.0000	0.0154	0.0152	0.1263	0.2197	0.2936
8	0.4000	0.5000	0.5000	0.0158	0.0155	0.1193	0.2113	0.2905
9	0.4000	0.5000	0.9500	0.0172	0.0156	0.2084	0.3906	0.2936
10	0.8000	0.5000	0.0000	0.0078	0.0078	0.0244	0.0375	0.2289
11	0.8000	0.5000	0.5000	0.0082	0.0080	0.0258	0.0400	0.2318
12	0.8000	0.5000	0.9500	0.0107	0.0084	0.0486	0.1072	0.2678
13	0.4000	0.1000	0.0000	0.0158	0.0152	0.3672	0.4546	0.3023
14	0.4000	0.1000	0.5000	0.0161	0.0155	0.3597	0.4588	0.2994
15	0.4000	0.1000	0.9500	0.0177	0.0156	0.6264	0.5423	0.2965
16	0.8000	0.1000	0.0000	0.0086	0.0084	0.0743	0.1537	0.2725
17	0.8000	0.1000	0.5000	0.0086	0.0085	0.0774	0.1580	0.2731
18	0.8000	0.1000	0.9500	0.0107	0.0086	0.1428	0.2354	0.2809

Table 16. Estimation of the SDM Model for Large *n*.

n	20,000	100,000	500,000	2,500,000
mean $\hat{\rho}_{\mathrm{ML}}$	0.3984	0.3999	0.3996	0.4002
s.d. $\hat{\rho}_{\mathrm{ML}}$	0.0090	0.0039	0.0019	0.0008
mean $\hat{\rho}_{\mathrm{B}}$	0.3792	0.3209	0.4087	0.3954
s.d. $\hat{\rho}_{\mathrm{B}}$	0.8791	0.1695	0.0723	0.0322
mean $\hat{\rho}_{\mathrm{IV}}$	0.8149	0.5815	0.4588	0.4073
s.d. $\hat{\rho}_{\mathrm{IV}}$	0.1613	0.1097	0.0585	0.0314
mean $\hat{\rho}_{\mathrm{OLS}}$	0.7060	0.7081	0.7074	0.7080
s.d. $\hat{\rho}_{\mathrm{OLS}}$	0.0132	0.0057	0.0028	0.0012
mean R^2	0.0972	0.0983	0.0985	0.0984
Delaunay secs	0.35	1.74	12.18	62.54
ML secs	4.95	9.22	44.51	227.53
Best IV secs	0.38	2.57	24.93	295.37
2SLS IV secs	0.47	3.08	14.46	71.07
OLS secs	0.10	0.65	3.16	15.54

estimator, due to computing $(I_n - \rho W)^{-1} X\beta$, actually requires more time than ML when $n = 2{,}500{,}000$. Nonetheless, all the techniques were computationaly feasible, even for large sample sizes.

Marginal Effects

Tables 17 and 18 report the total marginal effects for SAR and SDM. The true value of the total marginal effect for $\rho = 0.4$ is 1.667 and for $\rho = 0.8$ is 5.0. We begin with the total marginal effects for the SAR model as shown in Table 17. For all but OLS, the table shows the statistics calculated on all the trials that result in estimates of ρ within the interval [0,0.99]. First, the ML, REML, and Best IV estimator usually display low biases across the cases. However, 2SLS displays a positive bias which becomes serious for some of the cases when $R^2 = 0.5$, and all of the cases when $R^2 = 0.1$. In almost every situation the variability of the estimates of the total effects from all the estimators rises with φ. In some situations, the variability rises more than double.

For the SDM, Table 18 displays the total effect results across the estimators and cases. In this case REML displays the lowest unadjusted total effect biases, although it often displays slightly more variability than ML. In the cases where $R^2 = 0.9$, the Best IV slightly underestimates the total effects. This underestimation becomes somewhat more prominent as the R^2 falls. However, 2SLS displays very serious biases when R^2 equals 0.5 and especially 0.1. In case 18, on average the parameter estimates for θ exceeded those for β and therefore 2SLS reversed the sign of the total effect. Note that using some other statistic such as the median may lead to fewer differences among the estimators. However, even the use of robust statistics does not alter the qualitative results.

Sensitivity to W

The above analysis always use a contiguity-based weight matrix. A common concern in spatial econometrics is that the results may vary across different specifications of W. To investigate this, we examine the case with a SDM DGP where $n = 3{,}000$, $\varphi = 0.95$, $\rho = 0.4$, and $R^2 = 0.9$. This is similar to the case used in analyzing models with large n in section "Performance by Sample Size." However, the sample size is smaller and the R^2 is raised from

Table 17. Estimated Total Effects from the SAR Model via Instrumental Variables and Maximum Likelihood.

Case	ρ	R^2	φ	$\tilde{T}_{\mathrm{ML}}$	$\tilde{T}_{\mathrm{RE}}$	$\hat{T}_{\mathrm{B}}$	$\hat{T}_{\mathrm{IV}}$	$\hat{T}_{\mathrm{OLS}}$
1	0.4000	0.9000	0.0000	1.6681	1.6679	1.6686	1.6733	1.7566
1	0.4000	0.9000	0.0000	0.0489	0.0489	0.0503	0.0501	0.0538
2	0.4000	0.9000	0.5000	1.6672	1.6677	1.6705	1.6731	1.7270
2	0.4000	0.9000	0.5000	0.0494	0.0494	0.0519	0.0519	0.0520
3	0.4000	0.9000	0.9500	1.6671	1.6689	1.6692	1.6709	1.7078
3	0.4000	0.9000	0.9500	0.0641	0.0642	0.0619	0.0623	0.0673
4	0.8000	0.9000	0.0000	5.0043	5.0029	5.0057	5.0414	5.8271
4	0.8000	0.9000	0.0000	0.1794	0.1793	0.1812	0.1902	0.2214
5	0.8000	0.9000	0.5000	5.0021	5.0021	5.0145	5.0424	5.7265
5	0.8000	0.9000	0.5000	0.2007	0.2010	0.2085	0.2129	0.2378
6	0.8000	0.9000	0.9500	4.9878	4.9959	5.0173	5.0371	5.9029
6	0.8000	0.9000	0.9500	0.3458	0.3471	0.3402	0.3389	0.4629
7	0.4000	0.5000	0.0000	1.6715	1.6706	1.6723	1.7135	2.2256
7	0.4000	0.5000	0.0000	0.1324	0.1323	0.1510	0.1507	0.2022
8	0.4000	0.5000	0.5000	1.6743	1.6763	1.6776	1.7014	2.0627
8	0.4000	0.5000	0.5000	0.1419	0.1421	0.1550	0.1563	0.1917
9	0.4000	0.5000	0.9500	1.6705	1.6748	1.6741	1.6881	1.8484
9	0.4000	0.5000	0.9500	0.1901	0.1906	0.1869	0.1916	0.2333
10	0.8000	0.5000	0.0000	5.0073	5.0019	5.0098	5.3103	17.3840
10	0.8000	0.5000	0.0000	0.4533	0.4529	0.5465	0.5937	2.9525
11	0.8000	0.5000	0.5000	5.0047	5.0053	5.0368	5.2735	15.8394
11	0.8000	0.5000	0.5000	0.5542	0.5541	0.6199	0.6736	2.9666
12	0.8000	0.5000	0.9500	4.9912	5.0058	5.0601	5.2461	−11.7811
12	0.8000	0.5000	0.9500	1.0044	1.0109	1.0211	1.0778	13.1279
13	0.4000	0.1000	0.0000	1.6853	1.6840	1.6740	2.0528	3.0537
13	0.4000	0.1000	0.0000	0.3648	0.3642	0.4548	0.5255	0.7371
14	0.4000	0.1000	0.5000	1.7011	1.7041	1.6927	1.9238	2.7350
14	0.4000	0.1000	0.5000	0.4218	0.4226	0.4661	0.5236	0.7255
15	0.4000	0.1000	0.9500	1.6849	1.6905	1.6975	1.8340	1.9817
15	0.4000	0.1000	0.9500	0.5557	0.5583	0.5687	0.6500	0.7606
16	0.8000	0.1000	0.0000	5.0357	5.0290	4.9641	7.9957	−16.9445
16	0.8000	0.1000	0.0000	1.1136	1.1121	1.6774	2.8672	5.3322
17	0.8000	0.1000	0.5000	5.1088	5.1097	5.0499	7.5818	−14.0975
17	0.8000	0.1000	0.5000	1.5626	1.5629	1.9185	3.0750	6.0601
18	0.8000	0.1000	0.9500	5.0829	5.1015	5.2099	7.1609	−0.6701
18	0.8000	0.1000	0.9500	2.9592	2.9814	3.2568	6.7314	3.9017

0.1 to 0.9 to allow better performance of the IV estimators for this sample size of 3,000 observations.

We examine the aforementioned contiguity weight matrix, weight matrices based on 6, 20, and 30 nearest neighbors (NN-6, NN-20,

Table 18. Estimated Total Effects from the SDM Model via Instrumental Variables and Maximum Likelihood.

Case	ρ	R^2	φ	$\tilde{T}_{\mathrm{ML}}$	$\tilde{T}_{\mathrm{RE}}$	$\hat{T}_{\mathrm{B}}$	$\hat{T}_{\mathrm{IV}}$	$\hat{T}_{\mathrm{OLS}}$
1	0.4000	0.9000	0.0000	1.6507	1.6574	1.6499	1.7651	2.6580
1	0.4000	0.9000	0.0000	0.0681	0.0681	0.1669	0.1768	0.2171
2	0.4000	0.9000	0.5000	1.6443	1.6564	1.6453	1.7552	2.6058
2	0.4000	0.9000	0.5000	0.0730	0.0736	0.1584	0.1751	0.2147
3	0.4000	0.9000	0.9500	1.6329	1.6581	1.6537	2.0091	2.9685
3	0.4000	0.9000	0.9500	0.1783	0.1820	0.3085	0.3970	0.4300
4	0.8000	0.9000	0.0000	4.9459	4.9490	4.9654	5.0811	9.0384
4	0.8000	0.9000	0.0000	0.2441	0.2439	0.3131	0.3278	0.6777
5	0.8000	0.9000	0.5000	4.9195	4.9380	4.9683	5.0943	9.4189
5	0.8000	0.9000	0.5000	0.2814	0.2826	0.3455	0.3578	0.8043
6	0.8000	0.9000	0.9500	4.8010	4.9138	4.9465	5.4273	66.8883
6	0.8000	0.9000	0.9500	0.9072	0.9334	1.0107	1.0857	59.7624
7*	0.4000	0.5000	0.0000	1.6483	1.6560	1.6226	2.5829	3.1540
7	0.4000	0.5000	0.0000	0.1360	0.1365	0.5040	0.9034	0.4073
8	0.4000	0.5000	0.5000	1.6425	1.6567	1.6009	2.5267	3.1150
8	0·4000	0.5000	0.5000	0.1576	0.1589	0.4745	0.7849	0.4553
9	0.4000	0.5000	0.9500	1.6192	1.6466	1.5711	4.0734	3.1670
9	0.4000	0.5000	0.9500	0.5031	0.5132	0.9757	3.3751	1.0860
10	0.8000	0.5000	0.0000	4.9039	4.9105	4.8895	5.9812	−32.2970
10	0.8000	0.5000	0.0000	0.4460	0.4459	0.9282	1.1946	13.6989
11	0.8000	0.5000	0.5000	4.8703	4.9035	4.9010	6.0701	−29.0026
11	0.8000	0.5000	0.5000	0.6060	0.6090	1.0187	1.3284	11.8103
12	0.8000	0.5000	0.9500	4.6716	4.8153	4.7548	10.1375	−14.4369
12	0.8000	0.5000	0.9500	2.6279	2.7193	2.9596	7.5898	8.0974
13	0.4000	0.1000	0.0000	1.6379	1.6454	1.2985	4.2001	3.2351
13	0.4000	0.1000	0.0000	0.3594	0.3609	1.3675	4.6443	0.7975
14	0.4000	0.1000	0.5000	1.6272	1.6424	1.2711	4.2402	3.2010
14	0.4000	0.1000	0.5000	0.4500	0.4540	1.2751	5.1848	0.9702
15	0.4000	0.1000	0.9500	1.6103	1.6463	0.6335	2.8419	3.1254
15	0.4000	0.1000	0.9500	1.4948	1.5232	1.7459	12.0521	3.0639
16	0.8000	0.1000	0.0000	4.8442	4.8508	4.6500	12.2291	−12.7589
16	0.8000	0.1000	0.0000	1.1063	1.1079	2.7628	12.9961	3.3146
17	0.8000	0.1000	0.5000	4.7894	4.8211	4.5625	12.0907	−12.5116
17	0.8000	0.1000	0.5000	1.6429	1.6566	3.0707	15.9700	4.8945
18	0.8000	0.1000	0.9500	4.6336	4.7955	2.6696	−7.4605	−11.5743
18	0.8000	0.1000	0.9500	7.8888	8.1310	8.0018	39.7988	19.6848

NN-30), weight matrices where the 3, 20, and 30 diagonals closest to the main diagonal are positive (D-6, D-20, and D-30), a random matrix where approximately *6/n* proportion of the off-diagonal elements are positive (Random), and a block diagonal matrix composed of a 50 by 50 submatrix

S_d with the 6 diagonals nearest the main diagonal are positive (and 0 otherwise) so that the candidate weight matrix equals $I_{60} \otimes S_d$. All candidate weight matrices are symmetricized and scaled to be doubly stochastic.

Table 19 shows that the weight matrices make a difference in the average estimates of ρ across the 1,000 trials. In this section we report the usual arithmetic averages, RMSEs, and standard deviations of the various estimates (no medians). For all the weight matrices, REML gave answers close to the true parameter value of 0.4. The Best IV estimator also came close to this value with the most bias appearing in the case with a matrix with 20 positive diagonals (D-20) where the average estimate of ρ equals 0.3705. Interestingly, ML shows considerable bias in the case of the D-30 weight matrix with an average estimate of 0.3316. Also, 2SLS shows its highest bias in this case with an estimate of 0.745. OLS shows its lowest bias in the case of NN-30. Typically, the determinant plays a smaller role; the denser the weight matrix, the less clustered neighbors are to the main diagonal.

Table 20 shows the RMSE for ρ estimates for the different estimators across the different weight matrices normalized by the REML RMSE. Interestingly, ML performs much worse than REML for some of the weight matrices. In the case of a weight matrix with positive entries in the 15 closest off-diagonals on either side of the main diagonal (D-30), ML had 67.2 percent more RMSE than REML. Almost all of that, however, is bias. For the Best IV estimator, it has RMSE worse than OLS for D-20, D-30, and Random weight matrices. For 2SLS, it has RMSE worse than OLS for the NN-20, D-20, NN-30, D-30, and Random weight matrices.

Table 19. Average Estimates of $\tilde{\rho}$ for the SDM Model Across W.

W	ML	REML	BEST	2SLS	OLS
Contiguity	0.3881	0.3980	0.3923	0.5065	0.6672
NN-6	0.3870	0.3989	0.3847	0.5570	0.6617
D-6	0.3871	0.3984	0.3760	0.6143	0.6563
NN-20	0.3565	0.3961	0.3880	0.6080	0.5910
D-20	0.3560	0.3959	0.3705	0.7291	0.5963
NN-30	0.3357	0.3949	0.3883	0.6374	0.5525
D-30	0.3316	0.3927	0.3740	0.7450	0.5546
Random	0.3318	0.3928	0.3755	0.7379	0.5539
Block Diagonal	0.3881	0.3985	0.3728	0.6315	0.6590

Table 20. Relative RMSE of $\tilde{\rho}$ for the SDM Model Across W.

W	ML	REML	BEST	2SLS	OLS
Contiguity	1.1184	1.0000	4.6860	5.8506	11.2825
NN-6	1.1356	1.0000	5.7661	7.6381	10.7066
D-6	1.1506	1.0000	7.2979	10.3208	11.1566
NN-20	1.3963	1.0000	3.3209	5.0693	4.4431
D-20	1.4788	1.0000	5.4840	8.3331	4.9435
NN-30	1.5477	1.0000	2.9276	4.6719	3.1370
D-30	1.6719	1.0000	4.1952	6.9101	3.3118
Random	1.6705	1.0000	4.0901	6.7866	3.3025
Block Diagonal	1.1384	1.0000	8.4173	11.5782	11.6931

SUGGESTIONS FOR SPATIAL MONTE CARLO EXPERIMENTS

Our main focus lies on the interaction between spatial dependence in the explanatory variables and estimation performance. Based on theory and empirical measurements, both explanatory variables and the dependent variable may exhibit spatial dependence.

Therefore, it seems unrealistic to use explanatory variables that arise from a non-spatial DGP or one that is vastly different from the one generating the dependent variable. We offer some examples. First, an *iid* DGP for the explanatory variables could arise if an explanatory variable is idiosyncratic and does not influence other observations. If one home uses interior beige paint while another uses taupe, these choices might not show spatial dependence and may have no effect on surrounding homes. However, important characteristics such as defective drywall usually result in spillovers. Since defective dry-wall building products may be popular at various points in time and with various builders, these may show spatiotemporal dependence that can produce a spatial equilibrium. Empirically, it seems difficult to find truly *iid* observations on many economic variables. Second, use of empirical explanatory variables in conjunction with y or disturbances generated using a spatial weight matrix not based on the location of explanatory variables tacitly destroys these possible interactions. Third, spatial weight matrices with random non-zero entries may "appear" spatial, but may exhibit very little simultaneity. For example, one could randomly permute a strictly lower triangular matrix from time series to make it appear spatial, but this matrix would not exhibit spatial simultaneity. Fourth, block diagonal weight matrices may also "appear" spatial, but spatial dependence is restricted to each of the blocks. If the block sizes are small relative to the sample size, this becomes more of an independence problem than a spatial problem. Fifth, assuming band matrices implies a regular grid for the underlying units, so Monte Carlo results may not extend to more irregular configurations. In general, it is better practice to use location coordinates for the variables, explanatory or dependent, which can be used to produce a corresponding W, in lieu of simply assuming a W that no possible configuration of observations could generate.

Our second focus is on the use of reasonable levels of goodness-of-fit. Some applications may have low signal-to-noise (e.g., finance) while others have very high signal-to-noise (e.g., real estate). If Monte Carlo experiments are used to test techniques, they should range over signal-to-noise levels found in applied practice. Historically, in the spatial literature there has been an emphasis on negative values of spatial dependence, which seem to occur rarely in practice, but not on examining low signal-to-noise situations which are more commonly encountered.

CONCLUSION

Much of the research conducted through Monte Carlo experiments in spatial econometrics assumes that regressors are *iid*. This is not a realistic assumption given the very high level of measured spatial dependence found in common economic variables. Although dependence in the regressors may not be an important issue in all situations, this study shows that performance of IV techniques when estimating SAR and especially SDM models can be sensitive to spatial dependence in the regressors, even for sample sizes involving thousands of observations.

In the spatial context, 2SLS has often been used as a means of solving a numeric problem of the *functional* simultaneity created by the transformation of y by $(I_n - \rho W)$. It has been popular partly due to economist familiarity with 2SLS and its ease of implementation. However, using 2SLS to solve a numeric problem differs fundamentally from the usual role of 2SLS in addressing *economic* simultaneity. Since many problems also involve right-hand-side endogenous variables, using instruments to address a numeric problem may impede the performance of also dealing with endogenous variables connected with the economics of the problem. Insofar as the numeric problems have largely been overcome by the progress in computational hardware and software, individuals should carefully weigh whether they wish to use their instruments to solve a numeric rather than an economic problem. If researchers do choose to use 2SLS with spatial data, they should carefully examine the sensitivity of the results to the instrument set, and calculate partial R^2 as well as the degree of dependence in the residuals of the first-stage regression to ensure their results have some robustness to small specification changes. Ideally, they should compare IV estimates to those from REML or ML. If there is no compelling reason to stay with the 2SLS framework, researchers should just present the results from REML or ML.

NOTES

1. In spatial econometrics the acronym SAR usually refers to an autoregression in the dependent variable while in spatial statistics it is usually an autoregression in the disturbances Ripley (1981).
2. A row-stochastic matrix in linear algebra is a non-negative matrix with rows that sum to 1. The individual elements are fixed and exogenous. Similarly, a doubly stochastic matrix has both rows and columns that sum to 1. The different meanings of stochastic in linear algebra versus statistics sometimes creates confusion.

REFERENCES

Anselin, L. (1988). *Spatial econometrics: Methods and models.* Dordrecht: Kluwer Academic.

Barry, R. P., & Pace, R. K. (1999). A Monte Carlo estimator of the log determinant of large sparse matrices. *Linear Algebra and its Applications, 289*, 41–54.

Bound, J., Jaeger, D. A., & Baker, R. (1995). Problems with instrumental variables estimation when the correlation between the instruments and the endogenous explanatory variable is weak. *Journal of the American Statistical Association, 90*, 443–450.

Bowden, R. J., & Turkington, D. A (1984). *Instrumental variables.* Cambridge: Cambridge University Press.

Byme, P. F. (2005). Strategic interaction and the adoption of tax increment financing. *Regional Science and Urban Economics, 35*(3), 279–303.

Cressie, N. (1993). *Statistics for spatial data* (rev. ed.). New York, NY: Wiley.

Gilley, O. W., & Pace, R. K. (1996). The Harrison and Rubinfeld data revisited. *Journal of Environmental and Economic Management, 31*, 403–405.

Harrison, D., & Rubinfeld, D. L. (1978). Hedonic housing prices and the demand for clean air. *Journal of Environmental Economics and Management, 5*, 81–102.

Kelejian, H., & Prucha, I. (1998). A generalized spatial two stage least squares procedure for estimating a spatial autoregressive model with autoregressive disturbances. *Journal of Real Estate Finance and Economics, 17*, 99–121.

Kelejian, H., Prucha, I., & Yuzefovich, Y. (2004). Instrumental variable estimation of a spatial autoregressive model with autoregressive disturbances: Large and small sample results. In J. P. LeSage & R. K. Pace (Eds.), *Advances in econometrics: Volume 18: Spatial and spatiotemporal econometrics* (pp. 163–198). Oxford: Elsevier Ltd.

Klotz, S. (2004). *Cross sectional dependence in spatial econometrics: With an application to German start-up activity data.* Munster: LIT Verlag Munster.

Land, K., & Deane, G. (1992). On the large-sample estimation of regression models with spatial or network-effects terms: A two stage least squares approach. In P. Marsden (Ed.), *Sociological methodology* (pp. 279–303). San Francisco, CA: Jossey-Bass.

Lee, L. F. (2003). Best spatial two-stage least squares estimators for a spatial autoregressive model with autoregressive disturbances. *Econometric Reviews*, *22*, 307–335.

Lee, L. F. (2007). GMM and 2SLS estimation of mixed regressive, spatial autoregressive models. *Journal of Econometrics*, *137*, 489–514.

Lee, L. F., Liu, X., & Lin, X. (2010). Specification and estimation of social interaction models with network structures. *The Econometrics Journal*, *13*, 145–176.

LeSage, J. P., & Pace, R. K. (2009). *Introduction to spatial econometrics.* Boca Raton: CRC Press/Taylor & Francis.

Millimet, D. L., & Rangaprasad, V. (2007). Strategic competition amongst public schools. *Regional Science and Urban Economics*, *37*(2), 199–219.

Ord, J. K. (1975). Estimation methods for models of spatial interaction. *Journal of the American Statistical Association*, *70*, 120–126.

Pace, R. K., & Barry, R. (1997). Quick computation of spatial autoregressive estimators. *Geographical Analysis*, *29*, 232–246.

Pace, R. K., & LeSage, J. P. (2009). Biases of OLS and spatial lag models in the presence of an omitted variable and spatially dependent variables. In A. Pa'ez, J. Gallo, R. N. Buliung & S. Dall'erba (Eds.), *Progress in spatial analysis: Methods and applications.* Berlin: Springer-Verlag.

Patterson, H. D., & Thompson, R. (1971). Recovery of inter-block information when block sizes are unequal. *Biometrika*, *58*, 545–554.

Richards, T. J., & Padilla, L. (2009). Promotion and fast food demand. *American Journal of Agricultural Economics*, *91*(1), 168–183.

Ripley, B. (1981). *Spatial statistics.* New York, NY: Wiley.

Staiger, D., & Stock, J. H. (1997). Instrumental variables regression with weak instruments. *Econometrica*, *65*(3), 557–586.

PART IV
APPLIED MACROECONOMETRICS

SECTORAL EFFECTS OF AGGREGATE SHOCKS☆

Nathan S. Balke

ABSTRACT

In this chapter, using a combination of long-run and sign restrictions to identify aggregate monetary and productivity factors, I find that the monetary factor is responsible for long swings in nominal variables but has little effect on fluctuations in output, real wage, or labor input growth. The productivity factor in addition to increasing output growth and real wage growth in the short and long run, also results in increases in labor input and decreases in prices, but the quantitative effect of the productivity factor on labor input is relatively small. These results are robust to the number of factors included in the model and to alternative priors about the short-run effects of the monetary factor, and to the inclusion of oil prices. Oil prices, in fact, appear to be largely driven by the other aggregate factors.

Keywords: Common factor model; Bayesian; sectoral data; long-run restrictions; monetary neutrality

☆This is a much revised version of a previous paper titled "Sectoral Effects of Monetary Shocks."

30th Anniversary Edition
Advances in Econometrics, Volume 30, 299–357

ISSN: 0731-9053/doi:10.1108/S0731-9053(2012)0000030015

INTRODUCTION

Much of the recent literature on accounting for the sources of economic fluctuations have focused on monetary and productivity shocks. Quantifying the effect of money shocks is important for evaluating theories of price stickiness and for determining the efficacy of monetary policy.[1] Productivity or technology shocks have featured prominently in stochastic general equilibrium models of business cycles although recent empirical work has cast doubt on the ability of technology shock driven models to explain business cycle fluctuations.[2]

Most of the literature use VARs to describe the data generating process and assume a recursive structure to identify economic shocks. These recursive schemes are not always well motivated and often rely on ad hoc timing restrictions. A few papers have applied non-recursive identification schemes. For example, Gali (1992) and others use long-run restrictions to identify monetary shocks while Gali (1999) and Francis and Ramey (2005) use long-run restrictions to identify productivity shocks. These long-run relationships are often thought to be more robust across models and policy regimes.[3] Sign restrictions rather than zero restrictions for impulse responses have also been used to identify monetary shocks (Canova & De Nicolo, 2002; Faust, 1998; Uhlig, 2005) as well as productivity shocks (Peersman & Straub, 2009). Again, these sign restrictions might be viewed as less stringent and more robust to alternative economic structures than traditional recursive zero restrictions.

In this chapter, I depart from the standard approach taken in most of the literature in several key respects. First, instead of using aggregate data, I use matched sectoral data on prices, output, wages, and labor input to identify monetary and aggregate productivity shocks. Using disaggregated data can bring additional information to bear in identification beyond that in aggregate variables alone. Not only are more series used, but economic theory often has implications for sectoral variables as well as for aggregate variables. For example, I use long-run monetary neutrality to identify a common monetary factor; monetary neutrality implies that relative prices and individual sector outputs, real wages, or labor input are unaffected by a one-time change in the stock of money. Using only an aggregate price index or an aggregate measure of output would miss this sectoral implication of monetary neutrality. Similarly, aggregate productivity shocks should have effects across many sectors, in particular, raising real wages and output across sectors in the long run; the timing and the extent of these effects, however, may vary across sectors. Here, I capture these sectoral effects of aggregate productivity shocks in the form of non-negativity constraints on the long-run response of sectoral real wages and output.[4]

Second, rather than estimate a VAR I use a common factor model to estimate monetary and aggregate productivity shocks. A common factor model is a natural way to model the effect of a few important shocks on a wide cross-section of observables. As the primary focus in the literature has been on a small number of aggregate shocks, focusing on a few common factors does not lessen the economic interest of the empirical analysis. In addition to aggregate monetary and productivity factors, I include a third factor that while not having a clear a priori interpretation will end up looking like a traditional aggregate demand factor. Also, I consider a model that includes oil prices and an additional factor that is interpreted as an oil factor.

Third, I use Bayesian methods to estimate the common factor model. While Bayesian methods are not crucial for the implementation of long-run or sign restrictions, in this application they greatly simplify the empirical analysis. Furthermore, a Bayesian approach allows one to set priors over short-run responses and use those, in addition to long-run restrictions, in the identification of aggregate shocks.

I find that the estimated monetary factor is responsible for much of the long swings in sector price growth, M2 growth, and the nominal interest rate, suggesting that this factor is capturing underlying inflationary pressure in nominal variables. Taking the unweighted average across sectors to be a measure of aggregate activity, the monetary factor, on the other hand, explains little of the historical fluctuations in "aggregate" output growth, real wages, or labor growth. Impulse response analysis for the monetary factor is somewhat sensitive to the prior assumptions that one makes about the short-run responses of sectoral variables to a monetary shock, suggesting that the data are not too informative about these short-run effects.

Productivity shocks, on the other hand, are estimated to increase aggregate output, real wages, and labor input in both the short run and long run but the size of the increases varies substantially across sectors. These results are robust to alternative priors about the short-run effect of monetary shocks. That productivity shocks are associated with increases in labor input both at the aggregate and sectoral level are in contrast to the results of Gali (1999) and Francis and Ramey (2005). Nevertheless, while historical decompositions for output and real wages suggest that the aggregate productivity factor is an important source of fluctuations in these variables, the aggregate productivity factor has only a small impact on labor fluctuations over our sample.

When I include oil prices and an "oil factor" in the analysis, the findings for the monetary or productivity factors are not substantially changed. The "oil factor" contributes only modestly to fluctuations in aggregate output, inflation, labor, and real wages. Interestingly, the "oil factor" seems to affect aggregate and sectoral economic activity primarily through a demand channel rather than through a supply-side channel. Finally, a large fraction of oil price fluctuations appears to be driven by the aggregate demand factor, suggesting that oil prices are responding endogenously to overall economic activity.

Several previous papers have applied factor models to disaggregated data in a macroeconomic context. Explicit dynamic common factor models have been employed to explain cyclical behavior of sectoral output (Forni & Reichlin, 1998; Quah & Sargent, 1994), world and regional growth rates (Kose, Otrok, & Whiteman, 2002), and disaggregated prices (Bryan & Cecchetti, 1993; Nath, 2004).[5] Stock and Watson (2002), Bernanke, Boivin, and Eliasz (2005), among others, employed factor models where the factors are based on the principal components of disaggregate data.

Four papers that come closest to the analysis in the current chapter are Boivin, Giannoni, and Mihov (2009), Ahmadi and Uhlig (2007), Reis and Watson (2010), and Foerster, Sarte, and Watson (2011). Boivin et al. (2009) use a large number of disaggregated variables to estimate a monetary factor. In fact, in one of their specifications they impose a sort of long-run monetary neutrality restriction for relative prices. However, they do not consider implications of long-run neutrality for other real variables. Ahmadi and Uhlig (2007) examine a Bayesian FAVAR and employ sign restrictions to identify monetary shocks, but do use long-run monetary neutrality. Neither paper tries to identify aggregate productivity shocks. Reis and Watson (2010) look at large cross-section of price indices and extract a common "pure inflation" factor in which all prices move one-to-one with this factor (i.e., no relative price effects). However, they do not examine disaggregate real variables nor do they examine the impact of productivity shocks. Foerster et al. (2011) estimate a structural factor model to sectoral industrial production and use a multi-sector growth model to capture the input–output relationships across sectors. They find that the aggregate factor explains a substantial portion of sectoral industrial production fluctuations. They do not, however, examine sectoral price, wage, or labor data, nor do they examine aggregate nominal factors.

The rest of this chapter is organized as follows. In the second section, I discuss the sectoral data used in the chapter and present some evidence about the degree to which this sectoral data can be described by a few common factors. In the third section, I discuss the specification and estimation of the structural dynamic common factor model. The fourth section contains the empirical results for the benchmark factor model. In the fifth section, I examine the effects of shocks to the aggregate factor on individual sectors. The sixth section examines the robustness of the benchmark model to alternative short-run priors and the number of factors. The seventh section adds oil prices to the analysis. The eighth section concludes.

SECTORAL CO-MOVEMENT

In this section, I examine the degree of co-movement in sector price growth, output growth, real wage growth, and labor input growth for the U.S. economy using principal components. The primary source of data is an extended version of the Jorgenson KLEM data set originally compiled by Jorgenson, Gollop, and Fraumeni (1987) that consists of annual observations on gross output, as well as the price and various inputs for 35 sectors of the U.S. economy. These 35 sectors roughly correspond to 2-digit SIC level for U.S. industries.[6] Table 1 lists the 35 sectors by sector number and the fraction of their output used for consumption, investment, and materials input. Two data sets, one covering the period 1947–1989 and another covering the more recent period from 1960 to 2005, are spliced together. I divide sector price of labor input by sector output price to obtain sector real wage.

Table 2 displays fraction the total sectoral variance that can be explained by the five largest principle components of the cross-sector covariance matrix. In addition to examining variance/covariance matrices of sectoral price growth, output growth, real wage growth, and labor input growth separately, the joint covariance matrix of all four sectoral variables is also examined. From Table 2, one observes that the first five principal components can explain a large fraction of the cross-sector variances. For sectoral prices and output, a single principal component can explain over 40 percent of the total variance. Even when one examines the covariance matrix for all four sectoral variables jointly, the first three principal components explain over 46 percent of the sum of sectoral variances.[7] In fact, the Bai and Ng (2002) IC_1 and IC_2 criteria for selecting the number of factors in a static factor model both suggest two factors.

If one looks at each sectoral variable individually, the average contribution across sectors of the first three principal components is close to 30 percent for all but sectoral real wages (see Table 3). If one takes unweighted averages of sector variables to construct aggregate variables, these same three principle components account for over 80 percent the

Table 1. Intermediate and Final Uses of Sectoral Output in 1992.

Sector	Fractions of Output in Different Uses				
	Total intermediate uses (1)	Personal consumption expenditures (2)	Private fixed investment (3)	Government expenditures (4)	Net exports (5)
1 Agriculture	0.84	0.11	0.00	0.00	0.02
2 Metallic ores mining	0.77	0.00	0.00	−0.01	0.24
3 Coal mining	0.88	0.00	0.00	0.00	0.11
4 Crude oil and natural gas	1.42	0.00	0.00	0.00	−0.42
5 Nonmetallic minerals	1.01	0.00	0.00	0.00	−0.02
6 Construction	0.23	0.00	0.53	0.23	0.00
7 Food and kindred products	0.38	0.61	0.00	0.02	−0.01
8 Tobacco products	0.08	0.76	0.00	0.00	0.11
9 Textile	0.85	0.13	0.04	0.01	−0.04
10 Apparel	0.31	1.03	0.00	0.03	−0.41
11 Lumber and wood products	0.92	0.03	0.05	0.00	−0.01
12 Furniture and fixtures	0.13	0.50	0.38	0.07	−0.08
13 Paper	0.85	0.12	0.00	0.03	−0.01
14 Printing	0.60	0.29	0.00	0.08	0.02
15 Chemicals	0.66	0.26	0.01	0.05	0.01
16 Petroleum refining	0.56	0.36	0.00	0.10	−0.02
17 Rubber	0.89	0.13	0.00	0.02	−0.06
18 Footwear and leather products	0.41	1.73	0.00	0.02	−1.19
19 Stone	0.98	0.06	0.00	0.01	−0.06
20 Primary metal	1.08	0.00	0.00	0.00	−0.08
21 Fabricated metal	0.92	0.04	0.04	0.02	−0.02
22 Machinery	0.46	0.03	0.47	0.06	−0.02
23 Electrical machinery	0.68	0.21	0.17	0.07	−0.13
24 Motor vehicle	0.33	0.48	0.35	0.05	−0.20
25 Transportation equipment	0.25	0.09	0.14	0.36	0.22
26 Instruments	0.27	0.09	0.39	0.27	−0.01
27 Misc. manufacturing	0.34	0.92	0.14	0.06	−0.48
28 Trans. and warehousing	0.57	0.24	0.02	0.06	0.11
29 Communication	0.46	0.44	0.02	0.06	0.02
30 Electric utility service	0.51	0.41	0.00	0.09	0.00
31 Gas utility service	0.64	0.31	0.00	0.04	0.01
32 Wholesale and retail trade	0.26	0.63	0.05	0.01	0.05
33 FIRE	0.36	0.59	0.02	0.02	0.02
34 Personal and busi. services	0.42	0.56	0.01	0.00	0.01
35 Govt. enterprises	0.58	0.37	0.00	0.06	0.00

Note: These ratios have been calculated from BEA's 1992 Benchmark I-O Table. The shares may not add up to 1 because changes in business inventories are not included in investment.

Table 2. Percentage of the Sectoral Variances Explained by First Five Principal Components.

Principal component	Sectoral Price Growth		Sectoral Output Growth		Sectoral Real Wage Growth		Sectoral Labor Input Growth		All Sectoral Data	
	Percent	Cumulative	Percent	Cumulative	Percent	Cumulative	Percent	Cumulative	Percent	Cumulative
1	0.48	0.48	0.44	0.44	0.33	0.33	0.29	0.29	0.21	0.21
2	0.11	0.59	0.12	0.56	0.11	0.44	0.14	0.43	0.17	0.38
3	0.08	0.67	0.08	0.64	0.09	0.53	0.11	0.54	0.08	0.46
4	0.05	0.72	0.06	0.70	0.07	0.60	0.09	0.63	0.06	0.52
5	0.05	0.77	0.05	0.75	0.07	0.67	0.07	0.70	0.05	0.57

Table 3. Average (Across Sectors) Contribution to Individual Sector Variables of the First Five Principal Components from All Sectoral Data.

Principal Component	Sectoral Price Growth (%)	Sectoral Output Growth (%)	Sectoral Real Wage Growth (%)	Sectoral Labor Input Growth (%)
1	0.04	0.30	0.04	0.22
2	0.22	0.05	0.14	0.03
3	0.02	0.04	0.03	0.05
4	0.04	0.02	0.03	0.04
5	0.02	0.03	0.03	0.06
Sum of 1st five components	0.34	0.44	0.27	0.42

Table 4. Percentage of Variance of the Aggregate (Unweighted Mean Across Sectors) Variables Explained by First Five Principal Components from All Sectoral Data.

Principal Component	Aggregate Price Growth (%)	Aggregate Output Growth (%)	Aggregate Real Wage Growth (%)	Aggregate Labor Input Growth (%)
1	0.04	0.87	0.00	0.81
2	0.71	0.00	0.70	0.09
3	0.02	0.06	0.02	0.00
4	0.02	0.00	0.00	0.00
5	0.01	0.00	0.00	0.03
Sum of 1st five components	0.80	0.93	0.72	0.93

aggregate variables' variances (Table 4). Interestingly, the first principal component accounts for the majority of the variance of sectoral and aggregate output and labor input while the second principal component accounts for most of the variance of sectoral and aggregate prices and real wages. This suggests that the aggregate factors that affect wages and prices may be different from those that affect output and labor input.

COMMON FACTOR MODEL WITH AGGREGATE MONETARY AND PRODUCTIVITY SHOCKS

While the principal components analysis is suggestive of strong co-movement across sectoral variables, I build a common factor model to delve more deeply into the nature of the co-movement and to provide an economic interpretation to these co-movements. In addition to the sectoral variables examined above, I include M2 growth and three-month T-bill as observation variables in the system as these have typically played a role in studies of aggregate shocks and to some extent reflect monetary conditions in the economy.

The relationship between the sectoral variables and aggregate factors is given by

$$Y_t = A(L)Y_{t-1} + H(L)S_t + E_t \tag{1}$$

where Y_t is the vector of sectoral data (plus M2 growth and the nominal interest rate), S_t is the vector of aggregate common factors, and E_t is a vector of sector specific or idiosyncratic shocks. $H(L)$ is the factor loading matrix on current and lagged values of the common factors; in the empirical analysis I set the number of lags in $H(L)$ to four. As S_t reflects the co-movement of variables across sectors, the $A(L)$ matrix is restricted so that lags of variables in other sectors are zero. Similarly, the elements of E_t are uncorrelated across sectors but not necessarily with in a sector; that is, sector specific shocks affect variables in that sector but not variables in other sectors. A low order VAR describes the intra-sector relationship of output, prices, real wages, and labor inputs. Individual sectoral variables are assumed to have no direct effect on the aggregate money stock or interest rate.

The evolution of the common factors is governed by

$$S_t = F(L)S_{t-1} + V_t \tag{2}$$

where V_t is a vector of shocks to the common factors. The common factors are orthogonal to one another so that $F(L)$ consists of stationary univariate AR(2) processes for the common factors while the elements of V_t are uncorrelated with each other and with the elements of E_t. Based in part on the principal components analysis in the previous section, I will specify three common factors in the benchmark model: a monetary factor ($s_{m,t}$), an aggregate productivity factor ($s_{z,t}$), and a third factor ($s_{f,t}$) that a priori does not have a specific economic interpretation.[8] I assume the monetary factor shock is distributed: $v_{m,t} \sim N(0, \sigma^2_{s_m})$, while to fix the scale of the aggregate productivity factor and the third common factor I assume $v_{z,t}$ and $v_{f,t} \sim N(0, 1)$. Finally, the factor loading in the interest rate equation for $s_{f,t}$ is constrained to be positive.

Identifying Monetary and Aggregate Productivity Shocks

The monetary factor is assumed to satisfy long-run monetary neutrality; that is, an increase to the monetary factor is assumed to have no effect on real economic activity and that all prices move proportionally to the money stock in the long run. This means that the level of sectoral output, real wages, and labor input are unaffected in the long run by an exogenous money shock while log sectoral prices, nominal wages, and the money stock move one-for-one in the long run in response to a monetary shock.[9] While long-run monetary neutrality at the macro level has previously been used to identify monetary shocks (Gali, 1992; Lastrapes & Selgin, 1995), I impose those restrictions on sectoral prices and outputs.[10]

Why long-run neutrality? The short-run effects of monetary shocks are still of substantial debate within the profession; long-run neutrality is less controversial (Lucas, 1996). Also, many New Keynesian models such as Christiano, Eichenbaum, and Evans (1999, 2005), Gali and Gertler (1999), and Mankiw and Reis (2002) and nominal misperceptions models such as Lucas (1972) and Barro (1976) imply long-run monetary neutrality but do not necessarily have the same predictions for the short run.

In order to identify the aggregate productivity factor, I assume that an aggregate productivity shock will result in an increase in sector outputs and real wages in the long run.[11] This increase need not be the same across sectors, and there is no restriction on the long-run effect of the productivity shock on sectoral prices, labor input, and the stock of money. Unlike monetary neutrality which implies a one-for-one change in prices and money and a zero response for outputs and real wages, the restriction on aggregate productivity takes the form of a sign restriction on the long-run response of sector output and real wages to an aggregate productivity shock. Note that it is not just the long-run restrictions that matter in the identification but also the common factor restrictions. These cross-section restrictions are particularly important here because the cross-section dimension is substantially larger than the time dimension (142 variables in the observation vector vs. 57 observations).

For these identification schemes to be valid, the (logarithms) of money stock, sectoral prices, and sectoral wages must be integrated of at least order one (or $I(1)$). That is, there must be permanent shocks to the money stock, prices, and wages in order to use long-run restrictions to identify a factor. In order to use long-run restrictions to identify the aggregate productivity factor, output and real wages must be $I(1)$. For our data, the assumption of nonstationary money stock, prices, and wages is uncontroversial. While it is not necessary for our identification scheme, we model sectoral labor input as $I(1)$ as well.[12] The nominal interest rate (here 3 month T-Bill) is arguably $I(0)$ and is modeled as such. This means that the interest rate does not have a direct effect on the identification of the monetary factor. We do use the nominal interest rate to normalize the third common factor.

To see the restrictions imposed by the proposed long-run restrictions, write the common factor model implied by Eq. (1) for a single sector, i

$$\begin{pmatrix} \Delta p_{i,t} \\ \Delta q_{i,t} \\ \Delta w_{i,t} \\ \Delta l_{i,t} \end{pmatrix} = A_i(L) \begin{pmatrix} \Delta p_{i,t-1} \\ \Delta q_{i,t-1} \\ \Delta w_{i,t-1} \\ \Delta l_{i,t-1} \end{pmatrix} + \begin{pmatrix} H_{i,m}(L) & H_{i,z}(L) & H_{i,f}(L) \end{pmatrix} \begin{pmatrix} s_{m,t} \\ s_{z,t} \\ s_{f,t} \end{pmatrix} + \begin{pmatrix} e_{i,p,t} \\ e_{i,q,y} \\ e_{i,w,t} \\ e_{i,l,t} \end{pmatrix} \tag{3}$$

where $(\Delta p_{i,t} \Delta q_{i,t} \Delta w_{i,t} \Delta l_{i,t})'$ is the vector of observable variables in sector i, $(s_{m,t} s_{z,t} s_{f,t})'$ is the vector of unobserved common factors, $A_i(L)$ is a 4×4 lag polynomial matrix, $H_{i,m}(L)$, $H_{i,z}(L)$, and $H_{i,f}(L)$ are 4×1 vectors that reflect the effect of the current value and four lags of the common factors on the sectoral variables, and $(e_{i,p,t} e_{i,q,t} e_{i,w,t} e_{i,l,t})'$ is the vector of sector specific shocks. I allow these sector specific shocks to be correlated within sectors but not across sectors. For M2 growth (first difference of log of M2) and interest rate, the equations have the form

$$\Delta m2_t = A_{m2}(L) \Delta m2_{t-1} + \begin{pmatrix} H_{m2,m}(L) & H_{m2,z}(L) & H_{m2,f}(L) \end{pmatrix} \begin{pmatrix} s_{m,t} \\ s_{z,t} \\ s_{f,t} \end{pmatrix} + e_{m2,t} \tag{4}$$

and

$$r_t = A_r(L) r_{t-1} + \begin{pmatrix} H_{r,m}(L) & H_{r,z}(L) & H_{r,f}(L) \end{pmatrix} \begin{pmatrix} s_{m,t} \\ s_{z,t} \\ s_{f,t} \end{pmatrix} + e_{r,t} \tag{5}$$

The errors $e_{m2,t}$ and $e_{r,t}$ are uncorrelated with each other and with sector specific shocks.

Long-run neutrality for the monetary factor implies the following restriction

$$\begin{pmatrix} 1 \\ 0 \\ 0 \\ 0 \end{pmatrix} = (I - A_i(1))^{-1} H_{i,m}(1) \tag{6}$$

for every sector. That is, the long-run multiplier of the monetary factor is one for all sectoral prices and zero for sectoral output, real wages, and labor input. For M2, the long-run multiplier for monetary factor must satisfy

$$1 = (1 - A_{m2}(1))^{-1} H_{m2,m}(1) \tag{7}$$

Thus, the money stock and all prices and nominal wages move one-for-one in the long run with the monetary factor. Because interest rates are stationary, long-run neutrality has no implications for interest rates.

For aggregate productivity the long-run multiplier for sector i variables is

$$\begin{pmatrix} \eta_{i,p} \\ \eta_{i,q} \\ \eta_{i,w} \\ \eta_{i,l} \end{pmatrix} = (I - A_i(1))^{-1} H_{i,z}(1) \tag{8}$$

where $\eta_{i,p}$, $\eta_{i,q}$, $\eta_{i,w}$, and $\eta_{i,l}$ are the long-run multipliers for sector i price, output, real wage, and labor input, respectively. I impose the sign restrictions, $\eta_{i,q} > 0$ and $\eta_{i,w} > 0$, to ensure that the aggregate productivity factor results in increases in sector output and real wages in the long run. However, these effects can differ across sectors. The long-run responses of sector i price, $\eta_{i,p}$, and labor input, $\eta_{i,l}$, are unrestricted. Similarly, the long-run response of M2 (not shown) is unrestricted.

Other than the long-run effect of the monetary factor and the long-run sign restrictions for the aggregate productivity factor, there are few other restrictions. Most importantly, no "hard" constraints are placed on the short-run responses of prices, output, real wages, and labor input to a money shock nor are constraints placed on the short-run response of sectoral variables to an aggregate productivity shock. Note that the specification does not rule out the possibility that shocks to the other aggregate factor ($s_{f,t}$) can have permanent effects on sectoral variables. As one is primarily interested in identifying the effects of aggregate shocks, the sectoral shocks, e_i's, do not have clear economic interpretations. They do, however, allow the model to capture any sector specific co-movement in output, price, real wage, and labor input not captured by the common (across sectors) factors.

Bayesian Estimation of Common Factor Model

I take a Bayesian approach rather than the more standard maximum likelihood approach to estimate the common factor model. By using Markov Chain Monte Carlo methods, here the Gibbs sampler, we can break up a rather large problem into a number of smaller, but computationally tractable, problems by sequentially drawing a parameter from its conditional distribution given all the other parameters (and unobserved states). Furthermore, by treating unobserved common factors like parameters, we can obtain the joint posterior distribution of the parameters and unobserved states. In fact, the linear structure of the model and standard conjugate priors make it very easy to write posterior conditional distributions. Even the restrictions implied by Eqs. (6) and (7) can be handled fairly easily. If the common factors were known, it is straightforward to rewrite Eqs. (3) and (4), imposing the restrictions given by Eqs. (6) and (7), so that the resulting equations are linear equations with no restrictions on parameters (see section "Imposing Long-Run Neutrality Restrictions" in the appendix). The inequality restrictions for the long-run response of real wages and output to a productivity shock are also easy to implement in this context as parameter draws that do not satisfy the sign restrictions are rejected. The empirical estimate of the posterior distribution is constructed by taking every fifth draw from the Gibbs sample, 10,000 times, after running the sampler for a burn-in period of 100,000 draws. See section "Bayesian Estimation" in the appendix for a detailed discussion of the Gibbs sampler.

Prior Distributions

For most of the parameters of the model, relatively economically neutral and diffuse priors are employed (see section "Prior Distributions of Parameters" in the appendix for details). Setting priors over the parameters does have implications for the short-run response of variables to monetary shocks. In our benchmark model, the prior distribution is centered around short-run money neutrality with a relatively large spread of outcomes for the prior distribution. The mean and interior 80 percent of the implied prior distribution for the impulse responses to a money shock are plotted in Fig. 1a.[13] While the prior distributions of the short-run responses are

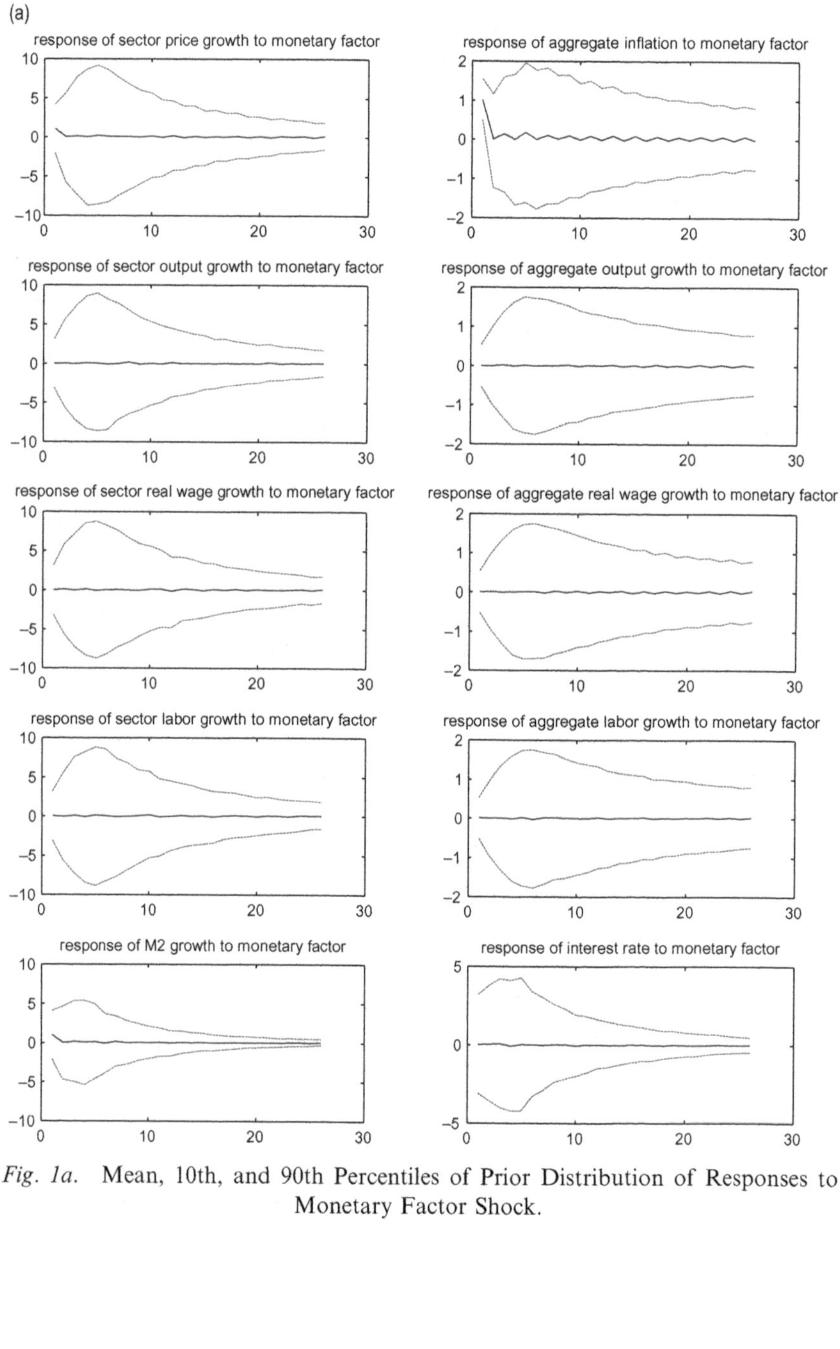

Fig. 1a. Mean, 10th, and 90th Percentiles of Prior Distribution of Responses to Monetary Factor Shock.

centered around monetary neutrality, one can see that the prior distributions support a wide range of non-neutral outcomes. This is particularly true for individual sectoral responses. For aggregated responses, here just the unweighted average of sectoral responses, the prior distribution is tighter around monetary neutrality but still suggests reasonable likelihood of non-neutral behavior in the short run. In section "Robustness", we discuss an alternative set of priors for which the prior distribution of impulse responses to a monetary shock implies short-run money non-neutrality and coincides with responses typically found in the traditional VAR literature.

Fig. 1b displays the implied prior distribution for the response of sector and aggregate variables to an aggregate productivity shock. The mean of the prior distribution is positive for the response of output and real wages, although the individual sector prior is much more diffuse than that for the aggregate prior. For the response of prices, labor input, M2 growth, and interest rates to a productivity shock, the prior distributions are centered around zero but with substantial dispersion. With the exception of the nominal interest in the current period, the prior distributions for the response of individual sector and aggregate variables to a third aggregate factor shock (not shown) are all centered about zero. This reflects the lack of an a priori interpretation for the third aggregate factor.

EMPIRICAL RESULTS: IMPLICATIONS FOR AGGREGATE VARIABLES

Fig. 2 displays the mean of the posterior distribution along with the 10th and 90th percentiles for the monetary factor, the aggregate productivity factor, and the third aggregate factor which I will argue below looks like an aggregate demand factor. From the figure, the monetary factor displays substantially more persistence than either the productivity or the third factor. The monetary factor shows a long swing upward in the mid-1970s and early 1980s and a decline in the late 1980s and through 1990s. These movements mirror the general movements in inflation over the sample. The aggregate productivity and demand factors feature movements that often, but not always, coincide with aggregate business cycles. For example, the productivity factor displays declines in the mid-1970s and early 1980s that coincide with recessions while third aggregate factor displays declines in the early 1990s and 2000s.

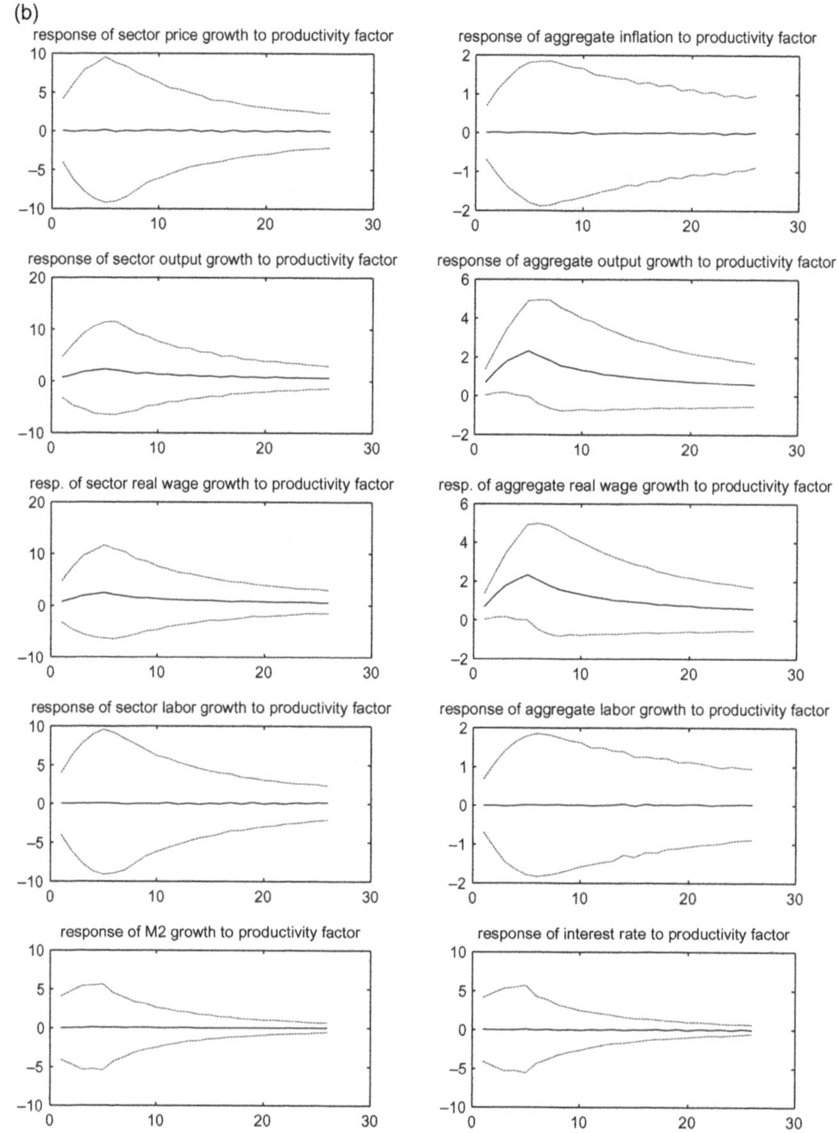

Fig. 1b. Mean, 10th, and 90th Percentiles of Prior Distribution of Responses to Productivity Factor Shock.

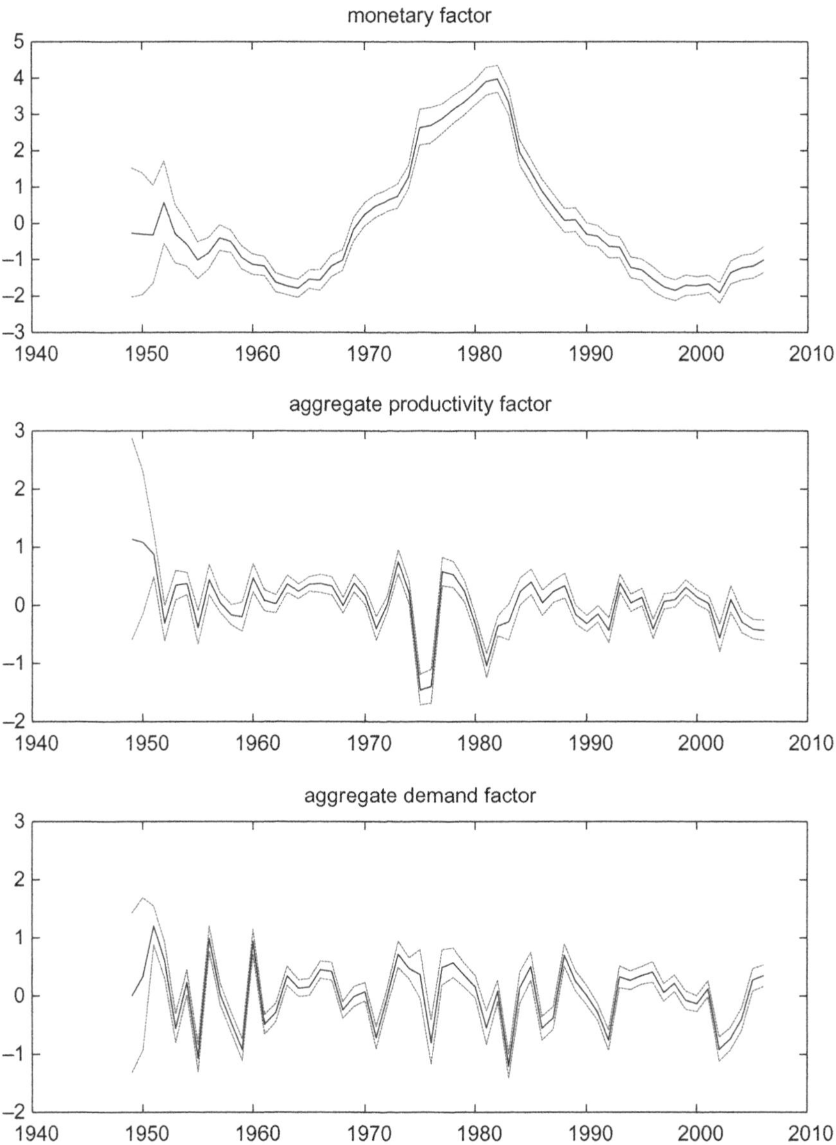

Fig. 2. Mean, 10th, and 90th Percentiles of the Posterior Distribution of Aggregate Factors.

Before turning to the sectoral details, it will be instructive to examine the aggregate implications of the estimated model. Doing so will help evaluate whether the estimated aggregate factors make sense in light of more aggregated data. With that in mind, I examine impulse responses and historical decompositions for unweighted averages (across sectors) of sector price, output, real wage, and labor.

Aggregate Impulse Response Analysis

Fig. 3 displays the mean, 10th, and 90th percentiles of posterior distribution for the impulse responses of "aggregate" variables to a shock in the monetary factor. The mean of the cross-section distribution of price growth (inflation) rises in response to an expansionary monetary shock, hence, there is no "price puzzle." Perhaps this is not too surprising given the identifying assumptions, but recall these assumptions did not preclude finding a price puzzle in the short run. On the other hand, mean output, real wage, and labor growth across sectors do not appear to respond significantly even in the short run. Again, even though the short-run prior implied monetary neutrality, it was diffuse enough to allow non-neutral responses. Both M2 growth and the nominal interest rate increase in response to a monetary factor shock but do so with some delay.

Fig. 4 displays the response of "aggregate" variables to an aggregate productivity shock. Output, real wage, and labor input growth all increase in response to a productivity shock. Furthermore, their responses display strong "humped" shapes. Interest rates also tend to rise after a productivity shock. Inflation, on the other hand, falls in response to a productivity shock. This decline while not as quantitatively large as the output response is long lasting. The strong, procyclical response of labor to a productivity shock contrasts with the results of Gali (1999) and Francis and Ramey (2005) who found a negative response of labor to an identified technology shock. My results are, however, consistent with results of Peersman and Straub (2009) whose sign restriction identification yielded a positive response of labor to a technology shock.

Fig. 5 displays the "aggregate" responses to the third common factor. Here the mean price, output, and labor input growth rates across sectors increase as does the nominal interest rate. On the other, hand real wage growth falls. These aggregate responses, at least in the short run, would be consistent with an increase in aggregate demand. In fact, the response of inflation, output, labor, and real wages are reminiscent of a shift in the "IS"

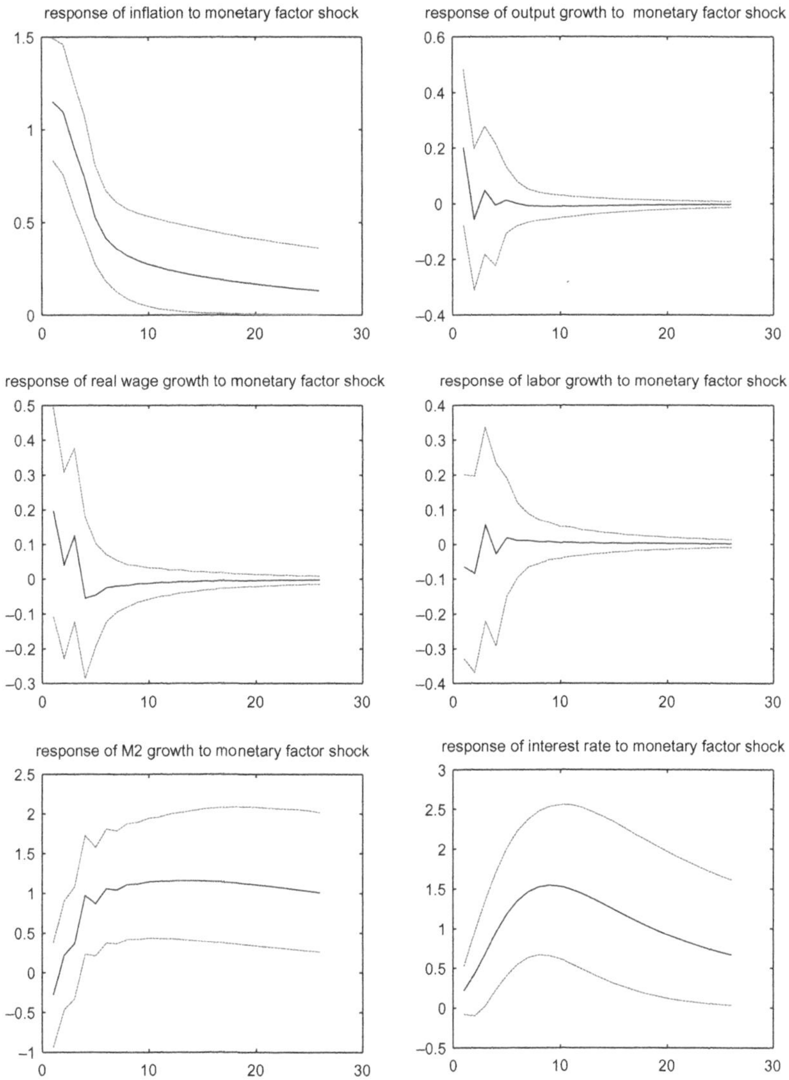

Fig. 3. Mean, 10th, and 90th Percentiles of Posterior Distributions for Impulse Responses to a Monetary Factor Shock.

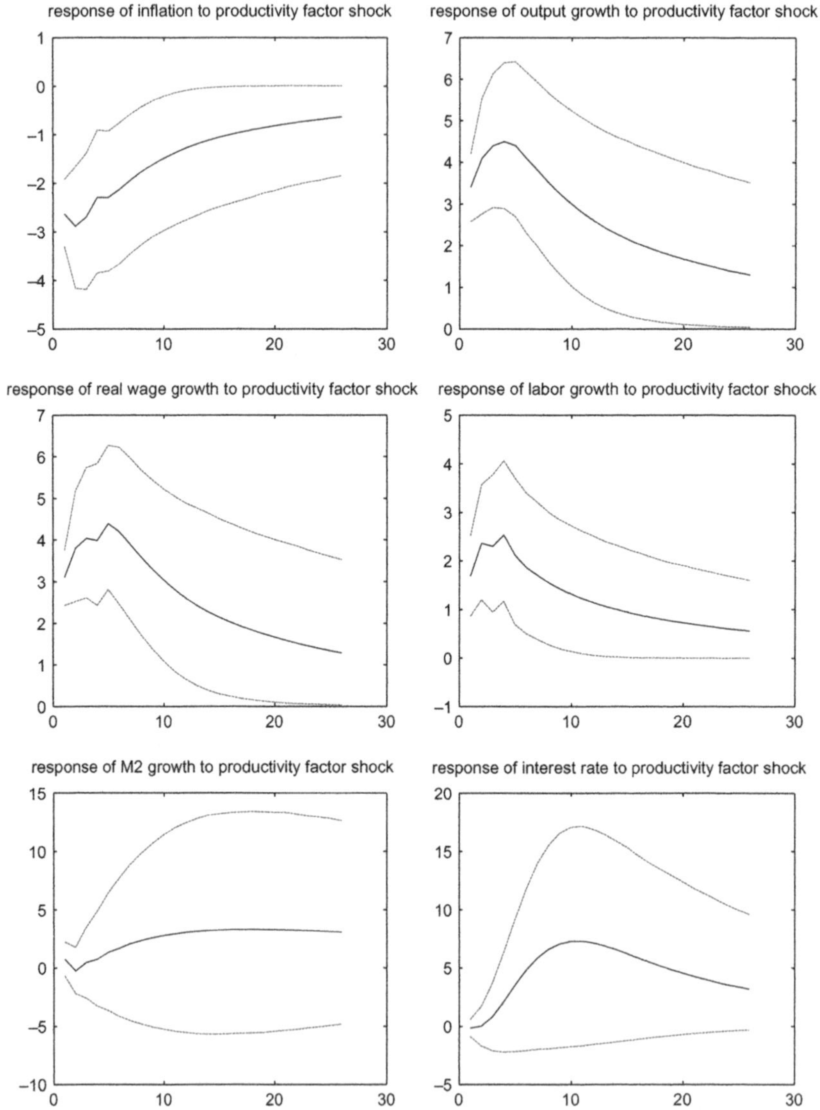

Fig. 4. Mean, 10th, and 90th Percentiles of Posterior Distributions for Impulse Responses to a Productivity Factor Shock.

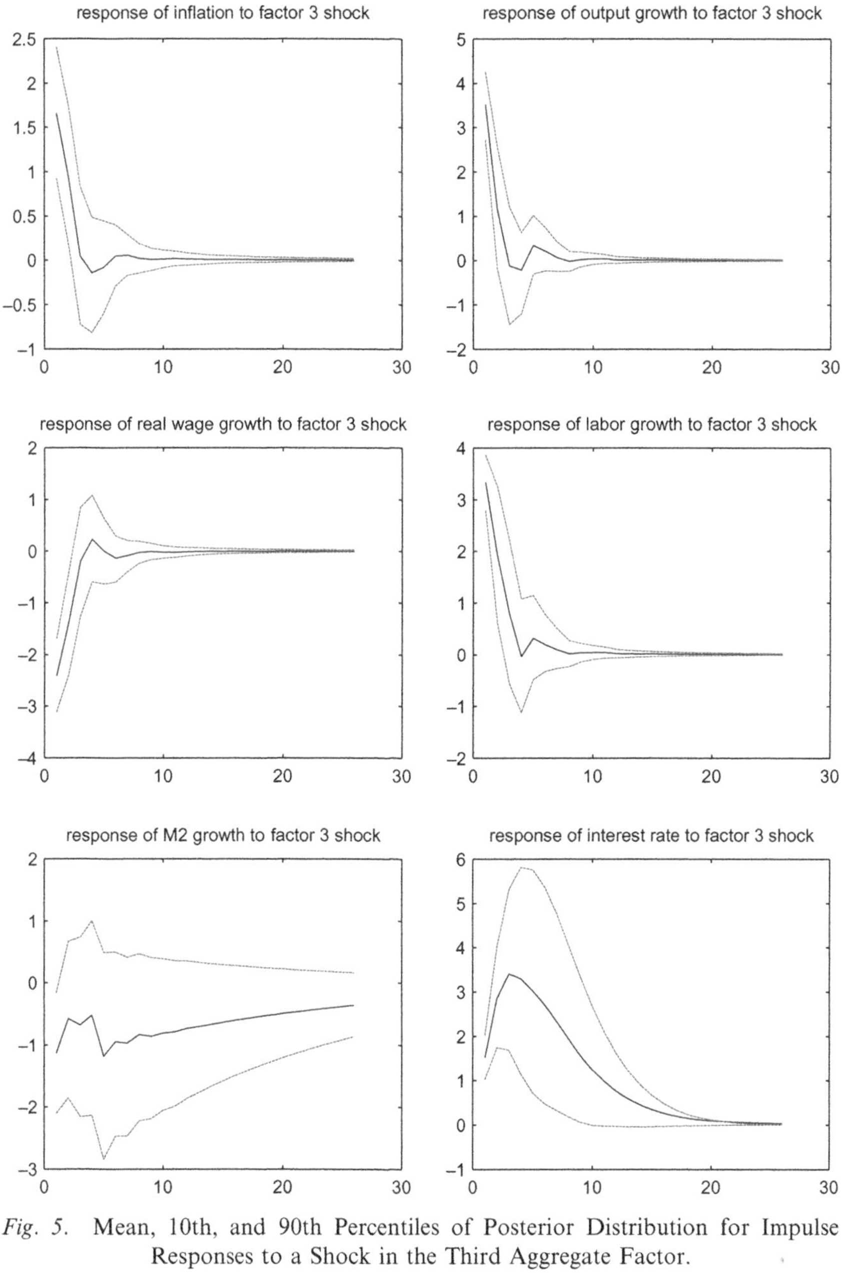

Fig. 5. Mean, 10th, and 90th Percentiles of Posterior Distribution for Impulse Responses to a Shock in the Third Aggregate Factor.

curve in a traditional IS/LM/AS model with sticky nominal wages. The slightly negative response of M2 growth and positive response of interest rates might also reflect a "lean against the wind" stance for monetary policy in response to positive aggregate demand shocks.

Historical Decompositions of Aggregated Sectoral Variables

Figs. 6–9 present posterior distribution for the historical decompositions of the unweighted average (across sectors) of sector price, output, real wage, labor input growth as well as M2 growth and three-month T-Bill rate. In addition to the mean contribution, the figures display the 10th and 90th percentiles of the posterior distribution. Fig. 6 displays the contribution of

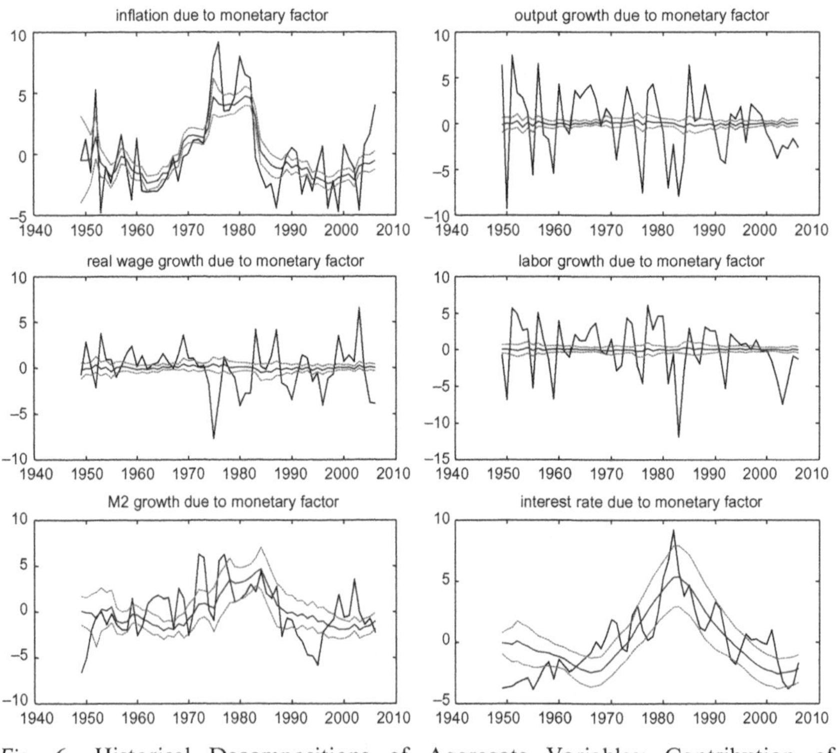

Fig. 6. Historical Decompositions of Aggregate Variables: Contribution of Monetary Factor. Actual (Sample Means Removed) and Mean, 10th, and 90th Percentiles of the Posterior Distribution.

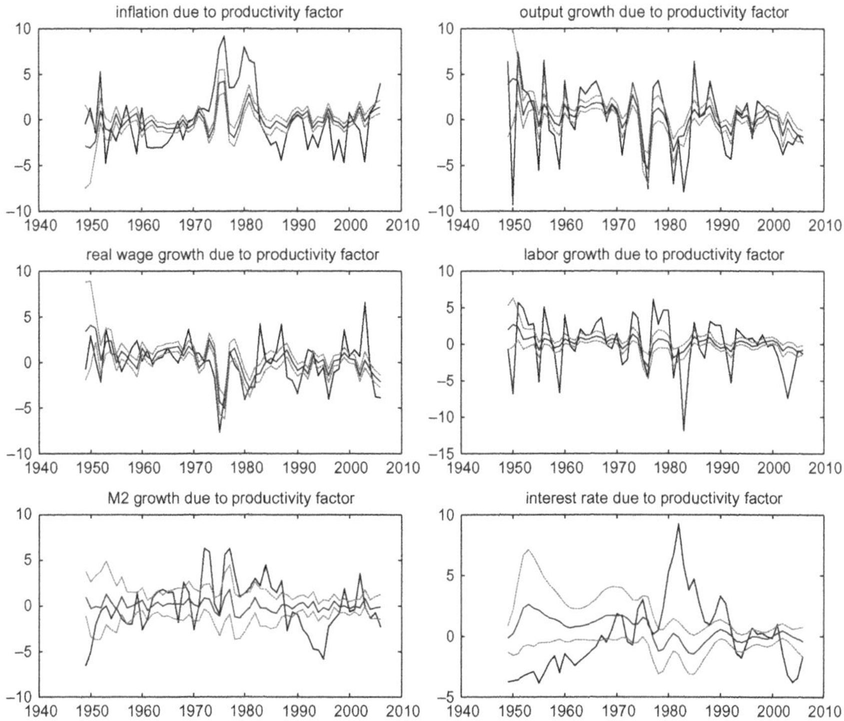

Fig. 7. Historical Decompositions of Aggregate Variables: Contribution of Productivity Factor. Actual (Sample Means Removed) and Mean, 10th, and 90th Percentiles of the Posterior Distribution.

the monetary factor to movements in these "aggregate" variables. From Fig. 6, one can observe that the monetary factor contributes very little to movement in aggregate output, real wage, and employment growth, but it does contribute substantially to the long swings in sectoral price growth, M2 growth, and the nominal interest rate. Clearly, the monetary factor reflects lower frequency movements in nominal variables – prices, money stock, interest rates (and implicitly nominal wages as well) – over the sample. In this respect, the monetary factor is quite similar to the "pure inflation" factor in Reis and Watson (2010). However, unlike Reis and Watson, the monetary factor does not imply a one-to-one movement in prices in every period, only in the long run. Recall that the common factor model did not impose the restrictions implied by long-run money neutrality on nominal interest rates, yet, nonetheless the monetary factor explains the run up in nominal interest rates in the 1970s and early 1980s and the run down of nominal rates in the late 1980s and 1990s.

Fig. 7 displays the contributions of the aggregate productivity factor to fluctuations in the aggregated sectoral variables. Here, the productivity factor explains a substantial fraction of the movements in the average output growth across sectors. In fact, declines in the aggregate productivity factor contribute to declines in output growth in nearly every single recession, particularly in 1974 and 1980. The productivity factor also contributes substantially to fluctuations in real wage growth, although perhaps not to the extent of its contribution to output growth. The effect of aggregate productivity on labor input growth, on the other hand, is much more modest than its contribution to output growth with perhaps only the decline in labor growth in 1974–1975 largely the result of aggregate productivity. Here the results are more in line with those of Gali (1999) in which technology shocks do not contribute much to employment fluctuations. While the productivity factor is not nearly so important in fluctuations of the nominal variables (prices, M2 growth, and interest rate), it does help contribute to the jump up in "aggregate" price growth in 1974–1975 and 1980–1981.

Fig. 8 displays the contribution of the third aggregate factor, which I will interpret as representing an aggregate demand factor. This demand factor contributes to a substantial amount of the fluctuations in output growth and most of the fluctuations in labor input. In fact, with the exception of 1974–1975 and 1980–1982 periods, this factor plays a key role in contributing to declines in output and labor input during recessions. On the other hand, the demand factor's contributions to real wage growth and short-run fluctuations in inflation are more modest than the contributions of the aggregate productivity factor. While the demand factor contributes little to the fluctuations in M2 growth, it does contribute to short-run swings in the nominal interest rate.

Fig. 9 shows the posterior distribution of the total contribution of the aggregate factors. The resulting unexplained fluctuations are the result of idiosyncratic factors for the sectors. Together the three aggregate factors contribute to a large portion of the fluctuations in the unweighted average of sector price, output, real wage, and labor input growth. Of course, taking the average of the sectoral variables tends to average away the contribution of the sectoral idiosyncratic factors for prices, output, real wages, and labor input tend leaving the bulk of "aggregate" fluctuations to be explained by the common factors. This does, however, suggest that the three common factors are sufficient to capture nearly all of the cross-sectoral co-movement. For M2 growth and interest rates, the idiosyncratic factors are more evident leaving a substantial amount of their fluctuations unexplained by the three aggregate factors.

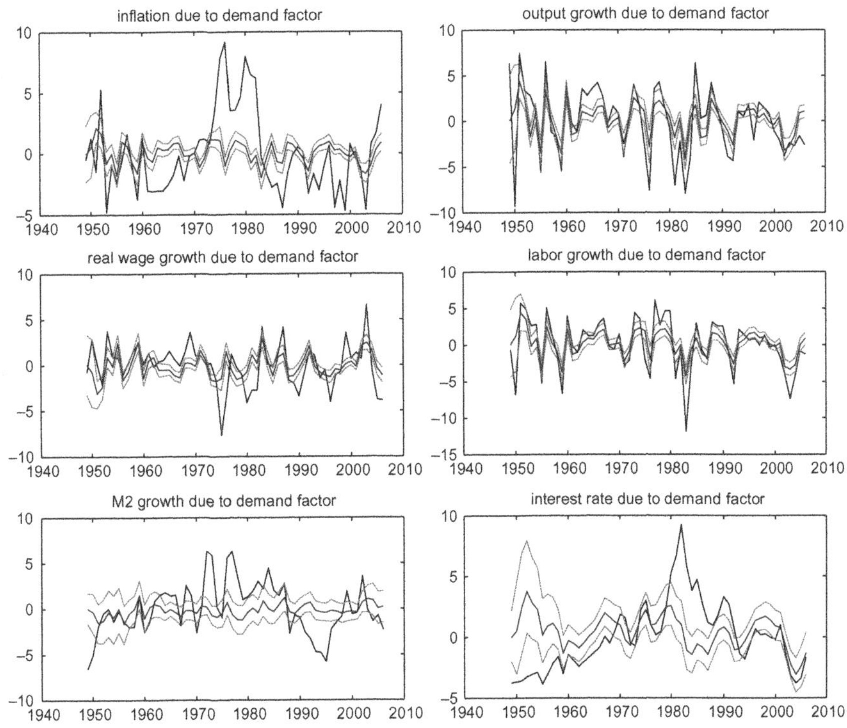

Fig. 8. Historical Decompositions of Aggregate Variables: Contribution of Third Factor. Actual (Sample Means Removed) and Mean, 10th, and 90th Percentiles of the Posterior Distribution.

Interpretation of Aggregate Factors

The estimated monetary factor appears to capture the long swings in nominal variables over the sample. As such, shocks to this monetary factor probably do not reflect so-called monetary policy shocks as typically discussed in the VAR literature. These "monetary policy shocks" typically identified in the VAR literature can be thought of as surprises or mistakes in the monetary authority's reaction function. In this chapter, shocks to the monetary factor might be better thought of as shocks to the long-run inflation target in something like a Taylor rule. Shocks in this long-run target will over time change inflation, interest rates, and the money growth rate by similar proportions. In the short run, however, these variables may respond differently to change in the long-run inflation target.

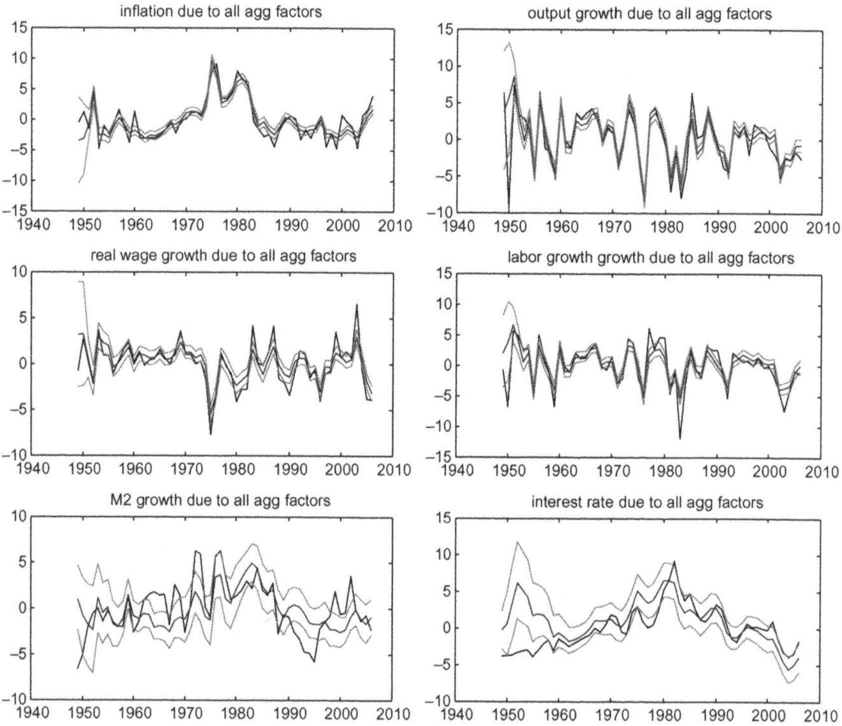

Fig. 9. Historical Decompositions of Aggregate Variables: Contribution of all Aggregate Factors. Actual (Sample Means Removed) and Mean, 10th, and 90th Percentiles of the Posterior Distribution.

The aggregate productivity factor was identified by assuming that a positive shock to this factor increased all sector real wages and outputs in the long run. Could this aggregate factor just be picking up business cycles driven by other phenomenon rather than aggregate productivity shocks? To get a sense whether the productivity factor is truly capturing productivity, we examine the actual unweighted average across sectors of the logarithm of labor productivity, $\log(Q_{i,t}/L_{i,t})$. Economic theory suggests that shocks to productivity should move sector (log) real wages and sector (log) average products of labor by similar amounts. In fact, if technology is Cobb–Douglas, the response of log real wages and log average products should be equal.[14] While nominal frictions and adjustment costs may prevent real wages and average products from moving together in the short run, there is

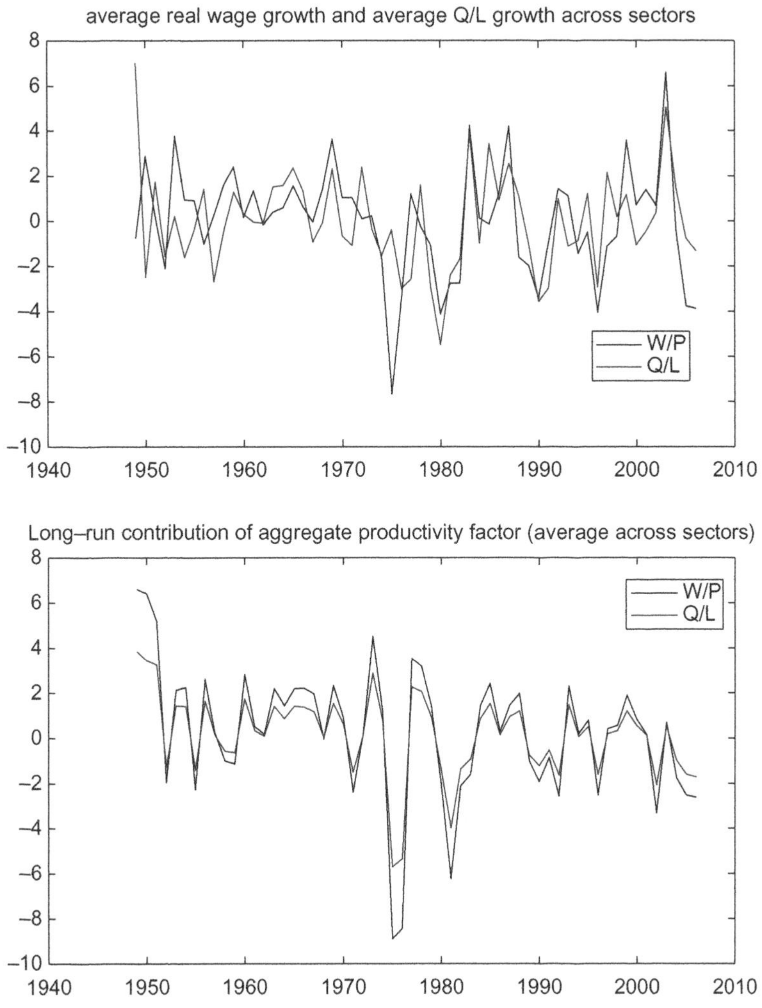

Fig. 10. Average Real Wage and Labor Productivity Across Sectors and Long-Run Contribution of Aggregate Productivity Factor.

strong a priori reasons to believe they are likely to move together in the long run in response to shocks in productivity.

The top panel of Fig. 10 displays the average growth rate across sectors of real wages and output per labor. While over the sample there are periods in which the two series move together, there is also a substantial portion of the sample in which to two series do not move in the same direction. The bottom panel of Fig. 10 plots the long-run contribution of the aggregate productivity factor to real wage growth, $\sum_{i=1}^{N}\eta_{i,w}s_{z,t}/N$, and output per labor growth, $\sum_{i=1}^{N}(\eta_{i,q}-\eta_{i,l})s_{z,t}/N$, where $\eta_{i,w}$, $\eta_{i,q}$, and $\eta_{i,l}$ are the long-run multipliers for sector i of a productivity shock for real wage, output, and labor given in Eq. (8). Unlike actual real wages and average products, the long-run contribution of the productivity factor to real wages and average products move together quite closely.[15] Furthermore, the correlation of the productivity factor and the unweighted average of sector KLEM Solow residual growth is 0.63. Taken together, it appears that the estimated productivity factor generates movements in real wages and average products consistent with aggregate productivity shocks.

EMPIRICAL RESULTS: SECTORAL VARIABLES

Sectoral Impulse Responses

Turning to what happens at the sectoral level, Fig. 11 displays the mean of the posterior distribution of the response to a monetary shock for every sector at various horizons. The horizontal axis displays the sector numbers referenced in Table 1. Unlike previous impulse response diagrams, these figures display the (log) level of sector variables so that one can better discern the effect of shocks on relative sectoral behavior. Looking at log level of sector prices first, one observes that a monetary shock affects sector prices broadly – all sector prices rise; there is no evidence of a price puzzle at the sectoral level. All prices, however, do not move one-for-one, suggesting that monetary shocks have relative price effects, at least in the short run. These relative price effects are small initially, rise as the horizon lengths, and then fall again as the horizon gets large (horizon = 20). Among the sectors with the largest price increases are: coal mining, crude oil and natural gas, gas utility services, and petroleum refining. Among the sectors with the smallest price increases are: communications, personal and business services, and printing.

Monetary factor shocks have different real effects across sectors in the short run as well, as some sectors increase while others decrease. Again the sectoral differences are modest at first, rise as the horizon lengthens, and then fall again as the horizon gets large. The sectors with the largest positive

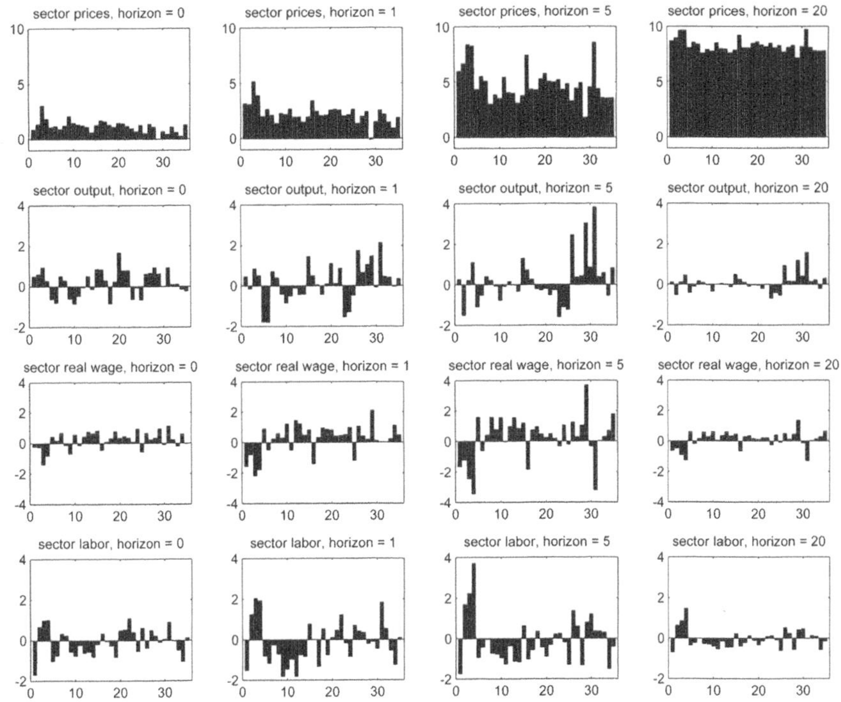

Fig. 11. Sectoral Impulse Responses to Monetary Factor Shock: Mean of the Posterior Distributions.

output response include: primary metals in the initial period, gas utility service, instruments, and communication in later periods. The sectors with the largest negative responses are apparel and footwear initially and electrical machinery, transportation equipment, and metallic ore mining at longer horizons. Interestingly the sectors with the largest real wage declines are typically those with the largest labor input increases (crude oil and natural gas, coal mining, gas utilities).

Fig. 12 displays the sectoral responses to an aggregate productivity shock. By construction, aggregate productivity shocks have a positive effect on all sectors' outputs and real wages in the long run. Nonetheless, there are substantial differences across sectors. Sectors like transportation equipment, fabricated metal, machinery, electrical machinery, and rubber show large increases in their outputs while agriculture, FIRE, and food show only small increases. Prices in most sectors decline with agriculture, lumber and wood

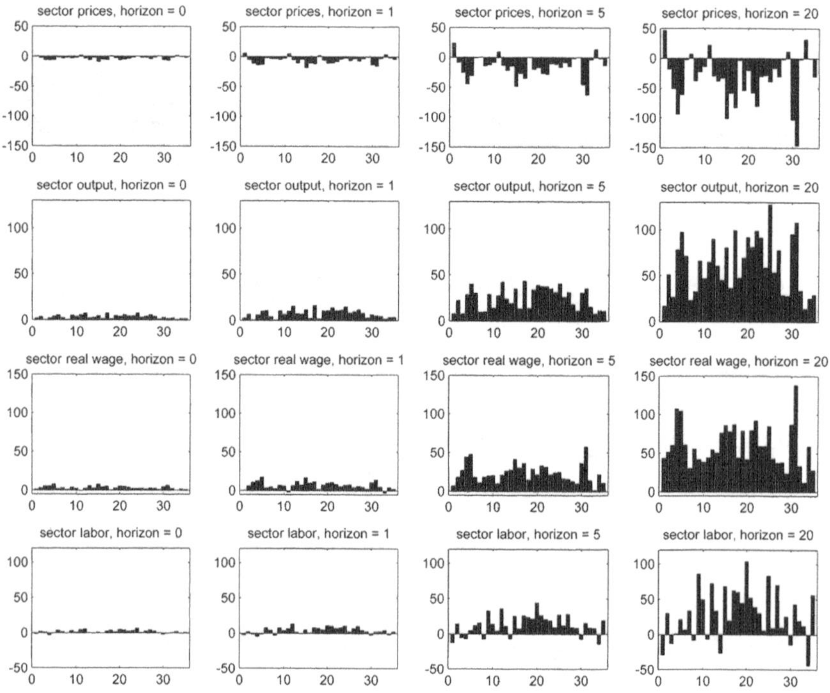

Fig. 12. Sectoral Impulse Responses to a Productivity Factor Shock: Mean of the Posterior Distributions.

products, and FIRE being notable exceptions. Labor input generally increases across sectors but falls in five sectors, the largest declines occurring in personal and business services, agriculture, crude oil and natural gas, and printing sectors.

Fig. 13 displays the sectoral responses to a shock in the third aggregate factor. Recall this factor's aggregate impulse responses were reminiscent of an increase in goods demand in an IS/LM model. Again the effect of a shock to this "aggregate demand" factor varies substantially across sectors. Among the sectors with the largest price responses are: primary metals and petroleum refining at short horizons and petroleum refining, chemicals, and nonmetallic minerals at longer horizons. The sectors with the largest output response include: machinery, electrical machinery, fabricated metals and primary metals while agriculture, FIRE, lumber and wood products, and construction display the smallest responses. The sectors with the largest

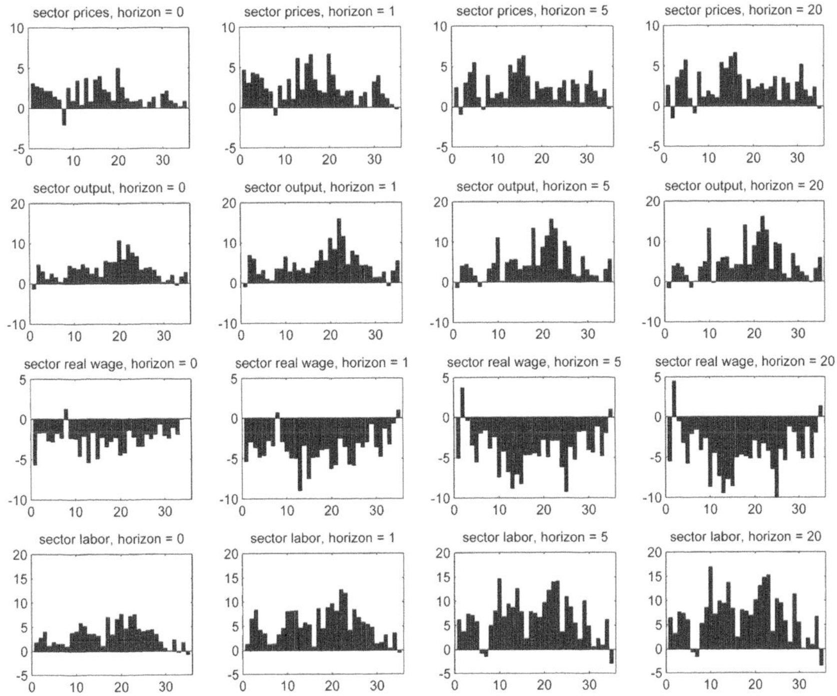

Fig. 13. Sectoral Impulse Responses to a "Demand" Factor Shock: Mean of the Posterior Distributions.

labor responses include many of the sectors with large output responses plus the apparel and printing sectors.

To get a sense of whether sector responses vary according to how the sectors' goods are used, I classified sectors according to whether they are used primarily as a consumption good, an investment good, or an intermediate good (see Table 1). Figs. 14–16 display the average sectoral responses (across good type) to the three types of aggregate shocks. From Fig. 14, it appears that the response to monetary shocks does not differ substantially across types of goods. Of the three types of goods, the investment good sector's output and labor input respond the most to a shock in the productivity factor (see Fig. 15). While the responses of output, real wages, and labor input do not vary much across good type for a demand shock (see Fig. 16), prices of intermediate goods appear to respond more than do the prices of consumption or investment goods.

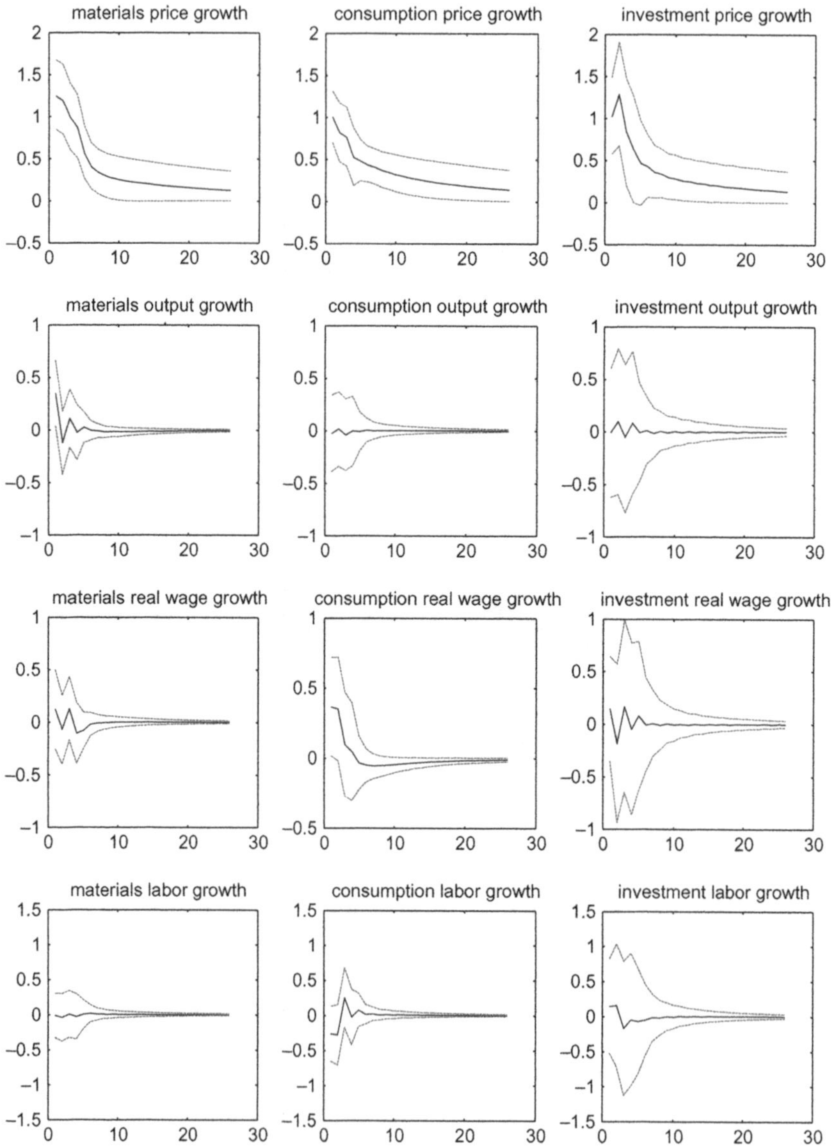

Fig. 14. Response to a Monetary Factor Shock Across Good Types: Mean, 10th, and 90th Percentiles of the Posterior Distributions.

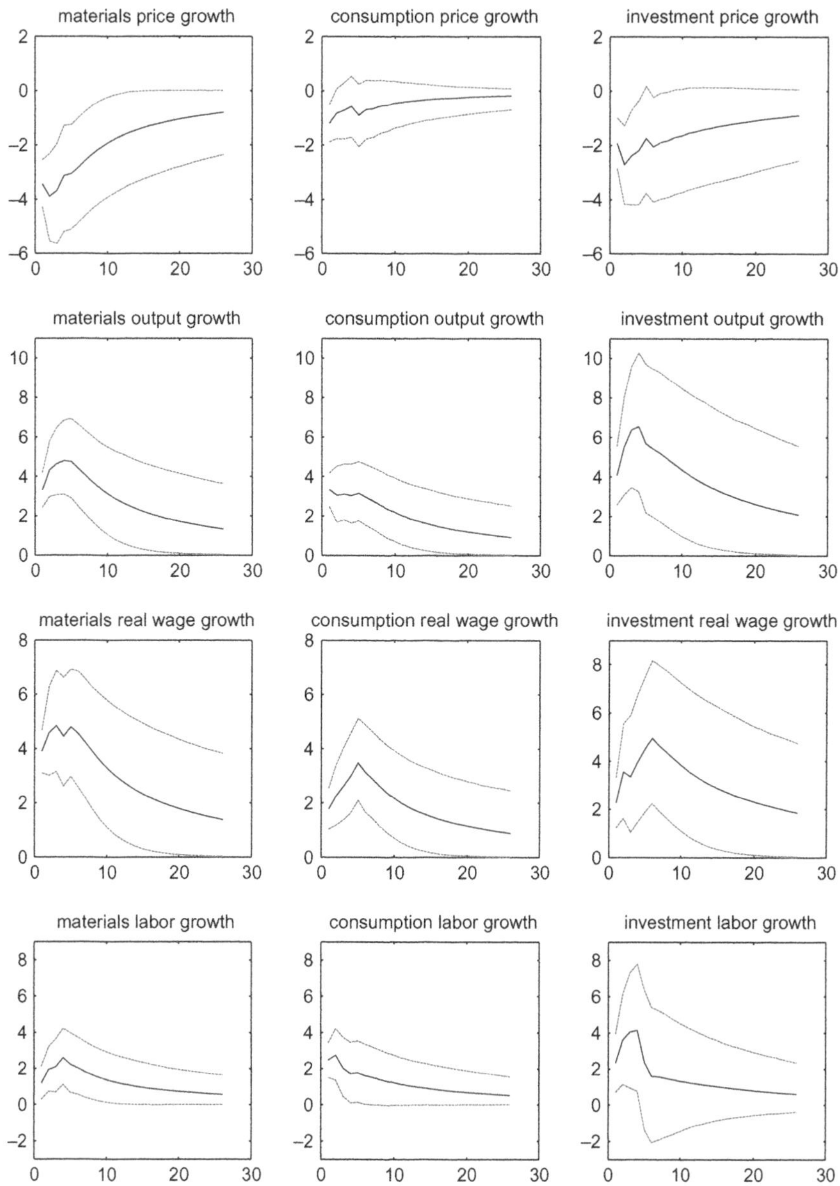

Fig. 15. Response to a Productivity Factor Shock Across Good Types: Mean, 10th, and 90th Percentiles of the Posterior Distributions.

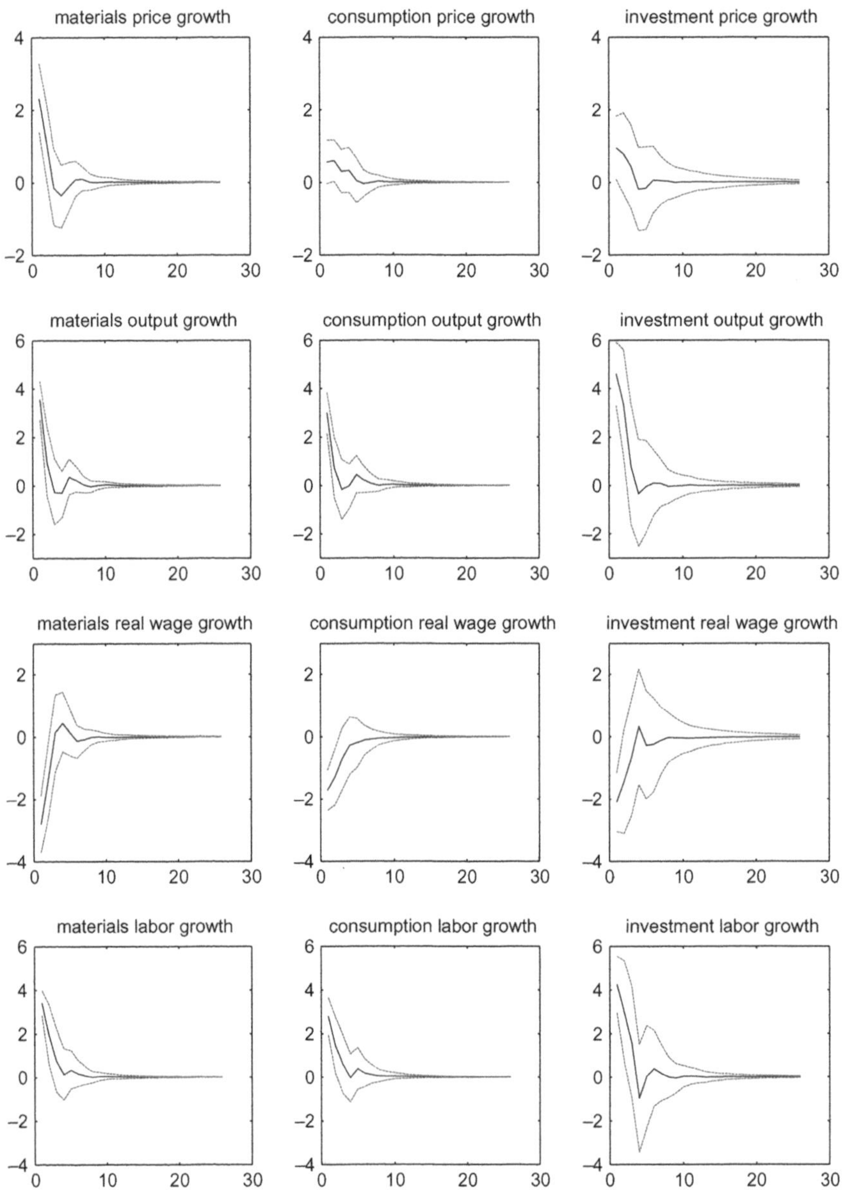

Fig. 16. Response to a "Demand" Factor Shock Across Good Types: Mean, 10th, and 90th Percentiles of the Posterior Distributions.

ROBUSTNESS

Alternative Short-Run Priors

While the identification of monetary shocks imposes long-run neutrality, one may want to examine the sensitivity of the results to alternative priors about the short-run responses of monetary shocks. In particular, I consider a prior distribution in which the responses of output, price, labor input, and interest rates to a monetary shock correspond, at least in the short run, to those typically employed in the VAR literature. In this literature, output responds positively and interest rates negatively to an expansionary monetary shock. There is less consensus about the initial response of prices to a monetary shock, witness the discussion in the literature of the "price puzzle." Here, I set the mean of the prior distribution for prices so that they do not respond in the initial period to a monetary shock as suggested by the VAR analysis of Christiano et al. (1999). Unlike the benchmark model in which the prior responses are fairly diffuse, I assume that the prior distribution around the so-called traditional VAR prior is relatively tight. For the response of real wages to a monetary shock, I continue to assume the prior distribution is centered around zero and is fairly diffuse.

Fig. 17 presents the means and 10th and 90th percentiles for the "traditional VAR" prior distribution. From Fig. 17, the mean of the prior distribution for the initial response of sectoral and aggregate output and labor input are positive while the mean of the prior distribution for the nominal interest rate response is negative. For individual sectors, the prior distribution is set so that a wide range of sectoral responses is still possible – for example, it is quite possible for an individual sectoral output to fall in response to an expansionary monetary shock. The aggregate (or unweighted mean response across sectors) responses, however, suggest the "traditional VAR" responses to a monetary shock – initially aggregate output (and labor input) rise and the nominal interest rate falls while the aggregate price level remains unchanged. Note that the prior distribution implies a "zig-zag" shape for the output and labor responses. The reason for this shape is that long-run neutrality implies that positive output and labor growth responses initially must be offset by subsequent negative responses.

Fig. 18 displays the posterior distribution of the impulse responses of aggregate variables to a monetary factor shock. Unlike the benchmark model, setting fairly tight priors results in both output and labor input growth rising in response to a monetary factor shock. This positive response

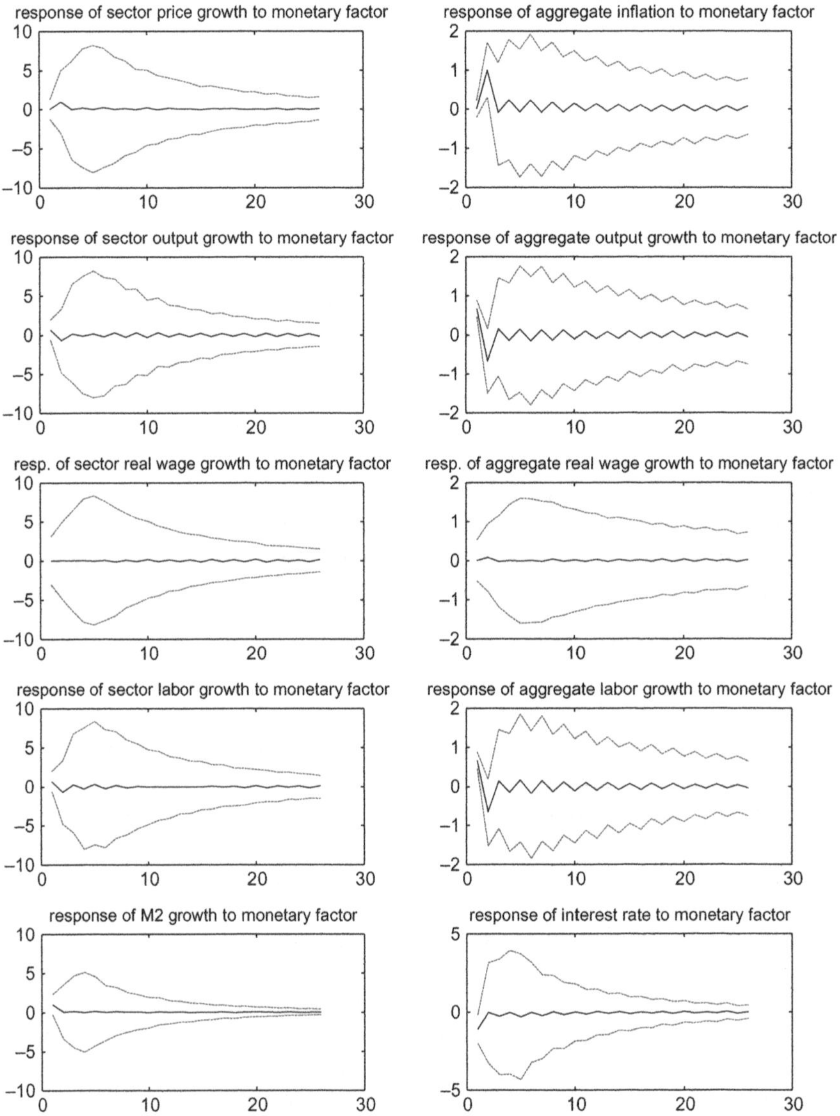

Fig. 17. Mean, 10th, and 90th Percentiles of Prior Distribution of Responses to Monetary Factor Shock for "Traditional" VAR Prior.

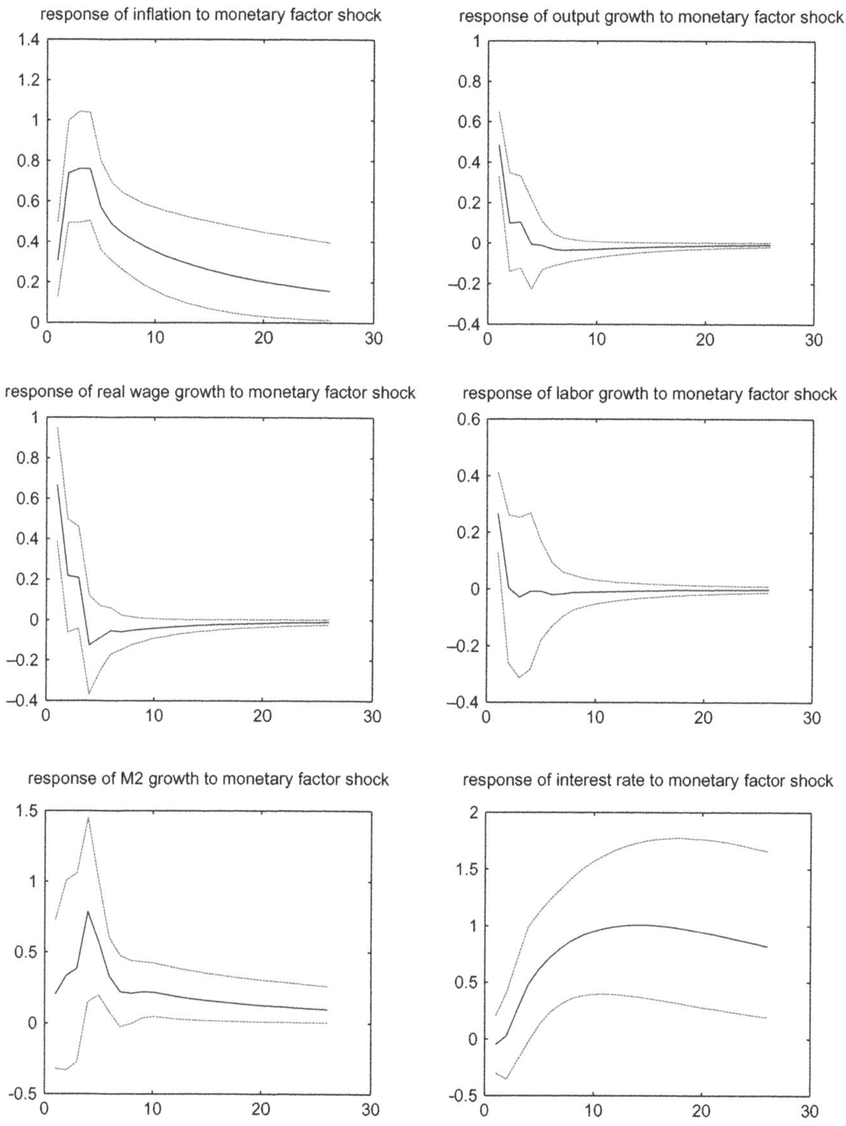

Fig. 18. Mean, 10th, and 90th Percentiles of Posterior Distributions for Impulse Responses to a Monetary Factor Shock for "Traditional" VAR Prior.

is short-lived as long-run neutrality requires this initial positive response in growth to be followed by negative responses. Given that the results for the short-run monetary priors are quite different, this suggests that the data alone are not too informative about the short-run response of output to a long-run neutral monetary shock. On the other hand, despite the relatively tight prior that the inflation response is zero, the posterior distribution implies a strong positive response of inflation to a monetary factor shock. Clearly, the data suggest a strong initial price response to a shock in the monetary factor. Similarly, despite a relatively tight prior around a negative interest rate response, the posterior distribution for nominal interest rate response is close to zero for the initial response and positive shortly thereafter. Finally, for the initial response of real wage growth, the mean of the posterior distribution is negative, although the interior 80 percentiles of the posterior distribution includes zero. The general tendency for real wages to fall is in line with what traditional models of wage stickiness would predict for a monetary shock.

Perhaps not surprisingly, the alternative prior distribution does not substantially change the impulse responses to a productivity shock (not shown). Similarly, the "traditional VAR" prior has only a small effect on the means of the posterior distribution for the responses to a shock to the third aggregate factor (demand). However, under the "traditional VAR" prior, the posterior distributions of responses to the "demand" factor are more diffuse.

Fig. 19 displays the contribution of the monetary factor to the historical decomposition of the aggregated sectoral variables when the "traditional VAR" prior is used. Comparison of these decompositions with those of the monetary neutral prior (Fig. 6 above) suggests that assuming non-neutral short-run priors does not change the interpretation of the main sources of fluctuations over our sample. Regardless of whether short-run priors imply monetary non-neutrality or neutrality, most of the long swings in prices, money stock, and interest rates are due to the monetary factor. On the other hand, the monetary factor does not contribute much to fluctuations in real output, real wage, or labor growth over the sample regardless of the short-run monetary prior. Likewise, assuming non-neutral priors does not appreciably change the contributions of the other factors. This suggests that it is the long-run identifying assumptions and the data rather than the short-run priors that are primarily responsible for the historical decomposition results.

Fig. 20 displays the mean of the posterior distribution for all the individual sector variables' responses to a monetary factor shock when

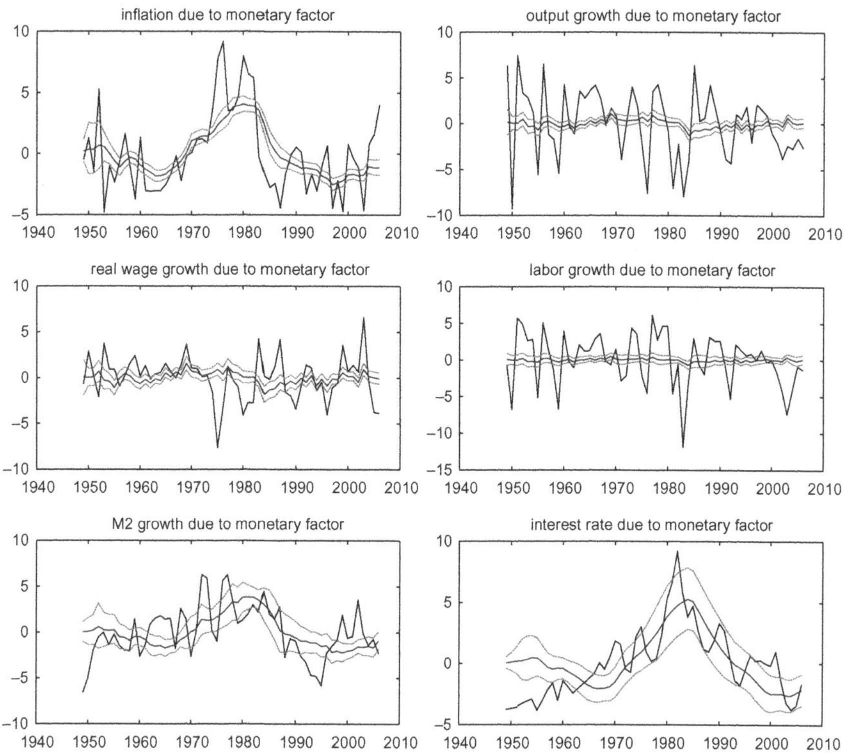

Fig. 19. Historical Decompositions of Aggregate Variables for "traditional" VAR Prior: Contribution of Monetary Factor. Actual (Sample Means Removed) and Mean, 10th, and 90th Percentiles of the Posterior Distribution.

non-neutral priors are assumed. As in the benchmark model, there are substantial quantitative differences in the sectoral responses to a monetary shock when the alternative, non-neutral monetary priors are used. Unlike the benchmark model, the alternative priors result in widespread increases in output and labor input across sectors during the initial period after a monetary factor shock. Much like the benchmark model, the sectors with greatest output growth increases were primary metal manufacturing and transportation initially and gas utility service and chemicals later. While prior distributions for the initial sectoral price responses are centered around zero, the posterior mean for nearly all the sectors is positive. Thus, the data (and not the priors) suggests that the inflationary effect of the monetary factor is widespread across sectors. The sectoral effects of

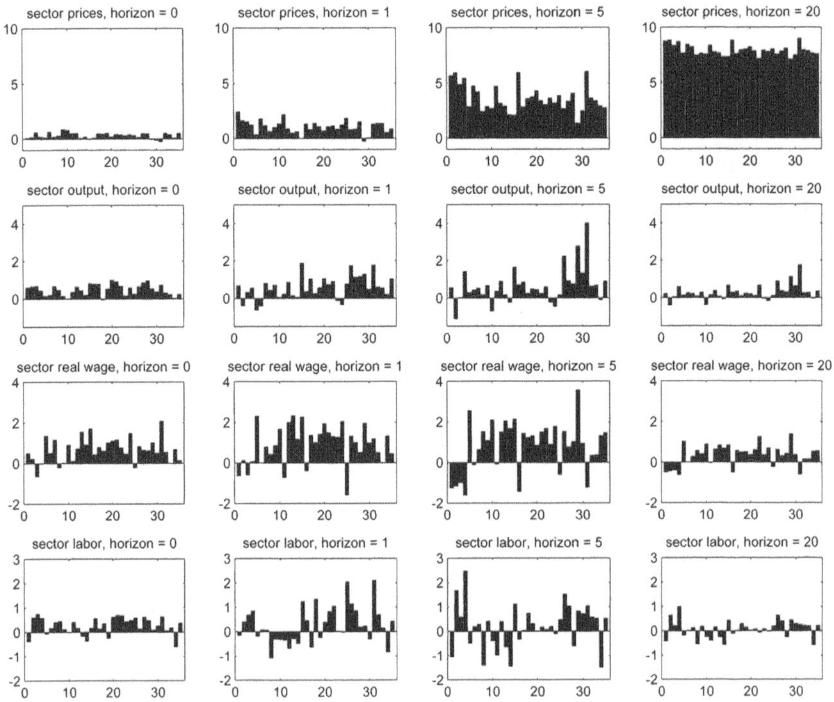

Fig. 20. Sectoral Impulse Responses to a Monetary Factor Shock for "traditional" VAR Prior: Mean of the Posterior Distributions.

productivity factor shocks and the demand factor shocks for the alternative short-run monetary prior is similar to the benchmark model.

Adding Additional Aggregate Factors

To test the sensitivity of the initial assumption of three aggregate factors, I added an additional aggregate factor to the common factor model. This factor was normalized so that it resulted in an increase in M2 growth in the initial period, but was otherwise unrestricted. Like the demand factor in the three-factor model, the priors of the other parameters were centered around zero and relatively diffuse. It turns out adding this factor does not alter the previous results for the monetary and productivity factors. It does seem to reduce the effect of what was termed the demand factor in the previous

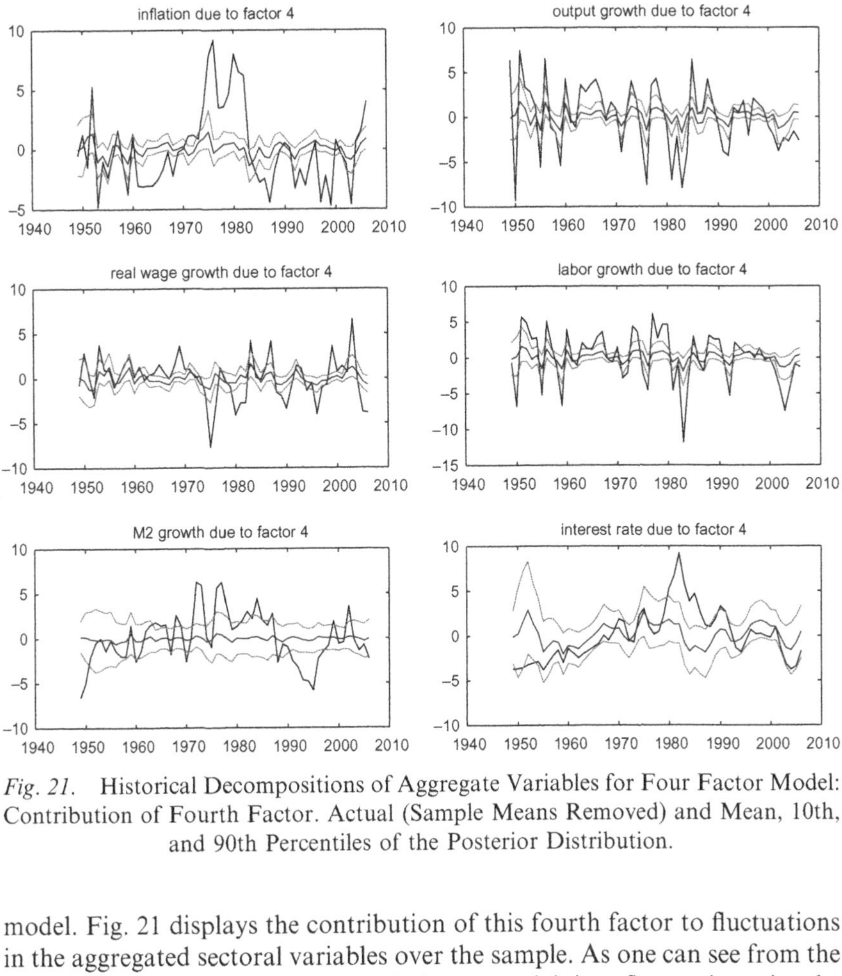

Fig. 21. Historical Decompositions of Aggregate Variables for Four Factor Model: Contribution of Fourth Factor. Actual (Sample Means Removed) and Mean, 10th, and 90th Percentiles of the Posterior Distribution.

model. Fig. 21 displays the contribution of this fourth factor to fluctuations in the aggregated sectoral variables over the sample. As one can see from the figure, this factor adds only a little to explaining fluctuations in the aggregated sectoral variables.

OIL PRICES, SECTORAL, AND AGGREGATE FLUCTUATIONS

The previous results suggested that the aggregate productivity factor had important contributions to both sectoral and aggregate output fluctuations.

In particular, the decline in output during the recessions of 1974–1975 and 1980 show up largely as due to fluctuations in the aggregate productivity factor. It is possible that this result might actually reflect the effect of oil price changes that also occurred during these time periods. Indeed, oil prices have been considered to be an important source of fluctuations in their own right.[16] To determine whether the estimated aggregate productivity factor was in fact picking up the effect of oil prices on sectoral economic activity, I add oil prices to the observation vector and add an additional factor to the benchmark model.

To be consistent with the other sectoral data, I include the growth rate in nominal oil prices to the vector of observables. Identification of the monetary and productivity factors is the same as the benchmark model above. Like the other nominal variables, I assume that the monetary factor affects the (log) level of nominal oil prices one-for-one in the long run so that the real oil price is unaffected in the long run by the monetary factor. Also like the other nominal prices, I place no restrictions on the long-run effect of the aggregate productivity factor on nominal oil prices. In addition to monetary and productivity factors, I include two additional common factors. One factor is normalized so that it has a positive effect on the interest rate in the current period; this will turn out to be similar to the so-called demand factor in the benchmark model. The other is normalized so that it has a positive effect on nominal oil prices in the current period; I will interpret this factor as an oil factor. As in the benchmark model, there are no additional restrictions on the factor loadings aside from these normalizations. I present the results from model with "traditional VAR" priors about the impact effects of the monetary factor; these results are largely unaffected if short-run money neutrality priors were used instead.

First, how are the results for monetary and productivity factors affected by including oil prices and another factor into the benchmark model? Figs. 22 and 23 display the contribution of the monetary and productivity factors to the historical decompositions of "aggregate" inflation, output, real wage, and labor growth for the model with oil prices. Adding oil prices and an "oil factor" to the model, does not appreciably change the relative contributions of the monetary and productivity factors. Monetary factor still contributes to the long swings in nominal variables over the sample and the productivity factor is an important contributor to output and real wage fluctuations. Even in the 1974 and 1980 recessions, the aggregate productivity factor is still an important source of output declines. Thus, adding oil prices and an oil factor do not appreciably affect the previous results for the monetary and productivity factors.[17] Fig. 24 displays the contribution of

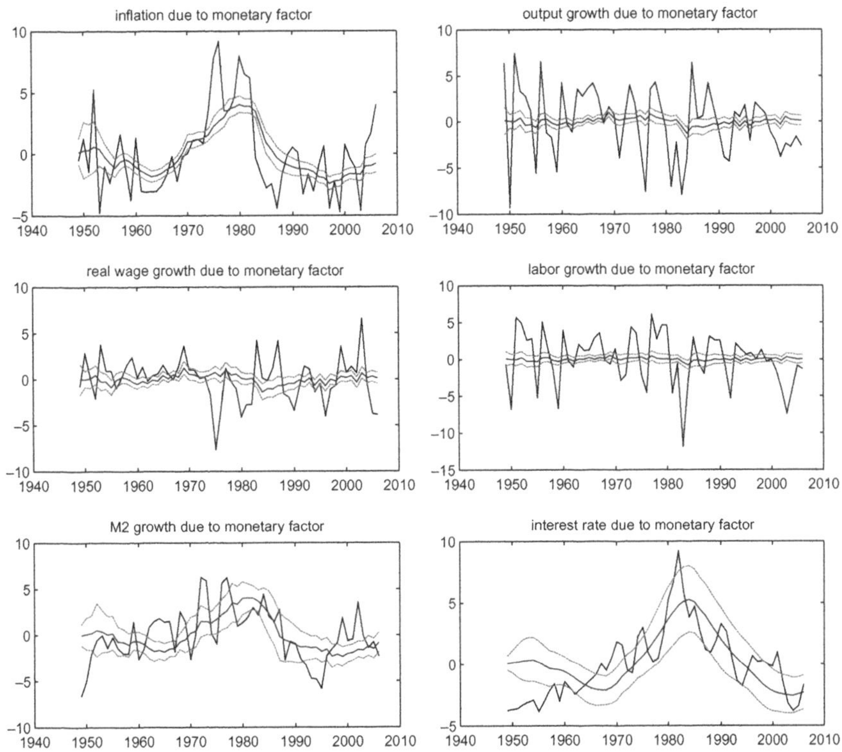

Fig. 22. Historical Decompositions of Aggregate Variables for Four Factor Model that Includes Oil Prices and Traditional VAR Prior. Actual (Sample Means Removed) and Mean, 10th, and 90th Percentiles of the Posterior Distribution.

the oil factor to historical fluctuations in the aggregated sectoral variables. From Fig. 24 one observes that the oil factor has only a moderate contribution to output and labor fluctuations and an even smaller contribution to inflation and real wage growth.

Does the oil factor shock look like an "oil shock"? Typically, oil shocks are envisioned to have a negative affect on output and employment and a positive effect on inflation. Fig. 25 displays the response of the aggregated sectoral variables to a shock in the oil factor. As is typical in the oil shock literature, a shock to the oil factor does result in a decrease in output and labor growth (remember, by assumption, an oil factor shock has a positive effect on current oil prices). However, unlike the literature, inflation

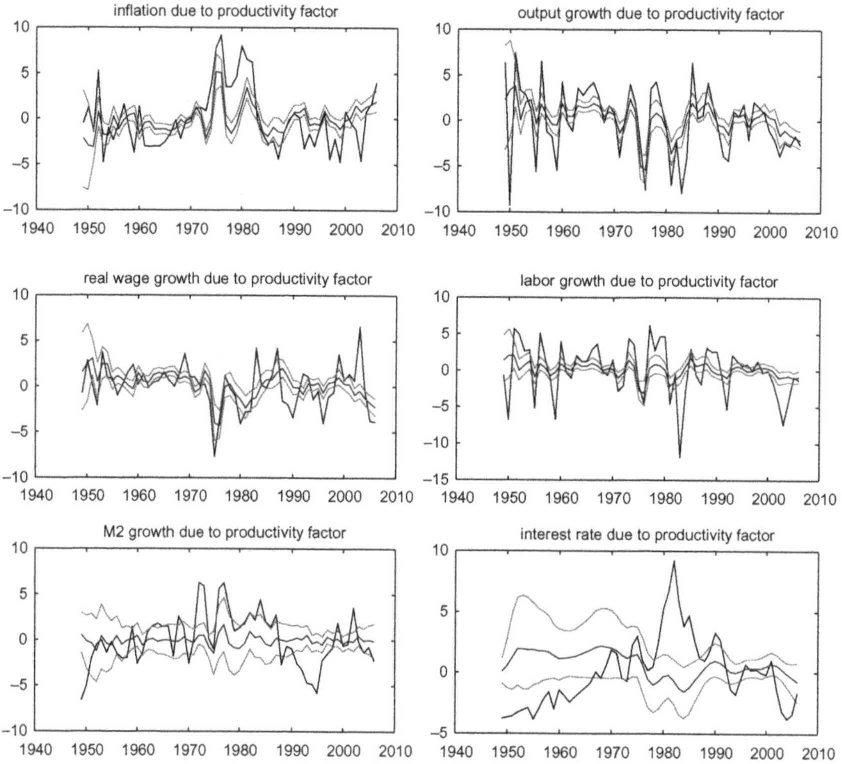

Fig. 23. Historical Decompositions of Aggregate Variables for Four Factor Model with Oil Prices and Traditional VAR Prior. Actual (Sample Means Removed) and Mean, 10th, and 90th Percentiles of the Posterior Distribution.

responds negatively to an oil price shock. This suggests a shock to the oil factor acts more like an aggregate demand shock rather than an aggregate supply shock.

Fig. 26 displays the individual sector responses to a shock in the oil factor. The sector responses also suggest that an oil factor shock acts more like a demand shock than a typical supply shock. Prices, output, and labor across a wide group of sectors fall in response to a shock in the oil factor. Only prices in the crude oil, petroleum, and gas utility sectors rise in response to an oil factor shock while output and labor in the crude oil, FIRE, and, interestingly, construction sectors rise.[18] These results are consistent with

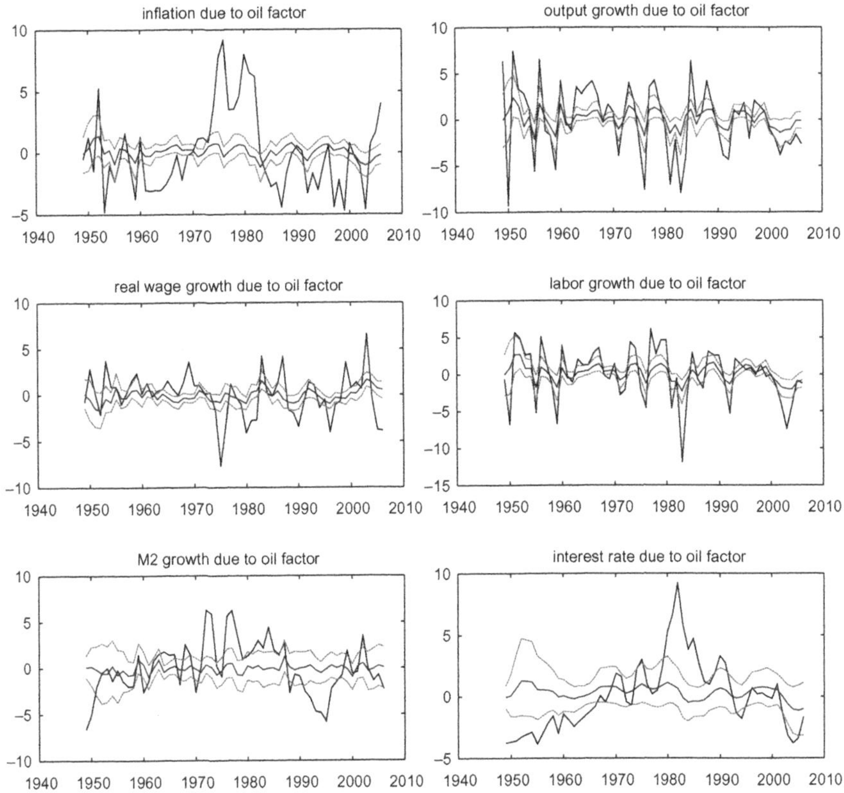

Fig. 24. Historical Decompositions of Aggregate Variables for Four Factor Model with Oil Prices and Traditional VAR Prior. Actual (Sample Means Removed) and Mean, 10th, and 90th Percentiles of the Posterior Distribution.

the demand channel for transmission of oil price increases emphasized recently by Edelstein and Kilian (2009) and Herrera (2009).[19] My sectoral results do not suggest a widespread supply channel for oil prices. The sectors hardest hit, in terms of output and labor declines, by an oil factor shock include primary metals manufacturing, machinery, and autos at short to medium horizons and textiles and apparel at longer horizons.

Given that the oil factor is only a modest contributor to fluctuations in "aggregate" variables, to what extent are oil prices themselves are driven by fluctuations in other factors? Fig. 27 displays the response of oil price

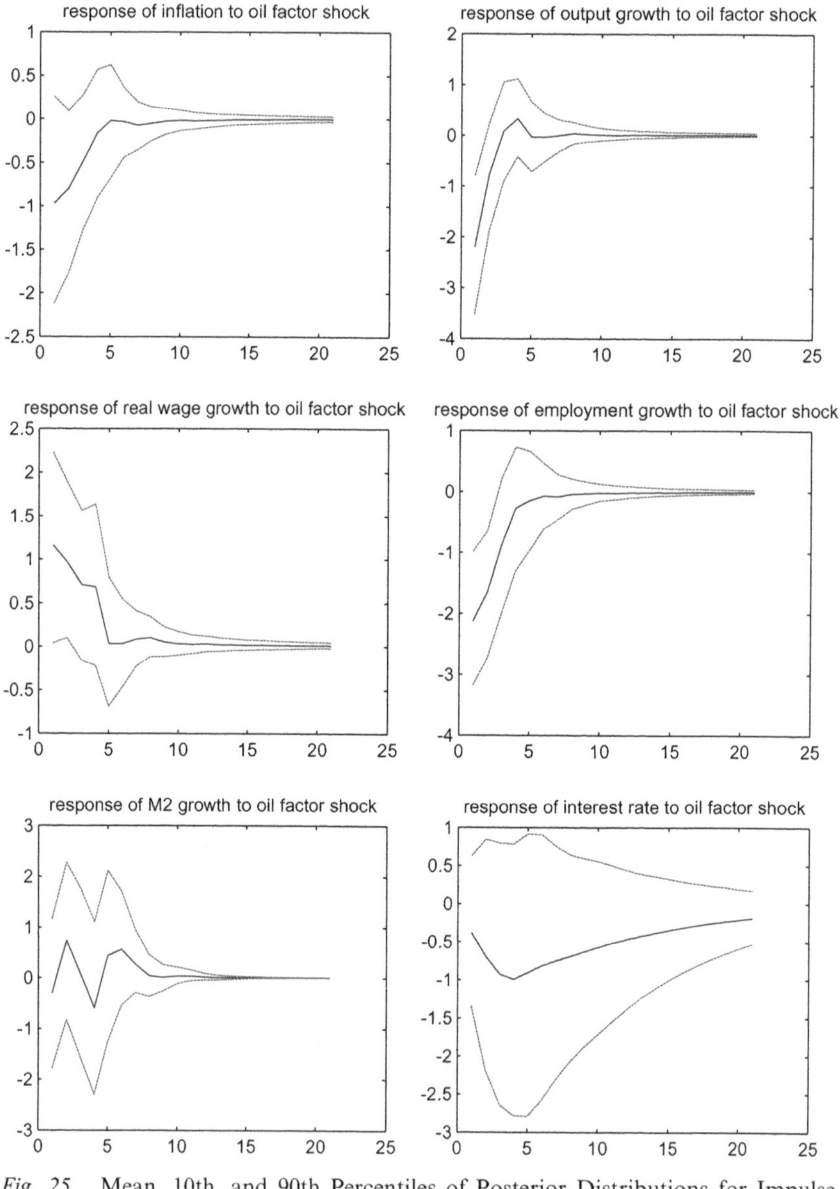

Fig. 25. Mean, 10th, and 90th Percentiles of Posterior Distributions for Impulse Responses of Aggregate Variables to a Oil Factor Shock for "Traditional" VAR Prior.

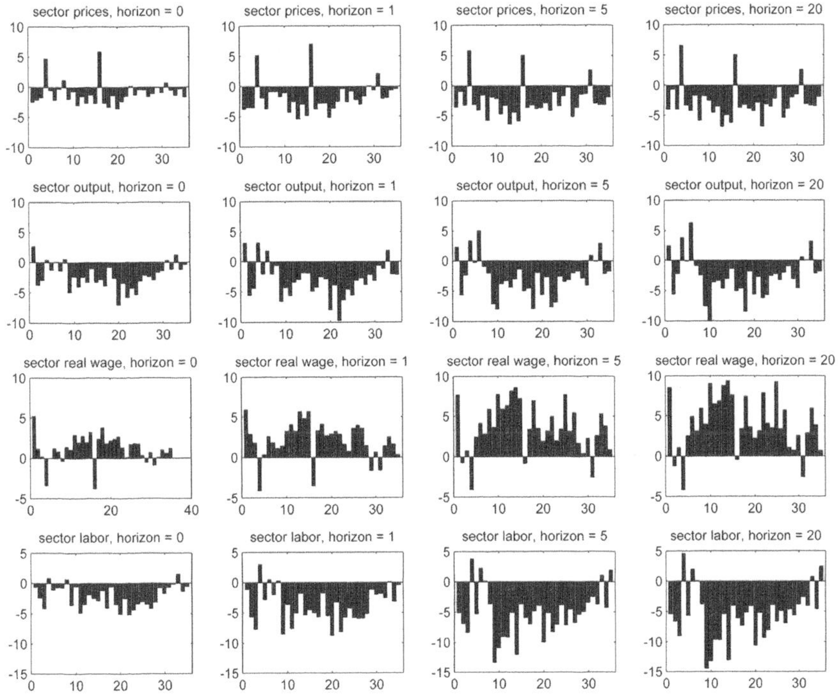

Fig. 26. Sectoral Impulse Responses to a Oil Factor Shock: Mean of the Posterior Distributions.

changes to shocks in the other aggregate factors. Here there is a strong response of oil price changes to monetary and particularly "demand" factor shocks, but little effect of productivity factor shocks on oil prices. Interestingly, the response of oil prices to an oil factor shock and to a demand factor shock look very similar (but the responses of the other variables look very different). Fig. 28 displays the historical decomposition of (nominal) oil price changes over the sample. While the monetary and productivity factors explain very little of the oil price fluctuations over the sample, the demand factor explains a large portion of the oil price fluctuations. In particular, the oil price increases at the end of our sample are driven primarily by the aggregate demand factor and not by the oil factor. On the other hand both the oil price and aggregate demand factors contribute to the 1974–1975 oil price spike while the oil price factor is largely responsible for the oil price increases in 1980 and in 1990. These results

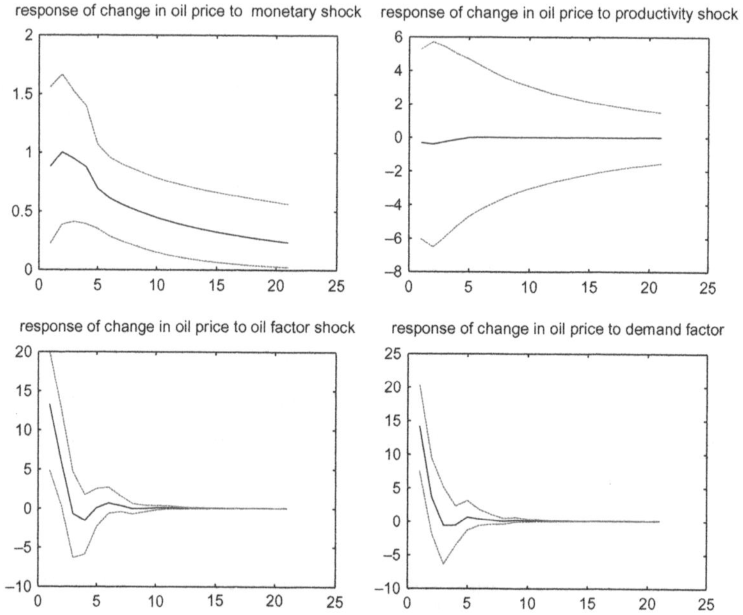

Fig. 27. Response of Change in Oil Prices to Factor Shocks (Traditional VAR Prior).

suggest that a large fraction of oil price fluctuations are the result of oil demand rather than oil supply fluctuations. This is consistent with results of Barsky and Kilian (2002) and Kilian (2009) who have argued for an important demand component in oil price fluctuations.

CONCLUSION

In this chapter, I used matched sectoral data on output, prices, real wages, and labor input to identify and quantify the effect of aggregate shocks on sectoral economic activity. Using the restrictions of long-run monetary neutrality to help identify monetary shocks and long-run sign restrictions on sector outputs and real wages for productivity shocks, I estimated aggregate monetary and productivity factors from sectoral data. Much of the common low frequency movements in sector price growth, M2 growth, and interest rates are captured by the estimated monetary factor. This factor, however,

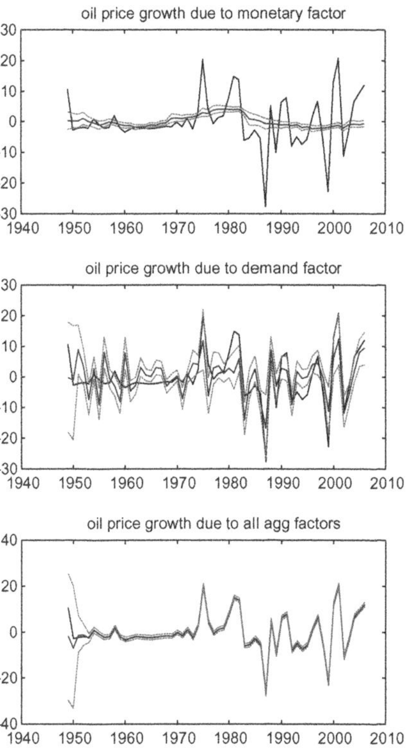

Fig. 28. Historical Decompositions of Oil Price Changes for Four Factor Model with Oil Prices and Traditional VAR Prior. Actual (Sample Means Removed) and Mean, 10th, and 90th Percentiles of the Posterior Distribution.

explains little of the fluctuations in output growth, real wages, or labor growth. Shocks to the productivity factor have widespread effects across sectors and, in contrast to the results of Gali (1999) and Francis and Ramey (2005), are associated with increases in labor input both at the aggregate and sectoral level. On the other hand, historical decompositions suggest that the productivity factor had only a modest contribution to labor fluctuations over the sample. Finally, results for monetary and productivity factors were robust to the inclusion of oil prices and an oil factor in the model.

Among the findings of this chapter is that what looked like an aggregate demand factor played an important role both in aggregate and sectoral fluctuations. The source of and the economic significance of this demand factor are largely unexplored in this chapter. To what extent this factor represents autonomous demand, fiscal policy, or financial and credit market effects is an interesting question for future research.

NOTES

1. See Christiano et al. (1999, 2005) for an extensive review of the VAR literature on quantifying the effect of monetary shocks on economic activity.
2. See for example, Gali (1999), Francis and Ramey (2005), and Basu, Fernald, and Kimball (2006).
3. Using long-run restrictions to identify structural shocks is not uncontroversial, see for example, the critique of Faust and Leeper (1998), Chari, Kehoe, and McGrattan (2008) and Dupor and Kiefer (2008).
4. Besides helping to identify aggregate shocks, the response of disaggregated variables to aggregate shocks is interesting in its own right. Many recent theories of nominal rigidities suggest that relative price effects are central to the effect of money on real economic activity and the welfare consequences of inflation (Woodford, 2001).
5. There is also an empirical literature that examines the relative price implications of monetary shocks. Hercowitz (1982) examined how the dispersion in cross-section distribution of prices changed in response to anticipated and unanticipated money changes. Bils, Klenow, and Kryvtsov (2003) examined how the response of prices differs depending on whether the price is classified as being in a flexible price versus a sticky price. Balke and Wynne (2007) find using 8 digit monthly PPI data that monetary shocks have substantial relative price effects. Lastrapes (2006), using more aggregated price indices than Balke and Wynne finds that monetary shocks have persistent effects on the cross-section distribution of prices.
6. The sum of the 35 sectors' output equals GDP.
7. There are a total of 140 sectoral variables yet only 57 observations for these variables. As a result, the variance/covariance matrix of all 140 variables is singular.
8. While the Bai and Ng (2002) criteria suggested two static factors, I start with three common factors in the benchmark dynamic factor model. I do this for two reasons. First, the monetary and productivity factors impose substantial restrictions across sectors and adding another factor might be useful capturing the remaining co-movement across sectors. Second, it seems plausible from economic grounds that there could be important aggregate sources of sector co-movement other than monetary or aggregate productivity shocks.
9. Long-run monetary neutrality can also imply that nominal interest rates are unchanged in the long run so long as a monetary shock does not change the long-run inflation rate.
10. Lastrapes and Loo (1998) examine the effect on money shocks on industry level output while Lastrapes (2006) examines the effect of money shock on relative prices. Both these papers use the long-run identifying assumption that a monetary shock has no effect on aggregate real output in the long run but no such restriction is implied for individual sector outputs or relative prices.

11. While an aggregate labor supply shock could also conceivably move real wages across sectors in the same direction, sectoral output would move in the opposite direction from real wages.

12. For all 35 sectors standard augmented Dickey–Fuller tests (with constant and time trend) fail to reject unit root in log prices and log real wages while for log output the unit root null is rejected in only 5 sectors and for log labor input in 6 sectors. Note that aggregate output and labor input (summing up gross output and labor input across sectors) also appear to contain a unit roots.

13. As the responses in Fig. 1a represent the response to a unit shock in the monetary factor ($v_{m,t} = 1$), these responses do not reflect uncertainty about the size of a typical monetary shock, σ_m. The responses below and their percentiles do reflect this uncertainty about the value of σ_m.

14. Recall, I am deflating nominal wages by its product price and not price of households' consumption goods.

15. Note that while $\eta_{i,w}$ and $\eta_{i,q}$ were restricted to be positive, there was no restrictions on $\eta_{i,l}$ nor on the relative sizes of these long-run multipliers in the estimation.

16. Many studies including Hamilton (1983), Mork (1989), and Hamilton (2009) have emphasized the importance of oil prices for aggregate fluctuations. Other papers have questioned the importance of oil price shocks on economic activity. These include Hooker (1996), Bernanke, Gertler, and Watson (1997), and Barsky and Kilian (2002) among others. Most of these papers examined the effect of oil prices on aggregate economic activity. Davis and Haltiwanger (2001), Lee and Ni (2002), and Herrera (2009), among others, examine the effect of oil prices on sectoral level data.

17. Adding oil prices and an oil factor also do not substantively change the impulse response functions for monetary, productivity, and demand factor shocks.

18. Output in the agriculture sectors rises but not employment, while labor rises in government enterprises but not output.

19. Herrera (2009) also found strong and quick responses of energy-intensive sectors and autos to oil price shocks.

ACKNOWLEDGMENTS

I thank an anonymous referee, participants at the Advances in Econometrics conference held in honor of Thomas B. Fomby and Carter Hill as well as seminar participants at Southern Methodist University, Washington University, The Federal Reserve Bank of Dallas, Texas Tech University, University of California at Riverside, and Sam Houston State University for comments. Thanks to Anna Marie Herrera, David VanHoose, Jeremy Piger, Mark Wynne, and Evan Koenig for useful conversations about this project. The views expressed are those of the author and do not necessarily reflect the views of the Federal Reserve Bank of Dallas or the Federal Reserve System.

REFERENCES

Ahmadi, A., & Uhlig, H. (2007). *Measuring the dynamic effects of monetary policy shocks: A Bayesian FAVAR approach with sign restrictions*. Unpublished manuscript, University of Chicago.

Bai, J., & Ng, S. (2002). Determining the number of factors in approximate factor models. *Econometrica, 70*(1), 191–221.

Balke, N. S., & Wynne, M. A. (2007). The relative price effects of monetary shocks. *Journal of Macroeconomics, 29*(1), 19–36.

Barro, R. J. (1976). Rational expectations and the role of monetary policy. *Journal of Monetary Economics, 2*(1), 1–32.

Barsky, R., & Kilian, L. (2002). Do we really know that oil caused the great stagflation? A monetary alternative. In B. S. Bernanke & K. Rogoff (Eds.), *NBER Macroeconomics Annual 2001* (pp. 137–183). Cambridge, MA: MIT Press.

Basu, S., Fernald, J. G., & Kimball, M. S. (2006). Are technological improvements contractionary? *American Economic Review, 96*(5), 1418–1448.

Bernanke, B., Boivin, J., & Eliasz, P. S. (2005). Measuring the effects of monetary policy: A factor-augmented vector autoregressive (FAVAR) Approach. *Quarterly Journal of Economics, 120*(1), 387–422.

Bernanke, B., Gertler, M., & Watson, M. (1997). Systematic monetary policy and the effects of oil price shocks. *Brookings Papers on Economic Activity, 1997*(1), 91–157.

Bils, M., Klenow, P. J., & Kryvtsov, O. (2003). Sticky prices and monetary policy shocks. *Quarterly Review – Federal Reserve Bank of Minneapolis, 27*(1), 2–8.

Boivin, J., Giannoni, M. P., & Mihov, I. (2009). Sticky prices and monetary policy: Evidence from disaggregated U.S. data. *American Economic Review, 99*(1), 350–384.

Bryan, M. F., & Cecchetti, S. G. (1993). The consumer price index as a measure of inflation. *Federal Reserve Bank of Cleveland Economic Review, 29*(4), 15–24.

Canova, F., & De Nicolo, G. (2002). Monetary disturbances matter for business fluctuation in the G-7. *Journal of Monetary Economics, 49*, 1131–1159.

Carter, C. K., & Koln, P. (1994). On Gibbs sampling for state space models. *Biometrika, 81*(3), 541–553.

Chari, V. V., Kehoe, P. J., & McGrattan, E. R. (2008). Are structural VARs with long-run restrictions useful in developing business cycle theory? *Journal of Monetary Economics, 55*(8), 1337–1352.

Christiano, L. J., Eichenbaum, M., & Evans, C. L. (1999). Monetary policy shocks: What have we learned and to what end? In J. B. Taylor & M. Woodford (Eds.), *Handbook of macroeconomics Volume 1A*. Amsterdam: Elsevier Science.

Christiano, L. J., Eichenbaum, M., & Evans, C. L. (2005). Nominal rigidities and the dynamic effects of monetary policy. *Journal of Political Economy, 113*(1), 1–45.

Davis, S. J., & Haltiwanger, J. (2001). Sectoral job creation and destruction responses to oil price changes and other shocks. *Journal of Monetary Economics, 48*(3), 465–512.

Dupor, B., & Kiefer, L. (2008). *Executing long-run restrictions*. Unpublished paper, Ohio State University.

Edelstein, P., & Kilian, L. (2009). How sensitive are consumer expenditures to retail energy prices? *Journal of Monetary Economics, 56*(6), 766–779.

Faust, J. (1998). The robustness of identified VAR conclusions about money. *Carnegie-Rochester Conference Series on Public Policy, 49*, 207–244.

Faust, J., & Leeper, E. (1998). When do long-run identifying restrictions give reliable results. *Journal of Business and Economic Statistics*, *15*, 345–353.

Foerster, A. T., Sarte, P.-D. G., & Watson, M. W. (2011). Aggregate shocks: A structural factor analysis of industrial production. *Journal of Political Economy*, *119*(1), 1–38.

Forni, M., & Reichlin, L. (1998). Let's get real: A factor analytical approach to disaggregated business cycle dynamics. *Review of Economic Studies*, *65*(3), 453–473.

Francis, N., & Ramey, V. A. (2005). Is the technology-driven real business cycle hypothesis dead? Shocks and aggregate fluctuations revisited. *Journal of Monetary Economics*, *52*(8), 1379–1399.

Gali, J. (1992). How well does the IS-LM model fit postwar U.S. data? *Quarterly Journal of Economics*, *107*(2), 709–738.

Gali, J. (1999). Technology, employment, and the business cycle: Do technology shocks explain aggregate fluctuations?. *American Economic Review*, *89*(1), 249–271.

Gali, J., & Gertler, M. (1999). Inflation dynamics: A structural econometric analysis. *Journal of Monetary Economics*, *44*(2), 195–222.

Hamilton, J. D. (1983). Oil and the macroeconomy since World War II. *Journal of Political Economy*, 228–248.

Hamilton, J. D. (2009). Causes and consequences of the oil shock of 2007–08. *Brookings Papers on Economic Activity*, *2009*(Spring), 215–259.

Hercowitz, Z. (1982). Money and price dispersion in the United States. *Journal of Monetary Economics*, *10*, 25–37.

Herrera, A. M., 2009, Oil Price shocks, inventories and macroeconomic dynamics, working paper. Wayne State University.

Hooker, M. (1996). What happened to the oil price-macroeconomy relationship? *Journal of Monetary Economics*, *38*(2), 195–213.

Jorgenson, D., Gollop, F. M., & Fraumeni, B. M. (1987). *Productivity and U.S. economic growth*. Cambridge, MA: Harvard University Press.

Kilian, L. (2009). Not all oil price shocks are alike: Disentangling demand and supply shocks in the crude oil market. *American Economic Review*, *99*(3), 1053–1069.

Kim, C.-J., & Nelson, C. R. (1999). *State-space models with regime switching*. Cambridge, MA: The MIT Press.

Kose, M. A., Otrok, C., & Whiteman, C. H. (2002). International business cycles: World, region, and country-specific factors. *American Economic Review*, *93*(4), 1216–1239.

Lastrapes, W. D. (2006). Inflation and the distribution of relative prices: The role of productivity and money supply shocks. *Journal of Money, Credit, and Banking*, *38*(8), 2159–2198.

Lastrapes, W. D., & Loo, C. M. (1998). Identifying the effects of money supply shocks on industry-level output. *Journal of Macroeconomics*, *20*(8), 431–449.

Lastrapes, W. D., & Selgin, G. (1995). The liquidity effect: Identifying short-run interest rate dynamics using long-run restrictions. *Journal of Macroeconomics*, *17*(3), 387–404.

Lee, K., & Ni, S. (2002). On the dynamic effects of oil price shocks: A study using industry level data. *Journal of Monetary Economics*, *49*, 823–852.

Lucas, R. E., Jr. (1972). Expectations and the neutrality of money. *Journal of Economic Theory*, *4*, 103–124.

Lucas, R. J. (1996). Nobel lecture: monetary neutrality. *Journal of Political Economy*, *104*(4), 661–682.

Mankiw, N. G., & Reis, R. (2002). Sticky information versus sticky prices: A proposal to replace the new Keynesian Phillips curve. *Quarterly Journal of Economics*, *117*(2), 1295–1328.

Mork, K. A. (1989). Oil and the macroeconomy, when prices go up and down: An extension of Hamilton's results. *Journal of Political Economy*, *97*, 740–744.

Nath, H. (2004). Relative importance of sectoral and aggregate sources of price changes. *Applied Economics*, *36*(16), 1781–1796.

Peersman, G., & Straub, R. (2009). Technology shocks and robust sign restrictions in a Euro area SVAR. *International Economic Review*, *50*(3), 727–750.

Quah, D., & Sargent, T. J. (1994). A dynamic index model for large cross sections. In J. Stock & M. Watson (Eds.), *Business cycle, indicators and forecasting*. Chicago, IL: NBER and University of Chicago Press.

Reis, R., & Watson, M. W. (2010). Relative goods' prices, pure inflation, and the Phillips correlation. *American Economic Journal: Macroeconomics*, *2*(3), 128–157.

Stock, J. H., & Watson, M. W. (2002). Macroeconomic forecasting using diffusion indexes. *Journal of Business and Economic Statistics*, *20*, 147–162.

Uhlig, H. (2005). What are the effects of a shock to monetary policy? Results from an agnostic identification procedure. *Journal of Monetary Economics*, *52*, 381–419.

Woodford, M. (2001). Imperfect common knowledge and the effects of monetary policy. Mimeo, Princeton University.

APPENDIX

State Space Representation of the Common Factor Model

The common factor model outlined in Eqs. (1) and (2) can be written in state space form as

$$Y_t = A(L)Y_t + HS_t + E_t \tag{A.1}$$

$$S_t = FS_{t-1} + V_t \tag{A.2}$$

where
$Y_t = (\Delta p_{1,t}\ \Delta q_{1,t}\ \Delta w_{1,t}\ \Delta l_{1,t}\ \cdots\ \Delta p_{35,t}\ \Delta q_{35,t}\ \Delta w_{35,t}\ \Delta l_{35,t}\ \ \Delta m2_t\ \ r_t)'$,
$E_t = (e_{1,p,t}\ e_{1,q,t}\ e_{1,w,t}\ e_{1,l,t}\ \cdots\ e_{35,p,t}\ e_{35,2q,t}\ e_{35,w,t}\ e_{35,l,t}\ \ e_{m2,t}\ \ e_{r,t})'$,
$S_t = (s_{m,t}\ s_{z,t}\ s_{f,t}\ \ s_{m,t-1}\ s_{z,t-1}\ s_{f,t-1}\ \cdots\ s_{m,t-N}\ s_{z,t-N}\ s_{f,t-N})'$,
$V_t = (v_{m,t}\ v_{z,t}\ v_{f,t}\ \ 0 \cdots 0\ \ \cdots\ \ 0 \cdots 0)'$,

$$A(L) = \begin{pmatrix} A_1(L) & & 0 \\ & \ddots & \\ 0 & & A_{35}(L) \end{pmatrix}$$

where $A_i(L)$ is a VAR(1) for $(\Delta p_{i,t}\Delta y_{i,t}\Delta w_{i,t}\Delta l_{i,t})'$, and

$$F = \begin{pmatrix} \rho_{m,1} & 0 & 0 & \rho_{m,2} & 0 & 0 & & \\ 0 & \rho_{z,1} & 0 & 0 & \rho_{z,2} & 0 & 0_{3,3(N-2)} & 0_{3,3} \\ 0 & 0 & \rho_{f,1} & 0 & 0 & \rho_{f,2} & & \\ & I_3 & & & 0_{3,3} & & 0_{3,3(N-2)} & 0_{3,3} \\ & 0_{3,3} & & & I_3 & & 0_{3,3(N-2)} & 0_{3,3} \\ & 0_{3(N-2),3} & & & 0_{(J+1)(N-2),J+1} & & I_{(J+1)(N-2)} & 0_{3(N-2),3} \end{pmatrix}$$

$$E_t \sim N(0, R)$$

where

$$R = \begin{pmatrix} R_1 & & & & \\ & \ddots & & 0 & \\ & & R_{35} & & \\ & 0 & & \sigma_{m2}^2 & \\ & & & & \sigma_r^2 \end{pmatrix}$$

and $R_i = E[(e_{i,p,t}e_{i,q,t}e_{i,w,t}e_{i,l,t})(e_{i,p,t}e_{i,q,t}e_{i,w,t}e_{i,l,t})]$. $v_{m,t} \sim N(0, \sigma^2_{s_m})$, $v_{z,t} \sim N(0, 1)$.

The factor loading matrix, H, reflects the restrictions given in Eqs. (6)–(8).

Imposing Long-Run Neutrality Restrictions

I impose the restrictions given by Eqs. (6) and (7) by rewriting Eqs. (3) and (4) so that the restrictions automatically hold. The resulting equations the sector variables are

$$\begin{pmatrix} \Delta p_{i,t} - s_{m,t} \\ \Delta q_{i,t} \\ \Delta w_{i,t} \\ \Delta l_{i,t} \end{pmatrix} = A_i(L) \begin{pmatrix} \Delta p_{i,t-1} - s_{m,t-1} \\ \Delta q_{i,t-1} \\ \Delta w_{i,t-1} \\ \Delta l_{i,t-1} \end{pmatrix} + (\tilde{H}_{i,m}(L) \quad H_{i,z}(L) \quad H_{i,f}(L)) \begin{pmatrix} \Delta s_{m,t} \\ s_{z,t} \\ s_{f,t} \end{pmatrix} + \begin{pmatrix} e_{i,p,t} \\ e_{i,q,y} \\ e_{i,w,t} \\ e_{i,l,t} \end{pmatrix} \tag{A.3}$$

$$\Delta m2_t - s_{m,t} = A_{m2}(L)(\Delta m2_{t-1} - s_{m,t-1}) + (\tilde{H}_{m2,m}(L) \quad H_{m2,z}(L) \quad H_{m2,f}(L)) \begin{pmatrix} \Delta s_{m,t} \\ s_{z,t} \\ s_{f,t} \end{pmatrix} + e_{m2,t} \tag{A.4}$$

The parameters in Eqs. (A.3) and (A.4) are unrestricted and can be estimated by OLS if the values of $s_{m,t}$, $s_{z,t}$, and $s_{f,t}$ were known. The restrictions implied by Eqs. (6) and (7) hold when Eqs. (A.3)–(A.4) are converted back to the original system.

Prior Distributions of Parameters

For the autogressive parameters in the F matrix, we set prior distribution to be $N(0,1)$. For lagged parameters in the $A(L)$ polynomials we also assume that the prior distribution of the parameters are independently distributed $N(0,.5)$. The factor loadings on the nonmonetary common factors, $H_{i,z}(L), H_{i,f}(L)$, are assumed to be normally distributed $N(0,10c^j)$, where $c=.9$ and j is the lag. The variance/covariance of sectoral shocks, R_i, is distributed $\mathrm{IW}(v\omega_i\hat{R}_i\omega'_i, v)$ where $v = 5$, $\hat{R}_i$ is the variance/covariance matrix of residuals from a VAR(1) fit to sector i data, and ω_i is a diagonal matrix with $\mathrm{sqrt}(1-\sum_{i=1}^{3}\alpha_{j,i})$ where $\sum_{i=1}^{3}\alpha_{j,i}$ is the proportion of the variance of variable j explained by the three largest principal components presented in Table 2. The prior distribution for σ_{m2}^2 and σ_r^2 is $\mathrm{IG}(5/2, 2.5/2)$ while the prior distribution for σ_{sm}^2 is $\mathrm{IG}(5/2, 5/2)$.

The parameters on the lagged polynomials $\tilde{H}_{i,m}(L)$ and $\tilde{H}_{m2,m}(L)$ in Eqs. (A.3)–(A.4) reflect the short-responses of sectoral variables to the monetary factor. For the benchmark model which assumes monetary neutrality, the prior distribution for parameters in $\tilde{H}_{i,m}(L)$ and $\tilde{H}_{m2,m}(L)$ were distributed $N(0,6c^j)$ where $c=.9$ and j is the lag.

Section "Alternative Short-Run Priors" presents alternative priors over the short-run response of sector output, prices, real wages, and labor input to a monetary factor shock. This prior assumes little adjustment in prices in the initial period, while output and employment rise and interest rates fall. To set this quantitatively, one can think of nominal money demand (in logs) as

$$m_t = p_t + q_t - \varepsilon r_t (\varepsilon = \text{semi-elasticity of money demand}) \tag{A.5}$$

or money, m_t, equals nominal income, $p_t + q_t$, less velocity, εr_t. An exogenous increase in money supply can be broken into a change in nominal income and a change in velocity (change in nominal interest rate). Nominal income changes can in turn be broken into a price change and an output change. Thus, the mean response in current period is: $\Delta p_t = \kappa\eta s_{m,t}$, $\Delta q_t = \Delta l_t = \kappa(1-\eta)s_{m,t}$, $\Delta r_t = -(1/\varepsilon)(1-\kappa)s_{m,t}$ where κ equals the fraction of the quantity of money demand increase due to nominal income, $1-\kappa$ equals the fraction of the quantity of money demand increase due to velocity, η equals the fraction of nominal income increase due to prices, while $1-\eta$ equals the fraction of nominal income increase due to output. To coincide with traditional VAR responses to money shock, I choose $\kappa=0.67$, $\eta=0.0$, and $\varepsilon=0.3$. Thus, the prior distribution for coefficient on current period monetary factor in Eq. (A.3) is $\tilde{h}_{i,p,m,0} \sim N(-1,1)$ and $\tilde{h}_{i,y,m,0}, \tilde{h}_{i,l,m,0} \sim N(0.67,1)$. The current period coefficient on the monetary factor in the interest rate equation: $h_{r,m,0} \sim N(-10/9,.5)$. The priors for the coefficients on the lagged monetary factor in Eq. (A.3) and in the interest rate equation are set to be the same distributions as in the benchmark model described in the previous paragraph.

Bayesian Estimation

The objective is to find the posterior distribution of both the parameters, Θ, and the unobserved state vector, **S**, given the data, **Y**, or $P(\Theta, \mathbf{S}|\mathbf{Y})$. For our problem, the posterior distributions $P(\Theta|\mathbf{S}, \mathbf{Y})$ and $P(\mathbf{S}, \Theta, \mathbf{Y})$ are fairly straightforward to derive (see, e.g., Kim & Nelson, 1999). By sequentially sampling $\mathbf{S}^{(i)}$ from $P(\mathbf{S}|\Theta^{(i-1)}, \mathbf{Y})$ and $\Theta^{(i)}$ from $P(\Theta|\mathbf{S}^{(i)}, \mathbf{Y})$ the resulting sample distribution of $(\mathbf{S}^{(i)}, \Theta^{(i)})$ converges to $P(\Theta, \mathbf{S}|\mathbf{Y})$.

The steps taken in the Gibbs Sampler are as follows:

1. Taking the parameters of the state space model as given, draw a realization of the state vector,S_t, from its conditional distribution. We use the multi-move approach of Carter and Koln (1994) (see also Kim & Nelson, 1999). The conditional distribution of the S_t given the data, parameters, and S_{t+1} is $N(S_{t|t,S_{t+1}}, P_{t|t,S_{t+1}})$ where

$$S_{t|t,S_{t+1}} = S_{t|t} + P_{t|t}F'(FP_{t|t}F' + Q)^{-1}(S_{t+1} - FS_{t|t}) \qquad \text{(A.2a)}$$

$$P_{t|t,S_{t+1}} = P_{t|t} - P_{t|t}F'(FP_{t|t}F' + Q)^{-1}FP_{t|t} \qquad \text{(A.2b)}$$

   $S_{t|t}$ and $P_{t|t}$ are the standard updating matrices from the Kalman filter. For $t=T$, the conditional distribution of S_T is given by $N(S_{T|T}, P_{T|T})$. Starting at $t=T$, the sampling proceeds backwards, drawing S_t from its conditional distribution using the previous period's draw, S_{t+1}, in Eq. (A.2a).
2. Given state vector,S_t, $t=1, ..., T$, and Q matrix, draw parameters of the state equation, i.e., the parameters in the F matrix. Let $\tilde{\rho}_m = (\rho_{m,1} \quad \rho_{m,2})'$

and $\tilde{\rho}_j = (\rho_{j,1} \quad \rho_{j,2})'$ for $j = 1, ..., J$. Given the prior distributions for $\tilde{\rho}_m$ and $\tilde{\rho}_j$ are N(0,I), their posterior distributions are given by $\tilde{\rho}_m \sim N(A_m, B_m)$ and $\tilde{\rho}_j \sim N(A_j, B_j)$ for $j = 1, ..., J$ where

$$B_m = (I + \sigma_m^{-2}(X'_m X_m))^{-1}, A_m = (I + \sigma_m^{-2}(X'_m X_m))^{-1}(\sigma_m^{-2} X'_m Y_m)$$

$$B_j = (I + (X'_j X_j))^{-1}, A_j = (I + (X'_j X_j))^{-1}(X'_j Y_j) \quad \text{for} \quad j = 1, ..., J$$

$$X_m = \begin{pmatrix} s_{m,0} & s_{m,-1} \\ \vdots & \vdots \\ s_{m,T-1} & s_{m,T-2} \end{pmatrix}, Y_m = \begin{pmatrix} s_{m,1} \\ \vdots \\ s_{m,T} \end{pmatrix}, X_j = \begin{pmatrix} s_{j,0} & s_{j,-1} \\ \vdots & \vdots \\ s_{j,T-1} & s_{j,T-2} \end{pmatrix}, Y_j = \begin{pmatrix} s_{j,1} \\ \vdots \\ s_{j,T} \end{pmatrix}$$

Draws of $\tilde{\rho}_m$ and $\tilde{\rho}_j$ which roots outside the stationary region were discarded and another draw was made.

3. Given state vector, and parameters of F matrix, draw variances in Q matrix. Here only σ_m^2 is drawn as the other common factors' variances are set to one. Posterior distribution of σ_m^2 given the other parameters and the states variables is IG$(\nu_m/2, \delta_m/2)$, where $\nu_m = T + 5$ and $\delta_m = (Y_m - X_m\tilde{\rho}_m)'(Y_m - X_m\tilde{\rho}_m) + 5$.
4. Given the state vector and the covariance matrix, R_i, draw parameters in Eq. (A.3) for the 35 sectors. As the explanatory variables are the same across regressions, the posterior distributions have similar forms as those in steps 2 and 3 with the linear regressions given by Eq. (A.3). Parameter draws which resulted in a nonstationary system for (A.3) were discarded as were draws in which the sign restrictions for $\eta_{i,q}$ and $\eta_{i,w}$ did not hold.
5. Given the state vector and the parameter vector for $A(L)$, variance/covariance matrix R_i is drawn from IW$(\nu\omega_i\hat{R}_i\omega'_i + \sum_{t=1}^{T} e_{i,t}e'_{i,t}, \nu + T)$ where $e_{i,t}$ is the vector of sector specific shocks in Eq. (A.3).
6. Repeat steps 1–5. This is done for a burn-in period of 100,000 draws after which every fifth draw is used to form the posterior distribution consisting of total of 10,0000 draws.

CYCLICAL CO-MOVEMENT BETWEEN OUTPUT, THE PRICE-LEVEL, AND THE INFLATION RATE

Joseph H. Haslag and Yu-Chin Hsu

ABSTRACT

In this chapter, we examine the relationship between the cyclical components of output, the price level and the inflation rate. During the post-war period, there is a negative correlation between output and the price level and a positive correlation between output and the inflation rate. A phase shift in the cyclical component between output and the price level can account for these two facts. The phase shift is consistent with movements in the price level Granger causes movements in output. In addition, we consider time-varying correlations between the two pairs of series. Spectral analysis suggest the price and output have different wavelengths, but the difference is not statistically significant.

Keywords: Countercyclical prices; procyclical inflation; phase shift

30th Anniversary Edition
Advances in Econometrics, Volume 30, 359–384

ISSN: 0731-9053/doi:10.1108/S0731-9053(2012)0000030016

INTRODUCTION

Over time, there has been a dramatic change in our understanding of the relationship between the price-level and output over the business cycle. For several decades, the conventional wisdom maintained that the price-level is procyclical. Arguably, the biggest development in our understanding came about because Lucas (1977) offered a transformative elegant definition of the business cycle itself. Armed with the definition that business cycles are deviations in output from trend, researchers applied new econometric techniques to reconsider key business-cycle facts.[1]

In this chapter, we concentrate on two related sets of business-cycle facts. More specifically, we consider the contemporaneous correlation between the price-level and output and between the inflation rate and output. Of course, the relationship between the price-level and inflation is tautological; the inflation rate is the time derivative of the log of the price-level. The existing evidence indicates a very interesting pair of observations; namely, that the price-level is countercyclical and the inflation rate procyclical.[2]

There is a simple explanation that can account for the countercyclical price-level and procyclical time derivative of the price-level. Consider the price-level and output are characterized by two sine waves with a given periodicity. Certainly, there is a phase shift that exists in which the price-level movements are negatively correlated with movements in output and the changes in the slope of the sine waves are positively correlated with movements in output. Thus, our first examination is to estimate two correlation coefficients; first, there is the price-level and output and second, there is the inflation rate and output. Over the full sample period, we verify that the price-level is countercyclical and the inflation rate is procyclical. Next, we use a numerical simulation to verify that a phase shift can account for these correlations. More concretely, if the price-level series and the output series are in phase, the estimated correlation between the price and output is negative and so is the estimated correlation between the inflation rate and output. We consider a positive phase shift applied to the price-level series. For an appropriate size positive phase shift, the correlation between the simulated price-level and simulated output is negative, while the correlation between simulated inflation and simulated output is positive. Note that a positive phase shift applied to the price-level is consistent with movements in the price-level temporally preceding, or Granger causing, output. We test for Granger causality and the results support that the price-level Granger causes output.

In addition, we examine whether the correlations are stable over the sample period. We present evidence that the correlation between the price-level and output changed over time. Based on a rolling set of sample period that is 180 quarters long, the evidence indicates that the price–output correlation is consistent and that countercyclical relationship is becoming stronger when estimated over the most recent 45-year sample than it was when estimated using the 45-year sample beginning in 1947. Meanwhile, the same rolling sample period indicates that the inflation–output correlation is consistent and that the procyclical relationship is weakening over time.

We conduct numerical simulations to study two possible explanations. One is that phase shift is varying over time. Indeed, we demonstrate that with a diminishing positive phase shift in the price-level, it is possible to observe a stronger countercyclical price–output correlation and a weaker procyclical inflation–output correlation. Another possible explanation is the wavelength of the price-level is different than the wavelength of output. We provide evidence that the wavelength of the price-level is slightly longer than the wavelength of output. However, when we test whether the difference between the two wavelengths is significant, we cannot reject the null hypothesis that the two wavelengths are equal.

The outline of the chapter is as follows. We provide a brief overview of the developments in the cyclical relationship between the price-level and output in Section "Views on the price-output relationship over time". The history helps to motivate why the time-varying relationship is considered. In Section "Full-sample results", we present the results of the statistical evidence for the full sample period. We examine the evidence on time-varying correlations in Section "Time-varying correlations". We offer a brief summary of our findings in Section "Summary and conclusion".

VIEWS ON THE PRICE–OUTPUT RELATIONSHIP OVER TIME

There is a huge literature that reports co-movements of variables at business-cycle frequencies. As different techniques emerge, researchers examine different questions. As such, there is an evolution of the stylized facts. In our view, one of the most interesting progressions involves the relationship between output and the price-level; in particular, the contemporaneous correlation between the price-level and output.

Friedman and Schwartz (1963) provided the rationale for a belief system that nakedly asserted that the price-level was procyclical. Their central thesis was that monetary disturbances were the most plausible source of cyclical fluctuations. After carefully documenting the relationship between money and output for every business cycle from the Civil War to 1970, Friedman and Schwartz (1963) concluded that monetary disturbances were the primary candidate capable of being the main source of business-cycle fluctuations. In part, they searched for sources that were "large enough" to shock a large, industrial economy. Armed with facts presented in the *Monetary History* and with the Quantity Theory, the monetary surprise literature sought to explain the procyclical price-level. In other words, the procyclical price-level was listed among the set of business-cycle regularities. Accordingly, models were judged on their ability to predict the procyclical price-level.[3]

Kydland and Prescott (1990) overturned this belief system, implementing a methodological improvement and using that methodology to report that the price-level is countercyclical. The methodological improvement owed to the definition put forward by Lucas (1977). Specifically, Lucas (1977) defined business cycles as the deviations of aggregate output from trend. Kydland and Prescott (1990) adopted this definition, estimated the deviations in trend for both output and the price-level (hereafter referred to as the cyclical components), and computed the contemporaneous correlation coefficient. For the sample period 1954 through 1989, they reported the correlation coefficient with the Implicit GNP Deflator was −0.57 and with the Consumer Price Index was −0.57. It is critical to note that the characterization of price–output correlation represents a change in measurement; that is, the definition of cyclical observations changed from characterizing leads and lags at business-cycle turning points to quarterly observations of detrended data.

Given that the price-level is countercyclical, the equilibrium models with monetary policy or price surprises had to be modified or simply ignore the price–output correlation. At the heart of the monetary surprise models was a Phillips Curve relationship. Rather than abandon the monetary surprise models, researchers moved to another cyclical relationship. Specifically, equilibrium models sought to explain the relationship between inflation and output. Cooley and Ohanian (1991) provided support for this modified approach, reporting that the relationship between output and inflation is positive during the postwar period.

The purpose of this chapter is to develop a simple framework that ties the disparate results between the cyclical component of output and the cyclical component of the price-level and inflation. We begin with a characterization of the cyclical relationship between the price-level and output and inflation and output using data for the post-World War II period. By examining these results from the perspective of a simulated deterministic relationship, it is possible to account for the observations as the result of a simple phase shift. By taking this approach, the countervailing cyclical relationships are viewed as part of an integrated approach. Once you get an equilibrium model that can account for the relationship between the price-level and output, it becomes much easier to see how the dynamic equilibrium incorporates the relationship between output and inflation. This integration is quite natural since there is a deterministic relationship between the price-level and inflation.

In addition, there are important questions about the robustness of the business-cycle relationships over time. For example, Cooley and Ohanian (1991) report the contemporaneous correlation between output and the price-level over three different postwar periods: 1948 through 1987, 1954 through 1973, and 1966 through 1987. The evidence indicates that correlation coefficient between output and the price-level is smaller in the 1966–1987 sample compared with the one estimated over the 1954–1973 sample In other words, the price-level exhibits a stronger countercyclical relationship in more recent data when compared with the relationship in the first two decades after World War II. Cooley and Ohanian (1991) extend their analysis to include annual data from Friedman and Schwartz (1963). After detrending, they report coefficients for four different sample periods: 1870–1900, 1900–1928, 1928–1946, and 1949–1975. They find that the output–price correlation is *positive* for sample periods before World War II and negative for the one sample period using postwar data. Together, these facts provide evidence that is consistent with the notion that the correlation has been changing over time between detrended output and the price-level.

The literature has also offered other ways to explain the developments in business-cycle facts.

FULL-SAMPLE RESULTS

Our first result is to report that the full-sample correlation between the detrended price-level and output is significantly negative, while the correlation between detrended inflation and output is significantly positive. Using sine functions to create simulated price and output series, we show that positive phase shift of the price-level series is consistent with our reported findings. In our case, a left-phase shift of the price-level series implies that movements in the cyclical component of the price-level would temporally precede or Granger cause the cyclical component of output. As a check, we test whether the detrended price-level Granger causes detrended output and vice-versa. The test results suggest that the price-level Granger causes output, but output does not Granger cause the price-level. The results are consistent with that there is a left-phase shift relationship between the price-level and output.

Following Lucas (1977), we define business cycles as deviations of aggregate output from trend. We use the Hodrick–Prescott filter to identify the trend component. Let $\{x_t: t = 1, 2, 3, \ldots, T\}$ denote a generic time series. At each date t, the realization of the random variable can be decomposed into its trend and cyclical components. Formally, let

$$x_t = x_t^g + x_t^c \tag{1}$$

where x_t^g denotes the trend component and x_t^c represents the cyclical component.[4] For both real GDP and the price-level measure, we use log transformations. The data are quarterly observations of real GDP for the period 1947 through 2010.[5] In addition, we compute the inflation rate for the period 1947:2 through 2010:4, where $\pi_t = p_t - p_{t-1}$ with p_t representing the (log of) the price-level.

There are two ways to compute the cyclical component of the inflation rate. We could apply the Hodrick–Prescott filter directly to the inflation rate computed from the source data; that is, $\pi_t = \pi_t^g + \pi_t^c$, where π_t^g is the result of decomposing the inflation time series. Note, however, that Eq. (1) implies that

$$\pi_t = p_t^g - p_{t-1}^g + p_t^c - p_{t-1}^c \tag{2}$$

Thus, based on Eq. (2), the cyclical component of the price-level could be used to compute a "consistent" cyclical component of the inflation rate by setting $\pi_t^c = p_t^c - p_{t-1}^c$. In small samples, the direct method of computing the cyclical component of the inflation rate is likely to commingle the trend component of the price-level with the cyclical component of the price-level. Since there is no guarantee that $\pi_t^c = p_t^c - p_{t-1}^c$, we perform a check, computing the correlation coefficient for $p_t^c - p_{t-1}^c$ from the consistent approach and π_t^c from the direct approach. The correlation coefficient is 0.99. Because the correlation coefficient is less than 1, though not significantly so, we recognize that the trend component of the price-level in Eq. (2) is commingled into the cyclical component of the inflation rate computed by the direct approach. We confirm that the correlation is close enough that the results reported in this chapter when using the direct approach are qualitatively identical to those obtained using the consistent approach. Therefore, throughout this chapter we will use the direct approach to computing the cyclical component of the inflation rate.

Table 1 reports two summary statistics for the cyclical components of output, the price-level and the inflation rate for the full sample. Figs. 1–3 plot the cyclical components for output, the price-level, and the inflation rate, respectively. In Table 1, we report the standard deviation of the cyclical components and the contemporaneous cross-correlation of the cyclical component of output to the cyclical component of both the price-level and the inflation rate separately.

The standard deviation tells us the average amplitude of the cyclical component. The average amplitude of the cyclical component of output is

Table 1. Key Summary Statistics.

Variable	Std. Dev.	Corr. Coeff. w/y
y	0.0167	1
p	0.0095	−0.216
π	1.583	0.293

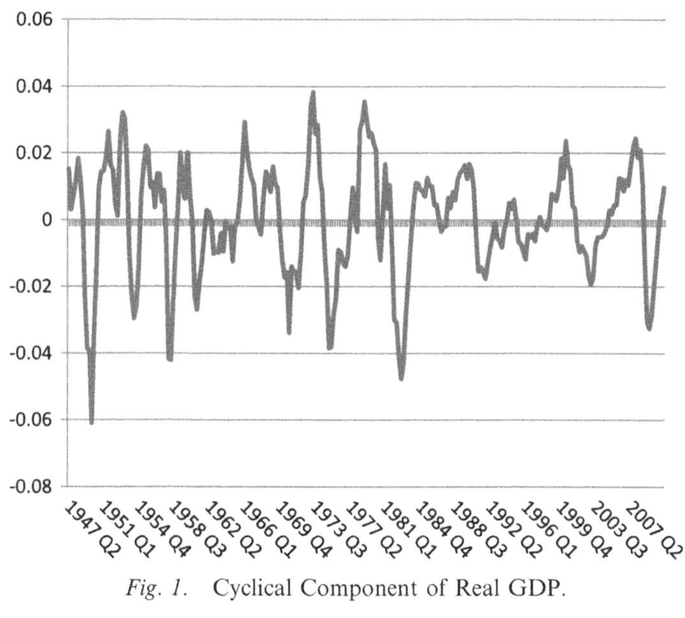

Fig. 1. Cyclical Component of Real GDP.

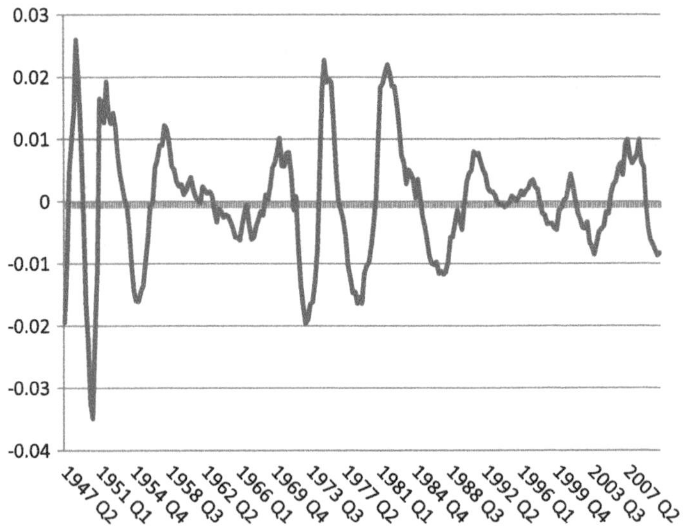

Fig 2. Cyclical Component of the Price-Level.

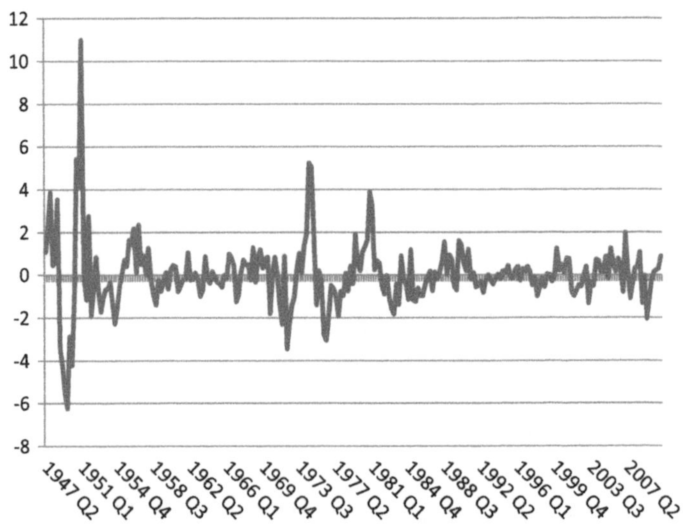

Fig. 3. Cyclical Component of the Inflation Rate.

1.7 percent, while the average amplitude of the cyclical component of the price-level is slightly less than 1 percent.

The correlation coefficient characterizes the average co-movement in the cyclical components. Because business cycles are defined by Lucas (1977) as deviations from trend output, we compute the correlation coefficients relative to real GDP. As Table 1 shows, the negative correlation coefficient between output and the price-level indicates that the price-level is mildly countercyclical. Interestingly, the evidence indicates that inflation is mildly procyclical. The standard error of the estimate of the correlation coefficient is 0.063.[6] Thus, in both cases, we can reject the null hypothesis that the correlation coefficient is equal to 0 at the 5 percent confidence level.

A Phase-shift explanation

The correlation coefficient between the cyclical components of the price-level and output has the opposite sign of the correlation coefficient between the cyclical components of the inflation rate and output. One way to account for these two results is to submit that there is a phase shift between the cyclical components of the price-level and output. Because the inflation rate is the slope of the price-level curve with respect to time, the evidence suggests that the slope of the cyclical component of the price-level is, on average, the opposite sign of the slope of the cyclical component of the price-level. Moreover, the change of the slope of the cyclical component of the price-level – that is, the cyclical component of the inflation rate – on average, is the same sign as the slope of the cyclical component of output.

To illustrate how a phase shift can account for this pair of results, we use a simple numerical example. Consider two sine waves. Let $\tilde{y} = \sin(x)$ for $x \in \mathbb{R}$. Similarly, let $\tilde{p} = -\sin(x + \gamma)$ where $2\pi > \gamma > 0$. In comparing $\tilde{y}$ and $\tilde{p}$ it is straightforward to see that $\tilde{p}$ is the inverse of $\tilde{y}$ and graphically is phase shifted to the left by γ. In other words, the graphs of the two variables would correspond to a correlation coefficient of -1 for $\gamma = 0$. Let $\rho(\tilde{y}, \tilde{p})$ denote the correlation coefficients for $\tilde{y}$ and $\tilde{p}$. Let $\tilde{\pi} = \frac{d\tilde{p}}{dx}$, or the slope of the sine function characterizing $\tilde{p}$. Formally, $\frac{d\tilde{y}}{dx} = \cos(x)$ and $\frac{d\tilde{p}}{dx} = -\cos(x + \gamma)$. Hence the slopes of the simulated price and simulated output are opposite. The change in the slope of the simulated price series is represented by $\frac{d^2\tilde{p}}{dx^2} = \sin(x + \gamma)$. Clearly, γ determines the correlation coefficient between simulated output and simulated price-level. Indeed, by shifting the price-level series "far enough" it is possible to have a positive correlation. Similarly, the correlation coefficient between the simulated output series and the simulated inflation rate depends on the size of the phase shift.[7]

Let $\gamma \in [0,\bar{\gamma}]$ where $\bar{\gamma}$ is defined as the value of γ for which $\rho(\tilde{y},\tilde{p}) = 0$. There exist values $\gamma \in [\gamma,\bar{\gamma}]$ for which $\rho(\tilde{y},\tilde{p})$ is negative and $\rho(\tilde{y},\tilde{\pi})$ is positive. In this sense, the cyclical relationship between the price-level and output can be characterized as a phase shift. Analytically, the degree of the phase shift is small enough to maintain the negative correlation between $\tilde{y}$ and $\tilde{p}$. The phase shift parameter essentially moves the slope of the simulated output series toward the change in the slope of the simulated price-level series for small enough values of γ.

In our numerical example, let $\gamma = 0.75$. We plot $\tilde{y} = \sin(x)$ and $\tilde{p} = -\sin(x + 0.75)$ for $x \in [-2\pi, 2\pi]$ in Fig. 4. To develop the intuition further, by shifting the sine curve for the simulated price-level to the left of the simulated output sine curve, we draw attention to the trough of the simulated price curve. In the neighborhood of the simulated price curve's trough, the slope of the curve is increasing. Meanwhile, just prior to the trough in the simulated price curve, the slope is negative and the slope of the simulated output curve is positive. Turning points are the easiest to illustrate the two correlations, but the sample statistics are in line with the intuition at the turning points. For $\gamma = 0.75$, $\rho(\tilde{y},\tilde{p}) = -0.732$, and $\rho(\tilde{y},\tilde{\pi}) = 0.26$. Thus, we demonstrate that a left-phase shift in the simulated price curve relative to the simulated output curve can account for the negative correlation between

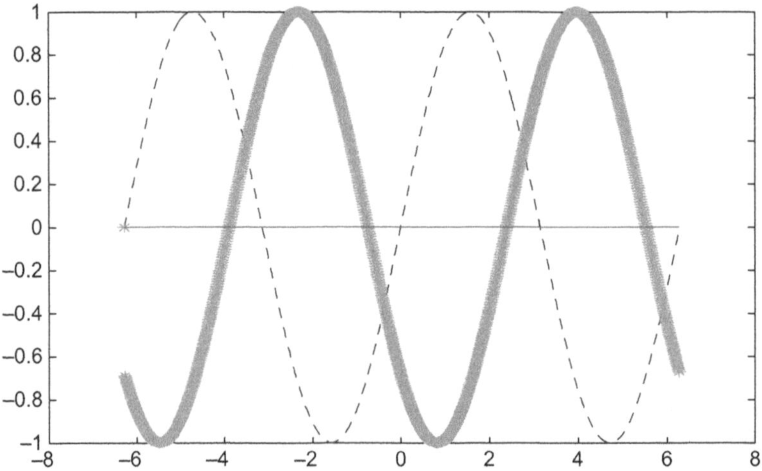

Fig. 4. Numerical Example of Positive Phase Shift.

the price-level and output and the positive correlation between inflation and output.[8]

Implications for Granger causality

By applying a left-phase shift to the simulated price-level series relative to the simulated output series, it is natural to test whether the actual time series exhibit Granger causality. Specifically, Granger causality indicates whether movements in the price-level help to predict movements in output. A left-phase shift corresponds to movements in the simulated price-level temporally preceding movements in the simulated output series. Here, temporal precedence is consistent with the ability of movements in one variable to help forecast movements in the other variable. Therefore, we can test whether movements in the cyclical component of the price-level temporally precede movements in the cyclical component of output. Such evidence is consistent with a left-phase shift characterizing the relationship between the two time series.

To check whether the temporal precedence is one-way or two-way, we assume that both series are with autoregressive lag length 4 as in [11.2.6] of Hamilton (1994) and estimate the following two regressions by ordinary least squares separately:

$$p_t = \mu_p + \sum_{i=1}^{4} \alpha_i p_{t-i} + \sum_{i=1}^{4} \beta_i y_{t-i} + \varepsilon_{pt} \qquad (3)$$

$$y_t = \mu_y + \sum_{i=1}^{4} \delta_i y_{t-i} + \sum_{i=1}^{4} \theta_i p_{t-i} + \varepsilon_{yt} \qquad (4)$$

where ε_{pt}'s are error terms that are assumed to be independent over time and so are ε_{yt}'s, but ε_{pt} and ε_{yt} might be correlated contemporaneously. Take the price question, for example. Granger causality occurs if the coefficients on the lagged output levels are significantly different from zero after taking into account the lagged price-levels. Therefore, the null hypothesis that output does not Granger cause price is formulated as $H_{p,0} : \beta_i = 0$ for $i = 1, 2, 3, 4$. If H_0 is rejected, then we conclude that movements in the cyclical component of output temporally precede, or Granger cause, movements in the cyclical component of the price-level. Instead of using an F test to test $H_{p,0}$ because the errors are potentially heteroskedastic, we use a Wald test and we use the heteroskedasticity-robust covariance matrix estimator.[9] Table 2 reports the Wald statistics for the null hypotheses that $H_{y0} : \beta_i = 0$ for $i = 1, 2, 3, 4$, and $H_{y,0} : \theta_i = 0$ for $i = 1, 2, 3, 4$.

Table 2. Granger Causality Tests.

Coefficient	Wald-Statistic
$\beta_i = 0$	2.91
$\theta_i = 0$	15.15

The Wald-statistic indicates that movements in the cyclical component of output do not temporally precede movements in the cyclical component of the price-level after accounting for lagged values of the price-level. In contrast, the evidence indicates that with 99 percent confidence, one can reject the null hypothesis that movements in the cyclical component of the price-level do temporally precede movements in the cyclical component of output.

TIME VARYING CORRELATIONS

Cooley and Ohanian (1991) use annual data to assess the relationship between output and the price-level during the 20th century. They present evidence consistent with the notion that the cyclical components exhibited different properties depending on the subsample. In particular, the contemporaneous correlation between output and the price-level was positive during the Interwar Period – between 1921 and 1940 – and negative when using data from the post-World War II period.

In this section, our aim is to assess whether the contemporaneous correlation coefficients change over time. In our assessment, the data are taken at quarterly frequencies. We compute the contemporaneous correlations for two separate estimation windows. First, we consider a fixed-window sample that is 45 years long – 180 quarters – in which the starting date begins in 1947:1 and moves forward so that the final 45-year window begins 1966:1. Thus, the initial 180-quarter sample covers the period 1947:1 through 1994:4 and the last 180-quarter sample spans the period 1966:1 through 2010:4. Second, we consider a sample period in which the number of observations increases; that is, the window length starts with a 45-year sample and then adds a quarterly observation to the sample. We refer to this latter approach as an expanding window. In the expanding-window approach, first the observations are computed for the sample period 1947:1 through 1994:1, while the last correlation coefficient is computed for the sample period 1947:1 through 2010:4.

In the fixed-window sample, the results of the contemporaneous correlation coefficient computed for each 180-quarter sample between the price-level and output are reported in Fig. 5. For inflation, the 180-quarter sample starts in 1947:2 and the correlation coefficient for the cyclical components of the inflation rate and output are reported in Fig. 6. Fig. 5 shows that the contemporaneous cross-correlation exhibits greater countercyclicality as the fixed-window moves forward through time. The correlation

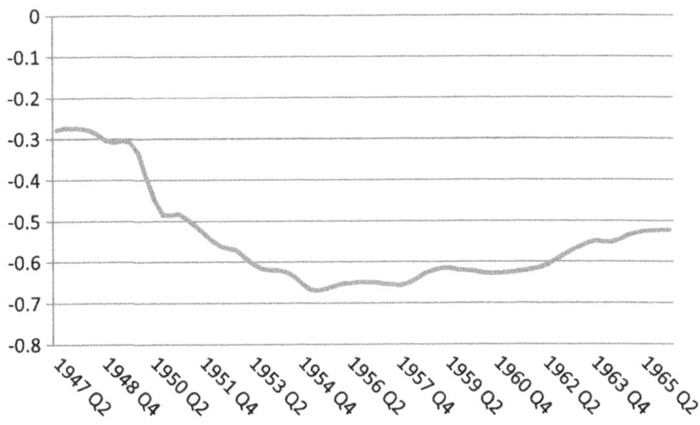

Fig. 5. Time-Varying Output-Price Correlation: Fixed-Window Case.

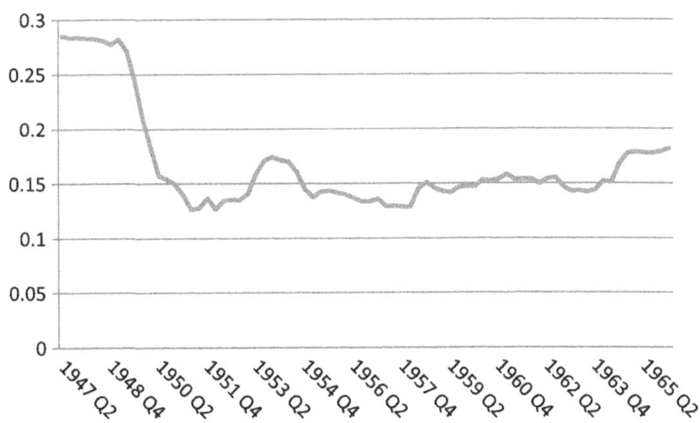

Fig. 6. Time-Varying Output–Inflation Correlation: Fixed-Window Case.

coefficients are initially around −0.25, but as the start date of the window moves toward the mid-1950s, the correlation coefficients approach −0.7. As the fixed-window sample spans periods that begin in the mid-1960s, the correlation coefficient increases to about −0.55. Meanwhile, in Fig. 6, the evidence suggests that the correlation between output and inflation is weakening as the fixed-window length moves forward in time. For windows starting in 1947, for example, the correlation coefficient is 0.28. The correlation coefficient declines as the sample dates begin in the 1950s, with the correlation coefficient falling to 0.12. As the sample periods move from the 1950s through to the 1960s, the coefficients increase slightly to around 0.16.

Two points emerge from the fixed-window length analysis. First, the evidence points to substantial variation in the contemporaneous relationship between the cyclical components of output and the price-level and between the cyclical components of output and the inflation rate. The correlation coefficients indicate that the relationship between output and the price-level is more intensely countercyclical, while the relationship between output and the inflation rate is less procyclical as the constant-length sample period omits observations closer to World War II.

In addition, we consider correlation coefficients through time with an expanding window length. Formally, we successively add one quarter to the sample period, computing the correlation coefficients for both output–price relationship and the output–inflation relationship. Fig. 7 reports the results from the expanding-window analysis for the correlation between the cyclical

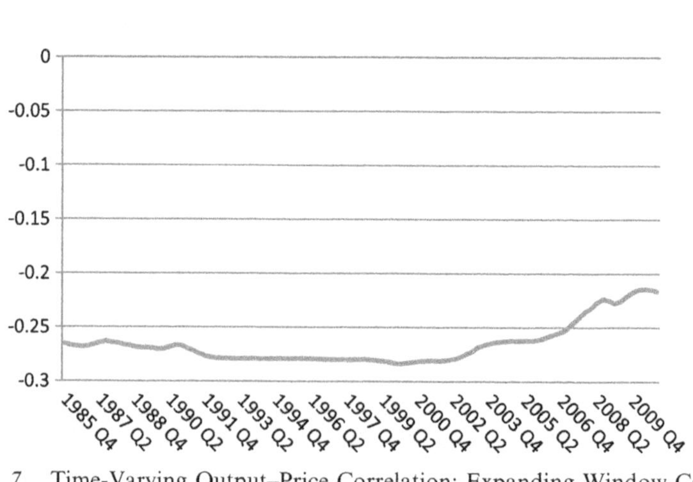

Fig. 7. Time-Varying Output–Price Correlation: Expanding-Window Case.

components of output and the price-level. As Fig. 7 shows, the contemporaneous correlation between output and the price-level is getting weaker as the window length increases. In the initial 45-year window, the correlation is −0.26 but falls to slightly greater than −0.22 when the full sample is used.

Fig. 8 plots the correlation coefficients for the cyclical components of output and the inflation rate. Here, Fig. 8 shows that there is very little movement in the correlation coefficient as the sample periods increase in size. The coefficients vary in a narrow range between 0.28 and 0.3.

We conduct Chow test to test if there is a structural change in the correlation coefficient of each pair of series. We assume that there is a known break-point at 1973:1 because the oil shocks in the 1970s represented distinguishable identifiable macroeconomic shocks that could have accounted for the differences in the correlation coefficients over time.[10] For a pair of time series random variables $\{(x_t, y_t) : t = 1, \ldots, T\}$, let ρ_{xy} denote the correlation between x_t and y_t which is estimated by

$$\hat{\rho}_{xy} = \frac{\hat{\mu}_{xy} - \hat{\mu}_x \hat{\mu}_y}{\hat{\sigma}_x \hat{\sigma}_y}$$

where $\hat{\mu}_{xy}$, $\hat{\mu}_x$ and $\hat{\mu}_y$ are sample averages of $x_t y_t$, x_t and y_t respectively. Let $\hat{\mu}_{xx}$ and $\hat{\mu}_{yy}$ be the sample averages of x_t^2 and y_t^2. We define $\hat{\sigma}_x = \sqrt{\hat{\mu}_{xx} - \hat{\mu}_x^2}$ and $\hat{\sigma}_x = \sqrt{\hat{\mu}_{yy} - \hat{\mu}_y^2}$ that are the sample standard deviations of x_t and y_t.

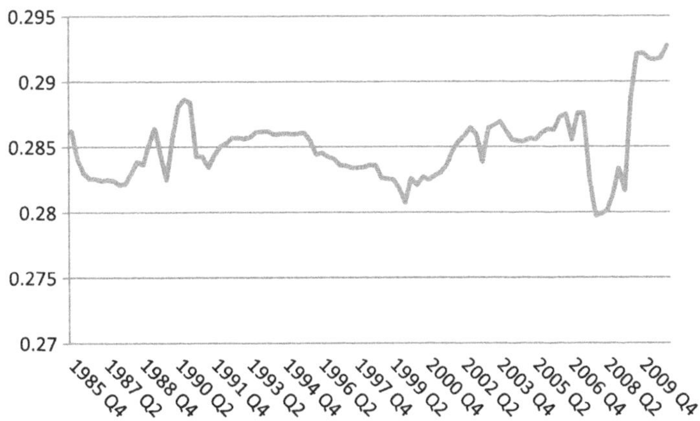

Fig. 8. Time-Varying Output–Inflation Correlation: Expanding-Window Case.

Note that it is true that $\hat{\rho}_{xy}$ is a function of $\hat{\mu}_{xy}$, $\hat{\mu}_x$, $\hat{\mu}_y$, $\hat{\mu}_{xx}$ and $\hat{\mu}_{yy}$ that are all $\sqrt{T}$-consistent and asymptotically normally distributed by a suitable central limit theorem. Therefore, by delta method, one can show that

$$\sqrt{T}(\hat{\rho}_{xy}-\rho_{xy}) = \frac{1}{\sqrt{T}}\sum_{t=1}^{T}\xi_t + o_p(1),$$
$$\xi_t = \frac{(x_t-\mu_x)(y_t-\mu_y)}{\sigma_x\sigma_y} - \rho_{xy} - \frac{\rho_{xy}}{2}\left(\frac{(x_t-\mu_x)^2}{\sigma_x^2} + \frac{(y_t-\mu_y)^2}{\sigma_y^2} - 2\right) \tag{5}$$

The derivation of Eq. (5) is straightforward but tedious, so we refer the readers to Chen (2012) for more details. Therefore, we have

$$\sqrt{T}(\hat{\rho}_{xy}-\rho_{xy}) \xrightarrow{d} N(0,\Sigma_{xy}),$$
$$\Sigma_{xy} = \lim_{T\to\infty} \; Var\left[\frac{1}{\sqrt{T}}\sum_{t=1}^{T}\xi_t\right] \tag{6}$$

We use the heteroskedasticity-autocorrelation-consistent (HAC) estimator by Newey and West (1987) to estimate Σ_{xy}. To be more specific, for a positive integer τ, Σ_{xy} is estimated by

$$\hat{\Sigma}_{xy} = \hat{\Gamma}(0) + \sum_{\ell=1}^{\tau-1} 2\left(1-\frac{\ell}{\tau}\right)\hat{\Gamma}(\ell),$$
$$\hat{\Gamma}(\ell) = \frac{1}{T-\ell}\sum_{t=1+\ell}^{T}\hat{\xi}_t\hat{\xi}_{t-\ell} \qquad \text{for } \ell = 0,\ldots,T-1 \tag{7}$$

where $\hat{\xi}_t$ is the estimated influence function of ξ_t by plugging in estimates of all unknown parameters in ξ_t. Let $\rho_{xy}(1)$ and $\rho_{xy}(2)$ denote the correlation between x_t and y_t before and after the break-point respectively. Define $\hat{\rho}_{xy}(1)$, $\hat{\rho}_{xy}(2)$, $\hat{\Sigma}_{xy}(1)$, and $\hat{\Sigma}_{xy}(2)$ similarly. Let T_1 and T_2 denote the sample sizes before and after the break-point respectively. Then we have

$$\frac{\hat{\rho}_{xy}(1)-\hat{\rho}_{xy}(2)-\rho_{xy}(1)+\rho_{xy}(2)}{\sqrt{(\hat{\Sigma}_{xy}(1)/T_1)+(\hat{\Sigma}_{xy}(2)/T_2)}} \xrightarrow{d} N(0,1) \tag{8}$$

Eq. (8) holds because $\hat{\rho}_{xy}(1)$ and $\hat{\rho}_{xy}(2)$ are asymptotically independent. Therefore, we can construct two-sided t-test based on Eq. (8). We set $\tau = 4$ when calculating the HAC variances.[11] The test results are presented in Table 3. From Table 3, we conclude that there is a structural change in the

Table 3. Structural Change in Correlations.

Series	t-Statistic
Price–output	2.979
Price-inflation	0.579
Output-inflation	0.933

correlation coefficient between the price and output series at 1973:1, but we accept that there are no structural changes in the correlation coefficients between the price and inflation series, and the output and inflation series.

Thus, the evidence indicates three things with respect to the time-varying nature of the contemporaneous relationships between output and the price-level and output and inflation. First, the evidence suggests that relationship between output and the price-level has not been constant over time. In the fixed-window sample, the correlation coefficients indicate a stronger negative correlation as the window becomes more modern. In other words, as the 45-year window excludes the period between World War II and the Federal Reserve-Treasury Accord, the evidence points to a strong countercyclical relationship. That relationship weakened only modestly as the sample period focused on the 45 year period beginning in the 1960s. In particular, the correlation coefficient begins to indicate slightly stronger countercyclicality somewhere after data of 5 years are added to the sample estimate. This is consistent with, though much less pronounced with, the evidence produced in the fixed-sample length estimation. As the sample length continues to increase, the correlation coefficient indicates that the countercyclical relationship is weakening. This also is consistent with the smaller correlation coefficients observed in the fixed-window length results that include more recent observations.

Second, the relationship between output and inflation weakens over time. This result is more pronounced in the fixed window where the procyclical relationship weakens as the sample period includes more recent observations.

Third, the time-varying nature of the contemporaneous correlation tells us something about the changes in cycles that can account for the time-varying cross-correlations we observe. A corollary holds; namely, we reported Granger causality results in the previous section. With time-varying correlations, the additional evidence is consistent with the notion that the Granger causality is a small-sample result. In the next section, we present a simple atheoretical depiction of two cycles. Using these cycles, we consider changes between these cycles that are consistent with our presentation of the time-varying changes in the cross-correlations. In other words, we begin our investigation by searching over changes in the properties of the time series of the price-level and output that can account for time-varying correlations. In particular, we will use simulated cycles to assess whether time-varying phase shifts or differences in the wavelengths can account for the time-varying correlation coefficients. In addition, the evidence is consistent with changes in the correlation coefficients for output and the price-level as the sample period lengthens.

Time-varying Phase Shifts

With a slight modification to our numerical example, it is possible to demonstrate that a time-varying phase-shift can account for the time-varying correlation coefficients. To formalize this notion, let $\tilde{p} = -\sin(x + \gamma t)$ so that the phase shift is changing over time. Consider, for example, a simple change to phase-shift parameter. With $\gamma = 0.35$, the correlation between the simulated price-level and the simulated output is -0.94, while the correlation between simulated inflation and simulated output is 0.15. (Recall that with $\gamma = 0.75$, the correlation coefficients were $\rho(\tilde{y}, \tilde{p}) = -0.732$ and $\rho(\tilde{y}, \tilde{\pi}) = 0.26$.) Thus, by letting the phase shift parameter decline over time in the simulated series, we see that the correlation between the price-level and output becomes more countercyclical, while the correlation between inflation and output becomes less procyclical. Hence, this reduction in the phase shift parameter over time is consistent with the observations we presented in the post war period in the fixed-window analysis.

Cycles of Different Wavelengths

In addition to a time-varying phase shift, the evidence is consistent with cycles of different wavelengths. In other words, the cyclical components of output and the price-level have different wavelengths. To extend our simulation, consider a case in which output is simulated as $\tilde{y} = \sin(x)$ and the simulated values for the price-level are determined by $\tilde{p} = -\sin(kx)$. Here, the parameter k denotes the wavelength so that $k > 1(< 1)$ represents a time series with a higher (lower) number of cycles per unit of time. Note that the slope of simulated output is $\cos(x)$ and the slope of the simulated price series is $-k\cos(kx)$. Note that as k deviates more from 1 in either direction, the correlation between the simulated price and output series decays. Not surprisingly, as the wavelength of one series increases relative to the other, the wavelength that is occurring more frequently will look more like noise compared with the other series.

To check whether the wavelengthes of the two cycles are identical, we assume that both series follow AR(m) processes:

$$p_t = \mu_p + \sum_{j=1}^{m} \rho_{pj} p_{t-j} + \varepsilon_{pt} \tag{9}$$

$$y_t = \mu_y + \sum_{j=1}^{m} \rho_{yj} y_{t-j} + \varepsilon_{yt} \tag{10}$$

where ε_{pt}'s are error terms that are assumed to be identical and independent over time, and so are ε_{yt}'s. Note that ε_{pt} and ε_{yt} are allowed to be correlated contemporaneously and the $Cov(\varepsilon_{pt}, \varepsilon_{yt})$ is allowed to vary over time, so it is compatible with the time-varying correlation we observed. If p_t is generated according to Eq. (9), then by [6.1.14] of Hamilton (1994), the spectral density $s_p(\omega)$ for $\omega \in [0,\pi]$ is defined as

$$s_p(\omega) = \frac{\sigma_p^2}{2\pi(1-\sum_{j=1}^{p}\rho_{pj}e^{-ij\omega})(1-\sum_{j=1}^{p}\rho_{pj}e^{ij\omega})}$$

where $i = \sqrt{-1}$ and $\sigma_p^2 = Var(\varepsilon_{pt})$. To estimate $s_p(\omega)$, we use the the system ordinary least squares estimator to estimate autoregressive parameters and get $\hat{\rho}_{pj}$ for $j = 1, \ldots, m$ and $\hat{\sigma}_p^2$. Hence, for $\omega \in [0,\pi]$, $s_p(\omega)$ is estimated by

$$\hat{s}_p(\omega) = \frac{\hat{\sigma}_p^2}{2\pi(1-\sum_{j=1}^{m}\hat{\rho}_{pj}e^{-ij\omega})(1-\sum_{j=1}^{p}\hat{\rho}_{pj}e^{ij\omega})}$$

We plot the spectral densities of the price and output series in Figs. 9 and 10. In Fig. 9, we set $m = 3$ and in Fig. 10, $m = 4$. In both figures, we normalize both spectral densities such that $\int_0^{\pi} \hat{s}_Y(\omega)d\omega = 1$, so the two densities are comparable. It is true that $2\int_0^{\pi} s_p(\omega)d\omega = \sigma_p^2$, see [6.1.17] of Hamilton (1994). Therefore, the normalization is not harmful.

To test if the normalized spectral densities for price and output series are identical, first note that according to Herglotz's Theorem, see for example, Corollary 4.3.2 of Brockwell and Davis (1991), two (non-normalized) spectral densities are identical if and only if the autocovariances functions of the stationary processes are identical. In our case, two normalized spectral densities are identical in AR processes iff all the autoregressive parameters

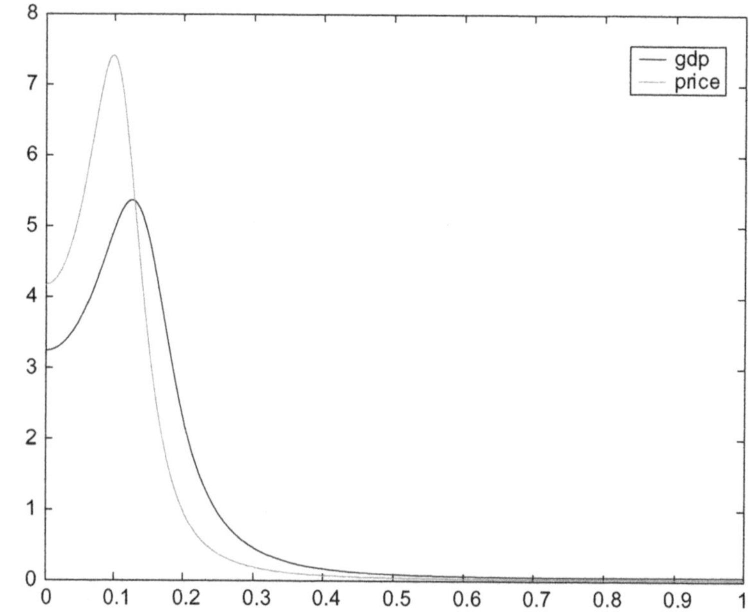

Fig. 9. Spectral Densities of the Price and Output Series: AR(3).

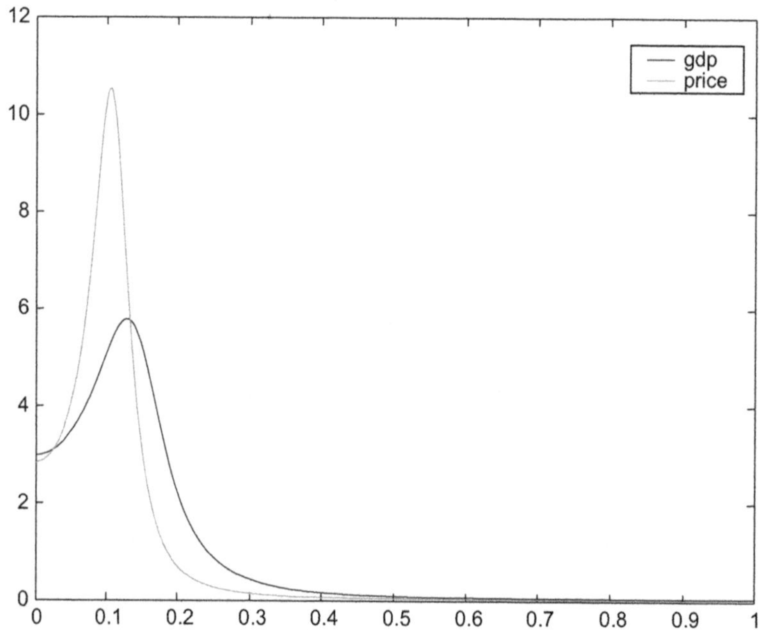

Fig. 10. Spectral Densities of the Price and Output Series: AR(4).

Table 4. Spectral Densities Equality Test.

Model	Wald-Statistic
AR(3)	5.774
AR(4)	6.715

are the same for both processes. As a result, it is equivalent to $H_0: \rho_{pj} = \rho_{yj}$ for all $j = 1, \ldots, m$. The Wald-statistics are summarized in Table 4. As we can see, we cannot reject the null hypotheses that the two spectral densities are identical.

Based on the simulations, a time-varying phase shift can account for the time-varying correlations between output and the price-level and between output and the inflation rate. More specifically, as the positive phase shift diminishes over time, the contemporaneous correlation between output and the price-level is more countercyclical over time, while the correlation between output and inflation is less procyclical over time. In effect, the smaller positive shift results in the two simulated series being more in phase.

We also consider the role that different wavelengths in the simulated series could have on the contemporaneous correlations. In the actual series, the evidence indicates that we cannot reject the null hypothesis that the wavelength of the price-level is the same as the wavelength of the output. As such, it does not appear that different wavelengths can explain why the correlation coefficients are changing over time.

Other Explanations

There are other explanations that can account for the time-varying correlations between the price-level and output and inflation and output. Here we consider two. First, there is the correlation between shocks to output and shocks to the price-level. More specifically, we modify the simulated series for output and the price-level as follows. Let $\tilde{y} = \sin(x) + \varepsilon_y$ for $x \in \mathbb{R}$ and let $\tilde{p} = -\sin(x + \gamma) + \varepsilon_p$. Note that we simulate ε_y such that it is drawn from a pseudo-normal distribution with mean zero and standard deviation equal to 0.167. Consider a case in which $\varepsilon_p = \eta\varepsilon_Y + u_t$, where $\eta > 0$ and u is a random variable with mean zero and variance $\sigma_u^2 > 0$. For our purposes, we examine a case in which $\rho(\varepsilon_y, \varepsilon_p) = 0.99$.

The results of these modified simulations indicate that allowing for more correlated shocks can account for the time-varying correlations we observe in the data. With this high degree of correlated shocks, we compute the correlation for the simulated output and price-level series for this case, finding $\rho(\tilde{y}, \tilde{p}) = -0.94$ and $\rho(\tilde{y}, \tilde{\pi}) = 0.16$. Compared with the baseline correlations of the simulated value,[12] the results are consistent with correlation between output and the price-level becoming more countercyclical over time and the correlation between output and inflation becoming less procyclical over time.[13]

SUMMARY AND CONCLUSION

In this chapter, we examine two sets of business-cycle facts. One is the contemporaneous correlation between output and the price-level. The other is the contemporaneous correlation between output and inflation. The first is countercyclical and the second is procyclical. Because the inflation rate is the time derivative of the log price-level, we are interested in examining properties of the time series that will account for the two facts. The Folk Theorem is that the price-level series is a positive phase shift of the output series. We verify that the Folk Theorem holds by simulating a positive phase shift. Because a positive phase shift is consistent with Granger causality, we test for that. Indeed, the price-level does Granger cause output in the actual time series.

In addition, we ask whether the contemporaneous correlation is stable over time. Two types of evidence are presented. First, we compute correlation coefficients for the price level–output and the inflation–output over different sample periods. In one the set of rolling 45-year windows is used, while the other starts with a 45-year sample and adds next time-series observation to each sample period. The evidence from this approach indicates that the price level–output correlation is becoming smaller algebraically over time; that is, the price-level is more countercyclical in the most recent sample when compared with the statistic computed with older data. The evidence also indicates that the inflation–output correlation is smaller, meaning that inflation is less procyclical when estimated over the most recent samples than it was when estimated with older data. We test for a structural break. The test statistic indicates that we reject the null hypothesis of constant correlation coefficients.

The simulations offer some clues that could explain the movements in the correlation coefficients over time. We present simulation results suggesting that a time-varying phase shift could account for the pattern in correlation coefficients that have evolved over time. Another possible explanation is that the phase shift is constant, but that price-level and output have different wavelengths. The power spectrum for the two variables are consistent with the wavelength for the price-level being longer than the wavelength for output. However, we cannot reject the null hypothesis that the two wavelengths are equal.

Lastly, we consider explanations that do not depend on changing the properties of the cycle relative to one another. For example, if there are correlated shocks present, we present simulations consistent with the notion that as the correlation between price-level and output shocks increases, the correlation between the price-level and output, not surprisingly, becomes more negative. Moreover, the correlation between inflation and output becomes less positive Table A.1.

In our view, the questions and answers developed in this chapter are a starting point. Indeed, a more ambitious project would be to tackle the frictions that would give rise to a positive phase shift between the price-level and output Table A.2.

NOTES

1. See Lucas (1977, p. 9).
2. See, for example, Kydland and Prescott (1990), Cooley and Ohanian (1991), and King and Watson (1996).
3. See, for example, Bernanke (1986) and Mankiw (1989). Both raised concerns about the King and Plosser model because it did not explain procyclical prices.
4. Formally, let x_t^g denote the trend component for $t = 1, 2, ..., T$. The Hodrick–Prescott filter minimizes

$$\sum_{t=1}^{T}(x_t - x_t^g)^2 + \lambda \sum_{t=2}^{T-1}[(x_{t+1}^g - x_t^g) - (x_t^g - x_{t-1}^g)]^2$$

Following the literature, we set $\lambda = 1,600$ for quarterly observations. See Hodrick and Prescott (1997) for more details.

Note that Cooley and Ohanian considered a wide variety of detrending approaches. In general, they showed that the results were not sensitive to the detrending method used.

5. The price-level is defined as the chain-type GDP deflator. All data are obtained from the Federal Reserve Bank of St. Louis FRED database with the Bureau of Economic Analysis as the source data.
6. See Ashley, Granger, and Schmalensee (1980) for the estimator of the variance of the correlation coefficient.
7. While this analysis presents the results for a left-phase shift applied to the price-level series, it is obviously equivalent to consider a right-phase shift to the simulated

output series. We normalize on the output cycle because business cycles are defined in terms of deviations in output from trend.

8. Alternatively, we could apply multivariate spectral methods to estimate the phase shift. The cross spectrum function $\Gamma_{yp}(\omega)$ for $\omega \in [-\pi, \pi]$ is defined as

$$\Gamma_{yp}(\omega) = \gamma_{yp}(0) + \sum_{j=1}^{\infty} \gamma_{yp}(j)\exp(-\omega \cdot j \cdot i) + \gamma_{yp}(-j)\exp(\omega \cdot j \cdot i)$$

and the phase shift information is

$$\phi_{yp}(\omega) = \arctan\left(\frac{\mathrm{Im}(\Gamma_{yp}(\omega))}{\mathrm{Re}(\Gamma_{yp}(\omega))}\right)$$

where Re(a) and Im(a) denote the real and imaginary parts of a complex number a. If $y_t = p_{t+\tau}$ for some τ that is an integer, then we have $\varphi_{yp}(\omega) = \tau \cdot \omega$. In our view, the phase shift parameter simulation approach allows for the phase shift to take on any real value and not just integer values. See von Storch and Zwiers (1999) for a more complete discussion.

9. The following test results on Granger causality tests remain the same when we use the covariance matrix estimator under homoskedasticity.

10. See Hamilton (1983) and (2011) for excellent discussions on the evolution of oil prices shocks on the macroeconomy. Hamilton applies regime-switching methods to examine the differences in cyclical fluctuations over time.

11. The testing results are not sensitive to the choice of τ between 3 and 10.

12. Recall that $\rho(\tilde{y}, \tilde{p}) = -0.73$ and $\rho(\tilde{y}, \tilde{\pi}) = 0.26$.

13. We conduct a Chow-test to see if there is a structure change in the correlations between the error terms from Eqs. (9) and (10), before and after 1973:1 and we fail to reject that there is a structure change.

REFERENCES

Ashley, R., Granger, C. W., & Schmalensee, R. (1980). Advertising and aggregate consumption: An analysis of causality. *Econometrica, July*, *48*(5), 1149–1167.

Bernanke, B. S. (1986). *Alternative explanations of the money-income correlation*. National Bureau of Economic Research Working Paper No. 1842. National Bureau of Economic Research, Cambridge, MA.

Brockwell, P. J., & Davis, R. A. (1991). *Time series: Theory and methods* (2nd ed.). New York, NY: Springer-Verlag.

Chen, Y.-T. (2012). *Exceedance correlation tests for financial returns*. Working Paper.

Cooley, T. F., & Ohanian, L. E. (1991). The cyclical behavior of the price level. *Journal of Monetary Economics, August*, *28*(1), 25–60.

Friedman, M., & Schwartz, A. (1963). *A monetary history of the United States, 1867–1960*. Princeton, NJ: Princeton University Press.

Hamilton, J. D. (1983). Oil and the macroeconomy since world war II. *Journal of Political Economy*, *91*, 228–248.

Hamilton, J. D. (1994). *Time series analysis*. Princeton, NJ: Princeton University Press.

Hamilton, J.D. (2011, February). *Historical oil shocks*. National Bureau of Economic Research Working Paper No. 16790. National Bureau of Economic Research, Cambridge, MA.

Hodrick, R. J., & Prescott, E. C. (1997). Postwar U.S. business cycles: An empirical investigation. *Journal of Money, Credit, and Banking, February*, *29*(1), 1–16.

King, R. G., & Watson, M. W. (1996). Money, the price level, interest rates and the business cycle. *Review of Economics and Statistics*, *78*, 35–53.

Kydland, F., & Prescott, E. C. (1990). Business cycles: Real facts and a monetary myth. *Federal Reserve Bank of Minneapolis Quarterly Review*, *14*(2), 3–18.

Lucas, R. E., Jr. (1977). Understanding business cycles. In K. Brunner & A. H. Meltzer (Eds.), *Stabilization of the domestic and international economy* (Vol. 5, pp. 7–29). Amsterdam: North Holland. Carnegie-Rochester Series on Public Policy.

Mankiw, G. N. (1989). Real business cycle: A new Keynesian perspective. *Journal of Economic Perspectives, Summer*, *3*(3), 79–90.

Newey, W., & West, K. (1987). A simple positive semi-definite heteroskedasticity and autocorrelation consistent covariance matrix. *Econometrica*, *55*, 703–708.

von Storch, H., & Zwiers, F. W. (1999). *Statistical analysis in climate research*. New York, NY: Cambridge University Press.

APPENDIX

Table A.1. Parameter Estimates for Output Regression.

Variable	Parameter Estimate	Std. Err.
Const	7.78e-06	4.95 e-04
δ_1	1.03	0.064
δ_2	−0.138	0.092
δ_3	−0.188	0.091
δ_4	−0.195	0.063
θ_1	0.105	0.165
θ_2	−0.35	0.263
θ_3	−0.005	0.263
θ_4	0.021	0.166

Table A.2. Parameter Estimates for Price Regression.

Variable	Parameter Estimate	Std. Err.
Const	−3.6e-05	1.89 e-04
α_1	1.234	0.063
α_2	−0.221	0.101
α_3	−0.019	0.10
α_4	−0.176	0.064
β_1	0.026	0.025
β_2	−0.006	0.035
β_3	−0.04	0.035
β_4	0.011	0.024

MONEY–INCOME GRANGER-CAUSALITY IN QUANTILES

Tae-Hwy Lee and Weiping Yang

ABSTRACT

The causal relationship between money and income (output) has been an important topic and has been extensively studied. However, those empirical studies are almost entirely on Granger-causality in the conditional mean. Compared to conditional mean, conditional quantiles give a broader picture of an economy in various scenarios. In this paper, we explore whether forecasting conditional quantiles of output growth can be improved using money growth information. We compare the check loss values of quantile forecasts of output growth with and without using past information on money growth, and assess the statistical significance of the loss-differentials. Using U.S. monthly series of real personal income or industrial production for income and output, and M1 or M2 for money, we find that out-of-sample quantile forecasting for output growth is significantly improved by accounting for past money growth information, particularly in tails of the output growth conditional distribution. On the other hand, money–income Granger-causality in the conditional mean is quite weak and unstable. These empirical findings in this paper have not *been observed in the money–income literature. The* new *results of this paper have an important implication on monetary policy, because they*

30th Anniversary Edition
Advances in Econometrics, Volume 30, 385–409

ISSN: 0731-9053/doi:10.1108/S0731-9053(2012)0000030017

imply that the effectiveness of monetary policy has been under-estimated by merely testing Granger-causality in conditional mean. Money does Granger-cause income more strongly than it has been known and therefore information on money growth can (and should) be more utilized in implementing monetary policy.

Keywords: Money – income Granger-causality; conditional mean; conditional quantiles; conditional distribution

INTRODUCTION

Granger-causality (GC), introduced by Granger (1969, 1980, 1988), is one of the important issues that has been much studied in empirical macro-economics and empirical finance. Particularly the causal relationship between money and income is one of the most widely studied subject in economics. In this paper, we extend the literature in two ways. The literature on money–income causality is studied for the conditional *mean* and most papers have used the *in-sample* significance of money variables in the output growth equation. (In this paper, the terms, income and output, will be used interchangeably.) First, we go beyond the conditional mean, and examine the conditional distribution and conditional *quantiles*. Second, we examine the *out-of-sample* predictive contents of money variables in forecasting output growth.

While GC is naturally defined in terms of conditional distribution (see Granger & Newbold 1986), almost all the papers in this literature have focused on GC-in-mean (GCM). The GC-in-distribution (GCD) has been less studied empirically perhaps because it is in fact about independence and so it may be too broad to be directly linked to a policy implication. More useful may be the particular quantiles of the conditional distribution because by inverting the conditional distribution we obtain the conditional quantiles. Hence, we may examine directly the GC in distribution, or indirectly via GC in conditional quantiles (GCQ). Granger (2003) notes that the study of the time series of quantiles is relevant as the predictive distribution can be expressed in terms of the CDF, the density, the characteristic function, or quantiles.

Vast empirical literature on the money–income causality has very mixed results on GCM – usually unstable and sensitive to the choice of sample periods, data sets, and variables (e.g., M1 or M2 for money, personal income (PI) or industrial production (IP) for income, with or without including some other variables such as interest rates and business cycle indicators in the regression, etc.). Different countries, sample periods, and variables are studied in those empirical research, but no consensus has been reached. The results in this paper for GCD and GCQ are much more stable and stronger.

The aim of this paper is to study the GC beyond the conditional mean between money and income, which is in line with the suggested directions of Granger (2003, 2005, 2006).[1] Forecasting conditional quantiles is important in economic policy when a particular scenario of the economy is concerned. For instance, in asset valuation, different scenarios of output growth are extremely useful in sensitivity analysis, scenario analysis, and risk management. For industries greatly influenced by overall macroeconomic conditions, forecasting output growth helps to evaluate the industry exposure in different scenarios. The fan chart of Bank of England for different quantile forecasts of inflation rate is another example.

Although it is not stable in the U.S. data on whether money growth helps to improve forecasting of the conditional mean of output growth, we find that there is much stronger evidence that it helps to improve forecasting of its conditional quantiles. Forecasting the conditional quantiles of output growth depends on its conditional distribution, and so we may also test for GCD. GCD implies GCQ in some quantiles, although GCD does not necessarily imply GCQ in each quantile. GCD is a necessary but not sufficient condition for GCQ in specific quantile. GCQ in a specific quantile exists if the lagged money variables help to improve forecasting the output growth at that quantile. Two quantile regression models for output growth with and without money growth information are estimated and the out-of-sample average of the "check" loss values of the two quantile models are compared. Because these two quantile forecasting models are nested, the unconditional predictive ability test proposed by Diebold and Mariano (1995) and West (1996) fails in that its asymptotic distribution degenerates. We therefore utilize the *conditional* predictive ability test proposed by Giacomini and White (GW henceforth, 2006).

Our empirical study uses several different data sets over various sample periods. We find the following results. First, for the causality in the conditional mean, differently from Chao, Corradi, and Swanson (2001) who test out-of-sample Granger-causality in mean using moment conditions, we compare the squared forecast error loss values of two conditional mean forecasts of output growth with and without money. The result is very weak for the GCM (as expected from the existing literature). We find that the predictive ability of a model with including money as a predictor for the conditional mean of output growth could be even worse than a model without money (due to parameter estimation error), and the result varies sensitively over time as pointed out by Eichenbaum and Singleton (1987), Stock and Watson (1996), Swanson (1998), and Thoma (1994). Second, for the money–income GC in the conditional distribution, we use a nonparametric copula function, and find a more stable and significant result for GCD in many subsamples even when there exists no significant GCM. Third, for the GCQ, two conditional quantile regression models with and without money are estimated and their quantile forecasts are compared for their out-of-sample check loss values. We find that GCQ is significant in tail quantiles in most subsamples and most data sets, while it is not significant in the center of the distribution. Fourth, comparing results across different data sets (which consist of different variables for money and income), it seems that GCQ between money and industrial production (IP) is more significant than between money and real personal income (PI).

The structure of this paper is as follows. In section "Granger-causality", we discuss GC in mean, GC in distribution, and GC in quantiles. Section "Empirical Analysis" reports the empirical findings. Section "Conclusions" concludes.

GRANGER-CAUSALITY

We use the following notation. Let R denote the sample size for estimation (for which we use a rolling scheme), P denote the size of the out-of-sample period for forecast evaluation, and $T = R + P$. Let x be money growth and y the output growth. Consider the distribution functions conditional on the information set $\mathcal{F}_t$ as $F_{t+1}(x|\mathcal{F}_t) = \Pr(x_{t+1} < x|\mathcal{F}_t)$, $G_{t+1}(y|\mathcal{F}_t) = \Pr(y_{t+1} < y|\mathcal{F}_t)$, and $H_{t+1}(x, y|\mathcal{F}_t) = \Pr(x_{t+1} < x$ and $y_{t+1} < y|\mathcal{F}_t)$. Let $f_{t+1}(x|\mathcal{F}_t)$, $g_{t+1}(y|\mathcal{F}_t)$, and $h_{t+1}(x, y|\mathcal{F}_t)$ be the corresponding densities. Let $u = F_{t+1}(x|\mathcal{F}_t)$ and $v = G_{t+1}(y|\mathcal{F}_t)$. Let $C_{t+1}(u, v|\mathcal{F}_t)$ and $c_{t+1}(u, v|\mathcal{F}_t)$ be the conditional copula function and the conditional copula density function respectively. Let $E(y_{t+1}|\mathcal{F}_t)$ be the conditional mean of y_{t+1}. Let $X_t = (x_t, \ldots, x_{t+1-q})'$ and $\mathcal{G}_t$ be the information set excluding X_t, that is, $\mathcal{G}_t = \mathcal{F}_t/\{X_t\}$.

Money–Income Granger-Causality in Mean

Starting with Friedman (1956) the debate about role of money on income attracts attention of a lot of economists. Numerous studies have been devoted to the interaction between money and income. Theoretical models are constructed to explore the roles of aggregate demand fluctuation and money demand fluctuation, such as in Kaldor (1970); Meltzer (1963); Modigliani (1977), among others. Along with the theoretical development, many empirical studies have been made following the seminal research of Sims (1972, 1980). Sims (1972) shows money Granger-causes income, but his results were criticized due to the bias caused by hidden factors. Sims (1980) applies a vector autoregressive (VAR) model to handle a vector of variables and reports that money does not Granger cause income after the World War II. After Sims, Granger-causality and VAR models have become generally accepted instruments for studying the money and income relationship. Stock and Watson (1989) contend that the deterministic trend plays important roles and use detrended money in the analysis. They find more significant money–income causality using the detrended money growth rate. Friedman and Kuttner (1992, 1993) and Thoma (1994) also report limited evidence for the money–income causality. However, they find money–income causality is time-varying with regard to different sample periods or with regard to different variables.[2]Swanson (1998) tests money–income Granger-causality in an error-correction model. Dufour and Renault (1998) and Dufour, Pelletier, and Renault (2006) test the long horizon causality.

Definition 1. (Non-Granger-causality in mean, NGCM): *X_t does not Granger-cause y_{t+1} in mean if and only if $E(y_{t+1}|X_t, \mathcal{G}_t) = E(y_{t+1}|\mathcal{G}_t)$ almost surely (a.s.).*

To test for Granger-causality in mean (GCM), we can utilize either an in-sample test or an out-of-sample test. In the literature, most tests of money–income causality focus on in-sample conditional mean in a VAR model. The in-sample Granger-causality test is to test the null hypothesis that coefficients of money are all insignificant in the output equation. A Wald-type test is often used in an in-sample test of GCM. Following Ashley, Granger, and Schmalensee (1980) who argue Granger-causality makes more sense in a predictive setting, we conduct an out-of-sample test for GCM based on two nested models. The first model does not account for money–income GCM (referred as Model 1 or "NGCM") and the second does (referred as Model 2 or "GCM"):

$$\text{Model 1 (NGCM)} \quad : \quad y_{t+1} = E(y_{t+1}|\mathcal{G}_t) + \varepsilon_{1,t+1} = V_t'\theta_1 + \varepsilon_{1,t+1} \tag{1}$$

$$\text{Model 2 (GCM)} \quad : \quad y_{t+1} = E(y_{t+1}|X_t, \mathcal{G}_t) + \varepsilon_{2,t+1} = W_t'\theta_2 + \varepsilon_{2,t+1} \tag{2}$$

where $V_t \in \mathcal{G}_t$ and $W_t = (X_t' V_t')$ are vectors of regressors. V_t includes a constant term. The parameters $\{\theta_i\}$ are estimated by minimizing the squared error loss using the rolling sample of the most recent R observations at time $t(t = R, \ldots, T-1)$

$$\hat{\theta}_{1,t} = \arg\min_{\theta_1} \sum_{s=t-R+1}^{t} (y_s - V_{s-1}'\theta_1)^2 \tag{3}$$

$$\hat{\theta}_{2,t} = \arg\min_{\theta_2} \sum_{s=t-R+1}^{t} (y_s - W_{s-1}'\theta_2)^2 \tag{4}$$

Denote $\hat{y}_{1,t+1}(\hat{\theta}_{1,t}) = V_t'\hat{\theta}_{1,t}$ and $\hat{y}_{2,t+1}(\hat{\theta}_{2,t}) = W_t'\hat{\theta}_{2,t}$, the forecasts of y_{t+1} from Model 1 and Model 2, respectively, and let $\hat{\varepsilon}_{i,t+1}(\hat{\theta}_{i,t}) = y_{t+1} - \hat{y}_{i,t+1}(\hat{\theta}_{i,t})$ be the forecast error of Model i. In the empirical analysis of section "Empirical Analysis", we choose $X_t = (x_t, \ldots, x_{t+1-q})'$ and $V_t = (Y_t', I_t, B_t)'$ where $Y_t = (y_t, \ldots, y_{t+1-q})'$, $q = 12$, I_t is the 3-month T-bill interest rate, and B_t is the business cycle coincident index. See Table 1, Panel A for details.

As the models are nested, we can not use the tests of Diebold and Mariano (1995) and West (1996). A test for Granger-causality is to compare the loss functions of forecasts conditional on two information sets, $\mathcal{G}_t$ and $\mathcal{F}_t$. As we are interested in comparing the loss of forecasting output growth y_{t+1} without and with using the information on past money growth X_t, we use the conditional predictive ability test of GW (2006). Let $L_{t+1}(\cdot)$ be a loss function. The null hypothesis of NGCM is therefore

$$H_0 : E[L_{t+1}(y_{t+1}, \hat{y}_{1,t+1}) - L_{t+1}(y_{t+1}, \hat{y}_{2,t+1})|\mathcal{F}_t] = 0, \qquad t = R, \ldots, T-1 \tag{5}$$

Under the H_0 the loss differential $\Delta L_{t+1} \equiv L_{t+1}(y_{t+1}, \hat{y}_{1,t+1}) - L_{t+1}(y_{t+1}, \hat{y}_{2,t+1})$ is a martingale difference sequence (MDS), which implies $E(h_t \Delta L_{t+1}) = 0$ for any h_t that is $\mathcal{F}_t$-measurable. Denoting $Z_{t+1} = h_t \Delta L_{t+1}$, the GW (2006) statistic is

$$GW_{R,P} = P\bar{Z}_{R,P}'\hat{\Omega}_P^{-1}\bar{Z}_{R,P} \tag{6}$$

Table 1. Description of Data Sets and Samples.

Panel A. Description of Data Sets				
	Income	Money	Interest Rate	Business Cycle Index
	y	m	I	B
Data Set 1	Real personal income	M2	3-Month T-bill rate	Coincident index
Data Set 2	Industrial production index	M2	3-Month T-bill rate	Coincident index

Panel B. Description of Subsamples in Out-of-Sample Tests					
	Starting Month	Ending Month	T	R	P
Subsample 1	1960:05	1991:04	360	240	120
Subsample 2	1962:05	1993:04	360	240	120
Subsample 3	1964:05	1995:04	360	240	120
Subsample 4	1966:05	1997:04	360	240	120
Subsample 5	1968:05	1999:04	360	240	120
Subsample 6	1970:05	2001:04	360	240	120
Subsample 7	1960:05	2001:12	500	380	120
Subsample 8	1960:05	2001:12	500	320	180
Subsample 9	1960:05	2001:12	500	260	240
Subsample 10	1960:05	2001:12	500	200	300

Notes: (1) To make these series stationary, we take log-difference of income and money variables, and take the first difference of interest rate.

(2) The business cycle index series are taken from James Stock's web page, http://www.economics.harvard.edu/faculty/stock/links.htm, while the other data are obtained from the Federal Reserve Economic Database (FRED) of Federal Reserve Bank of St. Luis.

(3) All data are monthly data, with sample period of 1959:04 to 2001:12, with 513 observations.

(4) Subsample 1 to 6 have a fixed window of 30 years, with 20 years as in-sample period and 10 years as out-of-sample period. Subsamples are moving forward by two years each time.

(5) Subsample 7 to 10 are the samples that contain all observations but with different combination of R and P. Due to the 12 lags used in the model and log-difference of money and income, there are 500 observations.

where $\bar{Z}'_{R,P} = \frac{1}{P}\sum_{t=R}^{T-1} h_t \Delta L_{t+1}$ and $\hat{\Omega}_P = \frac{1}{P}\sum_{t=R}^{T-1} Z_{t+1}Z'_{t+1}$. Under some regularity conditions, $GW_{R,P} \rightarrow^d \chi^2_q$ as $P \rightarrow \infty$ under H_0.[3] We choose the "test" function, h_t, such that it is $\mathcal{F}_t$-measurable but not $\mathcal{G}_t$-measurable. For simplicity, we choose $h_t = X_t = (x_t, \ldots, x_{t+1-q})'$.[4]

To test money–income Granger-causality in mean, we choose the squared error loss $L_{t+1}(y_{t+1}, \hat{y}_{i,t+1}) = \hat{\varepsilon}^2_{i,t+1} (i = 1, 2)$ for the out-of-sample forecast evaluation because the conditional mean is the optimal forecast under the squared error loss. We also minimize the same loss for in-sample parameter estimation as shown in Eqs. (3) and (4). Therefore, $Z_{t+1} = h_t \Delta L_{t+1} = h_t(\hat{\varepsilon}^2_{1,t+1} - \hat{\varepsilon}^2_{2,t+1})$. To be consistent with the literature using monthly series, we choose h_t using 12 lags of money growth rate, that is, $h_t = X_t = (x_t \ldots x_{t-11})'$ with $q = 12$.

Because the GW statistic is for equal conditional predictive ability test, the rejection of the null hypothesis only implies that the two models are not equal in conditional predictive ability. To choose one model over the other, we follow the decision rule suggested by GW (2006) to construct a statistic as

$$I_P = \frac{1}{P} \sum_{t=R}^{T-1} \mathbf{1}(\hat{\alpha}'_P h_t < 0) \tag{7}$$

where $\mathbf{1}(\cdot)$ is the indicator function and $\hat{\alpha}_P$ is the coefficient of h_t by regressing ΔL_{t+1} on $h_t (t = R, \ldots, T-1)$. As the rejection of H_0 occurs when the test function h_t can predict the loss difference ΔL_{t+1} in out-of-sample, $\hat{\alpha}'_P h_t \approx E(\Delta L_{t+1} | \mathcal{F}_t)$ will be the out-of-sample predicted loss differences. If I_P is greater than 0.5, Model 1 (NGCM) will be selected; otherwise Model 2 (GCM) will be selected.

Asymmetric GCM versus GCQ

Hayo (1999) nicely summarizes five stylized facts found in the empirical literature on the existence and strength of GCM between money and output using U.S. data: (a) In a model with only two variables, money Granger-causes output (Sims, 1972). (b) The statistical significance of the effect of money on output will be lower when including other variables in a multivariate test such as prices and interest rates (Sims, 1980). (c) The use of narrow money is less likely to support GC from money to output than broad money (King & Plosser 1984). (d) Assuming that variables are trend stationary and modelling them in (log-) levels with a deterministic trend is more likely to lead to significant test results than assuming difference stationary and employing growth rates (Christiano & Ljungqvist, 1988; Hafer & Kutan, 1997; Stock & Watson, 1989). (e) Allowing asymmetric effects of money on output growth and including the business cycle greatly influences results and strengthens the causal effect of money (Cover, 1992; Lo & Piger, 2005; Ravn & Sola, 2004; Thoma, 1994; Psaradakis, Ravn, & Sola, 2005; Weise, 1999).

Hayo (1999) revisited the above U.S. stylized facts using a broad data base of 14 EU countries plus Canada and Japan. It is found that very few of the above, particularly (b) and (d), can be sustained. Also found in the literature is that GCM is unstable, changing with sample periods, data to use (variables and frequency), and countries. Psaradakis et al. (2005) provide some summary on this instability evidence from the literature. Davis and Tanner (1997) also find the instability of the GCM across countries.

What appears to be robust is (e). Thoma (1994) shows for monthly data with M1 that the state of business cycle has considerable influence on the results and strengthens of the GCM of money. When real activity declines the effect of money on output becomes stronger, while the opposite takes place during a recovery. Numerous papers in the literature have found that the evidence for the GCM becomes more evident when some asymmetry has been introduced. Lo and Piger (2005) and Weise (1999) classify the three forms of asymmetry studies in a large body of empirical literature on money–income causality.

A1. (sign asymmetry): asymmetry related to the direction of the monetary policy action (Cover, 1992, Dolado, Pedrero, & Ruge-Murcia, 2004)

A2. (size asymmetry): asymmetry related to the size of the policy action. (Ravn & Sola 2004, Dolado et al., 2004)

A3. (business cycle asymmetry): asymmetry related to the existing business cycle business cycle phase (Garcia & Schaller 2002; Lo & Piger 2005; Thoma, 1994; Weise, 1999)

Weise (1999) finds no evidence for A1, some evidence for A2, and strong evidence for A3. Bernanke and Gertler (1995) and Galbraith (1996) explain A3 via credit rationing and its threshold effects on the relationship between money and output. Lo and Piger (2005) examine A3 using a regime-switching model in the response of U.S. output to monetary policy and find that policy actions during recessions have larger output effects than those taken during expansions. To deal with the instability and the asymmetry in GCM between money and income, many researchers have used split subsamples or rolling samples, or nonlinear models such as regime-switching models and threshold models.

The objective of this paper is to study GCQ, which is useful for scenario analysis in implementing monetary policy. Our empirical results (in section "Empirical Analysis") for GCQ are "symmetric", in that GCQ is insignificant in or near the center of the predicted distribution of the output growth while it is strongly significant in both tails. (The results of section "Empirical Analysis" show that GCQ is significant in both tails.) The difference between the asymmetric GCM and GCQ is that the former refers to the empirical fact that the predictive power of past money growth to predict the mean of output growth is stronger when the *past output growth* is negative (in recession), while the (symmetric) GCQ refers to the fact the predictive power of past money growth to predict the quantiles of output growth is stronger when the scenario of our interest is the *future output growth* in tails of its predicted distribution. Hence, the asymmetric GCM prescribes a monetary policy based on the past information, while the GCQ enables a monetary policy to be based on the forward looking scenarios of output growth. The GCQ can indicate how/whether the past and current money growth affects the various future states (that is, quantiles) of the output growth.

The GCQ is also different from the Granger-causality in Risk (GCR) proposed by Hong, Liu, and Wang (2009). They extend the causality in variance in the literature (volatility spillover) to causality in extreme distribution defined on the platform of Value-at-Risk. However, the focus of GCR is the risk spillover, that is, the causality from occurrence of one extreme event to the occurrence of another extreme event, while that for GCQ is the forecasting of quantiles based on all available information of the causal variable, not merely the tail distribution of the causal variable.

Money–Income Granger-Causality in Quantiles

Most empirical studies on money–income causality focus on Granger-causality in mean. As discussed above, in many cases, one may care about conditional distribution of output growth. Even without significant Granger-causality in mean, Granger-causality in distribution (GCD) may still be significant.

Definition 2. (Non-Granger-causality in distribution, NGCD): *X_t does not Granger-cause y_{t+1} in distribution if and only if* $\Pr(y_{t+1} < y | X_t, \mathcal{G}_t) = \Pr(y_{t+1} < y | \mathcal{G}_t)$ *a.s. for all y.*

Remark: Note that we can write for $y \in \mathbb{R}$,

$$G_{t+1}(y|\mathcal{F}_t) \quad = \quad \Pr(y_{t+1} < y | \mathcal{F}_t) = E[\mathbf{1}(y_{t+1} < y | \mathcal{F}_t)] = E(z_{t+1} | \mathcal{F}_t) \qquad (8)$$

$$G_{t+1}(y|\mathcal{G}_t) \quad = \quad \Pr(y_{t+1} < y|\mathcal{G}_t) = E[\mathbf{1}(y_{t+1} < y|\mathcal{G}_t)] = E(z_{t+1}|\mathcal{G}_t) \tag{9}$$

where $z_{t+1} = \mathbf{1}(y_{t+1} < y)$. Therefore, Definition 2 is equivalent to

$$E(z_{t+1}|\mathcal{F}_t) = E(z_{t+1}|\mathcal{G}_t) \ a.s. \text{ for all } y \tag{10}$$

Hong et al., (2009) use this to test for Granger-causality in risk for *a fixed* value of y between two financial markets (X_t and y_{t+1}). The GCD between X_t and y_{t+1} can be viewed as GCM between X_t and z_{t+1} for *all* y.

There is GCD if $\Pr(y_{t+1} < y|X_t, \mathcal{G}_t) \neq \Pr(y_{t+1} < y|\mathcal{G}_t)$ for *some* y. X_t does not Granger-cause y_{t+1} in distribution if $G_{t+1}(y|X_t, \mathcal{G}_t) = G_{t+1}(y|\mathcal{G}_t)$ *a.s.* or $g_{t+1}(y|X_t, \mathcal{G}_t) = g_{t+1}(y|\mathcal{G}_t)$ *a.s.*

As the conditional distribution can be inverted to conditional quantiles, we test for Granger-causality in conditional distribution via testing for Granger-causality in conditional quantiles. Let the conditional quantile of y_{t+1} be denoted $q_\alpha(y_{t+1}|\mathcal{F}_t)$ such that $G_{t+1}(q_\alpha(y_{t+1}|\mathcal{F}_t)|\mathcal{F}_t) = \alpha$. The conditional quantile $q_\alpha(y_{t+1}|X_t, \mathcal{G}_t)$ can be obtained by inverting the conditional distribution $G_{t+1}(y|\mathcal{F}_t) = \alpha$. Recall that $\mathcal{G}_t$ is the information set excluding X_t, that is, $\mathcal{G}_t = \mathcal{F}_t/\{X_t\}$. We now define GC in conditional quantile (GCQ).

Definition 3. (Non-Granger-causality in quantile): *X_t does not Granger-cause y_{t+1} in α-quantile if and only if $q_\alpha(y_{t+1}|X_t, \mathcal{G}_t) = q_\alpha(y_{t+1}|\mathcal{G}_t)$ a.s.*

GC in conditional quantile refers to the case that $q_\alpha(y_{t+1}|X_t, \mathcal{G}_t) \neq q_\alpha(y_{t+1}|\mathcal{G}_t)$. If X_t does not Granger-cause y_{t+1} in distribution, $q_\alpha(y_{t+1}|X_t, \mathcal{G}_t) = q_\alpha(y_{t+1}|\mathcal{G}_t)$ since $g_{t+1}(y|X_t, \mathcal{G}_t) = g_{t+1}(y|\mathcal{G}_t)$. Therefore, non-Granger-causality in distribution implies non-Granger-causality in conditional quantiles. On the contrary, GC in distribution does not necessarily imply GC in each quantile, while significant GC in any conditional quantile implies significant GC in distribution. For some quantiles, X_t may Granger-cause y_{t+1}, while for other quantiles it may not. Granger (2003, p. 700) notes that some quantiles may differ from other quantiles in time series behavior (such as long memory and stationarity). For example, different parts of the distribution can have different time series properties; one tail could be stationary and the other tail may have a unit root.

While the quantile forecast $q_\alpha(y_{t+1}|X_t, \mathcal{G}_t)$ can be derived from inverting the density forecast, in this paper we use linear quantile regression. An out-of-sample test for GCQ is based on two nested linear models. The first model does not account for money–income GC in α-quantile (referred as Model 1 or "NGCQ") and the second does (referred as Model 2 or "GCQ").

$$\text{Model 1}: \quad y_{t+1} = q_\alpha(y_{t+1}|\mathcal{G}_t) + e_{1,t+1} = V_t'\theta_1(\alpha) + e_{1,t+1} \tag{11}$$

$$\text{Model 2}: \quad y_{t+1} = q_\alpha(y_{t+1}|X_t, \mathcal{G}_t) + e_{2,t+1} = W_t'\theta_2(\alpha) + e_{2,t+1} \tag{12}$$

where $V_t \in \mathcal{G}_t$ and $W_t = (X_t' \;\; V_t')$ are vectors of regressors and V_t includes a constant term. The parameters $\theta_i(\alpha)$ are estimated by minimizing the "check" function discussed in Koenker and Bassett (1978) using the rolling sample of the most recent R observations at time $t (t = R, \ldots, T-1)$

$$\hat{\theta}_{1,t}(\alpha) \quad = \quad \arg\min_{\theta_1(\alpha)} \sum_{s=t-R+1}^{t} \rho_\alpha(y_s - V_{s-1}'\theta_1(\alpha)) \tag{13}$$

$$\hat{\theta}_{2,t}(\alpha) \quad = \quad \arg\min_{\theta_2(\alpha)} \sum_{s=t-R+1}^{t} \rho_\alpha(y_s - W_{s-1}'\theta_2(\alpha)) \tag{14}$$

where $\rho_\alpha(e) \equiv \alpha - \mathbf{1}(e<0)]e$. Denote $\hat{q}^1_{\alpha,t+1}(\hat{\theta}_{1,t}(\alpha))$, $\hat{q}^2_{\alpha,t+1}(\hat{\theta}_{2,t}(\alpha))$ for the α-quantile forecasts of y_{t+1} from Model 1 and Model 2 respectively, and let $\hat{e}_{i,t+1}(\hat{\theta}_{i,t}(\alpha)) = y_{t+1} - \hat{q}^i_{\alpha,t+1}(\hat{\theta}_{i,t}(\alpha))$.

Denote $\hat{\theta}_{2X,t}(\alpha)$ as parameters in the GCQ quantile regression model for X_t' and $\hat{\theta}_{2V,t}(\alpha)$ as parameters in that model for V_t'. Thus $\hat{\theta}_{2,t}(\alpha) = (\hat{\theta}'_{2X,t}(\alpha), \hat{\theta}'_{2V,t}(\alpha))$. A convenient test for NGCQ is therefore to test H_0 : $\hat{\theta}_{2X,t}(\alpha) = \mathbf{0}$. Koenker and Basset (1982) show that under some regularity conditions, the parameter estimates has asymptotic normal distribution, that is,

$$n^{1/2}(\hat{\theta}_{2,t}(\alpha) - \theta_{2,t}(\alpha)) \xrightarrow{d} N(\mathbf{0}, \omega^2(\alpha, G)\Omega^{-1}) \tag{15}$$

where $\Omega = \lim_{R\to\infty} R^{-1} \sum w_t w_t'$, $\omega^2(\alpha, G) = \alpha(\mathbf{1} - \alpha)/g^2(G^{-1}(\alpha))$ and $G(\cdot)$ and $g(\cdot)$ are the distribution function and density function of y respectively. The covariance matrix Ω can be estimated by its sample estimate $\hat{\Omega}$. $g(G^{-1}(\alpha))^{-1}$ is the reciprocal of the density function and is called the sparsity function. It can be estimated by the difference quotient of the empirical quantile function with a chosen bandwidth (such as the Bofinger bandwidth or the Hall–Sheather bandwidth).

Koenker and Machado (1999) propose two tests for $H_0 : \hat{\theta}_{2X,t}(\alpha) = \mathbf{0}$, a Wald-type test or a likelihood ratio test. Let the covariance matrix be partitioned accordingly in the GCQ model with Ω_{XX} and Ω_{VV} as the two diagonal terms and Ω_{XV} and Ω_{VX} as the two off-diagonal terms. The Wald-type test statistic is computed by

$$\text{Wald}_t(\alpha) = \hat{\theta}'_{2X,t}(\alpha)\hat{\Sigma}(\alpha)^{-1}\hat{\theta}_{2X,t}(\alpha) \tag{16}$$

where

$$\hat{\Sigma}(\alpha) = R^{-1}\hat{\omega}^2(\alpha, G)(\Omega_{XX} - \Omega_{XV}\Omega_{VV}^{-1}\Omega_{VX})^{-1}$$

and $\hat{\omega}(\alpha, G)$ is computed using the estimated sparsity function. The likelihood ratio test is based on the difference between the in-sample check loss. Denote $L_1(\alpha) = \sum_{s=t-R+1}^{t}\rho_\alpha(y_s - V'_{s-1}\theta_1(\alpha))$ and $L_2(\alpha) = \sum_{s=t-R+1}^{t}\rho_\alpha \times (y_s - W'_{s-1}\theta_2(\alpha))$. The LR test statistic is calculated by

$$\text{LR}_t(\alpha) = 2\left[\alpha(1-\alpha)\hat{g}(\hat{G}^{-1}(\alpha))^{-1}\right]^{-1}(L_1(\alpha) - L_2(\alpha)) \tag{17}$$

where $\hat{g}(\hat{G}^{-1}(\alpha)))^{-1}$ is the estimated sparsity function. Koenker and Machado (1999) show that under some regularity conditions, $\text{Wald}_t(\alpha)$ and $\text{LR}_t(\alpha)$ are equivalent and follow a χ^2_k distribution asymptotically where k is the dimension of X.

The above-mentioned two test statistics are relatively easy to compute and have good asymptotic property, but they have some deficiencies. For instance, the estimation of the sparsity function requires the Gaussian distribution and the nice asymptotic property of those test statistics is not robust with non i.i.d. cases. Some alternative methods have been put forward to deal with such deficiencies, for instance, the bootstrap resampling proposed by Parzen, Wei, and Ying (1994), or the Markov chain marginal bootstrap by He and Hu (2002), or the Huber sandwich local estimate of the sparsity function proposed by Koenker and d'Orey (1993). But no consensus has been reached on those issues. Moreover, it is merely an in-sample test and can not be used to compare the predictive power of different models, which, unfortunately, is exactly the essence of Granger-causality. Therefore, we propose to testing GCQ using a predictive ability test.

A test for Granger-causality is to compare the check loss functions of forecasts conditional on two information sets, $\mathcal{G}_t$ and $\mathcal{F}_t$. We again use the conditional predictive ability test of GW (2006) using the check loss function, that is, $L_{t+1}(y_{t+1}, \hat{y}_{i,t+1}) = \rho_\alpha(\hat{e}_{i,t+1}(\hat{\theta}_{i,t}(\alpha)))$. The null hypothesis of NGCQ is therefore

$$H_0 : E[\rho_\alpha(\hat{e}_{1,t+1}(\hat{\theta}_{1,t}(\alpha))) - \rho_\alpha(\hat{e}_{2,t+1}(\hat{\theta}_{2,t}(\alpha)))|\mathcal{F}_t] = 0, \qquad t = R, \ldots, T-1 \tag{18}$$

Under the H_0 the loss differential $\Delta L_{t+1} \equiv \rho_\alpha(\hat{e}_{1,t+1}(\hat{\theta}_{1,t}(\alpha))) - \rho_\alpha(\hat{e}_{2,t+1}(\hat{\theta}_{2,t}(\alpha)))$ is an MDS, which implies $E(h_t\Delta L_{t+1}) = 0$ for any h_t that is $\mathcal{F}_t$-measurable. Denoting $Z_{t+1} = h_t\Delta L_{t+1}$, the GW (2006) statistic is of the same form as in Eq. (6) with $\bar{Z}'_{R,P} = \frac{1}{P}\sum_{t=R}^{T-1} h_t \Delta L_{t+1}$ and $\hat{\Omega}_P = \frac{1}{P}\sum_{t=R}^{T-1} Z_{t+1} \times Z'_{t+1}$. Under some regularity conditions, $GW_{R,P} \to^d \chi^2_q$ as $P \to \infty$ under H_0. We choose the same test function $h_t = X_t = (x_t, \ldots, x_{t+1-q})'$ as before with q= 12. When the null hypothesis of the equal conditional predictive ability is rejected, the forecast model selection rule is the same as in Eq. (7) in Section "Money–Income Granger-Causality in Mean."

In Section "Empirical Analysis" for the empirical analysis, we choose $X_t = (x_t, \ldots, x_{t+1-q})'$ and $Y_t = (y_t, \ldots, y_{t+1-q})'$ with $q = 12$, and let $\mathcal{G}_t = \sigma(V_t)$ be the σ-field generated by $V_t = (Y'_t, I_t, B_t)'$ where I_t denotes the 3-month T-bill interest rate and B_t denotes the business cycle coincident index. See Table 1, Panel A.

EMPIRICAL ANALYSIS

In the literature, empirical studies of Granger-causality in mean commonly apply VAR models with exogenous variables. Different exogenous variables, such as treasury bill rates, federal funds rates, commercial paper rates, and business cycle indicators are used. Real personal income or industry production is used as the proxy for income, and M2 is used as the proxy for money stock. We report the results only with M2 for space as the results with M1 are similar. We use monthly data of real personal income, industrial production index, M2 money stock, 3-month T-bill interest rate, and the Stock and Watson experimental coincident index in the empirical study. The sample period is from 1959:04 to 2001:12 (513 observations). The source of the Stock and Watson experimental coincident index is the website of James Stock, while source for all other data is the Federal Reserve Economic Database (FRED) of Federal Reserve Bank of St. Luis.

We construct two data sets with two different income variables. Data Set 1 uses real personal income for income and M2 for money. Data Set 2 uses industrial production for income and also uses M2 for money. The description of those data sets is listed in Table 1, Panel A. Noting that output, money, and interest rate series are all non-stationary processes, we take the log-difference of output and money series and the first difference of interest rate series. Business cycle index (the Stock and Watson experimental coincident index) is a stationary process itself. Denote y_t as the output growth rate at time t,m_t as the money growth rate at time t, I_t as change of interest rate at time t, and B_t as the business cycle indicator at time t.

As discussed earlier in section "Introduction", it is well documented in the literature that the results for GCM are generally weak and sensitive over different sample periods. While we find that the results for GCQ in tails are much stronger, we also find that the GCQ results are robust over different sample periods. For the robustness check, we consider 10 different subsample periods constructed as follows. In each of Subsample 1 to 6, we set $T = 360$ (30 years), with $R = 240$ (20 years) and $P = 120$ (10 years). Forecasting horizon is 1, and a recursive method is used in each subsample. We shift the subsample period by two years to get another subsample. We also construct Subsample 7 to 10 with whole sample ($T = 500$), but with different combinations of R and P. A description of those subsamples is listed in Table 1, Panel B. As a referee pointed out, we can consider recent work by Rossi and Inoue (2012) and Hansen and Timmermann (2012) for testing predictive ability over a collection of sample splits. We leave this for other paper as the GCQ results presented in Table 2 are very strong and stable over all of the 10 subsamples, indicating the results are not due to data snooping over different sample splits.

Money–Income Granger-Causality in Mean

For the forecasting setting, as discussed in Section "Granger-causality", an out-of-sample Granger-causality test is more appropriate. We estimate two nested models, one model without money–income Granger-causality in mean (Model 1), and the other with money–income Granger-causality in mean (Model 2)

$$\text{Model 1}: \quad y_t = \beta_0 + \sum_{l=1}^{12} \beta_{y,l} y_{t-l} + \beta_I I_{t-1} + \beta_B B_{t-1} + \varepsilon_{1,t} \tag{19}$$

$$\text{Model 2}: \quad y_t = \beta_0 + \sum_{l=1}^{12} \beta_{y,l} y_{t-l} + \sum_{l=1}^{12} \beta_{m,l} m_{t-l} + \beta_I I_{t-1} + \beta_B B_{t-1} + \varepsilon_{2,t} \tag{20}$$

The unconditional out-of-sample mean quadratic losses of these two models are reported in Table 3, Panel A. In Data Set 1, the unconditional mean squared forecast error (MSFE) of Model 2 is less than that of Model 1, while in Data Set 2 the MSFE of Model 1 are generally smaller.

The p-values of $GW_{R,P}$ and I_P statistics are listed in Table 3, Panel B. The p-values of $GW_{R,P}$ indicate that the null hypothesis of the equal conditional predictive ability can not be rejected for all subsamples and for both data sets at a reasonable significance level.

Table 2. Out-of-Sample Test for Granger-Causality in Quantiles Test for Conditional Predictive Ability.

Panel A. Data Set 1

Sub- sample	α = 0.05		α = 0.10		α = 0.20		α = 0.30		α = 0.40		α = 0.50		α = 0.60		α = 0.70		α = 0.80		α = 0.90		α = 0.95	
	P_{GW}	I_{GW}	P_{GW}	I_{GW}	P_{GW}	I_{GW}	P_{GW}	I_{GW}	P_{GW}	I_{GW}	P_{GW}	I_{GW}	P_{GW}	I_{GW}	P_{GW}	I_{GW}	P_{GW}	I_{GW}	P_{GW}	I_{GW}	P_{GW}	I_{GW}
1	**0.0117**	**0.4500**	0.7422	0.4667	0.0471	0.6083	0.3107	0.5167	0.8232	0.6167	0.9025	0.4833	0.8333	0.5417	0.4207	0.6500	0.0002	0.6750	0.0085	0.6417	**0.0015**	**0.4000**
2	**0.0030**	**0.4667**	0.9390	0.4667	0.7655	0.5583	0.7137	0.6250	0.8008	0.7417	0.5908	0.6417	0.4173	0.5000	0.2084	0.5083	**0.0739**	**0.4667**	0.0192	0.5000	**0.0691**	**0.4583**
3	0.1779	0.3500	0.4385	0.4917	0.5595	0.4083	0.5189	0.5333	0.7671	0.5250	0.7387	0.6333	0.4010	0.4333	0.2000	0.4083	0.6170	0.3750	**0.0859**	**0.3917**	**0.0496**	**0.4083**
4	**0.0090**	**0.2833**	**0.0540**	**0.4583**	0.4654	0.4750	0.3460	0.5917	0.3853	0.6250	0.7581	0.5750	0.6841	0.4750	0.3531	0.4500	0.6698	0.5167	0.4106	0.3750	0.3230	0.3333
5	**0.0308**	**0.4333**	**0.0463**	**0.1583**	0.4546	0.4000	0.2683	0.5583	0.0689	0.5583	0.5733	0.6083	0.8052	0.5750	0.5605	0.4833	0.5908	0.4250	0.1066	0.3833	**0.0577**	**0.2917**
6	**0.0276**	**0.4167**	0.1218	0.1917	0.5791	0.3000	0.2310	0.5083	0.1766	0.6667	0.5141	0.4917	0.7996	0.4917	0.4245	0.4833	0.2632	0.3000	**0.0051**	**0.2667**	**0.0140**	**0.2417**
7	0.1685	0.4583	0.0846	0.4167	0.2196	0.3833	0.1963	0.5500	0.1008	0.6000	0.7883	0.6750	0.4584	0.6167	0.2115	0.4417	0.2124	0.4250	0.2393	0.2833	**0.0408**	**0.2500**
8	0.1136	0.5056	0.2827	0.4056	0.3981	0.5000	0.2287	0.6111	0.1810	0.6778	0.7713	0.6944	0.5418	0.6000	0.3520	0.4944	0.2462	0.4944	0.2197	0.3889	**0.0011**	**0.1889**
9	**0.0852**	**0.4958**	0.3348	0.4333	0.5132	0.4958	0.3477	0.5917	0.3066	0.6375	0.6190	0.6333	0.4214	0.5833	0.2017	0.4833	0.0854	0.5458	**0.0676**	**0.3917**	**0.0000**	**0.2375**
10	**0.0105**	**0.4567**	**0.0902**	**0.3967**	0.1426	0.5100	0.1308	0.6100	0.0696	0.6600	0.3022	0.5867	0.2714	0.6133	0.1773	0.4867	0.0378	0.5267	**0.0237**	**0.3300**	**0.0000**	**0.2433**

Panel B. Data Set 2

Sub-sample	α = 0.05		α = 0.10		α = 0.20		α = 0.30		α = 0.40		α = 0.50		α = 0.60		α = 0.70		α = 0.80		α = 0.90		α = 0.95	
	P_{GW}	I_{GW}	P_{GW}	I_{GW}	P_{GW}	I_{GW}	P_{GW}	I_{GW}	P_{GW}	I_{GW}	P_{GW}	I_{GW}	P_{GW}	I_{GW}	P_{GW}	I_{GW}	P_{GW}	I_{GW}	P_{GW}	I_{GW}	P_{GW}	I_{GW}
1	**0.0002**	**0.3417**	0.1628	0.3417	0.2214	0.4833	0.5189	0.6500	0.6532	0.4917	0.5541	0.6083	0.6286	0.5250	0.7982	0.4167	0.4725	0.5333	0.0689	0.5583	**0.0020**	**0.4333**
2	**0.0335**	**0.4083**	**0.1007**	**0.4083**	0.1330	0.4583	0.6568	0.3333	0.3921	0.3917	0.3471	0.5083	0.3360	0.5250	0.6631	0.2583	**0.0740**	**0.3750**	0.0083	0.5333	**0.0360**	**0.4000**
3	0.1249	0.2750	**0.1017**	**0.2417**	0.7646	0.6750	0.8068	0.4750	0.5166	0.5083	0.2172	0.6000	0.0282	0.5667	0.3202	0.5000	**0.0340**	**0.3667**	**0.0020**	**0.4250**	**0.0085**	**0.4167**
4	**0.0539**	**0.2083**	**0.1045**	**0.1417**	0.6332	0.4167	0.5413	0.4167	0.5070	0.4583	0.4126	0.6083	0.3790	0.5750	0.3531	0.4917	0.4377	0.3917	**0.0129**	**0.4417**	**0.0031**	**0.3333**
5	**0.0049**	**0.2333**	**0.0296**	**0.1583**	0.6305	0.3417	0.7290	0.5250	0.5032	0.4333	0.4176	0.5833	0.6161	0.4833	0.6840	0.4583	0.1679	0.4083	**0.0216**	**0.3833**	**0.0014**	**0.4500**
6	**0.0019**	**0.3000**	**0.0164**	**0.1333**	0.3927	0.3417	0.1993	0.4167	0.5847	0.6833	0.1255	0.4750	0.4161	0.4500	0.7979	0.4417	0.4001	0.3333	**0.0037**	**0.2417**	**0.1045**	**0.3833**
7	**0.0048**	**0.3167**	**0.0881**	**0.2167**	0.1320	0.3250	0.7178	0.4833	0.2517	0.5417	0.4681	0.4500	0.2813	0.4000	0.6677	0.4333	0.1654	0.4333	**0.0580**	**0.1833**	**0.0075**	**0.4333**
8	**0.0019**	**0.2556**	0.1205	0.2056	0.2170	0.3333	0.9851	0.4500	0.7164	0.6556	0.4882	0.5222	0.3393	0.5722	0.6736	0.4778	0.3259	0.4889	**0.0875**	**0.2722**	**0.0001**	**0.4111**
9	**0.0014**	**0.2042**	0.1190	0.1917	**0.0914**	**0.3917**	0.8946	0.5208	0.7946	0.6583	0.3457	0.4833	0.2288	0.4833	0.8215	0.4292	0.2571	0.5125	**0.0545**	**0.3167**	**0.0000**	**0.4375**
10	**0.0002**	**0.2167**	**0.0226**	**0.1933**	0.1229	0.3200	0.8991	0.4567	0.4114	0.6500	0.4170	0.5067	0.1150	0.5400	0.5778	0.5200	**0.0722**	**0.4800**	**0.0036**	**0.2533**	**0.0000**	**0.4333**

Notes: (1) The numbers in the first column is referring to the 16 subsamples. See Table 1, Panel B.
(2) P_{GW} refers to the asymptotic *P*-value of the nR^2 version of the Wald statistics of Giacomini and White (2006). We choose a linear test function which contains 12 lags of money growth rate. The asymptotic *P*-values of the Giacomini and White statistics are obtained from a chi-square distribution with 12 degrees of freedom.
(3) I_{GW} refers to the I_P statistic in Giacomini and White (2006). See Section 2.4.
(4) At 10% level, if $P_{GW} < 0.10$ and $I_{GW} < 0.5$, we prefer Model 2 (GCQ) over Model 1 (NGCQ). These cases are reported in bold font. If $P_{GW} < 0.10$ and $I_{GW} > 0.5$, we prefer Model 1 to Model 2.

Table 3. Out-of-Sample Test for Granger-causality in Quantiles Comparing Unconditional Predictive Ability (Check Loss).

Sub-sample			α=0.10		α=0.20		α=0.30		α=0.40		α=0.50		α=0.60		α=0.70		α=0.80		α=0.90		α=0.95	
	NGCQ	GCQ	NGCQ	GCQ	NGCQ	GCQ	NGCQ	GCQ	NGCQ	GCQ	NGCQ	GCQ	NGCQ	GCQ	NGCQ	GCQ	NGCQ	GCQ	NGCQ	GCQ	NGCQ	GCQ
Panel A. Data Set 1																						
1	0.0399	0.0277	0.0538	0.0535	0.0613	0.0641	0.0721	0.0715	0.0766	0.0743	0.0789	0.0745	0.0756	0.0738	0.0734	0.0733	0.0651	0.0690	0.0483	0.0491	0.0345	0.0301
2	0.0361	0.0259	0.0586	0.0573	0.0719	0.0720	0.0828	0.0851	0.0932	0.0930	0.0952	0.0920	0.0928	0.0899	0.0868	0.0832	0.0731	0.0713	0.0498	0.0463	0.0311	0.0296
3	0.0327	0.0300	0.0418	0.0415	0.0594	0.0566	0.0716	0.0738	0.0826	0.0793	0.0829	0.0833	0.0829	0.0782	0.0790	0.0710	0.0650	0.0615	0.0463	0.0390	0.0285	0.0223
4	0.0305	0.0277	0.0406	0.0394	0.0580	0.0573	0.0701	0.0733	0.0782	0.0829	0.0804	0.0821	0.0790	0.0800	0.0784	0.0752	0.0665	0.0655	0.0466	0.0420	0.0277	0.0218
5	0.0280	0.0274	0.0425	0.0379	0.0602	0.0572	0.0730	0.0764	0.0802	0.0838	0.0814	0.0846	0.0814	0.0826	0.0797	0.0792	0.0664	0.0653	0.0415	0.0392	0.0260	0.0173
6	0.0419	0.0339	0.0602	0.0537	0.0772	0.0701	0.0868	0.0891	0.0937	0.0986	0.0988	0.1006	0.0999	0.1010	0.0936	0.0950	0.0792	0.0762	0.0521	0.0478	0.0336	0.0259
7	0.0409	0.0346	0.0664	0.0639	0.0865	0.0869	0.0987	0.1010	0.1053	0.1093	0.1093	0.1099	0.1051	0.1077	0.1027	0.0993	0.0840	0.0838	0.0588	0.0483	0.0335	0.0256
8	0.0346	0.0324	0.0525	0.0503	0.0692	0.0705	0.0800	0.0821	0.0858	0.0896	0.0894	0.0898	0.0860	0.0879	0.0834	0.0824	0.0699	0.0707	0.0490	0.0450	0.0306	0.0241
9	0.0379	0.0320	0.0604	0.0593	0.0743	0.0748	0.0855	0.0865	0.0912	0.0922	0.0943	0.0922	0.0909	0.0908	0.0889	0.0859	0.0749	0.0757	0.0535	0.0503	0.0346	0.0280
10	0.0378	0.0326	0.0589	0.0572	0.0720	0.0728	0.0829	0.0845	0.0879	0.0904	0.0906	0.0895	0.0874	0.0880	0.0856	0.0836	0.0715	0.0728	0.0513	0.0470	0.0313	0.0255
Panel B. Data Set 2																						
1	0.0197	0.0155	0.0338	0.0306	0.0569	0.0550	0.0693	0.0718	0.0778	0.0779	0.0807	0.0811	0.0780	0.0764	0.0677	0.0664	0.0535	0.0543	0.0332	0.0342	0.0196	0.0174
2	0.0181	0.0149	0.0316	0.0278	0.0515	0.0506	0.0637	0.0607	0.0751	0.0708	0.0747	0.0731	0.0726	0.0722	0.0656	0.0625	0.0505	0.0500	0.0335	0.0356	0.0212	0.0191
3	0.0223	0.0157	0.0379	0.0313	0.0580	0.0607	0.0737	0.0744	0.0828	0.0825	0.0837	0.0859	0.0804	0.0822	0.0707	0.0713	0.0553	0.0541	0.0356	0.0346	0.0214	0.0192
4	0.0206	0.0165	0.0398	0.0310	0.0584	0.0564	0.0734	0.0728	0.0784	0.0784	0.0785	0.0801	0.0758	0.0786	0.0695	0.0687	0.0561	0.0536	0.0344	0.0331	0.0221	0.0201
5	0.0207	0.0168	0.0417	0.0337	0.0659	0.0619	0.0776	0.0774	0.0859	0.0841	0.0855	0.0870	0.0824	0.0833	0.0751	0.0743	0.0623	0.0601	0.0406	0.0373	0.0231	0.0231
6	0.0195	0.0167	0.0421	0.0323	0.0618	0.0578	0.0769	0.0731	0.0821	0.0855	0.0835	0.0827	0.0799	0.0792	0.0724	0.0702	0.0578	0.0556	0.0399	0.0324	0.0197	0.0187
7	0.0206	0.0172	0.0414	0.0331	0.0607	0.0553	0.0722	0.0708	0.0770	0.0789	0.0813	0.0792	0.0745	0.0723	0.0663	0.0645	0.0552	0.0535	0.0350	0.0297	0.0188	0.0163
8	0.0195	0.0164	0.0382	0.0318	0.0592	0.0553	0.0719	0.0703	0.0782	0.0799	0.0815	0.0810	0.0757	0.0761	0.0671	0.0663	0.0538	0.0531	0.0339	0.0312	0.0197	0.0176
9	0.0200	0.0162	0.0367	0.0312	0.0568	0.0543	0.0705	0.0701	0.0776	0.0796	0.0827	0.0813	0.0773	0.0758	0.0684	0.0667	0.0556	0.0560	0.0346	0.0325	0.0193	0.0175
10	0.0195	0.0161	0.0377	0.0321	0.0582	0.0550	0.0714	0.0702	0.0774	0.0793	0.0820	0.0810	0.0771	0.0759	0.0683	0.0676	0.0557	0.0557	0.0347	0.0317	0.0192	0.0173

Notes: (1) The numbers in the first column is referring to the 10 subsamples. See Table 1, Panel B.

(2) "NGCQ" refers to Model 1, the quantile forecasting model without money–income Granger-causality in quantile, that is, not including the lagged money growth rate as independent variables.

(3) "GCQ" refers to Model 2, the quantile forecasting model with money–income Granger-causality in quantile, that is, including the lagged money growth rate as independent variables.

(4) A check loss function proposed by Koenker and Bassett (1978) is used to evaluate the out-of-sample performance of the two quantile forecasting models. The out- of-sample average of the loss values are reported in this table. The loss value of Model 2 is shaded when it is smaller than that of Model 1.

Comparing Data Set 1 to Data Set 2, GCM remains insignificant whether real personal income or industrial production is used. (Similarly GCM is not significant with M1 or M2 although the results with M1 are not shown to save space.) The results of the different sample periods (Subsample 1–6) are very robust, showing that with the shift of the sample window, money–income causality in mean remains insignificant across all the data sets. With the increase of ratio of P/R from Subsamples 7–10, GCM still remains insignificant.

In a forecasting model, using so many lagged money variables in Model 2 may cause the "over-fit" of the model and damage the forecasting performance. Therefore, in order to reduce the number of parameters in the large model, we also check the robustness of our GCM results by using a weighed moving average of $(x_t, \ldots, x_{t+1-q})$ for estimation and forecasting, for example, $\sum_{l=1}^{q} w_l x_{t+1-l}$ with weights w_l such that $\sum_{l=1}^{q} w_l = 1$. We use three such different weight functions, namely a linear declining weight, a equal weight ($w_l = q^{-1}$), and a beta polynomial function which creates flexible nonlinear declining weights as introduced in Ghysels, Sinko, and Valkanov (2007). We use these weighted moving average (a scalar) in place of the q-vector X_t in estimation, forecasting, and testing. It is found that Model 2 (GCM) is still no better than Model 1 (NGCM) in terms of predictive ability. Hence, we find that out-of-sample GCM is not significant. Adding the information on lagged money growth rate is not very useful to improve the conditional mean forecasting of U.S. output growth over the various sample periods and different choices of the variables.

Money–Income Granger-Causality in Quantile

As discussed in section "Introduction," significant GCD does not imply GCQ in each conditional quantile. Therefore, in our empirical study, we choose 11 quantiles (α = 0.05, 0.1, 0.2, 0.3, 0.4, 0.5, 0.6, 0.7, 0.8, 0.9, 0.95). We check the GCQ in these different quantiles of the conditional distribution of output growth.

First, we use the check loss function to compare the *unconditional* predictive ability of the GCQ and the NGCQ models.[5] The unconditional check loss values of GCQ and NGCQ models are reported in Table 4. The ratios of the unconditional check loss values of GCQ to NGCQ can be easily obtained from Table 4 for each α. (The ratios are not presented for space.) The ratio less than 1 indicates the money–income Granger-causality in

Table 4. Out-of-Sample Test for Granger-Causality in Mean.

Panel A. Comparing Unconditional Predictive Ability (Squared Error Loss)

	Data Set 1		Data Set 2	
Loss	Model 1	Model 2	Model 1	Model 2
Subsample 1	0.0778	0.0744	0.0415	0.0419
Subsample 2	0.0927	0.0884	0.0378	0.0377
Subsample 3	0.0823	0.0780	0.0460	0.0468
Subsample 4	0.0820	0.0776	0.0435	0.0436
Subsample 5	0.0818	0.0768	0.0509	0.0516
Subsample 6	0.1124	0.1036	0.0452	0.0455
Subsample 7	0.1252	0.1162	0.0406	0.0405
Subsample 8	0.0926	0.0881	0.0418	0.0419
Subsample 9	0.1014	0.0956	0.0414	0.0420
Subsample 10	0.0993	0.0937	0.0416	0.0420

Panel B. Test for Conditional Predictive Ability

	Data Set 1		Data Set 2	
	P_{GW}	I_{GW}	P_{GW}	I_{GW}
Subsample 1	0.5427	0.5583	0.4873	0.5417
Subsample 2	0.5396	0.5250	0.5154	0.5000
Subsample 3	0.5588	0.4750	0.6943	0.5417
Subsample 4	0.6817	0.4583	0.8779	0.5583
Subsample 5	0.5015	0.4083	0.7118	0.6083
Subsample 6	0.4113	0.3250	0.6066	0.5500
Subsample 7	0.6436	0.4167	0.8561	0.4833
Subsample 8	0.5155	0.4500	0.8597	0.5778
Subsample 9	0.3409	0.4583	0.7694	0.5833
Subsample 10	0.2876	0.4233	0.8342	0.5900

Notes: Quadratic loss values for two models are reported. "Model 1" refers to the model without Granger- causality in mean, while "Model 2" refers to the model with Granger-causality in mean. The loss value of Model 2 is shaded when it is smaller than that of Model 1. P_{Gw} refers to the asymptotic *P*-value of the nR^2 version of the Wald statistics of Giacomini and White (2006). We choose a linear test function which contains 12 lags of money growth rate. The asymptotic *P*- values of the Giacomini and White statistics are obtained from a chi-square distribution with 12 degrees of freedom. I_{Gw} refers to the I_P statistic in Giacomini and White (2006). See Section "Money–Income Granger-causality in mean." At 10% level, if $P_{GW} < 0.10$ and $I_{GW} < 0.5$, we may prefer Model 2 (GCM) over the Model 1 (NGCM); if $P_{GW} < 0.10$ and $I_{GW} > 0.5$, we may prefer Model 1 to Model 2. None of the cases satisfies ($P_{GW} < 0.10$ and $I_{GW} < 0.5$) or ($P_{GW} < 0.10$ and $I_{GW} > 0.5$). In fact all P-values are very large.

quantile. In terms of check losses and loss ratios, the GCQ model performs better than the NGCQ model in almost all subsamples of both data sets in the tails. In the central region, however, the GCQ model has lower check losses than the NGCQ model only in a few subsamples. The same pattern is observed in the rolling subsamples (Subsample 1–6). This implies that GCQ is stable across the sample period and data sets. In the whole data subsamples (Subsample 7–10), we find more significant GCQ with the increase of P. The loss ratios are much smaller than 1 in the tails than in the central regions.

Next, to compare the *conditional* check loss values, the P-values of $GW_{R,P}$ and I_P statistics are reported in Table 2. According to the P-values of $GW_{R,P}$ and I_P for the conditional predictive ability test, the GCQ model is significantly better to the NGCQ model in the tails. After accounting for money–income Granger-causality, quantile forecasting of output is improved at tails. The Granger-causality in quantile seems to be more significant between money and industrial production than that between money and personal income. Money does significantly improve the forecasting of output/income tail quantiles. However, money does not improve forecasting of the output growth in conditional mean and the conditional quantiles close to median.

CONCLUSIONS

The relationship between money and income is a much-studied but controversial topic in the literature. This paper follows a VAR framework and applies an out-of-sample test for money–income Granger-causality. We find that money–income Granger-causality in mean is not significant for all data sets and all subsample periods that we considered.

We define Granger-causality in quantile and compare two quantile forecasts with or without money–income Granger-causality in quantile. Empirical results show the potential of improving quantile forecasting of output growth rate by incorporating information on money–income causality in quantile, especially in the tails. Causality between money and industrial production seems to be more significant than that between money and personal income (while M2 has stronger causality in quantiles to personal income than M1 does). However, money is not very useful for forecasting near the center quantiles of the conditional distribution of output growth.

The empirical findings of this paper on the money–income Granger-causality in quantiles is entirely new and have never been documented in the money–income literature. The new results on GCQ have an important implication on monetary policy, showing that the effectiveness of monetary policy has been under-estimated by merely testing Granger-causality in mean. Money does Granger-cause income more strongly than it has been known and therefore the information on money growth can (and should) be more utilized in implementing monetary policy.

What does the lack of GC in mean tell us that is different from the presence of GC in the extreme quantiles? As a referee points out, the answer may be "risk management" in the sense of Kilian and Manganelli (2008), who derive a generalization of the Taylor rule (that links changes in the interest rate to the balance of the risks implied by the dual objective of sustainable economic growth and price stability) which reconciles economic models of expected utility maximization with the risk management approach to central banking. The results of Kilian and Manganelli (2008) suggest that Fed policy decisions under Greenspan were better described in terms of the Fed weighing upside and downside risks to their objectives rather than simply responding to the conditional mean of output growth (or output gap) and inflation.

ACKNOWLEDGMENTS

We thank an anonymous referee, Clive Granger, Yongmiao Hong, Michael Owyang, Peter Phillips, Philip Rothman, and the seminar participants at the Symposium on Econometric Theory and Applications (SETA) and the Society of Nonlinear Dynamics and Econometrics (SNDE) conference for useful comments. All errors are our own. A part of the research was conducted while Lee was visiting the California Institute of Technology. Lee is thankful for their hospitality and the financial support during the visit. Yang is thankful for the Chancellor's Distinguished Fellowship from the University of California, Riverside.

NOTES

1. Granger (2006) remarks, in the 20th anniversary issue of the *Advances in Econometrics*, "For most of its history time series theory considered conditional means, but later conditional variances. The next natural development would be

conditional quantiles, but this area is receiving less attention than I expected. The last stages are initially conditional marginal distributions, and finally conditional multivariate distributions. Some interesting theory is starting in these areas but there is enormous amount to be done."

2. For instance, by replacing the 3-month T-bill interest rate by commercial paper rate, the money–income causality become less significant. In general, these empirical studies give us relatively controversial results on the money–income causality.

3. Chao et al. (2001) propose an out-of-sample test using following test statistic

$$CCS_{R,P} = \frac{1}{P}\sum_{t=R}^{T-1} \hat{\varepsilon}_{1,t+1} h(X_t)$$

which follows zero-mean normal distribution asymptotically with its asymptotic variance incoorporating estimation error.

4. Two possible ways to improve the power of the test are (i) to choose q in a way to maximize the test power and (ii) to choose h_t from transforms of X_t as suggested in Bierens (1990), Stinchcombe and White (1998), or Hong (1999). We do not consider these extensions in this paper for simplicity and also to match the choice of h_t with the vast literature on GCM. Following Lee, White and Granger (1993) and Stinchcombe and White (1998), h_t will be called a *test* function.

5. Besides the standard check loss, as a robust check, we also use the loss functions of the tick-exponential family introduced in Komunjer (2005). The results using these generalized check functions were essentially the same as those reported here with the standard check loss function and therefore not reported for space.

REFERENCES

Ashley, R., Granger, C. W. J., & Schmalensee, R. (1980). Advertising and aggregate consumption: An analysis of causality. *Econometrica, 48*, 1149–1167.

Bernanke, B. S., & Gertler, M. (1995). Inside the black box: The credit channel of monetary policy transmission. *Journal of Economic Perspective*, *9*, 27–48.

Bierens, H. B. (1990). A consistent conditional moment test of functional form. *Econometrica*, *58*, 1443–1458.

Chao, J., Corradi, V., & Swanson, N. (2001). An out of sample test for granger causality. *Macroeconomic Dynamics*, *5*, 598–620.

Christiano, L. J., & Ljungqvist, L. (1988). Money does Granger-cause output in the bivariate money-output relation. *Journal of Monetary Economics*, *22*, 217–234.

Cover, J. P. (1992). Asymmetric effects of positive and negative money supply shocks. *Quarterly Journal of Economics*, *107*(4), 1261–1282.

Davis, M. S., & Tanner, J. E. (1997). Money and economic activity revisited. *Journal of International Money and Finance*, *16*, 955–968.

Diebold, F. X., & Mariano, R. S. (1995). Comparing predictive accuracy. *Journal of Business and Economic Statistics*, *13*, 253–263.

Dolado, J., Pedrero, R., & Ruge-Murcia, F. J. (2004). Nonlinear monetary policy rules: Some new evidence for the U.S. *Studies in Nonlinear Dynamics and Econometrics*, *8*(3), Article 2.

Dufour, J., & Renault, E. (1998). Short run and long run causality in time series: Theory. *Econometrica, 66*, 1099–1125.
Dufour, J., Pelletier, D., & Renault, E. (2006). Short run and long run causality in time series: Inference. *Journal of Econometrics, 132*, 337–362.
Eichenbaum, M., & Singleton, K. J. (1987). Do equilibrium real business cycle theories explain postwar U.S. business cycles. *NBER Macroeconomics Annual, 2*, 91–134.
Friedman, M. (1956). The quantity theory of money: A restatement. In M. Friedman (Ed.), *Studies in the quantity theory of money*. Chicago: University of Chicago Press.
Friedman, B. M., & Kuttner, K. N. (1992). Money, income, prices, and interest rates. *American Economics Review, 82*, 472–492.
Friedman, B. M., & Kuttner, K. N. (1993). Another look at the evidence on money-income relationship. *Journal of Econometrics, 57*, 189–203.
Galbraith, J. W. (1996). Credit rationing and threshold effects in the relationship between money and output. *Journal of Applied Econometrics, 11*(4), 419–429.
Garcia, R., & Schaller, H. (2002). Are the effects on interest rate changes asymmetric? *Economic Inquiry, 40*, 102–119.
Ghysels, E., Sinko, A., & Valkanov, R. (2007). MIDAS regressions: Further results and new directions. *Econometric Reviews, 26*(1), 53–90.
Giacomini, R., & White, H. (2006). Tests of conditional predictive ability. *Econometrica, 74*, 1545–1578.
Granger, C. W. J. (1969). Investigating causal relations by econometric models and cross-spectral methods. *Econometrica, 37*, 424–438.
Granger, C. W. J. (1980). Testing for causality: A personal viewpoint. *Journal of Economic Dynamics and Control, 2*, 329–352.
Granger, C. W . J. (1988). Some recent developments in a concept of causality. *Journal of Econometrics, 39*, 199–211.
Granger, C. W . J. (2003). Time series concepts for conditional distributions. *Oxford Bulletin of Economics and Statistics, 65*, 689–701.
Granger, C. W . J. (2005). The past and future of empirical finance: Some personal comments. *Journal of Econometrics, 129*, 35–40.
Granger, C. W . J. (2006). The creativity process. In D. Terrell & T. B. Fomby (Eds.), *Advances in econometrics* (Vol. 20, pp. xxiii–xxv). Amsterdam: Elsevier.
Granger, C. W. J., & Newbold, P. (1986). *Forecasting economic time series* (2nd ed.). New York: Academic Press.
Hafer, R. W., & Kutan, A. M. (1997). More evidence on the money-output relationship. *Economic Inquiry, 35*, 48–58.
Hansen, P.R. and Timmermann, A. (2012). *Choice of sample split in out-of-sample forecast evaluation*. Working Paper. EUI and UC San Diego.
Hayo, B. (1999). Money-output Granger causality revisited: An empirical analysis of EU countries. *Applied Economics, 31*(11), 1489–1501.
He, X., & Hu, F. (2002). Markov chain marginal bootstrap. *Journal of the American Statistical Association, 97*, 783–795.
Hong, Y. (1999). Hypothesis testing in time series via the empirical characteristic function: A generalized spectral density approach. *Journal of American Statistical Association, 84*, 1201–1220.
Hong, Y., Liu, Y., & Wang, S. (2009). Granger causality in risk and detection of extreme risk spillover between financial markets. *Journal of Econometrics, 150*(2), 271–287.

Kaldor, N. (1970). The new monetarism,. *Lloyds Bank Review*, *97*(July), 1–18.

Kilian, L., & Manganelli, S. (2008). The central banker as a risk manager: Estimating the Federal Reserve's preferences under Greenspan. *Journal of Money, Credit and Banking*, *40*(6), 1103–1129.

King, R. G., & Plosser, C. I. (1984). Money, credit, and prices in a real business cycle. *American Economic Review*, *74*, 363–380.

Koenker, R., & Bassett, G. (1978). Regression quantiles. *Econometrica*, *46*, 33–50.

Koenker, R., & Bassett, G. (1982). Robust tests for heteroscedasticity based on regression quantiles. *Econometrica*, *50*, 43–61.

Koenker, R., & d'Orey, V. (1993). A remark on computing regression quantiles. *Applied Statistics*, *36*, 383–393.

Koenker, R., & Machado, J. (1999). Goodness of fit and related inference processes for quantile regression. *Journal of the American Statistical Association*, *94*, 1296–1310.

Komunjer, I. (2005). Quasi-maximum likelihood estimation for conditional quantiles. *Journal of Econometrics*, *128*(1), 137–164.

Lee, T.-H., White, H., & Granger, C. W. J. (1993). Testing for neglected nonlinearity in time-series models: A comparison of neural network methods and standard tests. *Journal of Econometrics*, *56*, 269–290.

Lo, M. C., & Piger, J. (2005). In the response of output to monetary policy asymmetric? Evidence from a regime-switching coefficients model. *Journal of Money, Credit, and Banking*, *37*(5), 865–886.

Meltzer, A. H. (1963). The demand for money: The evidence from time series. *Journal of Political Economy*, *71*, 219–246.

Modigliani, F. (1977). The monetarist controversy: Or should we forsake stabilization policies? *American Economic Review*, *67*, 1–19.

Parzen, M. I., Wei, L., & Ying, Z. (1994). A resampling method based on pivotal estimating functions. *Biometrika*, *81*, 341–350.

Patton, A. J. (2006). Modelling asymmetric exchange rate dependence. *International Economic Review*, *47*(2), 527–556.

Psaradakis, Z., Ravn, M. O., & Sola, M. (2005). Markov switching causality and the money-output relationship. *Journal of Applied Econometrics*, *20*, 665–683.

Ravn, M. O., & Sola, M. (2004). Asymmetric effects of monetary policy in the United States. *Federal Reserve Bank of St. Louis Review*, *86*(5), 41–60.

Rossi, B., & Inoue, A. (2012). Out-of-sample forecast tests robust to the choice of window size. *Journal of Business and Economic Statistics*, *30*(3), 432–453.

Sims, C. A. (1972). Money, income and causality. *American Economic Review*, *62*, 540–542.

Sims, C. A. (1980). Comparison of interwar and postwar business cycles: Monetarism reconsidered. *American Economic Review*, *70*, 250–259.

Stinchcombe, M., & White, H. (1998). Consistent specification testing with nuisance parameters present only under the alternative. *Econometric Theory*, *14*, 295–324.

Stock, J. H., & Watson, M. W. (1989). Interpreting the evidence on money-income Causality. *Journal of Econometrics*, *40*, 161–181.

Stock, J. H., & Watson, M. W. (1996). Evidence on structural instability in macroeconomic time series relations. *Journal of Business and Economic Statistics*, *14*, 11–30.

Swanson, N. R. (1998). Money and output viewed through a rolling window. *Journal of Monetary Economics*, *41*, 455–474.

Thoma, M. A. (1994). Subsample instability and asymmetries in money-income causality. *Journal of Econometrics*, *64*, 279–306.
Weise, C. L. (1999). The asymmetric effects of monetary policy: A nonlinear vector autoregression approach. *Journal of Money, Credit and Banking*, *31*, 85–108.
West, K. (1996). Asymptotic inference about predictive ability. *Econometrica*, *64*, 1067–1084.

COPULA–GARCH TIME-VARYING TAIL DEPENDENCE

Jiaqi Chen and Jeffery W. Gunther

ABSTRACT

Tail-dependence evolution for the symmetrized Joe–Clayton copula is proposed to depend on an exponentially weighted moving average (EWMA) of the absolute difference in probability integral transforms. Using these dynamics, time-varying tail dependence between bank and insurance equity prices is assessed in a parametric copula, generalized autoregressive conditional heteroscedastic framework. The results suggest a relatively long lag and support the EWMA lag structure as an effective estimation vehicle. Tail dependence is shown often to tend higher during periods of market stress.

Keywords: Copula; dependence; time-varying

Financial correlation and return dependence are general issues with important implications in a variety of market contexts, and they become particularly important in the area of risk management. Both practitioners and the public policy overseers charged with systemic stability take keen interest in the amplifying effects that rising dependence among financial instruments and asset classes can exert on the volatility of financial-market

30th Anniversary Edition
Advances in Econometrics, Volume 30, 411–425

ISSN: 0731-9053/doi:10.1108/S0731-9053(2012)0000030018

and investment-portfolio returns. Of particular interest is the notion of tail dependence and the susceptibility to crashes in a given asset return, given that another has crashed, as this type of risk can contribute to systemic crises, characterized by widespread, severe losses.

The example we address here involves bank returns and their tail dependence with the returns to other types of financial businesses, in this case the business of insurance. A high degree of such dependence could impair the ability of banking organizations to diversify effectively through conglomeration. More broadly, such dependence could impair the stability of the financial sector as a whole, if component businesses were susceptible to common shocks. Because the financial sector and banks in particular carry out the financial intermediation function necessary for overall economic growth, these issues become all the more important from a public policy perspective.

Yet, for present purposes, the problem of bank-insurance return dependence serves mainly as a setting in which to address related econometric issues. For the measurement and estimation of financial dependence, the theory of copulas can play a natural role, especially as it relates to tail dependence. However, financial dependence has proven anything but constant; hence, a crucial role for the theory of time-varying copulas.

In that regard, we pick up on the work of Patton (2002, 2006a, 2006b) in modeling time-varying dependence via copulas. Econometrically, our focus is on the evolution equation used to describe time-varying dependence, and in our application parametric copula, generalized autoregressive conditional heteroscedastic (GARCH) modeling is used to estimate time-varying tail dependence between bank and insurance daily stock returns over the 20-year period of 1992–2011. The sample encompasses both prosperous times and intervals of financial turmoil, allowing us to estimate the cyclical behavior of return dependence.

A related literature focuses on the potential for return linkages to become magnified during a crisis. Longin and Solnik (2001) and Ang and Chen (2002) support the view that equity market correlation is amplified for bear markets and downside moves. Forbes and Rigobon (2002) find comovements do not increase in a crisis, whereas Bartram and Wang (2005) question the generality of that result. Corsetti, Pericoli, and Sbracia (2005) find evidence of international contagion, as do Caporale, Cipollini, and Spagnolo (2005). Along these lines, Fomby, Gunther, and Hu (2012) show return dependence between banks and other financial businesses increased during the financial crisis of 2007–2009, relative to the preceding 3-year period.

In this chapter, we extend the work of Fomby, Gunther, and Hu by employing the previously mentioned longer dataset, covering several financial cycles, and also by incorporating a different formulation for time-varying tail dependence. Specifically, we estimate a dependence evolution equation for Patton's (2006b) symmetrized Joe–Clayton (SJC) copula using an exponentially weighted moving average (EWMA) of the absolute difference in probability integral transforms. Our primary econometric conclusion is that in attempting to model time-varying dependence, much is to be gained from careful consideration and specification of the formula through which time variation is driven, especially in terms of the lag structure between dependence and its forcing terms. In our results, a relatively long EWMA receives support over shorter and equally weighted lags. The findings also suggest tail dependence often moves higher during periods of market stress. The following section describes the copula–GARCH approach, after which the specification of time-varying tail dependence is addressed, followed by a presentation of our empirical results.

COPULA–GARCH

The bivariate framework presents two random variables Y_1 and Y_2 with joint distribution function $F(y_1, y_2) = \Pr(Y_1 \leq y_1, Y_2 \leq y_2)$ and continuous marginal distributions $F_1(y_1) = \Pr(Y_1 \leq y_1)$ and $F_2(y_2) = \Pr(Y_2 \leq y_2)$. Sklar (1959) demonstrates that a unique copula exists such that:

$$C(u, v) = C(F_1(y_1), F_2(y_2)) = F(y_1, y_2) \tag{1}$$

As such, the joint distribution $F(y_1, y_2)$ is decomposed into marginal distributions $F_1(y_1)$ and $F_2(y_2)$ together with a dependence function $C(u, v)$. The probability integral transforms $U = F_1(Y_1)$ and $V = F_2(Y_2)$ are uniformly distributed (Fisher, 1932; Rosenblatt, 1952). We have, then, the copula as a joint distribution function of uniform variates U and V.[1] Via this decomposition, attention can focus on the dependence structure, as formulated in $C(u, v)$. Patton (2002, 2006a, 2006b) extends these results to encompass conditional copulas, facilitating an examination of time-varying conditional dependence.

Depending on the specific dependence function, the copula framework can be used to estimate tail dependence, or the likelihood of an extreme event in one variable, given a similarly extreme event in the other. Pushing this conditional probability far into the tail region gives rise asymptotically to the tail index, the dependence measure with which we work here. Upper tail dependence is defined as:

$$\tau^U = \lim_{\varepsilon\to 1} \Pr[U > \varepsilon | V > \varepsilon] = \lim_{\varepsilon\to 1} \Pr[V > \varepsilon | U > \varepsilon] = \lim_{\varepsilon\to 1} \frac{1 - 2\varepsilon + C(\varepsilon,\varepsilon)}{1-\varepsilon} \quad (2)$$

If this limit exists, a copula demonstrates upper tail dependence when $\tau^U > 0$. Similarly, lower tail dependence is defined as:

$$\tau^L = \lim_{\varepsilon\to 0} \Pr[U \leq \varepsilon | V \leq \varepsilon] = \lim_{\varepsilon\to 0} \Pr[V \leq \varepsilon | U \leq \varepsilon] = \lim_{\varepsilon\to 0} \frac{C(\varepsilon,\varepsilon)}{\varepsilon} \quad (3)$$

Different copula specifications can be used to restrict τ^U and τ^L. In contrast, the SJC copula used here keeps them open to estimation. In the context of financial markets, lower tail dependence can serve as a measure of downside portfolio or sector risk. For the present application, the degree of lower tail dependence between bank and insurance returns is important, as a tendency for insurance returns to crash, when the returns to banking have crashed, would work against any diversification benefits of conglomeration.

To focus on τ^U and τ^L, we employ the SJC copula, a modification of the original Joe–Clayton copula.[2] The SJC copula is defined as:

$$\begin{aligned} C_{SJC}(u, v|\tau^U, \tau^L) = 0.5 \cdot (&C_{JC}(u, v|\tau^U, \tau^L) \\ &+ C_{JC}(1-u, 1-v|\tau^L, \tau^U) + u + v - 1) \end{aligned} \quad (4)$$

where C_{JC} represents the original Joe–Clayton copula, as found in Joe (1997):

$$\begin{aligned} C_{JC}(u, v|\tau^U, \tau^L) = 1 - \{1 - [&(1-(1-u)^\kappa)^{-\gamma} \\ &+ (1-(1-v)^\kappa)^{-\gamma} - 1]^{-1/\gamma}\}^{1/\kappa} \end{aligned} \quad (5)$$

where $\kappa = 1/\log_2(2-\tau^U)$, $\gamma = -1/\log_2(\tau^L)$, $\tau^U \in (0, 1)$, and $\tau^L \in (0, 1)$.

The SJC copula is especially advantageous because it allows for asymmetry in upper and lower tail dependence. This copula is symmetric when $\tau^U = \tau^L$.

The first step in our parametric approach is to specify the marginal distribution functions F_1 and F_2. We model the marginal distribution for each financial return series we investigate using an AR(1)-GARCH(1,1) formulation. An autoregressive AR(1) specification is used for the conditional mean, a GARCH(1,1) model specifies the conditional variance, as pioneered by Bollerslev (1986, 1987) in an extension of Engle (1982), and the underlying independent and identically distributed innovations are modeled using an appropriate choice for F_i:

$$\begin{aligned} r_t &= \alpha + \beta r_{t-1} + e_t \\ e_t &= \sigma_t y_t \\ \sigma_t^2 &= a + b\sigma_{t-1}^2 + ce_{t-1}^2 \\ y_t &\sim st(\xi, \upsilon) \end{aligned} \tag{6}$$

Here Y is modeled as a Student-t distribution with potential skewness, as developed in Fernández and Steel (1998) and Lambert and Laurent (2001). Specifically, Y follows a standardized skewed Student-t distribution with skewness parameter ξ, which in the absence of skewness is equal to 1, and degrees of freedom υ.

In estimation, we employ a two-stage procedure, whereby the univariate margins are first estimated via maximum likelihood, followed by maximum likelihood estimation of the tail dependence parameters, with the first-stage univariate parameters held fixed.[3] For a given pair of return series R_1 and R_2, the above AR-GARCH framework is estimated separately for each, resulting in Y_1 and Y_2, to which we apply the probability integral transforms $F_1(Y_1)$ and $F_2(Y_2)$ to generate U and V as inputs to the dependence function.[4]

EVOLUTION DYNAMICS

The SJC copula can be extended to allow dependence to evolve over time. Several alternative frameworks have been proposed for the estimation of such time-varying copulas. Here we employ a modification of the dependence evolution equation of Patton (2006b).[5]

We begin with Patton's evolution equations for τ^U and τ^L based on an ARMA(1,l)-type process with an autoregressive term and forcing term:

$$\tau_t^U = \Lambda\left(\omega_U + \beta_U \cdot \tau_{t-1}^U + \alpha_U \cdot \frac{1}{l}\sum_{j=1}^{l}\left|u_{t-j} - v_{t-j}\right|\right) \tag{7}$$

$$\tau_t^L = \Lambda\left(\omega_L + \beta_L \cdot \tau_{t-1}^L + \alpha_L \cdot \frac{1}{l}\sum_{j=1}^{l} |u_{t-j} - v_{t-j}|\right) \tag{8}$$

where $\Lambda = (1 + e^{-x})^{-1}$ is the logistic transformation. With close positive association, $|u_{t-j} - v_{t-j}|$ tends to zero, whereas when association diminishes, suggesting the possibility of reduced tail dependence, this term increases. Given the inverse relationship between this term and association, we might expect α_U, $\alpha_L < 0$.

However, rather than the equally weighted mean shown in the last terms of Eqs. (7) and (8), we employ an EWMA of the absolute difference in probability integral transforms:

$$\tau_t^U = \Lambda(\omega_U + \beta_U \cdot \tau_{t-1}^U + \alpha_U \cdot o_t) \tag{9}$$

$$\tau_t^L = \Lambda(\omega_L + \beta_L \cdot \tau_{t-1}^L + \alpha_L \cdot o_t) \tag{10}$$

with

$$o_t = \delta \cdot o_{t-1} + (1 - \delta) \cdot |u_{t-1} - v_{t-1}| \tag{11}$$

where $0 < \delta < 1$. The geometric lag specification is intuitively motivated by the view that in determining the current level of dependence, measures of association from the distant past should be less important than more recent measures. Moreover, the inclusion of δ as an additional parameter allows direct estimation of the speed of decay and effective lag length.[6] In contrast, the original specification in Eqs. (7) and (8) incorporates only equally weighted lags up to the point l, after which the direct informational effect of additional lags is reduced in binary fashion to zero. Also, in the original specification the method to be used for determining l is left undetermined.

A geometric lag specification has been used in a wide variety of econometric contexts, but not so much for time-varying copulas. In estimating a time-varying Gaussian copula, Bartram, Taylor, and Wang (2007) use a similar exponential moving average structure for the absolute difference in cumulative distribution functions. However, they choose an autoregression for the dependence parameter, without Λ the logistic transformation, instead relying on parameter restrictions and boundary conditions to maintain the dependence parameter within its required range. As such, their formulation differs fundamentally from the one used here, where as in Patton (2006b) the parameters may operate more widely.

EMPIRICAL APPLICATION

For an assessment of how stock return dependence might vary over the financial cycle, we examine the dependence between bank stock returns and those to insurance companies, based on daily log returns from January 3, 1992 through January 3, 2012, multiplied by 100. To focus on the returns to these two types of financial businesses, as opposed to individual companies, we employ corresponding sector indexes constructed by Dow Jones Indexes. *DJUSBK* is the Dow Jones U.S. Banks Index, designed to measure the stock performance of U.S. companies in the banking sector. *DJUSIR* is the Dow Jones U.S. Insurance Index, measuring the performance of U.S. companies in the insurance sector. Both are weighted by float-adjusted market capitalization (subject to some caps). Both series are obtained from Bloomberg L.P.

Descriptive statistics and test results for the two series are shown in Table 1. Normality is strongly rejected in each case, based on the Jarque–Bera test, reflecting substantial kurtosis and some skewness, as well. Serial correlation and conditional heteroscedasticity are also present, based on Ljung–Box and Engle tests, respectively.

Table 2 displays estimation results for the AR(1)-GARCH(1,1) marginal distribution models. We note substantial GARCH (b) and ARCH (c) effects

Table 1. Descriptive Statistics for the Log Return Series (*P-Values*).

	Mean	Std. Dev.	Kurtosis	Skewness	*JB*	Q(*20*)	ARCH(*20*)
DJUSBK	0.011	2.078	22.303	0.108	78,272	115.849	1,243.9
					(*0.000*)	(*0.000*)	(*0.000*)
DJUSIR	0.019	1.590	16.190	−0.163	36,564	79.685	1,399.0
					(*0.000*)	(*0.000*)	(*0.000*)

Note: Row headings refer to return series, as defined in the text. *JB* is the Jarque–Bera test of normality. *Q(20)* is the Ljung–Box test of serial correlation, using 20 lags. *ARCH(20)* is the LM test of conditional heteroscedasticity from Engle (1982), conducted on each centered series.

Table 2. Marginal Distribution Models (Standard Errors).

	α	β	a	b	c	υ	ξ
DJUSBK	0.059	0.021	0.011	0.909	0.091	7.785	0.991
	(0.014)	(0.015)	(0.003)	(0.008)	(0.009)	(0.821)	(0.020)
DJUSIR	0.048	0.021	0.011	0.900	0.098	6.681	0.952
	(0.012)	(0.014)	(0.003)	(0.009)	(0.010)	(0.606)	(0.019)

and also kurtosis, as reflected in the estimates for the shape parameter, υ. The estimated skewness parameter ξ falls below 1 by more than two standard deviations for *DJUSIR*, indicating a relatively long left tail.

When a marginal distribution is correctly specified, the resulting sequence of probability integral transforms should be independent and identically distributed as $U(0, 1)$. We follow Diebold, Gunther, and Tay (1998) to check the adequacy of the estimated marginal distribution models by assessing whether our estimates for the copula inputs – the probabilities $u_t = F_1(y_{1,t})$ and $v_t = F_2(y_{2,t})$ – are *iid* $U(0, 1)$. For this purpose, we apply the Ljung–Box test of serial correlation to the probability integral transform of the underlying error term from each AR(1)-GARCH(1,1)-*st* process and also to the transform's second, third, and fourth centered moments. We also employ Pearson's chi-square test of departures from the uniform distribution.

The results of these tests are shown in Table 3, and they indicate the AR(1)-GARCH(1,1) model is adequate for both *DJUSBK* and *DJUSIR*. The Ljung–Box results suggest a lack of serial correlation. The Pearson's results are consistent with the uniform distribution.

The stage is now set for copula estimation. Here we progress in steps, with the primary goal of illustrating the importance of careful consideration and specification of the formula through which time variation in tail dependence is driven, especially in terms of the lag structure between dependence and its forcing terms.

One approach would involve setting the lag length l in the equally weighted specification shown in Eqs. (7) and (8) to a reasonable, yet essentially arbitrary, value. In applied work, this appears sometimes to be the case. Table 4 then displays time-varying SJC copula estimates, and the

Table 3. Specification Tests for the Probability Integral Transforms (*P-Values*).

	$Q(20)$	$Q(20)^2$	$Q(20)^3$	$Q(20)^4$	Pearson's
DJUSBK	25.313	10.405	29.158	11.874	42.920
	(0.190)	*(0.961)*	*(0.085)*	*(0.921)*	*(0.307)*
DJUSIR	26.218	24.719	19.703	26.572	42.952
	(0.159)	*(0.213)*	*(0.477)*	*(0.148)*	*(0.306)*

Note: $Q(20)$ is the Ljung–Box test of serial correlation, using 20 lags. Q^k applies to the kth centered moment. *Pearson's* is the usual goodness-of-fit chi-square statistic using 40 equal-length intervals on [0, 1] and comparing the proportions of observations found in the bins to the expected value under the null.

Table 4. Time-Varying Symmetrized Joe–Clayton Copula.

Lag Weights	ω_U	β_U	α_U	ω_L	β_L	α_L	δ	AIC	BIC
Equal (10)	0.089	1.284	−4.033	1.021	0.635	−6.326	–	−0.9350	−0.9273
Equal (27)	3.092	−1.376	−14.470	5.227	−3.651	−17.131	–	−0.9535	−0.9445
Exponential	1.906	0.219	−12.550	3.050	−1.182	−12.260	0.954	−0.9547	−0.9457

first row corresponds to the equally weighted lag specification in Eqs. (7) and (8), with l arbitrarily set equal to 10.

Next, we continue with the equally weighted specification in Eqs. (7) and (8), but this time estimate an optimal lag length l, rather than assigning one arbitrarily. Specifically, we allow l to range from 1 to 100 and select from these possibilities the value resulting in the highest copula log likelihood. While we have not noticed this approach in applied work, it would seem to attempt to make the best of the equally weighted specification. However, because every lag length considered requires a separate numerical optimization, the approach does seem a bit cumbersome. This procedure suggests a lag length of 27, and the corresponding estimation results are shown in the second row of Table 4.

Finally, estimates for the specification we have proposed, the EWMA model in Eqs. (9)–(11), are shown in the third row. Of special note are the α estimates, representing the effects of the forcing terms on dependence evolution. As expected, $\alpha_U, \alpha_L < 0$, in each of the three specifications. For the EWMA model, interest also focuses on the decay parameter. The results indicate $\delta = 0.954$, implying an effective lag length of $2/(1-\delta)-1 = 42$ trading days.

To allow us to rank the alternative time-varying copulas we estimate, Table 4 also displays the Akaike information criterion (AIC), which has been widely used in the copula literature, and also the Bayesian information criterion (BIC), both based on the copula log likelihood. The model with the lowest information criteria is considered to provide the best fit. Both of these metrics point to the EWMA specification as superior.

The estimated time paths for upper and lower tail dependence generated by the three alternative specifications are shown in Figs. 1–3. Beginning with the equally weighted, 10-lag model, Fig. 1 displays noisy paths, with wide and rapid variation and muted cyclical features. With the lag length extended to 27, as suggested by our search, the level of noise is reduced and cyclical patterns become more pronounced, as shown in Fig. 2. Yet, the EWMA specification shown in Fig. 3 achieves substantially more noise reduction, leaving relatively clean time paths and more distinct cyclical patterns.

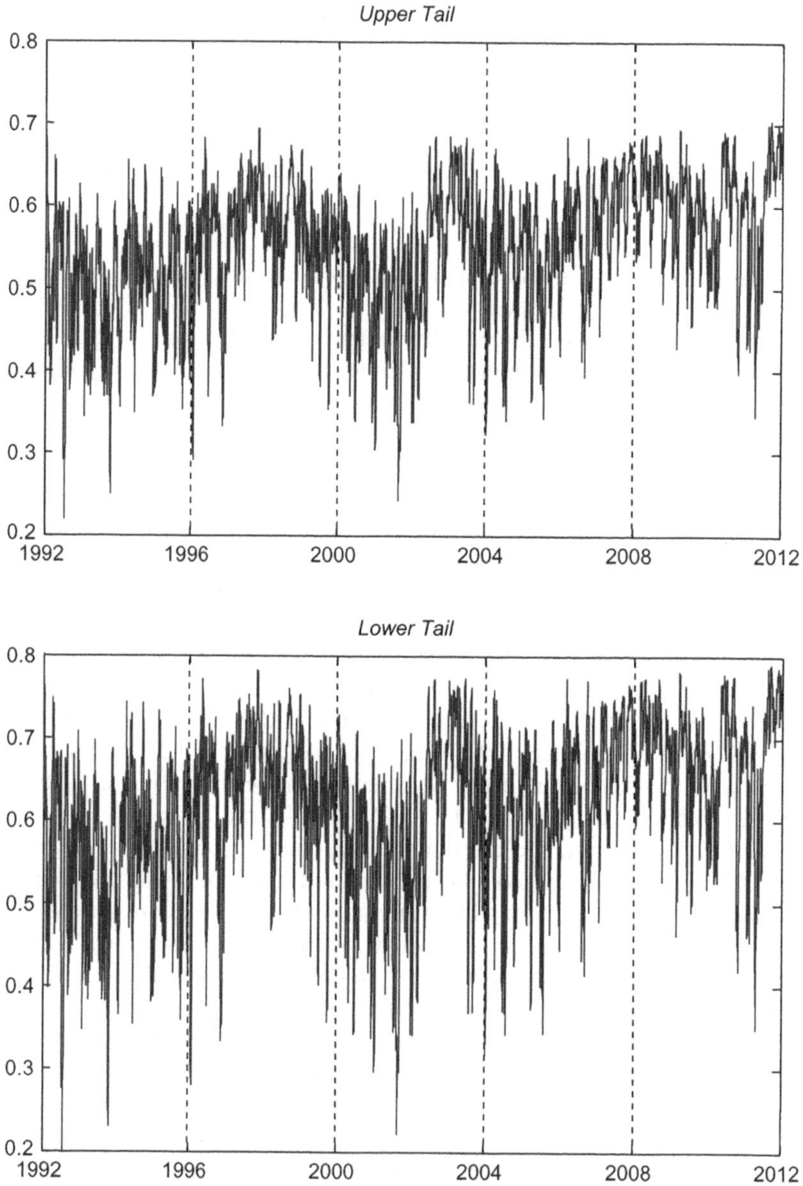

Fig. 1. Symmetrized Joe–Clayton Copula: Equally Weighted, 10 Lags.

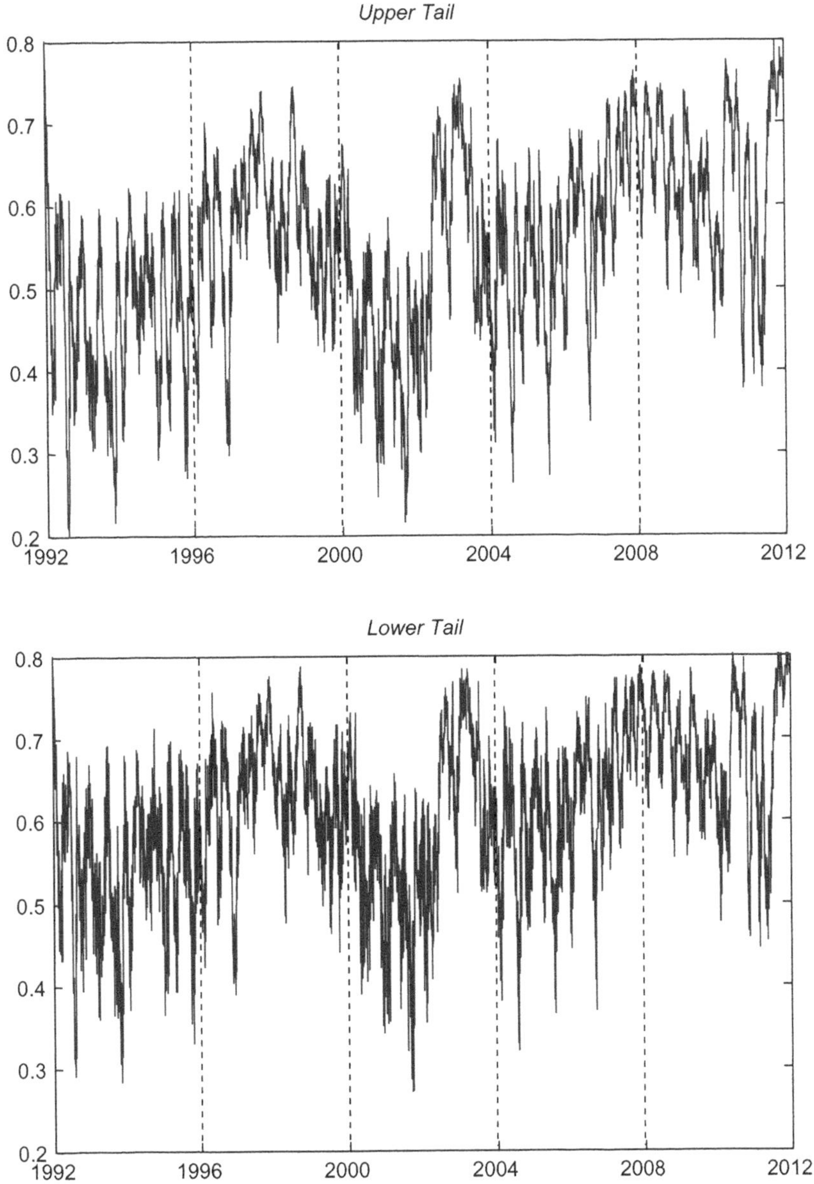

Fig. 2. Symmetrized Joe–Clayton Copula: Equally Weighted, 27 Lags.

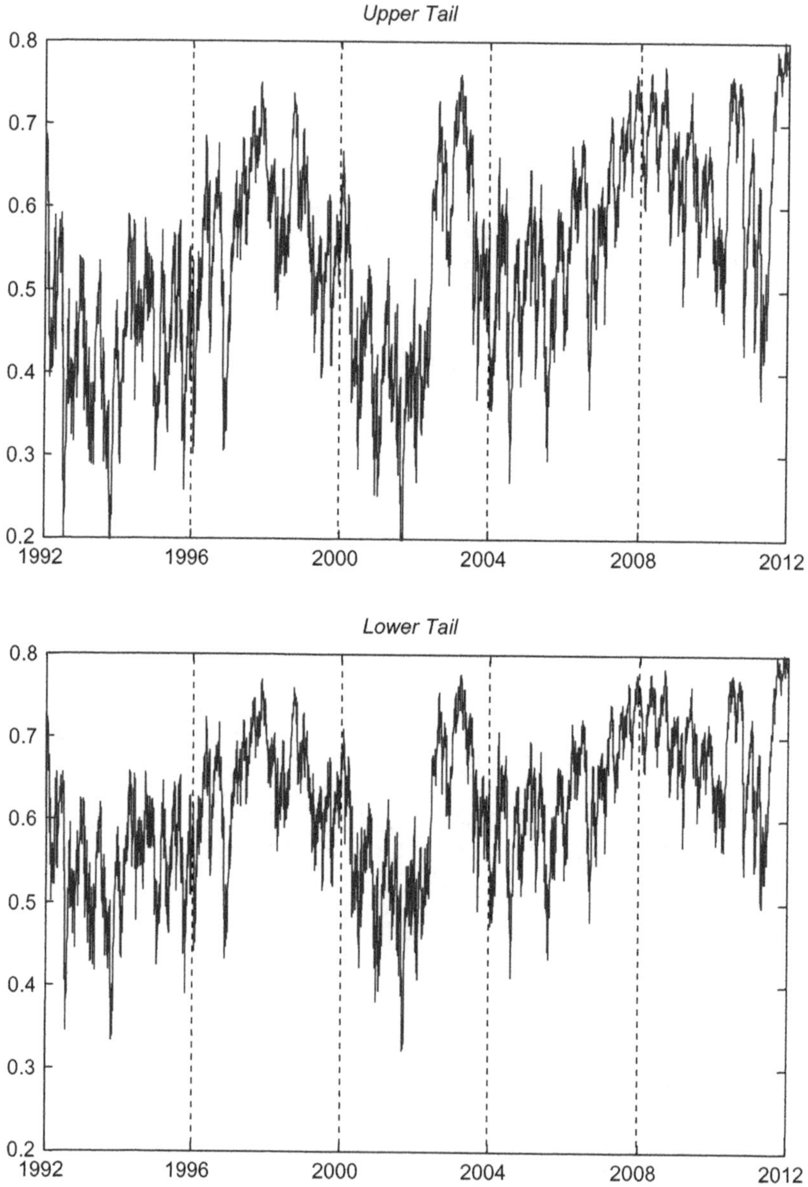

Fig. 3. Symmetrized Joe–Clayton Copula: Exponentially Weighted.

Interestingly, the three episodes of heightened dependence, as shown in Fig. 3, correspond to well-known periods of financial stress. Dependence tracks higher during the 1997–1998 period, encompassing the East Asian and Russian financial crises. Similarly, dependence also increases in 2002, a year of large corporate accounting scandals. Finally, dependence is elevated in 2008, the peak of the recent financial crisis, and tends to remain high thereafter, in association with ongoing financial tremors.

CONCLUSION

We have developed an SJC copula based on the EWMA of the absolute difference in probability integral transforms for use in the estimation of time-varying upper and lower tail dependence. We apply the model to 20 years of daily stock index data covering the banking and insurance industries.

In our application, considerable support is found for the proposed specification. It performs well based on information criteria and is relatively easy to implement. For this model, an appropriate effective lag length is estimated simultaneously with the other copula parameters. Moreover, the resulting estimates of time-varying tail dependence appear more well defined, revealing relatively distinct cyclical patterns.

Finally, the estimation results for bank and insurance returns suggest tail dependence often moves higher during periods of market stress. This element of heightened dependence would be expected to contribute to the systemic nature of financial crises and degrade any diversification benefits prevailing under more favorable operating conditions.

ACKNOWLEDGMENTS

We thank an anonymous referee for helpful suggestions. This chapter reflects the authors' views and not necessarily those of the Federal Reserve Bank of Dallas or Board of Governors of the Federal Reserve System.

NOTES

1. Two-dimensional copulas are grounded, 2-increasing functions with uniform margins and domain $\mathbf{I}^2$. For details, see Joe (1997) and Nelsen (2006).

2. A variety of methods have been used to estimate tail dependence. For a nonparametric approach, see Coles, Heffernan, and Tawn (1999) and Poon, Rockinger, and Tawn (2004).

3. If interest extends to hypothesis testing, Joe (1997, 2005) lays out the asymptotic covariance matrix of the two-stage estimator, facilitating Wald tests. Corresponding adjustments to the log likelihood ratio test are addressed by Gong and Samaniego (1981), Liang and Self (1996), and Chen and Liang (2010).

4. A related approach involves the semiparametric copula, as in Genest, Ghoudi, and Rivest (1995), Chen and Fan (2005, 2006), and Chan, Chen, Chen, Fan, and Peng (2009), using the empirical distributions of the residuals from the first stage, rather than their parametric distribution functions.

5. Alternative approaches include the autoregressive score model of Creal, Koopman, and Lucas (forthcoming) and the stochastic dependence model of Hafner and Manner (2012).

6. In principle, estimation of potentially different rates of decay for the upper and lower tail could be important for some applications, but for simplicity and ease of comparison with Patton's original specification using a single lag length we do not pursue that possibility here. Also, if it were to prove useful, one could of course simply fix the decay parameter to a calibrated value.

REFERENCES

Ang, A., & Chen, J. (2002). Asymmetric correlations of equity portfolios. *Journal of Financial Economics*, *63*, 443–494.

Bartram, S. M., Taylor, S. J., & Wang, Y.-H. (2007). The Euro and European financial market dependence. *Journal of Banking and Finance*, *31*, 1461–1481.

Bartram, S. M., & Wang, Y.-H. (2005). Another look at the relationship between cross-market correlation and volatility. *Finance Research Letters*, *2*, 75–88.

Bollerslev, T. (1986). Generalized autoregressive conditional heteroskedasticity. *Journal of Econometrics*, *31*, 307–327.

Bollerslev, T. (1987). A conditional heteroskedastic time series model for speculative prices and rates of return. *Review of Economics and Statistics*, *69*, 542–547.

Caporale, G. M., Cipollini, A., & Spagnolo, N. (2005). Testing for contagion: A conditional correlation analysis. *Journal of Empirical Finance*, *12*, 476–489.

Chan, N.-H., Chen, J., Chen, X., Fan, Y., & Peng, L. (2009). Statistical inference for multivariate residual copula of GARCH models. *Statistica Sinica*, *19*, 53–70.

Chen, X., & Fan, Y. (2005). Pseudo-likelihood ratio tests for semiparametric multivariate copula model selection. *The Canadian Journal of Statistics*, *33*, 389–414.

Chen, X., & Fan, Y. (2006). Estimation and model selection of semiparametric copula-based multivariate dynamic models under copula misspecification. *Journal of Econometrics*, *135*, 125–154.

Chen, Y., & Liang, K.-Y. (2010). On the asymptotic behavior of the pseudolikelihood ratio test statistic with boundary problems. *Biometrika*, *97*, 603–620.

Coles, S., Heffernan, J., & Tawn, J. (1999). Dependence measures for extreme value analyses. *Extremes*, *2*, 339–365.

Corsetti, G., Pericoli, M., & Sbracia, M. (2005). Some contagion, some interdependence: More pitfalls in tests of financial contagion. *Journal of International Money and Finance, 24*, 1177–1199.

Creal, D., Koopman, S. J., & Lucas, A. (forthcoming). Generalized autoregressive score models with applications. *Journal of Applied Econometrics*.

Diebold, F. S., Gunther, T. A., & Tay, A. S. (1998). Evaluating density forecasts with applications to financial risk management. *International Economic Review, 39*, 863–883.

Engle, R. F. (1982). Autoregressive conditional heteroscedasticity with estimates of the variance of U.K. inflation. *Econometrica, 50*, 987–1008.

Fernández, C., & Steel, M. F. J. (1998). On bayesian modeling of fat tails and skewness. *Journal of the American Statistical Association, 93*, 359–371.

Fisher, R. A. (1932). *Statistical methods for research workers*. Edinburgh: Oliver and Boyd.

Fomby, T. B., Gunther, J. W., & Hu, J. (2012). Return dependence and the limits of product diversification in financial firms. *Journal of Money, Credit, and Banking, 44*, 1151–1183.

Forbes, K. J., & Rigobon, R. (2002). No contagion, only interdependence: Measuring stock market comovements. *Journal of Finance, 57*, 2223–2261.

Genest, C., Ghoudi, K., & Rivest, L.-P. (1995). A semiparametric estimation procedure of dependence parameters in multivariate families of distributions. *Biometrika, 82*, 543–552.

Gong, G., & Samaniego, F. J. (1981). Pseudo maximum likelihood estimation: Theory and applications. *The Annals of Statistics, 9*, 861–869.

Hafner, C. M., & Manner, H. (2012). Dynamic stochastic copula models: Estimation, inference and applications. *Journal of Applied Econometrics, 27*, 269–295.

Joe, H. (1997). *Multivariate models and dependence concepts* (Vol. 73). Monographs in statistics and probability. London: Chapman & Hall.

Joe, H. (2005). Asymptotic efficiency of the two-stage estimation method for copula-based models. *Journal of Multivariate Analysis, 94*, 401–419.

Lambert, P., & Laurent, S. (2001). *Modelling financial time series using GARCH-type models with a skewed Student distribution for the innovations*. Discussion paper 0125, Institut de Statistique, Université Catholique de Louvain, Louvain-la-Neuve, Belgium.

Liang, K.-L., & Self, S. G. (1996). On the asymptotic behavior of the pseudolikelihood ratio test statistic. *Journal of the Royal Statistical Society Series B (Methodological), 58*, 785–796.

Longin, F., & Solnik, B. (2001). Extreme correlation of international equity markets. *Journal of Finance, 56*, 649–676.

Nelsen, R. B. (2006). *An introduction to copulas*. New York, NY: Springer.

Patton, A. J. (2002). *Applications of copula theory in financial econometrics*. Ph.D. dissertation, University of California, San Diego, CA.

Patton, A. J. (2006a). Estimation of copula models for time series of possibly different lengths. *Journal of Applied Econometrics, 21*, 147–173.

Patton, A. J. (2006b). Modeling asymmetric exchange rate dependence. *International Economic Review, 47*, 527–556.

Poon, S.-H., Rockinger, M., & Tawn, J. (2004). Extreme value dependence in financial markets: Diagnostics, models, and financial implications. *Review of Financial Studies, 17*, 581–610.

Rosenblatt, M. (1952). Remarks on a multivariate transformation. *The Annals of Mathematical Statistics, 23*, 470–472.

Sklar, A. (1959). Fonctions de Repartition à *n* Dimensions et Leurs Marges. *Publications de l'Institut Statistique de l'Universite de Paris, 8*, 229–231.

PART V
SIMULATION

MONTE CARLO EXPERIMENTS USING STATA: A PRIMER WITH EXAMPLES

Lee C. Adkins and Mary N. Gade

ABSTRACT

Monte Carlo simulations are a very powerful way to demonstrate the basic sampling properties of various statistics in econometrics. The commercial software package **Stata** *makes these methods accessible to a wide audience of students and practitioners. The purpose of this chapter is to present a self-contained primer for conducting Monte Carlo exercises as part of an introductory econometrics course. More experienced econometricians that are new to Stata may find this useful as well. Many examples are given that can be used as templates for various exercises. Examples include linear regression, confidence intervals, the size and power of* t*-tests, lagged dependent variable models, heteroskedastic and autocorrelated regression models, instrumental variables estimators, binary choice, censored regression, and nonlinear regression models. Stata do-files for all examples are available from the authors' website http://learneconometrics.com/pdf/MCstata/.*

Keywords: Monte Carlo; econometrics; Stata; simulations

30th Anniversary Edition
Advances in Econometrics, Volume 30, 429–477

ISSN: 0731-9053/doi:10.1108/S0731-9053(2012)0000030019

INTRODUCTION – USING MONTE CARLO SIMULATIONS IN ECONOMETRICS

Kennedy (2003, p. 24) defines a Monte Carlo study as "a simulation exercise designed to shed light on the small-sample properties of competing estimators for a given estimating problem." More importantly, it can provide "a thorough understanding of the repeated sample and sampling distribution concepts, which are crucial to an understanding of econometrics." The basic idea is to model the data generation process, generate sets of artificial data from that process, estimate the parameters of the model using one or more estimators, and then use summary statistics or plots to compare or study the performance of the estimators. Monte Carlo methods are used extensively by econometricians to study the finite sample performance of statistics, compare the power of tests, and determine the effects of various statistical designs on the statistical properties of estimators that are commonly used.

Davidson and MacKinnon (2004, p. ix) are enthusiastic proponents of simulation methods in econometrics. They state:

> Ten years ago, one could have predicted that [personal computers] would make the practice of econometrics a lot easier, and of course that is what has happened. What was less predictable is that the ability to perform simulations easily and quickly would change many of the directions of econometric theory as well as econometric practice.

It has also changed the way we teach econometrics to undergraduates and graduate students (e.g., see Barreto & Howland, 2006; Day, 1987; Judge, 1999). Kiviet (2012) has written a thorough discussion of Monte Carlo methods that is suitable for classroom use, employing EViews as the computing vehicle. In this chapter, the use of Monte Carlo simulations to learn about the sampling properties of estimators in econometrics will be discussed and the usefulness of **Stata** will be demonstrated using a number of examples. As such, this paper can serve a tutorial suitable for classroom use.

Stata 12 (StataCorp, 2011) provides a powerful means of conducting Monte Carlo studies. **Stata** contains many estimators and its programming structure allows functions to be looped rather easily, which greatly expands the types of things that can be studied via simulation. For the most part, programming simulations in **Stata** is not difficult, but there are a few tricks and idiosyncracies to contend with. Many of these are documented below.

The best introduction to simulations using **Stata** is found in Cameron, Colin, and Trivedi (2009, ch. 4). They concentrate on the use of **Stata**'s `simulate` command due to its simplicity. **Stata**'s `simulate` command runs a specified command or user written function (which **Stata** refers to as a *program*) a given number of times. The results can be output to another dataset. However, when the simulation finishes, the **Stata** dataset that is currently in memory is overwritten with the results of the simulation. This may not always be desirable; for instance, you may want to conduct additional loops of the simulation using different model parameters and collect the simulated statistics into a single dataset.

In this chapter, the slightly more basic, but versatile `postfile` command is used. The `postfile` command in **Stata** allows users to write the value of a computed statistic (or set of statistics) at each iteration of a simulation to a dataset.

Although `simulate` allows one to write the statistics from a Monte Carlo to a specified file, it does not allow that file to be appended with the results of subsequent runs of the simulation. That means if you change the value of one or more parameters in the simulation and rerun it, the results have to be written to a new dataset; joint analysis of the different parameterizations requires manually changing filenames and parameters, rerunning the simulation, and then merging the differently named datasets. That requires a lot of user intervention. With `postfile` no user intervention is required until the joint results are analyzed.

In addition, the syntax of `postfile` routines is more transparent. In most cases you do not have to write separate functions as you do with `simulate`; this allows you to avoid creating *return class objects*[1] and to minimize the use of **Stata**'s *macro* structure. In **Stata**, a *macro* permits the transport of data and results in and out of functions (and loops); inexperienced programmers find their use confusing and programming errors within them are more difficult to diagnose.[2] It should be said, however, their use cannot be avoided altogether because they are the principal way **Stata** moves objects (data and computations) in and out of functions.

We find the `postfile` and loop method no more difficult to use than `simulate` for most problems, and it is more versatile and slightly faster in execution. Evidence of this is presented in the seventh section. In order to compare the two ways of simulating in **Stata**, the first example below is developed using both methods.

In the next section, we briefly review the concept of a fully specified statistical model as laid out by Davidson and MacKinnon (2004) and present their basic recipe for using it in a Monte Carlo exercise. In the third section we review some of the basics of **Stata**. In the fourth section the basics of Monte Carlo simulations are summarized. Then in the fifth section we go through a complete example of using **Stata** to study the coverage probability of confidence intervals centered at the least squares estimator. In subsequent subsections, we outline how to generate simulated models for some of the estimators used in a first econometrics course. In the sixth section an example is presented that automates the changing of model parameterizations. Finally, our conclusion includes a few remarks on speed and a comparison of **Stata** to the free software **gretl**, version 1.9.9 (Cottrell & Lucchetti, 2012).

FULLY SPECIFIED STATISTICAL MODEL

The first step in a Monte Carlo exercise is to model the data generation process. This requires what Davidson and MacKinnon (2004) refer to as a fully specified statistical model. A **fully specified parametric model** "is one for which it is possible to simulate the dependent variable once the values of the parameters are known" (Davidson & MacKinnon, 2004, p. 19). First you need a regression function, for instance,

$$E(y_t|\Omega_t) = \beta_1 + \beta_2 x_t \tag{1}$$

where y_t is your dependent variable, x_t is the dependent variable, Ω_t is the current information set, and β_1 and β_2 are the parameters of interest. The information set Ω_t contains x_t as well as other potential explanatory variables that determine the average of y_t. The conditional mean of y_t given the information set could represent a linear regression model (e.g., as in Eq. (1)) or a discrete choice model (e.g., as in $E(y_t|\Omega_t) = F(\beta_1 + \beta_2 x_t)$ where F is a cumulative distribution function).

The actual values of $y_t|\Omega_t$ will differ from the mean by some amount, u_t, that is, $y_t|\Omega_t - E[y_t|\Omega_t] = u_t$.

$$y_t|\Omega_t = \beta_1 + \beta_2 x_t + u_t \tag{2}$$

To complete the model requires a description of how the unobserved or excluded factors, u_t, behave.

A fully specified model requires an "unambiguous recipe" for simulating the model on a computer (Davidson & MacKinnon, 2004, p. 17). This means one needs to specify a probability distribution for the unobserved components of the model, u_t, and then use a pseudo-random number generator to generate samples of the desired size. Again following Davidson and MacKinnon (2004), the recipe is as follows:

- Set the sample size, n.
- Choose the parameter values β_1 and β_2 for the deterministic conditional mean function (1).
- Obtain n successive values x_t, $t = 1, 2, \ldots, n$, for the explanatory variable. You can use real data or generate them yourself.
- Compute $\beta_1 + \beta_2 x_t$ for each observation.
- Choose a probability distribution for the error terms, u_t, and choose any parameters that it may require (e.g., the normal requires a mean, usually zero, and a variance, σ^2).
- Use a pseudo-random number generator to get the n successive values of the errors, u_t.
- Add these to your computed values of $\beta_1 + \beta_2 x_t$ to get y_t for each observation.
- Estimate the model using the random sample of y, the given values of x, and the desired estimator.
- Save the statistics of interest.
- Repeat this a large number of times.
- Print out the summary statistics from the preceding step.

In the last step it is common to evaluate the mean of the sampling distribution, bias, variance, and the mean square error. Kennedy (2003, pp. 26–27) gives explicit formulas for these.

STATA BASICS

Stata is a statistical package for managing, analyzing, and graphing data. It is available for a variety of computing platforms. It can be used either as a point-and-click application or as a command-driven package. **Stata**'s system of windows and dialog boxes provides an easy interface for newcomers to **Stata** and for experienced **Stata** users who wish to execute a command that they seldom use. The command language provides a fast way to communicate with **Stata** and to communicate more complex ideas. For Monte Carlo studies, the command language is a necessity.

Common Conventions

In the few instances when the pull-down menus are employed, the convention used will be to refer to menu items as `A>B>C` which indicates that you are to click on option A on the menu bar, then select B from the pull-down menu and further select option C from B's pull-down menu.

An important fact to keep in mind when using **Stata** is that its language is **case sensitive**. This means that lower case and capital letters have different meanings in **Stata**. The practical implication of this is that you need to be very careful when using the language. Since **Stata** considers x to be different from X, it is easy to make programming errors. If **Stata** gives you a programming error statement that you cannot decipher, make sure that the variable or command you are using is in the proper case.

A very powerful, but often confusing, feature of the **Stata** programming language is the macro concept. In **Stata**, macros can be thought of as the variables in a **Stata** program. A *macro* is a string of characters, called the *macroname*, that stands for another string of characters, called the *macro contents*. These can either be defined locally (exist only temporarily within a **Stata** program) or globally (permanently defined within the entire set of commands, which **Stata** calls a *do-file*). To access the contents of a macro, special syntax is required that identifies the name of the macro and tells **Stata** whether it is locally or globally defined. To substitute the macro contents of a global macro name, the macro name is typed (punctuated) with a dollar sign ($) in front. To substitute the macro contents of a local macro name, the macro name is typed (punctuated) with surrounding left and right single quotes (‘’). The main problem inexperienced users have with **Stata** macros is knowing when it is necessary to use them and then figuring out how to get the syntax correct – the ability to nest macros can make getting it right quite a trick. Since our chapter is intended to serve the less-experienced user, we try to limit their use in this chapter (though it is not entirely possible).

Ways to Work in Stata

There are several different ways to work in **Stata**. One can use the program through its built-in graphical user interface (GUI) where **Stata** collects input from the user through dialog boxes, delivered by mouse clicks and a few keystrokes, to generate a computer code that is executed in the background. The output is sent to the **Results Window**. After estimating a model, one can access a set of *postestimation* commands that will perform various functions on the output of your estimation The postestimation commands give one access to various graphs, tests, analyses, and a means of storing the results. Commands executed through the GUI are listed in the **Review window**, where they can be debugged and rerun using a few mouse clicks.

Stata commands can be collected and put into a file that can be executed at once and saved to be used again. **Stata** refers to these collections of code as *do-files* because they have .do filename extensions (e.g., MC.do). Even though you can run do-files from the command line, the **do-file editor** is the easiest way used to do Monte Carlo simulations in **Stata**. Start by opening a new do-file from the file menu. Select `Window>Do-file Editor>New Do-file Editor` from the pull-down menu to open the do-file editor. You can use "CRTL + 8" as a keyboard shortcut.

One of **Stata**'s great strengths is the consistency of its command syntax. Most of **Stata**'s commands share the following syntax, where square brackets mean that something is optional and a varlist is a list of variables.

```
1 [prefix:] command [varlist] [if] [in] [weight] [,options]
```

Some general rules:

- Most commands accept prefix commands that modify their behavior. One of the more common prefix commands is `by`, which tells **Stata** to repeat the given command on subsets of the data. See the example in the sixth section.
- If an optional `varlist` is not specified, all the variables are used.
- `if` and `in` restrict the observations on which the command is run.1
- `options` modify what the command does.
- Each command's syntax is found in the online help and the reference manuals.

Type the commands you want to execute in the tabbed editor box using one line for each command. If you have a very long command that exceeds one line, use the triple forward slash (///) as a **continuation command**.[3] Then, to save the file, use the "CRTL + S." If this is a new file, you will be prompted to provide a name for it.

MONTE CARLO BASICS

There are two main tools that can be used to do Monte Carlo simulation in **Stata**. It includes a special `simulate` command that will execute a program a given number of times and collect the outcomes in a dataset. It is specifically designed to make model simulation of the type suggested here very easy to do. The other method uses the `postfile` set of commands, which opens and posts observations to a dataset one at a time. Thus, with `postfile` one computes a statistic at each iteration of the Monte Carlo and adds it to a specified dataset. Used in conjunction with **Stata**'s flexible loop constructs, it is both versatile and powerful. We find the `postfile` and loop method just as easy to use as `simulate` and it is slightly more versatile; it is the main method used below in the series of examples. Cameron et al. (2009) provide a very nice chapter on the use of `simulate`, and the discussion of its use here is limited. For purposes of comparison, though, two examples are repeated using `simulate`.

Using the `postfile` command to run Monte Carlo experiments requires the programming of loops. There are three ways to loop in `Stata`. The fastest uses the syntax:

```
1       forvalues lname = range {
2               commands referring to 'lname'
3       }
```

The braces that appear in lines 1 and 3 must be specified in the `forvalues` loop (as well as in each of the methods used for looping), and

1. the open brace must appear on the same line as `forvalues`;
2. nothing may follow the open brace except, of course, comments; the first command to be executed must appear on a new line;
3. the close brace must appear on a line by itself.

Notice that '`lname`' in line 2 is enclosed in single quotes. It is a local macro that contains the values of `lname` in the preceeding line.

The `forvalues` loop has its limitations. If you need to create loops within loops (referred to as nested loops), you have to use the `while` looping construct.

```
1       while lname = range {
2               commands referring to 'lname'
3       }
```

There is also a `foreach` loop construct that can be used to loop over the elements of a list.

```
        foreach var of local varlist {
                do something based on 'var'
}
```

You can use lists of numbers rather than variables (numlist), which proves to be especially useful in the sixth section.

Each simulation follows a similar recipe. The basic structure is as follows:

(1) Open a dataset to use as a basis for the simulation or create values of the regressors using random numbers.
(2) Set a seed.
(3) Define global variables for the number of monte carlo trials (`nmc`) and sample size (`nobs`). For instance, using a samples of size 100 and doing a 1,000 simulations.
```
global nobs = 100
global nmc = 1000
```
(4) Set the values of the parameters.
(5) Initialize data that may need starting values in the loop, for example, errors and conditional mean.
(6) Create a temporary name, in this case we use the local macro name `sim`, using the `tempname` command. sim will hold statistics as they accumulate in the simulation.
```
tempname sim
```
(7) Initiate the `postfile` command. Put the macro name '`sim`' in single quotes. Assign variable names to the statistics that are being posted to `sim` and give a file name for placing the entire contents of sim once the simulation finishes. The `replace` option permits the `results.dta` file to be overwritten each time the simulation is executed.
```
postfile 'sim' vname1 vname2 .... vnamek using
results, replace
```
(8) Start a **quietly** block to enclose the loop. This suppresses output to the screen and speeds things up.
```
quietly {
```
(9) Start the **forvalues** loop, indicating the desired number of samples (`$nmc`).
```
forvalues i = 1/$nmc {
```

(10) Replace random errors.
(11) Replace conditional mean, y.
(12) Estimate the model.
(13) Compute the desired statistics (e.g., `b, se, pv`).
(14) Post the statistics into the local macro '`sim`' using variable names. The name of each computed statistic that is to be posted is contained within parentheses.

```
post 'sim' (b) (se) (pv)
```

(15) Close the loop with a right bracket.

```
}
```

(16) Close the temporary storage, '`sim`', using the `postclose` command.

```
postclose 'sim'
```

(17) Close the `quietly` block with a right bracket.

```
}
```

The `quietly` command suppresses the printing results to the screen at each iteration of the loop. In the `forvalues` command that appears in step 9, notice that the macro $nmc has a dollar sign in front. Assuming that the number of Monte Carlo simulations to use is stored in the global macro nmc as in step 3, it must be referred to by its global macro name, $nmc.

Once the samples of your statistics from the Monte Carlo experiments have been obtained, what do you do with them? For instance how can you tell whether an estimator is biased? The simple answer is to use statistics. The Monte Carlo mean, $\bar{x}$, for a statistic should be approximately normally distributed, e.g., $\bar{x} \overset{a}{\sim} N(\mu, \sigma^2/n)$. Therefore, $z = \sqrt{\text{NMC}}(\bar{x} - \mu)/\hat{\sigma}$ should be a standard normal. The statistic $\hat{\sigma}$ is simply the standard deviation that **Stata** prints for you when you use the `summarize` command; now, compute z and compare it to the desired critical value from the standard normal.

EXAMPLES

In this section of the chapter, a series of examples is given. Each example illustrates important features of model specification and estimation and is typical of models taught in introductory econometrics courses. The first example is based on the classical normal linear regression model. One thousand (NMC = 1000) samples are generated using Engel's food expenditure and income data (Koenker & Bassett, 1982) included with **gretl** (Cottrell & Lucchetti, 2012) and available from http://www.learneconometrics.com/pdf/MCstata/engel.dta. The slope and intercept parameters with each simulated set of data are computed using least squares, and 95% confidence intervals are constructed. The number of times the actual values of the parameters falls within the interval is counted. We expect that approximately 95% will fall within the computed intervals.

Subsequent examples are given, though with less explanation. These include estimating a lagged dependent variable model using least squares (which is biased but consistent). Autocorrelated errors are then added, making OLS inconsistent. A Breusch–Godfrey test to detect first-order autocorrelation is simulated and can be studied for both size and power. Heteroskedasticity Autocorrelation Consistent (HAC) standard errors are compared to the inconsistent least squares standard errors in a separate example.

Issues associated with heteroskedastic models are studied as well. The simple simulation shows that least squares standard errors are estimated consistently using the usual formula when the heteroskedasticity is unrelated to the model's regressors. The final examples demonstrate the versatility of the exercises. An instrumental variables example is used to study the error-in-variables problem, instrument strength, and other issues associated with this estimator. Binary choice and censored regression examples are given. The final example explores the properties of a nonlinear least squares estimator.

Although the errors in each of the examples are generated from the normal distribution, **Stata** offers other choices. These include uniform, Student's t, chi-square, beta, binomial, hypergeometric, gamma, and Poisson.

Classical Normal Linear Regression and Confidence Intervals

We start with the linear model:

$$\ln(y_t) = \beta_1 + \beta_2 \ln(x_t) + u_t \tag{3}$$

where y_t is total food expenditure for the given time period and x_t is income, both of which are measured in Belgian francs. Let $\beta_1 = .5$ and $\beta_2 = .5$ and assume that the error, u_t iid $N(0, 1)$.

The model's errors take into account the fact that food expenditures are sure to vary for reasons other than differences in family income. Some families are larger than others, tastes and preferences differ, and some may travel more often or farther making food consumption more costly. For whatever reasons, it is impossible for us to know beforehand exactly how much any household will spend on food, even if we know how much income it earns. All of this uncertainty is captured by the error term in the model.

Stata is used to generate sequences of random normals to represent these unknown errors. Distributions other than the normal could be used to explore the effect on coverage probabilities of the intervals when this vital assumption is violated by the data. Also, it must be said that computer-generated sequences of random numbers are not actually random in the true sense of the word; they can be replicated exactly if you know the mathematical formula used to generate them and the "key" that initiates the sequence. This key is referred to as a **seed**. In most cases, these numbers *behave as if* they are randomly generated by a physical process.[4]

A total of 1,000 samples of size 235 are created using the fully specified parametric model by appending the generated errors to the parameterized value of the regression. The model is estimated by least squares for each sample, and the summary statistics are used to determine whether least squares is biased and whether $1 - \alpha$ confidence intervals centered at the least squares estimator contain the known values of the parameters the desired proportion of the time.

The $(1 - \alpha)$ confidence interval is

$$P[b_2 - t_c\text{se}(b_2) \le \beta_2 \le b_2 + t_c\text{se}(b_2)] = 1 - \alpha \tag{4}$$

where b_2 is the least squares estimator of β_2, and that $\text{se}(b_2)$ is its estimated standard error. The constant t_c is the $\alpha/2$ critical value from the t-distribution and α is the total desired probability associated with the "rejection" area (the area outside of the confidence interval).

The complete do-file is found below. Start by opening a dataset. In the second line, a seed is chosen to initialize the stream of pseudo-random numbers. This ensures that the results can be replicated as shown here. In the next part of the do-file, the values of the parameters are set and then the natural logarithm of income, which is used as the independent variable in the simulation, is taken. In line 12 the systematic portion of the model is created. In lines 13 and 14 the error terms and the dependent variable are initialized and filled with missing values. This will allow us to use the `replace` statements in lines 20 and 21 to generate each new sample of y to use in the simulation. A temporary name called `sim` is created in line 15. This is the name that will hold the results posted in line 28 by the `post` command.

Performance of Confidence Intervals

```
use ''http://www.learneconometrics.com/pdf/
MCstata/engel.dta'', clear set seed 3213799


* Set the values of the parameters
scalar constant = .5
scalar slope = .5
scalar sigma = 1
scalar alpha = .05

* Take the natural log of income to use as x
gen x = log(income)
gen Ey = constant + slope*x    /* mean function */
gen u =.      /* initiate u - all missing values */
gen y =.      /* initiate y - all missing values */
tempname sim

postfile 'sim' b se coverage using results, replace
  quietly {
    forvalues i = 1/1000 {
    replace u = rnormal(0,sigma)
    replace y = Ey + u
    reg y x
    scalar b = _b[x]
    scalar se = _se[x]
    scalar lb = b - se*invttail(e(df_r),alpha/2)
    scalar ub = b + se*invttail(e(df_r),alpha/2)
    scalar pv = slope<ub slope>lb
   post 'sim' (b) (se) (pv)
  }
}
  postclose 'sim'

use results, clear
summarize
```

The heart of the simulation begins in line 17 with the `postfile` command. The set of `postfile` commands[5] was created to assist **Stata** programmers in performing Monte Carlo-type experiments. The `postfile` command declares the variable names and the filename of a (new) **Stata** dataset where results will be stored. In order to use the `postfile` construct, a temporary name has to be created that will store the returns generated by the program. In line 17 the `postfile` command tells **Stata** the names to be given to the computations being posted to the file in line 28 and provides the filename into which the temporary file will be written.

The `postclose` command in line 31 declares an end to the posting of observations. After `postclose`, the new dataset contains the posted results and may be loaded using `use`.

The loop itself uses the `forvalues` loop construct. The counter is given the name `i` which is instructed to loop from 1 to 1,000 using the syntax `1/1,000`. Any subsequent references to the counter `i` must be via its (local) macro name '`i`' (that is it must be enclosed in quotes). As a local macro the name '`i`' is temporary and exists only as long as the loop is executing. It becomes empty once the loop finishes. This allows you to reuse the name 'i' in other loops. In order to suppress the output from all of the iterations, the loop is actually initiated using the `quietly` command. Just as in the template given in the preceding section, notice how the loop actually sits within the `quietly` command.

The remainder of the program is very simple. The samples are created, the regression is estimated, and the desired results are computed using `scalar` computations. The logical statement in line 27 takes the value of 1 if the statements to the right of the equality are true. The ampersand (&) is the union of the two sets given on both of its sides. Therefore to be true, the slope must be less than the upper bound **and** greater than the lower. If not, `pv` is zero. Thus, the number of times the parameter falls within the interval can be counted and later averaged to estimate the coverage probability.

Running the do-file loads the data, initializes the parameters, and simulates the model. The results are stored in the dataset called `results`. Be sure to issue the `clear` option or the current dataset won't be cleared from memory and the new one containing the simulated values cannot be loaded. The `summarize` command will give you summary statistics. At this point you could do additional analysis, like test for normality or plot a kernel density.

When writing a do-file, it is always a good idea to add comments. In **Stata** the star sign (*) at the beginning of a line can be used to make comments, for example, lines 4 and 10. To insert a comment after a **Stata** command, you can enclose it in * *\, for example, lines 12–14.

The results from the simulation appear below.

```
summarize
```

Variable	Obs.	Mean	Std. Dev.	Min.	Max.
b	1000	.5023573	.1497425	−.0332597	1.021356
se	1000	.1488163	.0069646	.1289464	.1706571
coverage	1000	.956	.2051977	0	1

Note that the average value of the slope is about 0.502, which is very close to the true value set in line 6. If you were to repeat the experiments with larger numbers of Monte Carlo iterations, you will find that these averages get closer to the values of the parameters used to generate the data. This is what it means to be unbiased. Unbiasedness only has meaning within the context of repeated sampling. In your experiments, you generated many samples and averaged results over those samples to get closer to the truth. In actual practice, you do not have this luxury; you have one sample and the proximity of your estimates to the true values of the parameters is always unknown.

The bias of the confidence interval (or coefficients) can be tested using $z = \sqrt{\text{NMC}}(\bar{x} - \mu)/\hat{\sigma}$ which in this case of p1 is $\sqrt{1000}(0.956 - 0.95)/0.2052 = 0.9246$. This is less than 1.96 and not significantly different from zero at 5%. The two-sided p-value is 0.36, which just confirms this. Increasing the number of Monte Carlo samples to 4,000 produced $z = \sqrt{4000}(0.9485 - .95)/0.2205 = -0.3584$; this shows that increasing the number of samples increases the precision of the Monte Carlo results. Of course, we could let **Stata** do the arithmetic using

Monte Carlo test for size distortion of the 95% confidence interval

```
summarize coverage
scalar t = (r(mean)-.95)/(r(sd)/sqrt(r(N)))
scalar pval = 2*ttail(r(N),abs(t))
scalar list t pval
```

The saved results can be analyzed statistically or visualized graphically. From the command line typing

```
histogram b, normal kdensity
```

produces a histogram with a normal density plot. The kdensity option also adds a kernel density plot.

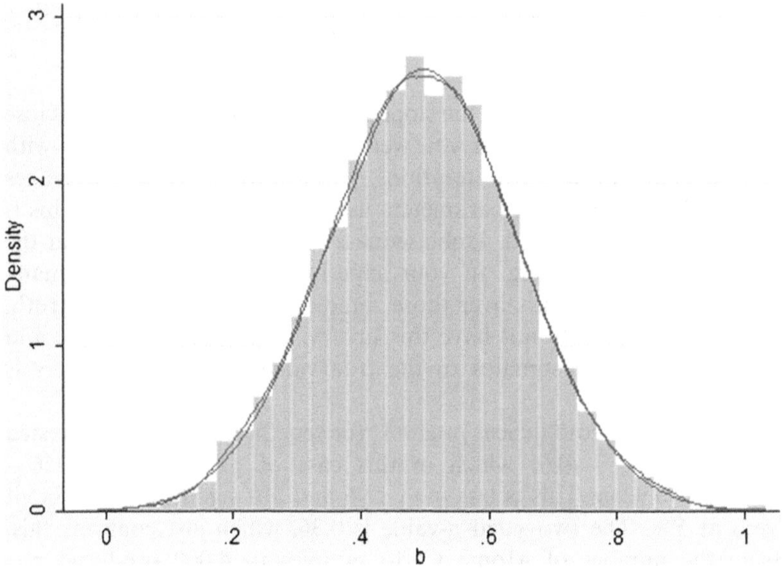

The kernel density plot and the normal plot are virtually indistinguishable.

Large sample properties like consistency and convergence in distribution can likewise be studied by increasing the number of observations, `nobs`.

Using simulate

In many cases you can use the `simulate` command to make things a bit easier. With `simulate`, you do not have to create loops and simple `return` statements are used to write results to the desired dataset. Otherwise, the two ways of working are very similar. Consider the confidence interval example done using `simulate`:

```
program drop _all
use ``http://www.learneconometrics.com/pdf/MCstata/engel.dta'', clear
set seed 3213799
* Set the values of the parameters
scalar constant = .5
scalar slope = .5
scalar sigma = 1
scalar alpha = .05
global scalar nmc = 1000
gen x = log(income)
gen Ey = constant + slope*x   /* mean function */
gen u =.    /* initiate u -- all missing values */
gen y =.    /* initiate y -- all missing values */
program regCI, rclass
   replace u = rnormal(0,sigma)
   replace y = Ey + u
   reg y x
   return scalar b = _b[x]
   return scalar se = _se[x]
   scalar lb = _b[x] - _se[x]*invttail(e(df_r), alpha/2)
   scalar ub = _b[x] + _se[x]*invttail(e(df_r), alpha/2)
   return scalar pv = (slope<ub) (slope>lb)
end
simulate b=r(b) s=r(se) pv = r(pv), reps ($nmc) ///
    saving (results, replace) nolegend nodots: regCI
use results, clear
summarize
summarize pv
scalar t = (r(mean) -.95)/(r(sd)/sqrt(r(N)))
scalar pval = 2*ttail(r(N), abs(t))
scalar list t pval
```

The top part of the program is unchanged. The data are loaded, a seed chosen, and parameters for the simulation set. We also initialized the variables that will be used in the simulation. The `simulate` command requires a program to host the set of commands to be simulated. In this example we create a program called `regCI`. The loop is no longer needed and there are no `postfile` and `post` commands. The loop is handled by `simulate` when the `regCI` function is called.

The `post` command is replaced by a series of `return scalar` commands in the body of the function. Each statistic that is to be sent to the dataset must be specified by a `return` statement. One thing requires caution, however. Notice that in lines 23 and 24 the accessors `_b[x]` and `_se[x]` were used rather than `b` and `se` that had been returned as scalars in the previous two lines. This is not accidental and it must be done this way. **The command `return scalar b` does not leave a copy of `b` in memory**; subsequent uses of `b` must refer to the accessor (or you need a separate line to define the scalar and then return it).

Running the simulation requires special syntax that can be tricky. Each of the statistics you want to collect in the simulation must be referred to by its `rclass` name; for instance, the `return` command in line 21 creates a return class object, a scalar named `b`, that is stored in `r(b)`. The simulate command in line 28 writes the `r(b)` object to a series called `b`. This is done similarly for the objects returned in `r(se)` and `r(pv)`. The options to `simulate` include the number of repetitions, `reps()`, a dataset name to which results are saved, a couple of options to reduce output to the screen, and finally a colon followed by the program name that is to be simulated.

Whether `simulate` is easier to use or more transparent then `postfile` is debatable. The fussiness of the `postfile` commands, which requires defining a temporary filename, `postfile`, `post`, the formal setup of a loop, and `postclose`, is replaced by another fussiness. With `simulate` one must write a program, include the properly named returns, call the program properly with `simulate` and its options, and populate the desired returns to match those in the program. The command `simulate` essentially collects decisions made when setting up a `postfile` and loop and puts those into the `simulate` calling function.

The postfile and loop commands can also be set up to use functions and in the examples below this is done. As you can see from the preceding examples, this is not necessary. For repeated use of the loop portion of the do-file, this can make things easier though.

Antithetic Variates

For the estimation of means and regression coefficients, anti thetic draws from the desired distribution can reduce the number of simulations needed to demonstrate unbiasedness. Some care has to be used though as Davidson and MacKinnon (1992) point out, antithetics can be harmful if one's interest lies elsewhere, for example, to study the variance of an estimator. See Train (2003, pp. 219–221) for an excellent discussion of antithetic variates and why they work.

Antithetic draws are perfectly negatively correlated with one another. For a symmetric distribution centered at zero like the standard normal, one simply draws a set of normal errors, *u*, and use these once to generate *y*. Then, reverse their signs, and use them again to generate another sample of *y*. The residuals in successive odd and even draws, *u* and *-u*, will be perfectly negatively correlated. This ensures that the regression errors are symmetrically centered around zero.

To use antithetic draws in the study of the least squares estimator of intercept and slope, replace line 22 of the preceding example with

Generating Antithetic Normal Samples

```
22 if mod(`i',2) != 0 {
23     replace u = rnormal(0,sigma)
24     }
25     else {
26     replace u = -u
27     }
```

First, notice that a new `if` command is used. This use of `if` should not be confused with the qualifier of the same name that is used in model statements (e.g., as in the syntax example in section "Ways to Work in Stata"). In this use, `if` evaluates the expression that appears to its right. If the result is true (nonzero), the commands inside the braces are executed. If the result is false (zero), those statements are ignored, and the statement (or statements if enclosed in braces) following the else is executed. The statement that is being checked is `mod(`i',2) != 0`. It uses the modulus command to produce the remainder of the division of the index `i` by 2. For odd integers the statement is true and the errors are generated and placed in u. When false (even integers), u is replaced by its negative value. u and −u are perfectly negatively correlated and hence make perfect antithetic variates for this example.

```
summarize
```

Variable	Obs.	Mean	Std. Dev.	Min.	Max.
b	10000	.5	.1492158	−.0806312	1.080631
se	10000	.148489	.0068318	.1252936	.174046
coverage	10000	.9486	.2208233	0	1

You can see that the least squares estimator has a mean equal to its theoretical value, 0.5.

Autocorrelation and Lagged Dependent Variables

In a linear regression with first-order autocorrelation among the residuals, least squares is consistent, provided there are no lags of the dependent variable used as regressors. Now, if the model contains a lagged dependent variable, least squares is inconsistent if there is autocorrelation. This is easy to demonstrate using a simple simulation.

Let the model be

$$y_t = \beta_1 + \beta_2 x_t + \delta y_{t-1} + u_t \quad t = 1, 2, \ldots, n \tag{5}$$

$$u_t = \rho u_{t-1} + e_t \tag{6}$$

where ρ is a parameter and e_t is random error with mean zero and variance, σ_e^2. This model encompasses several possibilities. If $\delta = 0$ the model is the usual AR(1) model. If $\rho = 0$ then it is a lagged dependent variable model. If both $\delta = 0$ and $\rho = 0$ the model reduces to the classical linear regression implied by Eq. (2). Stability requires $|\rho| < 1$. If $|\delta| = 1$, then the dependent variable is nonstationary and should be modeled as a change $(y_t - y_{t-1})$ rather than a level. A partial adjustment model is implied when $|\delta| < 1$.

The complete model can be written as an ARDL(2,1):

$$y_t = \beta_1(1 - \rho) + \beta_2 x_t + (\delta + \rho)y_{t-1} - (\rho\beta_2)x_{t-1} - (\rho\delta)y_{t-2} + e_t \tag{7}$$

There are many questions that can be answered by simulating this model. Among them, how biased is least squares when the LDV model is autocorrelated? Does this depend on the degree of autocorrelation? Suppose the model is autocorrelated but δ is small. Is least squares badly biased? Does HAC work effectively using standard bandwidth computations in various parameterizations? How powerful is the Breusch–Godfrey test of autocorrelation? How accurate is the approximation implied by use of the Delta theorem? and so on.

We have also introduced the ability to autocorrelate the independent variable, $x_t = \theta x_{t-1} + v_t$. This is useful because otherwise, x_t and y_{t-1} will not be contemporaneously correlated; hence, omitting the lagged dependent variable in the model will not create inconsistency in estimation of β_2. This can be verified quite easily by setting $\theta = 0$ in the simulation. As we have done in most of these simulations, we have set the value of the intercept parameter to be zero, which can be done without losing any generality.

The do-file in **Stata** is shown below.

generating samples from LDV model with Autocorrelation

```
global nobs = 200
global nmc = 1000
set seed 10101
set obs $nobs
gen time = _n
tsset time

scalar theta = .8     /* autocorrelation in x       */
scalar beta = 10      /* slope for x                */
scalar sigma = 20     /* variance of y              */
scalar delta = .5     /* coeff for lagged y         */
scalar rho = .8       /* autocorrelation in errors  */

gen x = rnormal()
replace x = theta*L.x + rnormal() in 2/$nobs
gen u = 0
gen y = 0

program regLDV, rclass
tempname sim
postfile `sim' b1 b2 b3 b4 se1 se2 se3 se4 using results, replace
```

```
22 quietly {
23   forvalues i = 1/$nmc {
24   * generate errors and samples of y
25     replace u = rho*L.u + rnormal(0,sigma) in
       2/$nobs
26     replace y = beta*x+delta*L.y + u in 2/$nobs
27   * run the regression
28     reg y x                     /* b1 OLS, w/o LDV    */
29   * save the estimated slopes and its std error of b
30     scalar b1 = _b[x]
31     scalar se1 = _se[x]
32   * repeat for other estimators
33     reg L(0/1).y x              /* b2 LDV w/o Prais   */
34     scalar b2 = _b[x]
35     scalar se2 = _se[x]
36     prais L(0/1).y x, twostep /* b3 LDV w/Prais     */
37     scalar b3 = _b[x]
38     scalar se3 = _se[x]
39     reg L(0/2).y L(0/1).x      /* b4 ARDL(2,1)       */
40     scalar b4 = _b[x]
41     scalar se4 = _se[x]
42 post 'sim' (b1) (b2) (b3) (b4) (se1) (se2)
   (se3) (se4)
43         }
44 }
45   postclose 'sim'
46 end
47
48 regLDV
49 use results, clear
50 summarize
```

There are a couple of new things in this example. First, in lines 1 and 2 global macro variables are created; `nobs` allows you to set the number of observations each sample will contain and `nmc` the number of simulated samples to draw. Recall that reference to a global variable contained in a macro requires the $ prefix as shown. In line 4 the `set obs $nobs` command opens an empty dataset with room for `nobs` observations. The autoregressive independent variable is created in lines 14 and 15. Rather than initialize the errors and dependent variables using missing observations, they are populated with zeros in lines 16 and 17. Essentially, this allows the initial values of the time series to be set to zero.

A program is created named `regLDV` and given a return classification using the `rclass` option; this allows something calculated within the `regLDV` program to be carried outside of `regLDV`. It is required in order to post the calculations of the coefficients and standard errors to the designated dataset.

The `replace` command makes it easy to construct the sequences of time series without resorting to recursion. However, some care must be exercised to limit the observations to those available. Hence in lines 24 and 25 the `in 2/$nobs` qualifier is required. In this example $u_1 = 0$ and $y_1 = 0$ due to the way these series were initialized above. Subsequent values will be computed using the formulas in 25 and 26. It is simple and elegant.

The rest of the example is straightforward. A model is estimated and statistics collected and posted to a temporary memory location called `sim`. The model is estimated in several ways: (1) least squares with y_{t-1} omitted; (2) least squares with y_{t-1} included, but ignoring the autocorrelation in the errors; (3) the LDV model with a two-step Prais–Winsten transformation; and (4) the correctly specified ARDL(2,1) model. Even though we did not include a constant in the data generation, one is estimated (the estimate of which should not be statistically significant from zero). When finished the contents of `sim` are written to permanent storage in the file *results.dta*, which is available for subsequent analysis.

The simulation reveals

Variable	Obs.	Mean	Std. Dev.	Min.	Max.
b1	1000	15.01713	9.167224	-12.40116	44.45594
b2	1000	5.307263	1.72269	-.9130358	10.61727
b3	1000	8.235005	1.635338	2.534937	12.85917
b4	1000	9.96528	1.445586	5.661244	14.54532

Based on 1,000 simulations, it is pretty clear that only the ARDL(2,1) estimator (`b4`) is getting anywhere near the correct value of 10. The high variance associated with `b1` is due to the high value of ρ in the simulation.

Breusch–Godfrey Test

To study the size or power of the Breusch–Godfrey test, a few more statements have to be added. These are shown below. Note, **Stata** can perform this test using `estat bgodfrey, lags(1)`, which is an `rclass` function. One of the results written to memory is the p-value of the test statistic and it is held in a return class object `r(p)`. The return, `r(p)`, is classified as a `matrix` rather than a `scalar` because additional lags may be specified when the function is called. If so, these will appear in a vector of results. The logical statement simply takes the value 1 when the p-value is less than α and zero otherwise. The mean will give you the overall rejection rate of the test. When $\rho = 0$ this should be close to α.

Size and Power of the Breusch-Godfrey Test for AR(1)

```
    reg y x L.y
    estat bgodfrey, lags(1)
        matrix p1 = r(p)
        scalar rej1 = (p1[1,1]<alpha)
    reg L(0/2).y L(0/1).x
        estat bgodfrey, lags(1)
        matrix p2 = r(p)
        scalar rej2 = (p2[1,1]<alpha)
post 'sim' (p1[1,1]) (rej1) (p2[1,1]) (rej2)
```

The errors of the ARDL(2,1) model are not autocorrelated, and the second use of the test verifies this. Letting $\rho = 0.4$ the rejection rate for the test for the regression $y_t = \beta_1 + \beta_2 x_t + \delta y_{t-1} + u_t$ is compared to that of the ARDL(2,1). The result is

Variable	Obs.	Mean	Std. Dev.	Min.	Max.
p1	1000	.0192006	.0658037	4.69e-10	.8686352
rej1	1000	.913	.2819761	0	1
p2	1000	.4883838	.2947545	.0012266	.9998515
rej2	1000	.057	.2319586	0	1

When the residuals are autocorrelated, the null hypothesis is rejected 91.3% of the time. The power is quite high. The residuals of the correctly specified ARDL(2,1) are not autocorrelated and the rejection rate is 5.7%, which is very close to the nominal level of the test.

HAC Standard Errors

The experimental design of Sul, Phillips, and Choi (2005) is used to study the properties of HAC. They propose a model $y_t = \beta_1 + \beta_2 x_t + u_t$, where $u_t = \rho u_{t-1} + e_t$ and $x_t = \rho x_{t-1} + v_t$ with $\beta_1 = 0$ and $\beta_2 = 1$ and the innovation vector (v_t, e_t) is independently and identically distributed (i.i.d.) $N(0, I_2)$ for $n = 10{,}000$. The do-file appears below.

Coverage probabilities of HAC and the usual OLS confidence intervals

```
global nobs = 10000
global nmc = 1000
set seed 10101
set obs $nobs

* Set the values of the parameters
scalar beta = 1
scalar sigma = 1
scalar rho = .8
scalar alpha = .05

* generate n observations on x and time. Set as time-series
gen x = rnormal()

gen u = 0
gen y = 0
gen time = _n
tsset time
global nwlag = ceil(4*($nobs/100)^(2/9))

program regLDV_hac, rclass
tempname sim
postfile 'sim' b1 se1 b2 se2 c_ls c_nw using results, replace
quietly {
forvalues i = 1/$nmc {
  * generate errors and samples of y
    replace u = rho*L.u + rnormal(0,sigma) in 2/$nobs
```

```
27      replace x = rho*L.x + rnormal(0,1) in 2/$nobs
28      replace y = beta*x + u
29    * run the regression
30      reg y x
31    * save the estimated slope and its std error of b
32      scalar b1 = _b[x]
33      scalar se1 = _se[x]
34    * Confidence Interval
35      scalar lb = b1 - se1*invttail(e(df_r),alpha/2)
36      scalar ub = b1 + se1*invttail(e(df_r),alpha/2)
37      scalar pv = beta<ub & beta>lb
38    * run the regression
39      newey y x, lag($nwlag)
40    * save the estimated slope and HAC std error of b
41      scalar b2 = _b[x]
42      scalar se2 = _se[x]
43    * HAC Confidence Interval
44      scalar lb2 = b2 - se2*invttail(e(df_r),alpha/2)
45      scalar ub2 = b2 + se2*invttail(e(df_r),alpha/2)
46      scalar pv2 = beta<ub2 & beta>lb2
47 post `sim' (b1) (se1) (b2) (se2) (pv) (pv2)
48      }
49 }
50 postclose `sim'
51 end
52
53 regLDV_hac
54 use results, clear
55 summarize
```

The commands are very similar to the ones in the preceding example. A notable difference is the creation of the global constant in line 18 to specify the desired lag length for the HAC kernel. The `replace` command is used as before, and once again the independent variable is autoregressive. Also, the `newey` command is used to estimate the model by least squares with the HAC standard errors in line 39.

Although some patience is required while the simulation runs, the results based on samples of size 10,000 are revealing.

Variable	Obs.	Mean	Std. Dev.	Min.	Max.
b1	1000	1.000417	.0215846	.9309496	1.068051
se1	1000	.0100025	.0002108	.0094095	.0108052
b2	1000	1.000417	.0215846	.9309496	1.068051
se2	1000	.0194082	.0007425	.0173165	.0216077
c_ls	1000	.633	.4822277	0	1
c_nw	1000	.924	.2651307	0	1

Even with a very large sample ($n = 10{,}000$), the confidence interval based on the HAC standard error is still too small on average to cover the slope at the nominal rate. It only covers the parameter 92.4% of the time. The usual least squares confidence interval is terrible, only covering the slope in 63.3% of the simulated samples. You can also see that the standard deviation of the least squares estimator, b1 or b2, is 0.0215846. The average value of the HAC standard error is slightly smaller, 0.0194982, which is consistent with coverage at less than the nominal rate. When $n = 1{,}000$, still a large sample in most people's minds, the coverage rate is only 0.9. The HAC standard errors are certainly an improvement over the usual ones, but one needs very large samples for them to closely approximate the actual LS standard error. This is essentially the same finding as Sul et al. (2005).

Heteroskedasticity

The data generation process is modeled $y_t = \beta_1 + \beta_2 x_t + u_t$, where u_t iid $N(0, \sigma_t^2)$ with $\sigma_t^2 = \sigma^2 \exp(\gamma z_t)$, $z_t = \rho_{xz} x_t + e_t$, e_t iid $N(0, 1)$ and $\beta_1 = 0$ and $\beta_2 = 1$ and the innovation vector (u_t, e_t) is independently distributed for $n = 100$.

In this example one can demonstrate the fact that heteroskedasticity is only a "problem" for estimating least squares standard errors when the model's error variances are correlated with the regressors. Setting $\rho_{xz} = 0$ leaves the errors heteroskedastic, but not with respect to the regressor, x. One can verify that the standard errors of OLS are essentially correct. As ρ_{xz} deviates from zero, the usual least squares standard errors become inconsistent and the heteroskedastic consistent ones are much closer to the actual standard errors.

Standard Errors Using HCCME

```
* set numbers of observations and MC samples
global nobs = 200
```

```
global nmc = 1000
set obs $nobs
set seed 10101

* Set the values of the parameters
scalar slope = .5
scalar sigma = 1
scalar rho_xz = .99
scalar gam = .5
scalar alpha = .05

* generate n observations on x and a correlated variable z
gen x = 10*runiform()
gen z = rho_xz*x+rnormal(0,sigma)
gen y =.
gen sig =.

program regHET, rclass
tempname sim
postfile `sim' b se se_r p_ls p_r using results, replace
quietly {
 forvalues i = 1/$nmc {
 replace sig = exp(gam*z)
    summarize sig
    replace sig = (1/(r(mean)))*sig
    replace y = slope*x + rnormal(0,sig)
reg y x
       scalar b = _b[x]
       scalar se = _se[x]
       scalar lb = b - se*invttail(e(df_r),alpha/2)
       scalar ub = b + se*invttail(e(df_r),alpha/2)
       scalar pv = slope<ub & slope>lb

       reg y x, vce(hc3)
       scalar ser = _se[x]
       scalar lb = b - ser*invttail(e(df_r),alpha/2)
       scalar ub = b + ser*invttail(e(df_r),alpha/2)
```

```
40        scalar pvr = slope<ub & slope>lb
41      post 'sim' (b) (se) (ser) (pv) (pvr)
42      }
43    }
44    postclose 'sim'
45 end
46
47 regHET
48 use results, clear
49 summarize
```

The only trick used here comes in line 27. As the parameter, γ changes, the overall variance of the model will change. Hence in this line the overall variance in the model is normalized around the mean.

Stata has options for the different versions of the Heteroskedasticity Consistent Covariance Matrix Estimator (HCCME). In line 36, `vce(hc3)` scales the least squares residuals $\hat{u}_t/(1-h_t)^2$.

The parameter ρ_{xz} controls the degree of correlation between the regressor, x, and the variable z that causes heteroskedasticity. If $\rho_{xz}=0$ then the degree of heteroskedasticity (controlled by γ) has no effect on estimation of least squares standard errors. That is why in practice z does not have to be observed. One can simply use the variables in x that are correlated with it to estimate the model by feasible GLS. This would make an interesting extension of this experiment. Whether this is more precise in finite samples than least squares with HCCME could be studied.

Instrumental Variables

The statistical model contains five parameters: $\beta, \sigma, \sigma_x, \gamma$, and ρ. The data generation process is modeled $y_t + \theta + \beta x_t + u_t$, where u_t iid $N(0, \sigma_t^2)$, $x_t = \gamma z_t + \rho u_t + e_t$, e_t iid $N(0, \sigma_e^2)$ and the innovation vector $(u_t,\ e_t)$ is independently distributed for $n = 100$. Without loss of generality, the intercept is set to zero, $\theta = 0$. The γ controls the strength of the instrument z_t, ρ controls the amount of correlation between x_t and the errors of the model u_t, σ_e^2 can be used to control the relative variability of x_t and u_t as in an error-in-variables problem.

Instrumental Variables

```
global nobs = 200
global nmc = 1000
set seed 10101
set obs $nobs

scalar slope = 1      /* regression slope */
scalar sigma = 10     /* measurement error in y */
scalar sige = .1      /* amount of measurement error in e */
scalar gam = 1        /* instrument strength */
scalar rho = .5       /* correlation between errors */
scalar alpha=.05      /* test size */

gen z = 5*runiform()
gen y = .
gen x = .
gen u = .

program regIV, rclass
tempname sim
  postfile `sim' b biv se se_iv p_ls p_iv using results, replace
  quietly {

 forvalues i = 1/$nmc {
    replace u = rnormal()
    replace x = gam*z+rho*u+rnormal(0,sige)
    replace y = slope*x + u
    reg y x
      scalar b = _b[x]
      scalar se = _se[x]
      scalar lb = b - se*invttail(e(df_r),alpha/2)
      scalar ub = b + se*invttail(e(df_r),alpha/2)
      scalar pv = slope<ub & slope>lb

 ivreg y (x=z)
      scalar biv = _b[x]
      scalar seiv = _se[x]
      scalar lb = biv - seiv*invttail(e(df_r), alpha/2)
```

```
   scalar ub = biv + seiv*invttail(e(df_r),alpha/2)
   scalar pvr = slope<ub & slope>lb
  post `sim' (b) (biv) (se) (seiv) (pv) (pvr)
  }
}
 postclose `sim'
end
regIV
use results, clear
summarize
```

Binary Choice

The statistical model contains only two parameters: β and σ. The latent variable is modeled $y_t^* = \mu + \beta x_t + u_t$, where u_t iid $N(0, \sigma_t^2)$. The observed indicator variable $y_t = 1$ if $y_t^* > 0$ and is 0 otherwise. By changing the parameters μ and β one can alter the proportion of 1s and 0s in the samples. Changing σ should have no effect on the substance of the results since it is not identified when the other two parameters are.

This simulation is designed to illustrate two things. First, parameter significance tests from maximum likelihood (ML) estimation of a probit and least squares estimation of the linear probability model usually agree. The exceptions occur, as Horrace and Oaxaca (2006) point out, if the least squares estimated index for any observation lies outside of the 0/1 interval. This is easy to check. If the only goal is to determine whether x affects the probability of observing $y = 1$, then the least squares estimator of the linear probability model may have better small sample properties than the probit MLE. Second, if estimating the average marginal effect is of interest, least squares of the linear probability model (LPM) offers a close approximation, is much easier to produce, and is unambiguous in interpretation. The following experiment demonstrates their similarity.

The do-file is:

Probit and Linear Probability Model

```
global nobs = 200
global nmc = 500
set seed 10101
```

```
set obs $nobs

scalar slope = 0   /* regression slope */
scalar sigma = 1   /* measurement error in y */
scalar alpha=.05   /* test size */

gen x = runiform()
gen u = .
gen y = .

program regBC, rclass
tempname sim
  postfile ‘sim’ t t_p pv pv_p b ame prop using results, replace
  quietly {
    forvalues i = 1/$nmc {
     replace u = rnormal(0,sigma)
     replace y = slope*x + u
     replace y = (y>0)
     summarize y
     scalar prop = r(mean)
     reg y x, vce(hc3)
       scalar b = _b[x]
       scalar t = _b[x]/_se[x]
       scalar pv = abs(t)>invttail(e(df_r),alpha/2)
     probit y x
       scalar tp = _b[x]/_se[x]
       scalar pv_p = abs(tp) > invnormal(1-alpha/2)
       margins, dydx(x)
       matrix m = r(b)
   scalar m = m[1,1]
       post ‘sim’ (t) (tp) (pv) (pv_p) (b) (m) (prop)
       }
   }

  postclose ‘sim’
end

regBC
use results, clear
summarize
```

In this exercise the data generation process is handled easily in lines 19, 20, and 21. The proportions of 1s to 0s is saved as a scalar in line 23 after computing its summary statistics in line 22. Then both least squares and the probit MLE are used to estimate the parameters of the binary choice model. In most econometrics courses, great effort is made to discuss the fact that marginal effects in probit models are different at each observation. This creates angst for many. In lines 31 and 32 the average marginal effects are computed and saved for the regressor, x. The results show that in fact the AME of the probit model (`ame`) and the slope from least squares (`b`) are nearly identical.

Variable	Obs	Mean	Std. Dev.	Min	Max
t	500	.0205409	.9701433	−3.010698	2.827217
t_p	500	.0209067	.9692976	−2.921262	2.780452
pv	500	.044	.2053005	0	1
pv_p	500	.044	.2053005	0	1
b	500	.0026487	.123441	−.3727096	.3533991
ame	500	.0026391	.1229774	−.3691716	.3499526
prop	500	.49916	.0348859	.4	.61

In 500 samples the size of the t-test is the same for the two estimators. The marginal effect in the LPM is 0.00265 and for probit it is 0.00264. There is slightly more variation in the t-ratio for the LPM than for probit, but the difference is very small. For this design, it is a toss-up. Increasing the slope to equal 1 increases the proportion of 1s in the sample to 0.69. The power of the t-test is the same to two decimal places, 0.87. The slope and AME are 0.3497 and 0.3404, respectively. The more unbalanced the sample, the larger the divergence in these estimates of the marginal effect. Still, with nearly 70% of the sample being observed as a 1, the two are very close (as are their standard deviations and range). Also, one could confirm the superiority of using the MLE when the index of the LPM strays outside of the logical (0,1) interval.

Tobit

To generate samples for tobit estimation requires an additional line in the probit code. Recall that in the tobit model, all of the latent variables that are less than zero are censored at zero. The continuous values above the threshold are actually observed. Therefore, the probit do-file can serve as a basis for this model. Lines 19–21 of the probit example are replaced by

```
19   replace ystar = const + slope*x + rnormal(0,sigma)
20       replace y = (ystar>const)
21       replace y = y*ystar
```

The constant in this model serves as the threshold for tobit. Computing the proportions of 1s in this model enables you to keep track of how much censoring occurs. Experiments could be conducted by changing the distance between actual threshold and zero and the tobit estimator could be compared to least squares. This is another poorly understood feature of this model. Choosing the proper threshold is critical for proper performance of the MLE. For instance, add a constant of 20 to the right-hand side of the equation in line 19, which generates the latent variable y_t^* (`ystar`), and change the actual threshold in line 20 from 0 to 20. Increase the value of `sigma` to 10, for instance, to create more variation in `y`. Re-run the simulation and see that least squares is now far superior to the tobit MLE, which erroneously assumes that the threshold is 0.

Nonlinear Least Squares

Stata can also be used to estimate more complicated models in a simulation. It includes generic routines for estimating nonlinear least squares and maximum likelihood estimators, the properties of which can also be studied in this way. In the following example, nonlinear least squares is placed within a loop and the estimates are collected in matrices and output to a data set. This simple example could be used as a rough template for more advanced problems.

Consider the nonlinear consumption function

$$C_t = \alpha + \beta Y_t^{\gamma} + u_t \tag{8}$$

where C_t is consumption and Y_t is output. Using data from Greene (2000), the model is estimated to obtain starting values, parameters are set, and simulated samples are drawn. In the first snippet of code, the data are opened in line 1, the parameter values set in lines 3–6, starting values are acquired by least squares (12–15). The `nl` command is used in lines 23 and

24 (using the /// line continuation). Parameters are enclosed in braces { } and given starting values using the assignment operator, =. The parameters in nonlinear models are accessible in postestimation, but the syntax is a little different. The `_b[varname]` convention used in linear models has been replaced by `_b[paramname:_cons]`. The `coeflegend` option can be used after the `nl` command to find the proper names for the parameters. To verify that you have identified the parameters correctly, run the nonlinear least squares regression again using the `coeflegend` option. This suppresses much of the output that you ordinarily want, but it does produce a legend that identifies **Stata's** names for each of the parameters.

```
 1 use ``http://www.learneconometrics.com/pdf/
      MCstata/greene11_3.dta'', clear
 2
 3 scalar A = 180
 4 scalar B = .25
 5 scalar G = 1
 6 scalar sigma = 2
 7
 8 global nmc = 1000
 9 gen c0 = A + B*y^G
10
11 * Starting values
12 reg c y
13 scalar alpha0 = _b[_cons]
14 scalar beta0 = _b[y]
15 scalar gamma0 = 1
16
17 program regNLS, rclass
18 tempname sim
19   postfile `sim' alpha beta gamma using results,
     replace
20    quietly {
21      forvalues i = 1/$nmc {
22      replace c = c0 + rnormal(0,sigma)
23      nl (c = {alpha=alpha0} + {beta=beta0} *
        y^{gamma=gamma0}), ///
24      variables(c y)
```

```
     post `sim' (_b[alpha:_cons]) (_b[beta:_cons])
     (_b[gamma:_cons])
       }
   }
   postclose `sim'
end

regNLS
use results, clear
summarize
```

The results of 1,000 iterations from the simulation are

Variable	Obs.	Mean	Std. Dev.	Min.	Max.
alpha	1000	179.8951	6.768631	157.7997	199.7664
beta	1000	.2527278	.0389655	.1565982	.4028383
gamma	1000	1.000128	.0183172	.9431589	1.056078

The results indicate that the nonlinear least squares estimator has done quite well, on average. It is statistically unbiased.

CHANGING THE DESIGN USING LOOPS

In some circumstances one wants to study the properties of an estimator or test statistic under different parameterizations of the model. The methods discussed here can easily be adapted to this and this is where the benefits of using the `postfile` commands can really be seen. Consider the instrumental variables example in section "Instrumental Variables" and suppose one wants to examine how instrument strength affects the bias of the IV estimator. One design consideration is to keep the overall variability in the endogenous regressor constant. This is desirable because it keeps the overall fit between y and x relatively constant as instruments get stronger. Note, $\text{var}(x) = \gamma^2\text{var}(z) + \rho^2\text{var}(u) + \text{var}(e)$. The revised program now includes a extra loop over which the simulation will execute.

In this version we add a computation for the F-statistic used to test the strength of the instruments, saving it as a scalar in line 11, and save the overall fit of the reduced form regression, R^2, as a scalar in line 14. The program follows.

```
 1 program regIV, rclass
 2 tempname sim
 3    postfile 'sim' gam r2 F b biv se se_iv p_ls p_iv
      using results, replace
 4    quietly {
 5    foreach gam of numlist 0.025 0.0375 0.05 0.1 0.15
      0.2 0.5 {
 6      forvalues i = 1/$nmc {
 7        replace u = rnormal()
 8        replace x = 'gam'*z+rho*u+rnormal(0,sige)
 9        replace y = slope*x + u
10        reg x z                          /* reduced form */
11        scalar F = (_b[z]/_se[z])^2 /* instrument
                                                 strength */
12
13      reg y x
14        scalar r2=e(r2)                  /* overall fit */
15        scalar b = _b[x]
16        scalar se = _se[x]
17        scalar lb = b - se*invttail(e(df_r),alpha/2)
18        scalar ub = b + se*invttail(e(df_r),alpha/2)
19        scalar pv = slope<ub & slope>lb
20
21      ivreg y (x=z)
22        scalar biv = _b[x]
23        scalar seiv = _se[x]
24        scalar lb = biv - seiv*invttail(e(df_r),
          alpha/2)
25        scalar ub = biv + seiv*invttail(e(df_r),
          alpha/2)
26        scalar pvr = slope<ub & slope>lb
27
28     post 'sim' ('gam') (r2) (F) (b) (biv) (se) (seiv)
       (pv) (pvr)
29        }
30      }
```

```
31     }
32   postclose 'sim'
33 end
```

The parameter gam changes instrument strength and we add it to the postfile statement in line 3; the reason for this will become apparent below. A foreach loop is initiated and populated in line 5. This type of loop is quite handy for this use. It will loop over the elements of the numlist (the numbers that follow numlist, i.e., 0.025, 0.0375, 0.05, 0.1, 0.15, 0.2, and 0.5). The loop is closed on line 30. Any reference to the contents of the local macro gam in the program must be enclosed in the now familiar single quotes. Also, note that line 28 now includes ('gam') (r2) and (F) which matches the declarations in the postfile command and ensures that the current values related to instrument strength are written to the *results.dta* file.

First, the output from the simulation can be summarized by the design parameter, γ.

```
regIV
use results, clear
by gam, sort: summarize r2 F biv b se_iv p_iv
by gam, sort: summarize p_ls p_iv
```

This produces the following set of summary statistics. Recall, that $\beta = 1$ and that the nominal coverage rate for the confidence interval is 95%.

-> gam = .025

Variable	Obs.	Mean	Std. Dev.	Min.	Max.
r2	1000	.7551756	.0127816	.7209752	.8055412
F	1000	2.027093	2.507469	4.73e-06	18.04113
biv	1000	.6928968	14.1146	−275.4101	126.6003
b	1000	1.401108	.0254189	1.30706	1.468426

-> gam = .0375

Variable	Obs.	Mean	Std. Dev.	Min.	Max.
r2	1000	.7539033	.0136158	.7106823	.7929032
F	1000	3.646357	3.672428	.0000266	23.84322

biv	1000	.7281356	6.266472	−108.6397	41.78983
b	1000	1.400052	.0254246	1.323858	1.478802

-> gam = .05

Variable	Obs.	Mean	Std. Dev.	Min.	Max.
r2	1000	.7538432	.0138031	.712468	.790554
F	1000	5.195976	4.309673	.0000954	32.33922
biv	1000	1.02068	4.401356	−37.72792	92.09246
b	1000	1.398645	.0251742	1.310623	1.47523

-> gam = .1

Variable	Obs.	Mean	Std. Dev.	Min.	Max.
r2	1000	.7535283	.0142405	.6980119	.7988308
F	1000	18.21511	8.546679	.7164295	55.45287
biv	1000	.969141	.2596705	−.4665824	1.800038
b	1000	1.39171	.0252075	1.280399	1.468121

-> gam = .15

Variable	Obs.	Mean	Std. Dev.	Min.	Max.
r2	1000	.7551459	.0131145	.7074705	.7903711
F	1000	39.52843	12.62963	8.662003	95.4943
biv	1000	.9862381	.1514629	.2889033	1.502262
b	1000	1.385531	.0236149	1.313096	1.467276

-> gam = .2

Variable	Obs.	Mean	Std. Dev.	Min.	Max.
r2	1000	.7560955	.0132708	.7115782	.7941926
F	1000	70.0239	16.88823	22.88008	136.5724
biv	1000	.9928045	.113901	.4058512	1.355679
b	1000	1.375369	.024529	1.297485	1.458363

-> gam = .5

Variable	Obs.	Mean	Std. Dev.	Min.	Max.
r2	1000	.7723736	.0122313	.7272431	.8121825
F	1000	432.9301	45.91504	304.8695	603.1799
biv	1000	.9993733	.0428512	.8498048	1.14667
b	1000	1.27941	.0201706	1.218947	1.355316

It is clear that one design goal has been met; `r2` is not changing as γ increases in value. The mean of the instrumental variable estimator is getting closer to 1 as the instrument strengthens. We can also see that least squares actually outperforms the instrumental variable estimator in terms of bias when instruments are very weak. Also, notice the very high standard deviation associated with the instrumental variable estimator when instruments are weak ($F < 18$). This is also supported by theory.

The coverage rate of the confidence intervals could likewise be studied. Below is a table that contains the coverage rates of 95% confidence intervals centered at the least squares and IV estimators, respectively.

Coverage rates for 95% confidence intervals

	gam						
Estimator	0.025	.0375	0.05	0.10	0.15	0.20	0.50
OLS	0	0	0	0	0	0	0
IV	.983	.966	.974	.965	.959	.945	.956

n=1,000 and 1,000 simulations per design point

Bias and precision of least squares may be lower when instruments are weak, but that does not translate into acceptable performance in the coverage rate of confidence intervals. Least squares never covers the parameter. The IV estimator may be erratic, but its confidence interval is doing a better job of covering the parameter. As instrument strength improves, it gets very close to the 95% nominal coverage rate.

Finally, we can explore the relationship between instrument strength and bias via the mechanism suggested by (Stock and Watson, 2011, p. 464). They suggest a rule of thumb that bias of an instrumental variable estimator is roughly proportional to the inconsistency of least squares scaled by the average F-statistic minus 1, for example, $E[\hat{\beta}_{\mathrm{IV}}] - \beta \approx [plim(\hat{\beta}_{\mathrm{OLS}}) - \beta]/(E(F) - 1)$, where we take F to be the average for a given design.

```
use results, clear
by gam: egen Fbar = mean(F)          /* Average F
                                          by gamma */
by gam: egen tslsbar = mean(biv)     /* Average biv
                                          by gamma */
```

```
by gam: egen olsbar = mean(b)       /* Average b
                                          by gamma */
by gam: egen r2bar = mean(r2)       /* Average r2
                                          by gamma */

gen tsls_bias = tslsbar-1           /* IV bias */
gen ols_bias = olsbar-1             /* OLS bias */
gen relb = tsls_bias/ols_bias       /* relative
                                          bias */
gen rot = 1/(Fbar-1)                /* rule of
                                          thumb */
```

This requires the use of the extended generation functions `egen` to populate the dataset with the average values of the *F*-statistic, R^2, and the averages of the IV and OLS estimators. With 1,000 observations in our sample, the average of the OLS estimator is hopefully close to its probability limit as required by the Stock and Watson result; increasing sample size will improve the approximation.

Based on our design with sample size $n = 1{,}000$ and on 1,000 simulations for each design, the results are

	F						
Variable	2.03	3.64	5.19	18.21	39.52	70.02	432.9
Rule of thumb	.97362	.37787	.23832	.05809	.02595	.01449	.00232
Relative bias	−.76564	−.67957	.05188	−.07878	−.03569	−.01917	−.00224

Keep in mind that the rule of thumb is always positive; the bias can be negative as is the case here. So, the magnitudes of the two indicators is what we are comparing. The approximations are reasonable, though the rule of thumb is much better when $F > 10$ it appears. A regression of relative bias onto the rule of thumb should yield a coefficient of -1. There are only seven actual design points; therefore, we need to eliminate all of the redundant observations.

```
gen t = _n                   /* obtain observation
                                      numbers */
keep if mod(t,1000) == 0     /* keep 1 observation
                                   per design */
```

A regression that includes the rule of thumb squared is used to test whether the relationship is actually linear. The coefficient on the squared term is not significant ($t = 0.72$) and we conclude that the relationship is linear. Next, we test whether the proportional relationship is one-to-one; the simple regression *relative bias* = β(*rule of thumb*) + *residual* is estimated and $\beta = -1$ is tested against $\beta \neq -1$. The t-ratio is approximately equal to 0.8, which is not significant at any reasonable level of significance.

```
reg relb rot, noconst
```

Number of obs = 7
R-squared = 0.8131

Relative bias:	Coef.	Std. Err.	t	$P>\|t\|$	[95% Conf.	Interval]
Rule of thumb	−.8644	.16921	−5.11	0.002	−1.278	−.4504

CONCLUSION

In this primer the basics of conducting a Monte Carlo experiment are discussed, and the concepts are put to use in **Stata**. As many authors have noted, Monte Carlo methods illustrate quite clearly what sampling properties are all about and reinforce the concepts of relative frequency in classical econometrics. The **Stata** software is well-suited for this purpose. The `postfile` construct makes it relatively easy to collect statistics of interest and to output them to a dataset. Once there, they can be recalled for further analysis without having to rerun the simulation.

A series of examples is given. These include linear regression, confidence intervals, the size and power of t-tests, lagged dependent variable models, heteroskedastic and autocorrelated regression models, endogenous regressor, binary choice, censored regression, and nonlinear regression models. Do-files and data for all examples are provided in the chapter and are available from the author's website http://learneconometrics.com/pdf/MCstata.

The large library of built-in estimators and tests makes **Stata** a rich platform for Monte Carlo analysis. The biggest drawback is that it can be relatively slow (see Table 1 and the discussion below), at least in the single processor versions of the software. For serious simulation work, one would want to consider a multiprocessor version. However, as demonstrated in the last example, Stata provides a fine tool for developing prototypes for

Table 1. Speed Comparison of **Stata**'s `simulate` and `postfile` Commands to **gretl** Measured in cpu Seconds Based on 4,000 Monte Carlo Samples.

		Stata[a]	
Program	**gretl**[b]	`postfile`	`simulate`
Confidence interval	0.608	2.86	3.26
Nonlinear least squares	1.81	50.01	52.35[c]

[a]**Stata** 11.2.
[b]**gretl** 1.9.9.
[c]The do-file for the simulate version can be found at http://www.learneconometrics.com/pdf/MCstata/index.htm

eventual translation to a faster computing platform. The huge library of built-in estimators and test statistics makes preliminary analysis a breeze.

Stata's `simulate` command allows you to put whatever you want to compute within a program and makes simulating the contents quite simple. The desired results are output to a **Stata** dataset, allowing for subsequent analysis. The advantages of using the `postfile` construction is that it gives you more control over what gets computed and saved. It also makes it easier to loop over different designs and have all of the output placed into the resulting dataset.

Below we compare time required to compute the first confidence interval example using three methods. The computations were performed using Windows 7 64-bit, equipped with an Intel i7 2600K processor (not over-clocked). The system has 16 gigabytes of RAM. A total of 4,000 Monte Carlo iterations were used in each example. These times were based on version 11.2 of **Stata**.[6] For purposes of comparison, we also used **gretl** (a native 32-bit program) to run comparable simulations. The **gretl** code appears in the appendix and examples similar to the ones in this chapter can be found in Adkins (2011a, 2011b). The simulation was repeated using the `postfile` commands and then using `simulate`. The numbers reported are cpu seconds. Smaller is faster.

As one can easily see, **gretl** is four times faster than **Stata** 11.2 in simulating the regression model and confidence interval coverage rates. The `simulate` command adds extra overhead that costs about a second. This extra time is not proportional to the number of iterations performed; the gap remains fairly constant as the number of simulations increases. If speed is important, **gretl** is an excellent choice. For nonlinear least squares, things get worse for **Stata. Gretl** manages to run 4,000 iterations of the simulation in less than 2 seconds while **Stata** takes over 50. **Stata** may not be very quick, but it contains a vast number of routines that can be easily simulated either via the `postfile` method or via the `simulate` command. Because of its wide availability and ease of use, it is an excellent vehicle for teaching sampling theory in econometrics and for exploring the properties of a wide variety of estimators.

NOTES

1. These are functions that return the result of a computation.
2. Experienced programmers love them because they conserve computer memory and when properly used, reduce the chances of making inadvertent programming errors, especially in very long programs.
3. If you are used to using software that requires explicit punctuation to indicate the end of a line of computer code, then you can set **Stata** to operate this way using the `#delimit` command. SAS programmers will be familiar with the convention that a line ends with a semicolon. To accomplish this in **Stata**, use `#delimit;` on the first line of your do-file and end each line with `;`.
4. If a seed is not specified, then **Stata** initializes it to be 123456789 each time the software is started.
5. `postfile, post, and postclose`
6. The timings of the programs in Stata 12.1 were also taken. Inexplicably, the do-files are much slower to execute in version 12.1. The `postfile` version of the confidence interval do-file takes 14.18 seconds and the simulate version 14.88. This suggests a potential problem with version 12.1 that is likely to be fixed in later updates. Hence, we have chosen to report the version 11.2 results in Table 1.

REFERENCES

Adkins, L. C. (2011*a*). *Monte Carlo experiments using gretl: A review with examples.* Retrieved from http://www.learneconometrics.com/pdf/MCgretl

Adkins, L. C. (2011b). Using gretl for Monte Carlo experiments. *Journal of Applied Econometrics*, *26*(5), 880–885.

Barreto, H., & Howland, F. H. (2006). *Introductory econometrics using Monte Carlo simulation with microsoft excel.* New York, NY: Cambridge University Press.

Cameron, A. Colin, & Trivedi, P. K. (2009). *Microeconometrics using Stata.* College Station, TX: Stata Press.

Cottrell, A., & Riccardo "Jack" Lucchetti (2012). *Gretl Users Guide.* Retrieved from http://sourceforge.net/projects/gretl/files/manual/

Davidson, R., & MacKinnon, J. G. (1992). Regression-based methods for using control variates in Monte Carlo experiments. *Journal of Econometrics*, *54*, 203–222.

Davidson, R., & MacKinnon, J. G. (2004). *Econometric theory and methods*. New York: Oxford University Press.

Day, E. (1987). A note on simulation models in the economics classroom. *Journal of Economic Education*, *18*(4), 351–356.

Greene, W. (2000). *Econometric analysis* (4th ed.). Upper Saddle River, NJ: Prentice-Hall.

Horrace, W. C., & Oaxaca, R. L. (2006). Results on the bias and inconsistency of ordinary least squares for the linear probability model. *Economics Letters*, *90*(3), 321–327.

Judge, Guy (1999). Simple Monte Carlo studies on a spreadsheet. *Computers in Higher Education Economics Review*, *13*(2). Retrieved from http://www.economicsnetwork.ac.uk/cheer/ch13_2/ch13_2p12.htm

Kennedy, Peter E. (2003), A guide to econometrics, 5th Ed. Cambridge, MA: MIT Press.

Kiviet, Jan F. (2012). Monte Carlo simulation for econometricians. *Foundations and Trends in Econometrics*, *5*(1–2), 1–181.

Koenker, R., & Bassett, G (1982). 'Robust tests of heteroscedasticity based on regression quantiles'. *Econometrica*, *50*, 43–61.

StataCorp (2011), *Stata statistical software: Release 12*. StataCorp LP, College Station, TX.

Stock, James H., & Watson, Mark W. (2011). *Introduction to econometrics* (3rd ed.). Boston, MA: Addison Wesley.

Sul, D., Phillips, P. C. B., & Chi-Young, C. (2005). Prewhitening bias in HAC estimation. *Oxford Bulletin of Economics and Statistics*, *67*(4), 517–546.

Train, Kenneth E. (2003). *Discrete choice methods with simulation*. Cambridge, UK: Cambridge University Press.

APPENDIX: *HANSL* SCRIPTS

To give the reader an idea about the speed of **Stata**, we compared it to a free software called **gretl. Gretl**'s matrix language is called *hansl*, and *hansl* refers to a collection of code as a *script*. The *hansl* scripts for studying confidence intervals and nonlinear least squares are found below in sections "Confidence Intervals" and "Nonlinear Least Squares," respectively.

Confidence Intervals

```
open engel.gdt
# set a seed if you want to get same results each time you run this
set seed 3213799

# Set the values of the parameters
scalar constant = .5
scalar slope = .5
scalar sigma = 1
scalar alpha = .025 # (size of one tail of a 2 sided CI)

# Take the natural log of income to use as x
genr x = log(income)

# start the loop, indicating the desired number of samples.
# -progressive is a special command for doing MC simulations (necessary)
# -quiet tells gretl not to print the iterations (highly recommended)
set stopwatch
loop 4000 -progressive -quiet
     # generate normal errors
     genr u = normal(0,sigma)
     # generate y, call it y1
     genr y1 = constant + slope*x + u
     # run the regression
     ols y1 const x
```

```
    # save the estimated coefficients
    genr b2 = $coeff(x)
    # save the estimated standard errors
    genr s2 = $stderr(x)
    # generate the lower and upper bounds for the confidence interval
    genr c2L = b2 - critical(t,$df,alpha)*s2
    genr c2R = b2 + critical(t,$df,alpha)*s2
    # count the number of instances when coefficient is inside inverval
    genr p2 = (slope>c2L && slope<c2R)
    # print the results
    print b2 p2
    # store the results in a dataset for future analysis is desired
    store cicoeff.gdt b2 s2 c2L c2R
endloop
scalar t = $stopwatch
```

Nonlinear Least Squares

```
open ''greene11_3.gdt''
setobs 1 1 -cross-section
# Set the actual values of the parameters
scalar A = 180
scalar B = .25
scalar G = 1

# Starting values
ols C 0 Y
genr alpha0 = $coeff(0)
genr beta0 = $coeff(Y)
genr gamma0 = 1


# Set the number of Simulated Samples
scalar NMC = 4000

```

```
# Create an empty matrix to store results
matrix coeffs = zeros(NMC, 3)
matrix vcvs = zeros(NMC, 6)

# Create systematic portion of the model and log(y)
series C0 = A + B*Y^G
series lY = log(Y)

set stopwatch
# The loop
set warnings off
loop i=1..NMC -quiet
# generate new sample
    genr C = C0 + normal(0,10)
    # Initialize parameters
    alpha = alpha0
    beta = beta0
    gamma = gamma0

    # Estimate parameters via NLS
    nls C = alpha + beta * Y^gamma
      deriv alpha = 1
      deriv beta = Y^gamma
      deriv gamma = beta * Y^gamma * lY
    end nls -quiet

    # Collect the coefficients into a vector and matrix
    matrix coeffs[i,] = {alpha, beta, gamma}
    matrix vcvs[i,] = vech($vcv)'
endloop

# open a new, empty dataset
nulldata NMC -preserve

# Convert the columns of matrix to data
series a = coeffs[,1]
series b = coeffs[,2]
series c = coeffs[,3]

```

```
# Print the summary Statistics
summary a b c
scalar t = $stopwatch
printf ''Monte Carlo vcv vs average estimated vcv\n''
MCV = mcov(coeffs)
AEV = unvech(meanc(vcvs)')
print MCV AEV
```